The Cytoskeleton in Health and Disease

Heide Schatten
Editor

The Cytoskeleton in Health and Disease

Springer

Editor
Heide Schatten
Department of Veterinary Pathobiology
University of Missouri-Columbia
Columbia, MO, USA

ISBN 978-1-4939-2903-0 ISBN 978-1-4939-2904-7 (eBook)
DOI 10.1007/978-1-4939-2904-7

Library of Congress Control Number: 2015945026

Springer New York Heidelberg Dordrecht London

Printed on acid-free paper

Springer Science+Business Media LLC New York is part of Springer Science+Business Media (www.springer.com)

Preface

New research on the cytoskeleton has led to exciting new data on its functions and dysfunctions in health and disease, and it has allowed a more complete understanding of cytoskeletal interactions in various cell organizations.

Cytoskeletal functions are essential for numerous cellular activities including cellular signaling, cell migration, cell division, transport of macromolecular complexes, and cell organelles that play a role in cellular metabolism, and various others, while cytoskeletal dysfunctions lead to a wide range of disorders and diseases such as cancer, Alzheimer's, neurodegenerative disorders, reproductive disorders, and a variety of metabolic disorders.

Much progress has been made in recent years to develop new strategies that target specific cytoskeletal dysfunctions with the goal to restore normal functions; other studies have been focused on inhibiting abnormal cytoskeletal activities to control cancer cell proliferation. New drugs are being developed to target cytoskeletal abnormalities on cell and molecular levels which include targeting cytoskeleton-associated proteins and posttranslational modifications.

This book covers recent advances in the cytoskeletal field and new discoveries that have emerged during the past few years related to the three major cytoskeletal components, microtubules, intermediate filaments, microfilaments, their associated proteins, interactions partners, and their regulation on cellular and molecular levels. It addresses cytoskeletal dynamics, regulation, posttranslational modification, and interactions with various cellular components in different cell systems including stem cells, reproductive cells, muscle cells, neuronal cells, bone cells, cancer cells, and others.

This book on the *Cytoskeleton in Health and Disease* features chapters written by experts in their specific areas of research who have been invited for their significant contributions to research on the cytoskeleton in health and disease. Chapters include: Brief Overview of the Cytoskeleton (Chap. 1); Cytoskeleton Dynamics in Health and Disease: Role of Molecular Switches and Rheostats (Chap. 2); Regulation of the Cytoskeleton by the Rho Family of GTPases in Hematopoietic Stem Cells in

Health and Disease (Chap. 3); The Role of the Cytoskeleton in Cell Migration, Its Influence on Stem Cells and the Special Role of GFAP in Glial Functions (Chap. 4); Centrosome–Microtubule Interactions in Health, Disease, and Disorders (Chap. 5); Cytoskeletal Elements and the Reproductive Success in Animals (Chap. 6); Cytoskeleton and Regulation of Mitochondrial Translocation in Mammalian Eggs (Chap. 7); Tubulin Detyrosination in Epithelial Cells (Chap. 8); Mutations in *Adenomatous Polyposis Coli*, Their Role in Cytoskeletal Dynamics and Cancer Onset (Chap. 9); Small GTPases Act as Cellular Switches in the Context of Cilia (Chap. 10); Desmin Plays Dual Structural and Regulatory Functions Through Its Interaction with Partners in Muscle (Chap. 11); Desmin Filaments and Desmin-Related Myopathy (Chap. 12); Possible Functions of Intermediate Filaments in Mammalian Ovarian Follicles and Oocytes (Chap. 13); Actin Organizing Proteins in Regulation of Osteoclast Function (Chap. 14); The Role of Drebrin-Binding Stable Actin Filaments in Dendritic Spine Morphogenesis (Chap. 15); The Role of the Actin Cytoskeleton in Cancer and Its Potential Use as a Therapeutic Target (Chap. 16).

It has been a privilege to edit this book on the cytoskeleton in health and disease, and I would like to thank all authors for their outstanding contributions and for sharing their special expertise with the cytoskeleton biology and biomedical community. I hope that the topics covered in this book will inspire further research and new approaches leading to new discoveries and new treatment possibilities for cytoskeletal diseases.

Columbia, MO, USA Heide Schatten

Contents

Contributors

Onnik Agbulut Sorbonne University, Institute of Biology Paris-Seine (IBPS), Biological Adaptation and Ageing, Paris, France

Diego Pinheiro Aguiar Institute of Biomedical Sciences, Federal University of Rio de Janeiro, Rio de Janeiro, Brazil

National Institute of Traumatology and Orthopedics, Division of Research, Rio de Janeiro, Brazil

Felipe Andreiuolo Department of Pathology, St. Anne Hospital, Paris Descartes University, Paris, France

Joana Balça-Silva Center for Neuroscience and Cell Biology, Institute for Biomedical Imaging and Life Sciences (CNC.IBILI), University of Coimbra, Coimbra, Portugal

Faculty of Medicine, University of Coimbra (FMUC), Coimbra, Portugal

Simon Brayford Oncology Research Unit, School of Medical Sciences, University of New South Wales, Sydney, New South Wales, Australia

Jose A. Cancelas Division of Experimental Hematology and Cancer Biology, Cincinnati Children's Hospital Medical Center, Cincinnati, OH, USA

Hoxworth Blood Center, University of Cincinnati College of Medicine, Cincinnati, OH, USA

Kyung-Hee Chang Division of Experimental Hematology and Cancer Biology, Cincinnati Children's Hospital Medical Center, Cincinnati, OH, USA

Hoxworth Blood Center, University of Cincinnati College of Medicine, Cincinnati, OH, USA

Alexander E. Davies Department of Molecular and Cellular Biology and the School of Veterinary Medicine, University of California, Davis, CA, USA

Juliana de Mattos Coelho-Aguiar Paolo Niemeyer State Brain Institute of Rio de Janeiro, Rio de Janeiro, Brazil

Institute of Biomedical Sciences, Federal University of Rio de Janeiro, Rio de Janeiro, Brazil

Anália do Carmo Center for Neuroscience and Cell Biology, Institute for Biomedical Imaging and Life Sciences (CNC.IBILI), University of Coimbra, Coimbra, Portugal

Alessandra Gallo Department of Biology and Evolution of Marine Organisms, Anton Dohrn Zoological Station, Naples, Italy

Jacqueline Gao-Li Sorbonne University, Institute of Biology Paris-Seine (IBPS), Biological Adaptation and Ageing, Paris, France

Henrike Gebhardt Department of Anatomy, University of Kiel, Kiel, Germany

Luiz Henrique Geraldo Institute of Biomedical Sciences, Federal University of Rio de Janeiro, Rio de Janeiro, Brazil

Peter Gunning Oncology Research Unit, School of Medical Sciences, University of New South Wales, Sydney, New South Wales, Australia

Jinghua Hu Department of Nephrology and Hypertension, Mayo Clinic, Rochester, MN, USA

Mayo Translational PKD Center, Rochester, MN, USA

Department of Biochemistry and Molecular Biology, Mayo Clinic, Rochester, MN, USA

Ralf Jacob Department of Cell Biology and Cell Pathology, Philipps University Marburg, Marburg, Germany

Kenneth B. Kaplan Department of Molecular and Cellular Biology, University of California, Davis, CA, USA

Noriko Koganezawa Department of Neurobiology and Behavior, Gunma University Graduate School of Medicine, Maebashi, Gunma, Japan

Beth S. Lee Department of Physiology and Cell Biology, The Ohio State University College of Medicine, Columbus, OH, USA

Yan Li Department of Nephrology and Hypertension, Mayo Clinic, Rochester, MN, USA

Department of Biochemistry and Molecular Biology, Mayo Clinic, Rochester, MN, USA

Zhenlin Li Sorbonne University, Institute of Biology Paris-Seine (IBPS), Biological Adaptation and Ageing, Paris, France

Maria Celeste Lopes Center for Neuroscience and Cell Biology, Institute for Biomedical Imaging and Life Sciences (CNC.IBILI), University of Coimbra, Coimbra, Portugal

Faculty of Pharmacy, University of Coimbra (FFUC), Coimbra, Portugal

Diana Isabel Lourenço Matias Paolo Niemeyer State Brain Institute of Rio de Janeiro, Rio de Janeiro, Brazil

Institute of Biomedical Sciences, Federal University of Rio de Janeiro, Rio de Janeiro, Brazil

Brooke K. McMichael Department of Physiology and Cell Biology, The Ohio State University College of Medicine, Columbus, OH, USA

Rolf Mentlein Department of Anatomy, University of Kiel, Kiel, Germany

Mathias Mericskay Sorbonne University, Institute of Biology Paris-Seine (IBPS), Biological Adaptation and Ageing, Paris, France

Vivaldo Moura-Neto Institute of Biomedical Sciences, Federal University of Rio de Janeiro, Rio de Janeiro, Brazil

Paolo Niemeyer State Brain Institute of Rio de Janeiro, Rio de Janeiro, Brazil

University of Rio Grande (UNIGRANRIO), Rio de Janeiro, Brazil

Ramesh C. Nayak Division of Experimental Hematology and Cancer Biology, Cincinnati Children's Hospital Medical Center, Cincinnati, OH, USA

Mahasin A. Osman Division of Biology and Medicine, Alpert Medical School, Brown University, Providence, RI, USA

Ara Parlakian Sorbonne University, Institute of Biology Paris-Seine (IBPS), Biological Adaptation and Ageing, Paris, France

Bruno Pontes Institute of Biomedical Sciences, Federal University of Rio de Janeiro, Rio de Janeiro, Brazil

Heide Schatten Department of Veterinary Pathobiology, University of Missouri, Columbia, MO, USA

Galina Schevzov Oncology Research Unit, School of Medical Sciences, University of New South Wales, Sydney, New South Wales, Australia

Tomoaki Shirao Department of Neurobiology and Behavior, Gunma University Graduate School of Medicine, Maebashi, Gunma, Japan

Qing-Yuan Sun State Key Laboratory of Reproductive Biology, Institute of Zoology, Chinese Academy of Sciences, Beijing, China

Hiroyuki Suzuki Faculty of Agriculture and Life Science, Hirosaki University, Hirosaki, Aomori, Japan

Elisabetta Tosti Department of Biology and Evolution of Marine Organisms, Stazione Zoologica Anton Dohrn, Naples, Italy

Julien Vos Oncology Research Unit, School of Medical Sciences, University of New South Wales, Sydney, New South Wales, Australia

Xuejun Wang Division of Basic Biomedical Sciences, Sanford School of Medicine of the University of South Dakota, Vermillion, SD, USA

Sabrina Zink Department of Cell Biology and Cell Pathology, Philipps University Marburg, Marburg, Germany

Part I
Overview of the Cytoskeleton

Chapter 1
Brief Overview of the Cytoskeleton

Heide Schatten

Introduction

The cytoskeleton consists of a complex network of fibers primarily including three families of protein molecules that assemble to form three main types of filaments: microtubules, intermediate filaments and microfilaments. Other components such as septins and centrosomes have been added in more recent years to be included under the larger umbrella of the cytoskeleton. These filaments are linked to each other and to different cellular components by hundreds of accessory proteins to allow intra- and intercellular communications and signal transductions for specific cellular functions. This chapter will briefly introduce the major cytoskeletal components. Several of these components and interactions are highlighted and discussed in more detail in specific chapters of this book.

Microtubules and Tubulins

Microtubules (MTs) are highly dynamic cytoskeletal fibers with an outer diameter of ca. 25 nm. They are composed of α/β subunit heterodimers that typically are assembled into laterally associated 13 protofilaments to compose one single cylindrical complete microtubule. The number of protofilaments can vary in different organisms. Microtubules display structural polarity with a slow growing minus end and a fast growing plus end. The minus end can be stabilized by attachment to cellular structures such as microtubule organizing centers (MTOC; centrosome),

H. Schatten (✉)
Department of Veterinary Pathobiology, University of Missouri, 1600 E Rollins Street, Columbia, MO 65211, USA
e-mail: SchattenH@missouri.edu

H. Schatten (ed.), *The Cytoskeleton in Health and Disease*,
DOI 10.1007/978-1-4939-2904-7_1

the Golgi apparatus, or cell membranes. Individual microtubules undergo phases of growth (polymerization) and shrinkage (depolymerization) in a process referred to as 'dynamic instability', allowing dynamic modulation of the microtubule cytoskeleton and its varied specific functions.

Gamma-tubulin is a member of the tubulin family whose functions include microtubule nucleation and organization. It is primarily associated with MTOCs at centrosomes but it has also been localized to other cellular compartments such as the Golgi and the plasma membrane where microtubule nucleation can take place. In non-polarized epithelial cells the majority of microtubules are nucleated by γ-tubulin from centrosomes and remain associated with centrosomes with their minus ends but the orientation changes in polarized epithelial cells when the majority of microtubules associate with their minus ends with γ-tubulin localized at the polarized cell membrane. The specific features of cellular polarization are discussed in Chap. 8 by Zink and Jacob.

Four more tubulins have been discovered in recent years, which are delta (Δ)-, epsilon (ε)-, zeta (ζ)- and eta (η)-tubulins [1]. Functions of these tubulins are mainly linked to eukaryotic centrioles and/or basal bodies.

Microtubules serve a variety of different functions including maintenance of cell shape, cellular transport of membrane vesicles, macromolecules and organelles such as mitochondria, cell motility, meiosis, mitosis, and cell division; they are critical for the formation of centrioles, immotile primary cilia, and motile cilia and flagella. The many different functions of microtubules are possible due to highly specific regulations that include a diversity of different factors such as Ca^{++}, pH, MT associated proteins, and posttranslational modifications (PTMs) which allows significant variability in dynamics and remodeling of the microtubule network to affect specific microtubule functions. Microtubule PTMs have recently been reviewed in detail [2] and include acetylation and detyrosination/tyrosination, PTMs that are oftentimes linked to microtubule stability. These PTMs also allow for specific associations of microtubules with the microtubule motor proteins dynein and kinesin; the plus-end directed microtubule motor protein kinesin and minus-end directed microtubule motor protein dynein are important for transport of cargo along microtubules to their functional destinations and therefore serve critical functions in cellular metabolism and intracellular signal transduction. The regulation of microtubule dynamics and stability further includes a role for a heterogeneous group of non-motor microtubule-associated proteins (MAPs) and microtubule-interacting molecules that provide additional functional diversity.

Several microtubule drugs are known to either inhibit microtubule polymerization (colcemid, colchicine, nocodazole, podophyllotoxin, and griseofulvin) or prevent depolymerization (taxol, paclitaxel). Several of these drugs have been validated as anti-cancer drugs to inhibit abnormal cell divisions based on their ability to interfere with microtubule functions in the mitotic apparatus, thereby preventing cancer cells from dividing.

Microtubule dysfunctions have been implicated in diseases and disorders with the best studied dysfunctions known for neuronal cells that is manifested in diseases such as Alzheimer's and Parkinson's. Microtubule dysfunctions are further observed in aging cells and in mitotic cells in which the highly labile microtubules become disorganized resulting in spindle abnormalities and aneuploidy that is associated with various diseases including cancer. These topics are specifically addressed in several chapters of this book.

Intermediate Filaments

Intermediate filaments (IFs) are composed of intermediate filament proteins that comprise a large and heterogeneous family of at least 65 different proteins that are subcategorized into six different types or classes in vertebrates. IFs are characterized as ropelike fibers with a diameter of ~10 nm. They are more stable to experimental treatments compared to microtubules and microfilaments. IFs provide cellular stability and allow intercellular communications in epithelial cells layers. IF are present in all cell types of the human body and they are developmentally regulated. Specific functions are seen in several cell types such as in epithelia (keratins), muscle (desmin), mesenchymal cells (vimentin), glia cells (GFAP), and neurons (neurofilament triplet proteins).

While IF dynamics and functions had been less explored compared to MTs and microfilaments (MFs), recent interest in IFs has resulted in significant new information on their structural and biochemical characteristics as well as IF regulation, assembly and disassembly, and their forms and functions [3,4].

Several diseases are associated with IF mutations or dysfunctions and include skin diseases with cytokeratin defects such as epidermolytic diseases. Vimentin and cytokeratin play a role in intra- and intercellular communications (reviewed in [5]). One type of IFs is organized into the nuclear lamins, a meshwork of filaments underlining the inner nuclear envelope. Defects in this organization, some caused by point mutations in IF proteins, is associated with laminopathies and other severe, inheritable multi-systemic diseases.

Microfilaments

Microfilaments (MFs) are composed of actin monomer molecules (g-actin; globular actin) to form filamentous (F) actin or microfilaments. The actin subunits form two-stranded helical polymers resulting in linear strands of polar filaments with a typical diameter of ca. 7–8 nm that are twisted around each other displaying an axial stagger of half a subunit.

Microfilaments have important functions in cell motility and can form a three-dimensional network to line the cell membrane or the lumen of microvilli. Other microfilament functions include membrane trafficking and shape changes. Actin filaments can be highly dynamic or they can be anchored, such as in muscle tissue. Cellular microfilaments can be organized in linear bundles, two-dimensional networks, and three-dimensional gels. The versatility of the actin cytoskeleton is achieved through different actin binding proteins that are discussed in detail in several specific chapters of this book. Microfilament-associated and microfilament-interacting molecules assure varied microfilament functions. The Arp2/3 (actin-related protein 2/3) complex serves as actin nucleation complex that is important for the formation of new actin filaments off the sides of existing microfilaments (reviewed by Sun and Kim [6]).

Centrosomes

Centrosomes are major Microtubule Organizing Centers (MTOCs) in cells that nucleate and organize microtubules during interphase and mitosis. In interphase, the centrosome complex is closely associated with the nucleus and organizes a radial microtubule formation that becomes reorganized into the mitotic apparatus after duplication and separation of the mitotic centrosome complex during S/G2 and early prophase, respectively.

As mentioned above, the major protein for microtubule nucleation is γ-tubulin that is integrated in the large γ-tubulin-ring complex (γTuRC) associated with the centrosome structure (reviewed in [7]). Hundreds of γTuRC-like rings may be associated with the centrosome matrix but the number changes with specific cell cycle stages. As will be discussed in more detail in Chap. 5, the large γ-TuRC contains 5-7 small complexes, the γTuSCs (around 280 kDa) that each comprises two molecules of γ-tubulin and one molecule each of GCP (γ-tubulin complex protein) 2 and 3 [8]. γTuRCs are enriched in mitotic and meiotic spindles where increased microtubule nucleation is important to generate microtubules for attachment to chromosomes, pole-to-pole microtubule organization and chromosome separation. As discussed in the specific chapter on centrosomes (Chap. x) centrosomes are further important for cytoskeletal coordination and functions. A typical somatic cell centrosome contains a pair of centrioles (reviewed in [7]) but acentriolar centrosomes are found in mammalian oocytes in which acentriolar centrosomes organize meiotic spindles (reviewed in [9,10]). Centrosomes have been best studied in somatic cells in which it has been shown that they are composed of a large number of centrosomal proteins, with at least 60 and perhaps 100 of them being directly associated with the interphase centrosome structure, and numerous others that are associated with centrosomes to perform cell cycle-specific functions (reviewed in [7]).

Septins

Septins are mentioned as an emerging class of cytoskeletal components that have been called a fourth novel unconventional component of the cytoskeleton [11]. Septins are a family of proteins that can form non-polar filaments or rings and they can interact with the actin and microtubule cytoskeleton. Septins had been discovered in budding yeast cells in which it was shown that they play a role in cytokinesis by recruiting different proteins to the contractile ring. It has been shown in mouse oocytes that Septin 2 is posttranslationally modified by SUMOylation and required for chromosome congression [12]. Other studies have shown that septin 1 is required for spindle assembly and chromosome congression [13] while Septin 7 is required for meiosis [14].

Together, the chapters presented in this book provide a selection of topics that are of current interest in the cytoskeleton field. They have been selected for their critical importance to research on the cytoskeleton and implications for cytoskeletal diseases and/or disorders. New discoveries are being made rapidly in this exciting field and the specific chapters have been chosen to stimulate new interest and further advances in the specific research areas.

References

1. McKean PG, Vaughan S, Gull K (2001) The extended tubulin superfamily. J Cell Sci 114:2723–2733
2. Schatten H, Sun Q-Y (2014) Posttranslationally modified tubulins and other cytoskeletal proteins: their role in gametogenesis, oocyte maturation, fertilization and pre-implantation embryo development. Adv Exp Med Biol 759:57–87. doi:10.1007/978-1-4939-0817-2_4
3. Goldman RD, Cleland MM, Murthy SN, Mahammad S, Kuczmarski ER (2012) Inroads into the structure and function of intermediate filament networks. J Struct Biol 177:14–23
4. Snider NT, Omary MB (2014) Post-translational modifications of intermediate filament proteins: mechanisms and functions. Nat Rev Mol Cell Biol 15:163–177
5. Goldman RD, Grin B, Mendez MG, Kuczmarski ER (2008) Intermediate filaments: versatile building blocks of cell structure. Curr Opin Cell Biol 20(1):28–34. doi:10.1016/j.ceb.2007.11.003
6. Sun SC, Kim NH (2013) Molecular mechanisms of asymmetric division in oocytes. Microsc Microanal 19:883–897
7. Schatten H (2008) The mammalian centrosome and its functional significance. Histochem Cell Biol 129:667–686
8. Murphy SM, Urbani L, Stearns T (1998) The mammalian γ-tubulin complex contains homologues of the yeast spindle pole body components Spc97p and Spc98p. J Cell Biol 141:663–674
9. Schatten H, Sun QY (2011) New insights into the role of centrosomes in mammalian fertilisation and implications for ART. Reproduction 142:793–801
10. Schatten H, Sun QY (2011) Centrosome dynamics during meiotic spindle formation in oocyte maturation. Mol Reprod Dev 78:757–768
11. Mostowy S, Pascale CP (2012) Septins: the fourth component of the cytoskeleton. Nat Rev Mol Cell Biol 13:183–194
12. Zhu JL, Lin SL, Li M, Ouyang YC, Hou Y, Schatten H, Sun QY (2010) Septin2 is modified by SUMOylation and required for chromosome congression in mouse oocytes. Cell Cycle 9(8):1607–1616
13. Zhu J, Qi ST, Wang YP, Wang ZB, Ouyang YC, Hou Y, Schatten H, Sun QY (2011) Septin1 is required for spindle assembly and chromosome congression in mouse oocytes. Dev Dyn 240(10):2281–2289. doi:10.1002/dvdy.22725, PMID: 21932310
14. Li S, Ou XH, Wei L, Wang ZB, Zhang QH, Ouyang YC, Hou Y, Schatten H, Sun QY (2012) Septin7 is required for orderly meiosis in mouse oocytes. Cell Cycle 11(17):3211–3218

Part II
Focus on Cytoskeletal Interactions

Chapter 2
Cytoskeleton Dynamics in Health and Disease: Role of Molecular Switches and Rheostats

Mahasin A. Osman

Introduction: Actin Dynamics Underlies Most Cellular Functions

Approximately 80 % of cellular proteins are of cytoskeletal nature and they provide the platform for nearly all cellular processes [1], ranging from DNA replication/repair, transcription, and cell motility to all aspect of vesicular traffic. An essential feature of the cytoskeleton is its dynamic nature, which is particularly important for the actin and the microtubule cytoskeleton. A plethora of specialized cellular proteins interact with the actin and microtubule cytoskeleton and mediate their spatio-temporal dynamics and thus facilitate the diversity and specificity of their numerous cellular functions. It is not surprising therefore that defects in the cytoskeleton associate with a wide range of human disease, including, but not limited to, cancer, heart disease, myopathies, skin disorders, neurodegenerative diseases like Alzheimer's disease (AD), Amyotropic lateral sclerosis (ALS), as well as cell death arising from disruption of mitochondrial function [2]. Most recently, downregulation of β-actin and oxidative posttranslational modifications has been suggested as an underlying factor in Rett syndrome, thus implicating cytoskeleton disorganization in the disease [3]. Accordingly, targeting the cytoskeleton therapeutically has potential for producing undesirable pleiotropic effects, but understanding the basic mechanisms that control its context-specific dynamics will be crucial to developing selective therapies. This chapter focuses on the regulation of the dynamic of filamentous actin and its crosstalk with the microtubule cytoskeleton through the Rho GTPase effector/regulator IQGAP1.

M.A. Osman (✉)
Division of Biology and Medicine, Alpert Medical School, Brown University, Providence, RI 02912, USA

Savannah State University, Savannah, GA 31404, USA
e-mail: Mahasin_Osman@Brown.edu; mo28@cornell.edu

H. Schatten (ed.), *The Cytoskeleton in Health and Disease*,
DOI 10.1007/978-1-4939-2904-7_2

Rho GTPases, N-WASPs and Arp2/3 Are Main Players of Actin Dynamics

Actin is a ubiquitous and widely conserved ~42 kDa protein present in all eukaryotic cell types [4]. Vertebrates have three isoforms; alpha, beta and gamma. Alpha actins are major components of the contractile apparatus such as muscle tissues whereas beta and gamma actins are main components of the cytoskeleton in most cell types. Beta-actin is a crucial player in a wide range of cellular functions, including the maintenance of cell shape and polarity, cell motility, protein trafficking and transcriptional regulation [5]. Actin exists in either a monomer globular form known as G-actin or a filamentous form known as F-actin that makes up large portions of the cytoskeleton. Polymerization of G-actin into F-actin is a highly regulated and reversible process, which provides the dynamic necessary for actin to influence multiple cellular functions. These functions include providing mechanical support that keeps cell shape and tracks for protein and organelle transport, facilitating cell motility and movement and regulating exocytosis and endocytosis [6, 7]. To carry out these functions, actin must interact with a diverse number of actin-binding proteins (ABPs) that cooperate in its cellular functions or most importantly regulate its polymerization–depolymerization cycles [8–10], commonly referred to as remodeling of the actin-cytoskeleton or actin dynamics. Assembly of F-actin into networks involves the actin branching regulator complex Arp2/3 [11], which nucleates new filaments from the sides of pre-existing filaments. This process requires activation of the Arp2/3 complex by the Neural Wiskott-Aldrich Syndrome Protein (N-WASP), or its relatives the hematopoietic WASP and the WASP-verprolin-homologous WAVE/Scar proteins [12]. WAVE, a subunit of the hetero-pentameric WAVE regulatory complex (WRC), is a downstream target of the Rho GTPase Rac1 [13–15] whereas N-WASP appears to be directly activated by the Rho GTPase Cdc42 and phosphatidylinositol 4,5-bisphosphate (PIP2) [16–19]. This differential interaction with two different GTPases suggests involvement of a complex crosstalk between Rac1 and Cdc42 in regulating actin dynamics and/or specialization in producing distinct actin-based structure–functions, which is discussed below.

A Complex Crosstalk Among the Rho GTPases in Actin Dynamics

The canonical Rho GTPases Cdc42, Rac1 and RhoA comprise a subfamily of the Ras superfamily of small GTPases, and are best known for their coordinate regulation of the actin cytoskeleton dynamics and vesicular traffic involved in both exocytosis and endocytosis [20, 21]. They also impact a plethora of other cellular functions that mainly depend on cytoskeleton dynamics. These include, but not limited to, cell-to-cell and cell-to-extracellular matrix adhesions, gene expression, morphogenesis, cell cycle, cytokinesis, cell migration, neurite outgrowth, and phagocytosis [6, 21–23]. In addition, Rho GTPases regulate microtubule-dependant cell polarity through actin-binding formin proteins like the diaphanous mDia [24, 25], supporting a role in regulating crosstalk between the actin and the microtubule cytoskeleton. Rho GTPases are molecular switches that, in response to a variety of upstream signals like nutrients, hormone and mitogens, cycle between a signaling-active GTP-bound and a signaling-inactive GDP-bound states. This cycling is accompanied by conformational changes in their GTP-binding switch 1 and effector-binding switch 2 regions [26]. This intrinsic cycling capacity is controlled by signal-dependent interactions with a set of regulatory proteins (reviewed in [27]). Guanine nucleotide exchange factors (Rho-GEFs) catalyze GDP dissociation and GTP binding, thus switching the GTPase to the active state, whereas GTPase-activating proteins (Rho-GAPs) catalyze GTP hydrolysis and thus returning the GTPase to the inactive state. Layered upon this, Rho GTPases activity also is regulated spatially by shuttling between membrane-bound and cytosolic states. In response to an activating signal, the GTPases undergo a series of carboxyl-terminal modifications resulting in covalent attachment of prenyl (farnesyl or geranyl) lipid anchors that attach them to membranes [28–30]. Dissociation from membrane-anchoring is catalyzed by guanine nucleotide-dissociation inhibitors (RhoGDIs) that bind the geranylgeranyl lipids and release inactive Rho proteins into the cytosol [31–33].

When activated, Rho GTPases interact with numerous effector proteins that help decode, amplify and transmit their signal to downstream proteins responsible for executing that signal into a specific cellular response. The consequence of this process is that small changes upstream in a Rho GTPase-pathway can translate into profound cellular responses downstream. Earlier studies using genetic analyses with classical GTPase dominant mutants revealed a level of specialization in formation of actin-based structures, where Cdc42 controls filapodia and microspike formation [34, 35], Rac1 promotes lamellipodia and membrane ruffles [36], and RhoA promotes actin stress fiber and focal adhesion in the back of migrating cells [37, 38]. Moreover, as initially revealed from studies in the budding yeast, all three GTPases also are required for G_1 to S phase transition of the cell cycle in mammals [39], and they induce gene transcription through the Jun N-terminal kinase (JNK) module of the mitogen-activated protein kinase (MAPK) pathways [40, 41]. Significantly, while earlier studies have implicated both Rac1 and Cdc42 in membrane traffic, Rac1 in particular was found to impact both exocytic and endocytic pathways [42]. This role may have significance in the hypothesis that Rac1 activity lies at the interface of coupling insulin and glucose homeostasis [43] that will be discussed under the exocytosis section below.

However, these earlier studies also revealed a level of complexity in Rho GTPase actions and hinted that different GTPases often coordinately regulate cellular functions via highly orchestrated crosstalk [44]. Indeed, recent advances in live cell imaging revealed exquisite spatiotemporal orchestration of the activity of these proteins in leading edge versus retracting back during directional cell migration [45] or during signal-driven exocytosis in different cell types [46, 47]. This mechanism likely applies to the myriad of their cellular functions in which they are involved. Adding to this complexity is their ability to bind a large variety of effecter proteins by which they gain selectivity to impact multiple functions at once, thus justifying their description as being multispecific [27]. This indicates that they gain both functional diversity and specificity via protein–protein interaction. Moreover, while their regulators, GEFs, GAPs and GDIs, can be Rho subfamily-specific, often they are not GTPase-specific, which means that a regulator can impart on multiple GTPases and conversely a GTPase can be regulated by multiple regulators of the same kind [31, 48–52]. Furthermore, Rho regulators themselves are regulated by complex mechanisms that include posttranslational modifications, auto-inhibition, subcellular localization as well as positive and negative feedback loops that determine specificity and range of signal durability (reviewed in [27]). Even more complexity is demonstrated by evidence that Rho regulators may have GTPase-independent functions [53], and that certain Rho GAPs may act as potential effectors of Rho GTPases [54, 55], reminiscent of p120RasGAP, which was implicated as an effector for Ras signaling in specific situations [56].

For all the reasons Stated above, studies utilizing overexpression of dominant mutants or knockdown of GTPases or their regulators will potentially not only globally interfere with multiple signaling pathways, but also likely will target more than one GTPase and produce confusing pleiotropic effects. Importantly, Rho GTPases do not bind F-actin directly neither in vitro nor in cells [57, 58], and thus must selectively choose from a menu of actin-binding effector proteins to execute, in a specific manner, the myriad of cellular functions involving actin dynamics. Furthermore, ensuring the selectivity, sensitivity and fidelity of signal transmission requires proper communication between a given Rho GTPase, its Rho regulator and cognate effector in time and space [27].

One of the best-known mechanisms to achieving this level of signal specificity is to isolate signaling components into modules via scaffold proteins [59]. Within a given module, scaffolds often play far more important roles than just simple scaffolding [60]. A relevant prototypical regulatory scaffold is IQGAP1 [43, 61], a RasGAP domain (GRD)-containing protein with selective affinity for the GTP-bound Rac1 and Cdc42 [62]. IQGAP1 binds to F-actin directly and cross-links actin filaments into interconnected bundles [63, 64]. For its potent regulatory activity of F-actin assembly, it has been dubbed the "master" regulator of actin dynamics [65]. It also binds other cytoskeletal elements [61] and is capable of not only mediating Cdc42 or Rac1 signaling to the cytoskeleton selectively, but also of regulating Cdc42 and Rac1 activities and crosstalk to execute a diversity of cellular functions with greater specificity. In addition, IQGAP1 also binds other GTPases like RhoA. As such and because of its modular feature, IQGAP1 is uniquely positioned both as an effector protein and as a context-specific signaling rheostat. These properties are discussed in the following sections.

Structural Basis of IQGAP1 as a GAP-Mimic and Molecular Rheostat

IQGAP1 is a modular protein that connects cell signaling and the cytoskeleton dynamic to diverse cellular functions. Rather than an inert scaffolding protein, IQGAP1 is best described as a rheostat of dynamic cellular functions that includes actin and microtubule dynamics. Whereas the budding yeast has one IQGAP ortholog; Iqg1p, humans have three isoforms; IQGAP1, IQGAP2 and IQGAP3. The three isoforms have similar primary sequence homology and modular structure, but they differ in tissue distribution and cellular functions [43, 61, 66, 67]. For relevance, this chapter is focused on IQGAP1, which is the most studied. Although, it has often been referred to as scaffold, its cellular activity can appropriately be described as a molecular rheostat [43]. This is primarily because while it serves as a scaffold to assemble a variety of subcellular modules, it modulates the activity of these modules in response to extracellular signals by directly or indirectly binding to a host of surface receptors. It appears that human cells utilize IQGAP1 where dynamic cellular responses must occur rapidly and with precise specificity. This function is facilitated not only by its modular nature that enables it to bind a large number of diverse signaling and structural proteins, but also by its ability to undergo phosphorylation-dependent conformational switch [43, 61]. Thus IQGAP1 serves both as a signaling as well as an effector molecule in many cellular functions, including in actin dynamic. Accordingly, IQGAP1 is a large protein containing 1657 residues that migrates at ~190 kDa on a 10 % polyacrylamide gel and it can also exist as an ~380 kDa homodimeric protein in solution [63]. To date, at least seven distinct functional domains and motifs have been identified on its protein structure (Fig. 2.1), which is discussed briefly below.

The N-terminal region of IQGAP1 contains a single calponin homology domain (CHD), which directly binds and bundles F-actin [63]. Downstream of the CHD are six putative coiled-coil specific IQGAP1 repeats (IR) followed by a Tryptophan-rich WW domain that mimics Src Homology (SH) domains in binding Proline rich domains. The IR-WW associates with the Exocyst complex, mTORC1, Akt [70–72] and with ERK1/2 [73]. Following the IR-WW domain are four IQ motifs that associate with calcium/calmodulin, myosin light chain (Mlc1), the MAPK b-Raf, MEK1/2 [74], and several surface receptors (see below). The RasGAP-related domain (GRD) of IQGAP1 (Fig. 2.1) is required, but not sufficient for binding activated forms of the Cdc42 and Rac1 [62, 63, 75–79]. Because of substitution of the catalytic arginine finger with a threonine in this GRD domain, IQGAP1 lacks enzymatic GAP activity [80]. This allows it to mimic a GAP and thus traps active Cdc42 at specific membrane sites without catalyzing GTPase activity, which explains previous observations why IQGAP1 expression increases the level of active Cdc42-GTP in cells [71, 81]. This feature may play an important unresolved role of IQGAP1 in dynamic cellular functions, such as allowing it to specify regions of polarized growth as in tubulogenesis, localized hot spots for signal-regulated exocytosis, or spatially activating signaling pathways for directed cell migration and invasion [23, 43, 61].

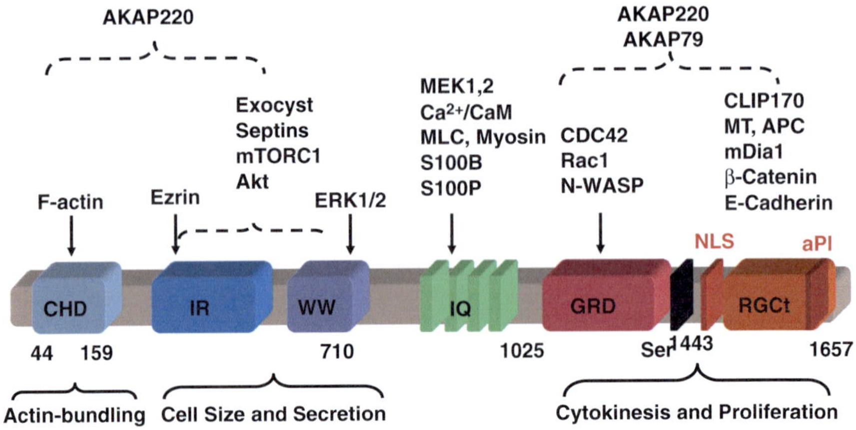

Fig. 2.1 Schematic diagram of human IQGAP1's domain structure, some binding-partners, and cellular function. *CHD* calponin homology domain, *IR-WW* IQGAP1-repeats (IR) and the tryptophan (WW) repeats, *IQ* four isoleucine and glutamine rich motifs, *GRD* Ras GTPase-activating protein-related domain, *RGCT* RasGAP-C terminus (RGCT) domain; the critical Ser-1443 is indicated, *NLS* nuclear localization signal, *aPI* C2/PH-like domain that binds the phospholipid PIP3. The A-kinase-anchoring protein 220 (AKAP220) appears to bind both the N-terminal and C-terminal regions of IQGAP1 with the exclusion of the IQ motifs in human mammary MCF-7 or HT1080 human fibrosarcoma cells [68], whereas AKAP79 was found to bind the C-terminal region of IQGAP1 in pancreatic β cells [69]

At the extreme C-terminal region of IQGAP1, resides the RGCt domain, which binds β-catenin and E-cadherin members of the Wnt signaling pathway, involved in modulating epithelial adhesion, and the formin mDia1, which catalyzes processive barbed end assembly of F-actin [82–85]. This domain also binds the cytoplasmic linker protein 170 (CLIP170) and the tumor suppressor adenomatous polyposis coli (APC) by which it anchors and modulates the microtubule cytoskeleton [86–88]. A nuclear localization signal (NLS) at the end of the RGCt domain mediates IQGAP1's cell cycle-dependent entry into the nucleus [89] where IQGAP1 may serve as transcriptional co-activator with β-catenin or in DNA replication. An atypical phosphoinositide (aPI)-binding domain, much similar to the Pleckstrin homology (PH) domains found in the PI3K-Akt and mTOR family, binds Phosphatidylinositol (3,4,5)-triphosphate (PIP3) [90], a phospholipid that activates signaling components such as Akt. PH domains also mediate binding to heterotrimeric G-proteins and protein kinase C (PKC) [91, 92], which are also binding-partners of IQGAP1 (see below). Altogether, these structures facilitate IQGAP1's spatiotemporal control of cell signaling events that impact aspects of cell proliferation such as size, division axis orientation, differentiation, epithelial polarity as well as cell migration and phagocytosis. The unifying role of IQGAP1's function in all these processes appears to be signal-controlled vesicular transport in which dynamic F-actin assembly plays a central role [23, 43, 61]. Accordingly, IQGAP1 has been implicated in multiple cellular functions, some of which are discussed below.

Cellular Functions of IQGAP1

A Master Regulator of Cytoskeleton Dynamics

A Modulator of F-Actin Dynamic

IQGAP1, the effector and regulator of Cdc42 and Rac1, has been established as a dynamic regulator of branched F-actin formation through direct binding to F-actin and activation of N-WASP.

Binding to F-actin and Cdc42 is a conserved feature of IQGAP1 across species [93, 94]. Co-immunoprecipitation [94] and co-sedimentation [95] experiments demonstrated that the budding yeast ortholog, Iqg1p, binds F-actin and influences cell polarity (bud-site selection and growth), cytokinesis and secretion [6, 94, 96]. In vitro experiments demonstrated that mammalian IQGAP1 cross-links F-actin into bundles and gels [63, 64], in a manner that is negatively regulated by calcium/calmodulin (Ca^{2+}/CaM) [97], which induces its dissociation from Cdc42 [76]. Thus two distinct signaling pathways, Ca^{2+}/CaM and Cdc42, appear to orchestrate IQGAP1's regulation of F-actin assembly. Subsequent assays showed that a single CHD of monomeric IQGAP1 binds F-actin with an ~47 μM affinity and that like full length IQGAP1, it localized with polymerized actin filaments at the cell cortex [98]. This evidence was the first to suggest IQGAP1's role in stimulating F-actin nucleation. Indeed, spectrofluorometric assays combined with direct visualization with total internal reflection (TIRF) microscopy, demonstrated that IQGAP1 binds N-WASP and mediates Arp2/3-N-WASP nucleation of F-actin side branches, independently of, but also cooperatively with, activated Cdc42 [83]. Because this activity occurred within a narrow concentration range, it hinted to interplay of stimulatory and inhibitory mechanisms by IQGAP1 [83].

Several independent studies revealed the existence of such mechanisms. First, kinetic assays of branched actin polymerization with N-WASP and Arp2/3 complex [84] and insulin exocytosis and cell migration assays using several mutants of IQGAP1 [71] revealed its positive and negative effects on F-actin assembly and exocytosis, respectively and suggested that it undergoes reversible auto-inhibition. Cell biological and biochemical assays showed that IQGAP1 controls N-WASP activity and localization to Arp2/3 complex at preexisting actin filaments in a Cdc42-dependant manner [84]. Whereas the C-terminal half of IQGAP1 activates N-WASP by binding to its BR-CRIB (basic region-Cdc42-Rac interactive binding) domain, the N-terminal half of IQGAP1 inhibited this activation by masking the C-terminal half of IQGAP1 through intra-molecular folding [84]. Second, cell transformation and migration assays demonstrated interplay between the N-terminal and the C-terminal regions of IQGAP1 [72]. Whereas the C-terminal domain promoted transformed phenotypes the N-terminal region inhibited such phenotypes, but both domains were required for efficient cell invasion in transwell assays and migration in wound-healing assays [72]. This evidence provided the functional framework for IQGAP1 dynamic auto-inhibition (see below), which is regulated by reversible phosphorylation of the C-terminal Serine 1443, as supported by functional analyses of phosphorylation-inhibitory and mimetic point mutant s–mutants [72]. It also suggested that these two functions converge at the plasma membrane where IQGAP1's regulation of F-actin dynamics provides a mechanism for regulating exocytosis [43].

However, the mechanics and the functional consequences of IQGAP1's binding and assembly of F-actin are far from being fully understood. Three lines of evidence support the notion that interplay of IQGAP1-Cdc42 and IQGAP1-Rac1 complexes plays a role in the stimulatory-inhibitory cycles of IQGAP1-mediated actin dynamics. First, N-WASP appears to bind to both C-terminal and N-terminal regions of IQGAP1 [83, 84], whereas both Cdc42 and Rac1 bind to sites in the C-terminal region of IQGAP1. Second, IQGAP1 and N-WASP bind different regions of Cdc42 [99]. Third, IQGAP1 binds either Cdc42 or Rac1 alone and influence F-actin nucleation. Together, this evidence hints to a GTPase-independent function of IQGAP1-N-WASP and likely differential actions of Rac1-bound and Cdc42-bound IQGAP1.

Additionally, IQGAP1 appears to regulate actin dynamics through another pathway that requires the formin mDia1 [82]. Formins are ubiquitous family of proteins that regulate the dynamics and organization of both the actin and the microtubule cytoskeletons [100]. IQGAP1 binds mDia1 and induces phagocytosis [82], which may represent a specialized function of an IQGAP1-mDia pathway in F-actin dynamics. However, it is conceivable that mDia1-IQGAP1 cooperate in stress-responsive actin networks assembly inside the nucleus where both proteins have been individually implicated [101, 102]. Yet another possibility involves actin-microtubule crosstalk, particularly in exocytic vesicle transport (see below). Evidence in support of this view is that while the mDia1-IQGAP1 interaction was dispensable for actin polymerization, it was required for microtubule capture at the leading edge of migrating breast carcinoma cells [103], a function previously demonstrated for IQGAP1 interaction with APC and CLIP170 [86, 87]. Thus much work lies ahead for sorting the mechanics and specific functions of the variable IQGAP1 complexes in a context-dependent manner.

A Modulator of Microtubule Dynamics

Several studies, both in yeast and mammals, have implicated IQGAP1 in microtubule dynamics associated with cell polarity and trafficking [6, 94], via binding to microtubule plus end tracking proteins [86–88]. The IQGAP1 RGCt directly binds the armadillo repeats of APC and the two proteins interdependently couple microtubules and F-actin network at the leading edge of migrating cells as revealed by pharmacologic inhibitors, RNAi, and mutant analyses [87]. A possible mechanism by which IQGAP1 couples microtubule and F-actin dynamics is that IQGAP1-APC complex is required for targeting CLIP170 to the microtubules and F-actin networks at the leading edge of migrating cells [86, 104]. CLIP170 can bind microtubule plus ends and nucleate microtubules from pure tubulin near the cell cortex [105]. In addition, IQGAP1 directly interacts with the CLIP associated protein 2 (CLASP2), albeit in a GSK-3β-dependent manner [88]. CLASP2, in turn, docks to the microtubule-plus End binding 1 (EB1) protein that links microtubules to the cell cortex [106]. These interactions suggest that IQGAP1 serves as a scaffold not only to regulate microtubule-dynamics, but also to mediate spatiotemporal coupling of microtubules and F-actin dynamics in different cell types. Interestingly, these interactions require the activity of both Rac1 and Cdc42, and overexpression or knockdown of either APC, IQGAP1 or CLIP170 affects both microtubules and F-actin dynamics, disrupts the localization of cognate partners and impairs directional migration [86, 87].

Collectively, this evidence suggests requirement of a fine balance between these proteins and existence of a regulatory crosstalk between Rac1-Cdc42 as well as involvement of additional signaling proteins such as mTOR and the MAPK-Akt axis, which control the activity of GSK-3β [43, 61, 70]. Evidence for this view is that expression of an IQGAP1 mutant defective in Rac1/Cdc42 binding induces multiple leading edges [86, 87]. Moreover, both IQGAP1 and CLIP170 are rapamycin-sensitive binding-partners of mTOR [70, 107]. Apparently cells adopt this fundamental IQGAP1-mediated mechanism universally not only during cell migration and likely vesicular transport, but also during developmental events where generating polarized structures require asymmetrical cell division, as discussed below.

A Mediator of Actin-Microtubule Crosstalk in Asymmetric Cell Growth

The role of IQGAP1 in asymmetric cell growth was first discovered in the budding yeast *Saccharomyces cerevisiae* (Fig. 2.2). Bud-site selection is a highly regulated asymmetrical cell growth and division event in which yeast Iqg1p exerts control both upstream and downstream of Cdc42 and determines the directional polarization of actin and microtubules [94, 96]. Likewise, lumen formation, which is required for tubulogenesis and cellular elongation, which is required for formation of structures like filapodia both depend on asymmetric organization of actin and microtubule networks [109]. Numerous actin-binding proteins, including the Ezrin-Radixin-Moesin (ERM) family of proteins, have been implicated in these processes [110], presumably guided by synaptotegmin [111]. Synaptotagmins are conserved SNARE-binding proteins that induce Ca^{2+}-dependent insulin exocytosis [112]. Given the conserved role of IQGAP1 in cell polarity, insulin exocytosis and binding to the SNARE Syntaxin 4 [23, 71] as well as to Ezrin [113], makes it plausible that it employs the same fundamental mechanism in tubulogenesis and lamillepodia-filapodia formation, as well as in regulated exocytosis. Much mechanistic analyses will be required for identifying the determinants of each of these distinct functions in different cell types.

Some of these mechanisms have begun to be unraveled. In several cell types, IQGAP1 localizes similar to centrosome (the microtubule-organizing centre or MTOC) proteins at the midbody (Fig. 2.2a) of mitotic cells [70]. Silencing or overexpressing IQGAP1 results in misorientation of MTOC and impairs directional cell migration [72, 87, 114]. This function is conserved in yeast, as null mutation of

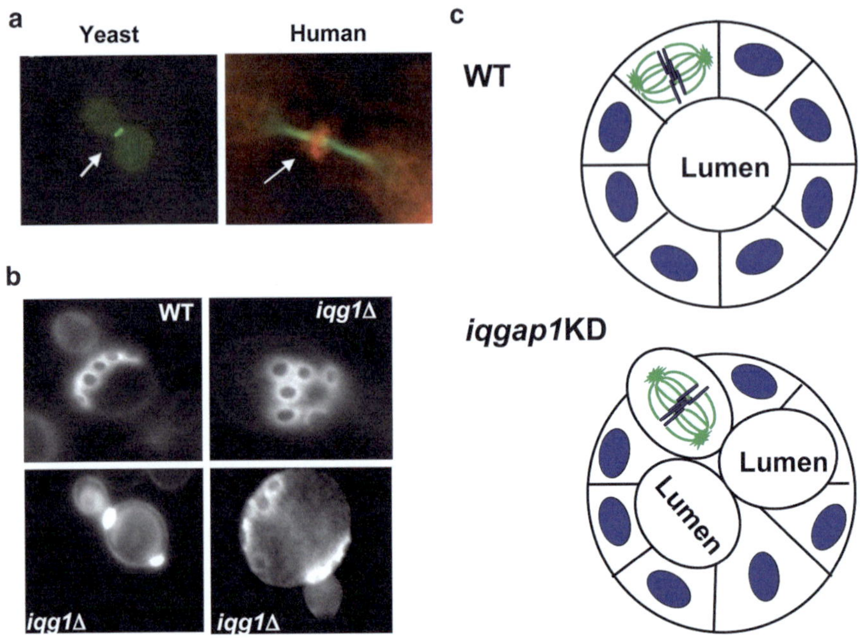

Fig. 2.2 IQGAP1 has a conserved role in directed cell division and morphogenesis. (**a**). Yeast and human IQGAP1 localize at the cytokinetic plate and influence cytokinesis ([96]; Osman unpublished; [70]). (**b**). Yeast Iqg1p controls directed cell division leading to budding. Wildtype (WT) haploid yeast bud in axial pattern. Deletion of *iqg1* leads to two types of budding patterns; random, indicating upstream effect like in *cdc42* null cells, and bipolar, a feature of diploid cells [96]. (**c**). Human IQGAP1 controls directed cell division leading to tubulogenesis. RNAi-mediated knockdown of IQGAP1 leads to misorientation of mitotic spindles and multilumen formation in epithelial cells growing in 3D cultures [108]

Iqg1p also misorients microtubules and directed cell division, leading to random budding (Fig. 2.2b, [94, 96]). Similarly, knockdown of mammalian IQGAP1 causes defective morphogenesis and tubulogenesis by misorienting the plane of cell division (Fig. 2.2c, [108]). In several epithelial cells, the IQ motifs of IQGAP1 directly bind the kinase domain of the EGFR [115] and both proteins have been shown to localize to the epithelial basolateral membrane [104, 116]. Recent evidence showed that EGFR directs the localization of IQGAP1 to the basolateral membrane in 3-D cell cultures, which in turn is required for the basolateral localization of the nuclear mitotic apparatus protein (NuMA) and formation of a single lumen [108]. NuMA, normally restricted to the interphase nucleus of non-mitotic cells, rapidly redistributes to the cytoplasm in early mitosis and form a complex with dynein/dynactin to tether microtubule minus ends to the spindle pole centrosome and persist in the cytoplasm until the onset of anaphase [117]. Overexpression of NuMA [117] or disruption of IQGAP1-EGFR binding [108] causes multipolar spindle formation, which leads to multilumen formation [108]. These findings are consistent with the evidence that overexpression of IQGAP1-IR-WW domain, which acts as a dominant negative mutant in cell proliferation, leads to multinuclear formation and cytokinesis arrest [72]. Altogether these findings demonstrate a universal role for IQGAP1 in anchoring both cortical microtubules in non-dividing cells and astral microtubules in mitotic cells, while mediating F-actin-microtubule crosstalk by cooperating with appropriate actin- or microtubule-binding proteins. This function provides more mechanistic understanding for its role in morphogenesis and the development of cancer. It remains unclear how Cdc42, which has also been implicated in restricting lumen formation, influences IQGAP1 function in this process. It is likely that unregulated IQGAP1-Cdc42 binding would promote multilumen formation since this binding requires active Cdc42 and phospho-IQGAP1 and disrupts epithelial adhesion and polarity [43, 61] similar to an effect brought about by EGF treatment of cells [108]. Moreover, these interactions likely are dynamic in time and space and involve differentially regulated crosstalk. An important role for an IQGAP1-mediated F-actin-microtubule interaction may occur during viral particles egress. Signaling through the RhoA-mDia1 axis was found to increase microtubule dynamics at the peripheries of cells infected with vaccinia virus and enhance the release of viral progenies [118]. Interactions of IQGAP1 with these components and its regulation of exocytosis and cytokinesis, support its role as effector in viral egress discussed later in this chapter.

Finally, although IQGAP1 appears to regulate many cell functions via combining negative and positive roles to modulate microfilaments dynamics, specialized cells may select one of the roles as suited for their specific function. One such example occurs during T-cell activation where IQGAP1 appears to modulate T-cell activity by interplaying microtubules and F-actin dynamics [119, 120]. However, unlike its dual roles in microfilament dynamics in cell motility or in exocytosis, in $CD8^{+}$ T cells, IQGAP1 negatively regulates the T-cell receptor (TCR)-mediated F-actin assembly during immunological synapse formation [120].

IQGAP1 Interacts with Several Other Cytoskeletal Elements

Besides F-actin and microtubules, IQGAP1 binds several other types of microfilaments. As this has been reviewed recently [61], a brief update will be provided in this section. IQGAP1 has a conserved role in organizing septin filaments and function. Septins comprise a family filament-forming GTPases that have been heavily studied for their involvement in cytokinesis [121–123]. Increasing evidence supports their participation in the various cellular functions involving IQGAP1, including adheren junctions remodeling [124], bacterial infection [125] and membrane dynamics at the cell cortex during cell motility [126]. Septin filaments polarization has been shown to require microtubule-dependent translation-coupled transport of septins on endosomes [127]. The microtubule-binding IQGAP1 also binds septins and is required for Sept2 localization and organization in mammalian cells [71]. Similarly, loss of yeast Iqg1p perturbs the localization of the yeast septin Cdc12 as well as the polarization of microtubules [94, 96]. This together with evidence for

IQGAP1 endosome function [128] and microtubule capture, support IQGAP1 participation in septin filaments polarized transport and/or anchor at the cell cortex.

Crosstalk between septin and actin filaments has also been demonstrated. It has long been observed that perturbation of septin organization in mammalian cells causes loss of actin bundles, implicating septin in actin bundles organization [129]. New evidence from studies in the fruit fly show that septins alone crosslink, bundle and bend actin filaments into rings, and thus participate in actin organization [130]. Given the conserved interactions of IQGAP1 with septins and role in cytokinesis [70–72, 96], it will be important to analyze how it mediates actin–septin function during cytokinesis and other cellular processes.

In addition, IQGAP1 couples to other types of cellular filaments; validated microarray analyses revealed a correlative stepwise upregulation of IQGAP1 and vimentin during development of hepatocellular carcinoma in a rat model [131]. Vimentin is an intermediate filament protein involved in the regulation of cell morphology primarily expressed in mesenchymal cells. Its expression in epithelial cells marks an epithelial-to-mesenchymal transition (EMT) during morphogenesis or cancer development [132–134]. While evidence for physical interaction between vimentin and IQGAP1 is lacking, bioinformatics-based network analyses supports such interaction [131]. Similarly, microarray-based data suggest correlation of IQGAP1 and cytokeratin (K19) in human hepatocellular cancer [135]. Taken together this evidence implicates IQGAP1 as a master cytoskeletal regulator controlling cell shape and polarity. An important question arises as to how might IQGAP1 bind so many partners and impact so many functions. It is clear that its modular nature allows for a variety of binding partners, modulated by a variety of posttranslational modifications in IQGAP1 as well as in the different partners. An important feature is that these modifications promote conformational changes, enabling IQGAP1 to act as a switch (Fig. 2.3, see below) and in turn allows for more interaction diversity. Such switch mechanism is prevalent in regulating the cytoskeleton, as it is adopted by many regulators, and is discussed below.

Autoinhibition as a Common Regulatory Mechanism in Actin Dynamics

A variety of mechanisms are involved in regulating the variable players of actin dynamics we discuss in this chapter. Prominent among these are signal-dependent recruitment to the plasma membrane and posttranslational modifications. However it appears that these mechanisms primarily input into autoregulation by intramolecular interactions, resulting in autoinhibition (Fig. 2.3). Biochemical and structural studies have uncovered autoinhibition as an emerging critical mechanism for proper response to upstream signals and appropriate selection of regulatory proteins and cognate downstream pathways. Furthermore, autoinhibition represents an efficient and reversible mechanism for providing tight "on-site" repression [136] of proteins involved in highly dynamic biological processes. Indeed, one of the best-known proteins regulated by this mechanism is the Ras GEF son of sevenless or Sos [137, 138]. Therefore, it appears to be the mechanism of choice adopted by many

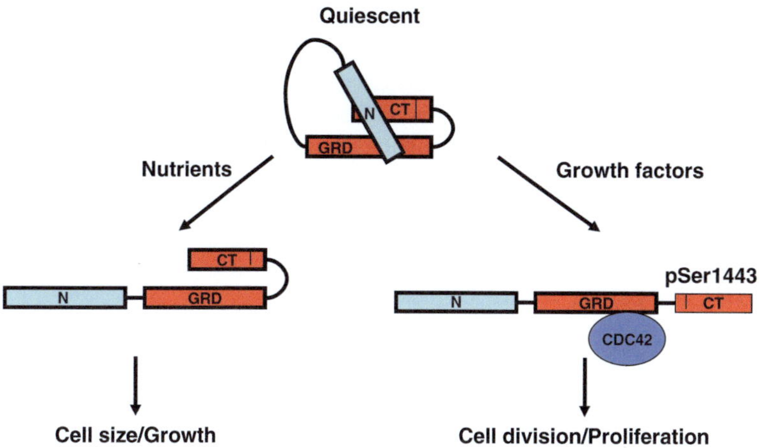

Fig. 2.3 IQGAP1 adapts auto-inhibition as a regulatory mechanism. Biochemical and functional analyses in multiple model cell lines revealed that the N-terminal region of IQGAP1 interacts with the C-terminal regions supporting the notion of a folded autoinhibited structure in quiescent cells [81]. *Left*, in nutrient-activated cells, opening of the N-terminal region leads to increased insulin secretion and cell growth [71]. *Right*, in growth factor-activated cells, phosphorylation of IQGAP1 on Ser^{1443} and binding of Cdc42 open the C-terminal fold leading to increased cell division and proliferation [72]. Cell motility requires dynamic conversion of the two forms [72]

factors in the IQGAP1's pathway, including regulators of small GTPases and the actin polymerization machineries.

Evidence for this mechanism was initially inferred from biochemical studies, and the definitive evidence was provided by structural studies of the RhoGEFs Vav1, which contains Dbl-homology (DH) domain [139]. A Vav1 N-terminal region, found in the GTPase interaction site and encompassing the Tyr174 Src-family kinase recognition site, masks the catalytic DH domain. Phosphorylation or truncation of this region results in activation of the GEF [139]. It is therefore not surprising that IQGAP1 activity is regulated by phosphorylation-sensitive autoinhibition (Figs 2.3 and 2.4 below) and that N-WASP represents a prototype of this mechanism. On this basis, N-WASP has been developed into fluorescence resonance energy transfer (FRET) biosensor that allows for in vivo visualization of its cellular activities [140]. Like IQGAP1, N-WASP is a multidomain protein, which remains autoinhibited by interaction between the CRIB (Cdc42- and Rac-interactive binding) domain and the catalytic VCA (Verprolin, cofilin, acidic) domain. Binding of active Cdc42 to the CRIB and phosphatidylinositol 4,5-bisphosphate (PIP2) to the basic region (BR), unmask the VCA domain, thus synergistically activate N-WASP [17, 141]. Similarly, Cdc42 binding to IQGAP1 opens its structure, which in turn opens N-WASP structure. Formins, such as the IQGAP1-binding partner mDia1, also are autoinhibited through intramolecular interactions and activated by Rho GTPases and additional proteins [100, 142]. Thus the factors involved in relieving the autoinhibition include Rho GTPAses, phosphorylation as well as interacting proteins that themselves might be regulated by autoinhibition. It is possible that this mechanism evolved to allow for functional versatility as well as quick response by preformed protein modules residing at specific cellular sites to different signals.

The following sections discuss how different cell types selectively utilize this property of IQGAP1 to regulate distinct functions like motility, invasion, protein traffic or immune response.

IQGAP1 Signaling to the Cytoskeleton

The cytoskeleton spans the cytoplasm, extends into the cell nucleus while anchoring the cell to the extracellular matrix and to its neighbors. Thereby it mediates the link with the environment, facilitates intercellular communications and impacts gene expression, by which the cell responds to these communications through signaling networks. As an actin-binding protein localized to the cell cortex, the cell–cell contacts and to the nucleus, IQGAP1 plays a central role in facilitating cell–cell, cell–matrix and cell–environment communications, as discussed below.

IQGAP1 Is a Rheostat of mTORC1 and MAPK Signaling

The mammalian target of rapamycin complex 1 (mTORC1) and complex 2 (mTORC2) both have been implicated in regulating the actin cytoskeleton [23]. Components like Rictor, the defining subunit of mTORC2, were reported to have an mTOR-independent function in actin dynamics in adipocytes [143]. The mTORC1 and mTORC2 are composed of shared and distinct subunits and are central regulators of cell growth and metabolism, and cell survival and proliferation respectively [144]. The catalytic "mechanistic" mTOR subunit, which is shared between the two complexes, is a serine/threonine protein kinase that belongs to a family of phosphatidylinositol kinase-related kinases that includes the Phosphoinositide 3-kinase (PI3K), and possibly IQGAP proteins. The effector of PI3K, Akt, also known as protein kinase B (PKB), is activated on the crucial S473 by mTORC2 (rictor-mTOR complex) in response to growth factor, and facilitates the activation of Akt on T308 by 3-phosphoinositide-dependent protein kinase 1 (PDK1) [145]. This sequential phosphorylation renders Akt full activity in cell proliferation, survival and metabolism [146]. However, several independent studies demonstrated that mutation of Rictor's Thr-1135, the site known to be responsible for activating Akt S473 and mediating the mTORC1-S6K-inhibition (Negative Feedback Loop, NFL) of Akt Ser473, does not alter the growth factor-dependent phosphorylation of Akt on Ser473, suggesting that mTORC2 or Thr-1135 site is dispensable for Akt activity at Ser473 [147–150]. Significantly, these data support the existence of alternative mechanism(s) for activating AktSer473 by an IQGAP1-dependent mechanism (Fig. 2.4, [70]). Thus, extensive functional overlap exists between the mTOR complexes and the PI3K-Akt pathway, owing to their regulation of one another by complex crosstalk

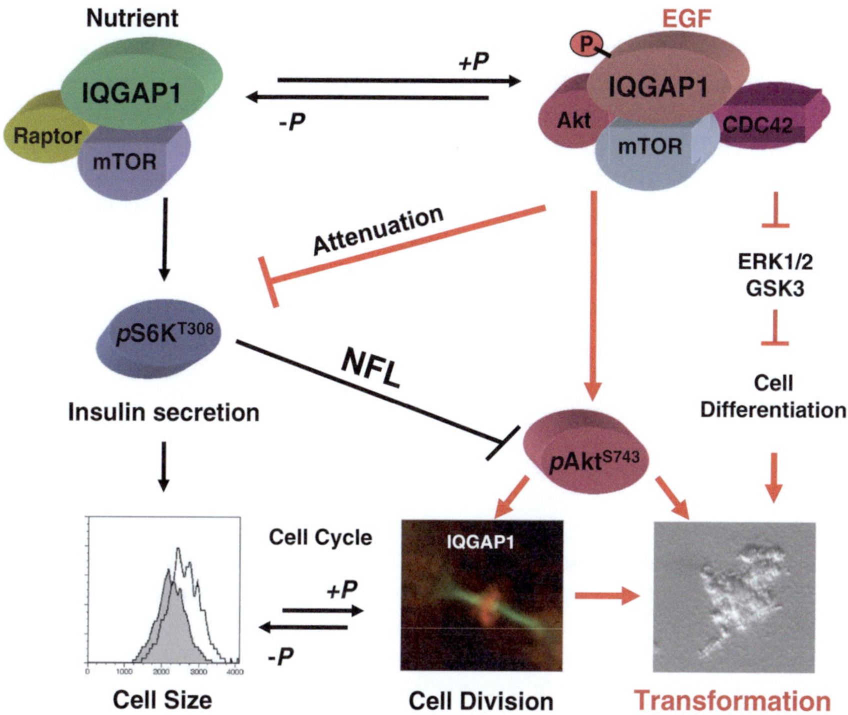

Fig. 2.4 IQGAP1 is a Rheostat of mTORC1-Akt signaling. A model summarizing IQGAP1's interactions with and regulation of mTORC1-Akt and functional outcomes [70–72]. In presence of nutrient signals, IQGAP1, in closed conformation, binds mTORC1 (mTOR-Raptor), activates S6K, increases insulin synthesis and secretion, and cell size, leading to obesity and possibly T2D. In response to mitogenic signals such as EGF, *p*IQGAP1, in open conformation, binds Cdc42, mTOR and Akt to suppress S6k activity, thus attenuating its negative feedback (NFL) on Akt. This in turn activates Akt^{S743}, suppresses differentiation and/or apoptosis by attenuating ERK1/2 and GSK3αβ, and activates mitosis/cell abscission and cell proliferation leading to transformed phenotypes and possibly cancer via $p\text{IQGAP1}^{S1443}\uparrow/\text{mTORC1/S6k1}\downarrow p\text{Akt1}^{S473}\uparrow$. *Black arrows* denote IQGAP1's regulation of cell size and a known mTORC1/S6k1 NFL. Thus IQGAP1 may act like a rheostat to modulate the NFL of $pS6K^{T389}\rightarrow p\text{Akt1}^{S473}$ and maintain cell homeostasis, dysregulation of which could underlie tumorigenesis and/or aberrant insulin signaling in diabetes. *Red arrows* denote the newly identified role of IQGAP1's action as a rheostat of the novel pathway Cdc42-mTORC1/S6K-Akt

and feedback mechanisms [151]. Adding to this complexity is that mTOR-PI3K-Akt crosstalk to regulate and be regulated by the mitogen activated protein kinase (MAPK) cascade, encompassing the Raf/MEK/ERK, and co-regulate downstream functions [152]. As a result targeting mTOR-PI3K-Akt and MAPK in human diseases has been challenging, leading to a paradigm shift to identifying their context-dependent regulators instead (reviewed in [43]). IQGAP1 represents an attractive target, as it binds both the lipid substrate (PIP2) and the product (PIP3) of PI3K, as well as mTOR, Akt and MAPK, and modulates their activities (reviewed in [43]).

In several cell types and model systems, IQGAP1 serves as a scaffold for the MAPK cascade by binding b-Raf, c-Raf, MEK1/2 and ERK1/2 [153, 154], and integrating their activity with mTORC1-Akt ([70], Fig. 2.4). Mechanistic analyses See notes with dominant mutants of IQGAP1 revealed that IQGAP1 binds Akt and mTORC1 (mTOR-Raptor) [70]. In presence of epidermal growth factor (EGF), expression of IQGAP1-IR-WW, the mTORC1-binding domain, attenuates ERK1/2/GSK3α/β and mTORC1/S6K1T389 and robustly elevates the level of phosphorylated AktSer473 [70]. These findings suggest that IQGAP1 relieves the mTORC1/S6K1T389 negative feedback (NFL) on pAktS473 [43, 61, 70] and position IQGAP1 as a rheostat that links cytoskeleton dynamics to metabolism. Significantly, IQGAP1 is more sensitive to rapamycin than TORC1 subunits both in yeast and mammals [70], suggesting its potential as predictor of therapeutic rapamycin-sensitivity in a subset of cancer and/or diabetes marked by deregulation of F-actin and IQGAP1.

Three important features in regulating this pathway that are likely to be intertwined must be considered. These are the role of signal dynamics, the interplay of kinase-phosphatase balance, and functional versatility gained through binding to different surface receptor. These features can be modulated by different signals to allow for a spatiotemporal ultra-sensitive system. The spatiotemporal Phosphatase-kinase balance imparting on IQGAP1 is yet to be analyzed. Signal dynamics appears to be an important factor in this pathway, as mall changes upstream seem to be translated into profound responses downstream. As powerful example, we observed that a small change in S6K activity level in response to IQGAP1-IR-WW expression, leads to a large change in AktS473 phosphorylation and promotion of robust cellular transformation [70]. Similarly significant but relatively modest changes in IQGAP1 level in human β-cells isolated from diabetic patients associate with frank type 2 diabetes (see below). IQGAP1's signaling through several surface receptors is considered below.

IQGAP1 Gains Functional Specificity by Relaying Signals from a Variety of Surface Receptors to the Cytoskeleton

The actin cytoskeleton plays an important role in receptor-mediated signaling. It has been noted that the WAVE regulatory complex (WRC) controls lamellipodia formation by activating Arp2/3 complex at distinct membrane sites downstream of Rac1 [15] and ERK [155]. It is not surprising therefore that a recent study in flies has identified a sequence motif present in a large number of membrane receptors that specifically binds to WRC and is required for actin dynamics at membrane sites [156]. In fact, much of the signaling through surface receptors has been shown to input into modulation of the cytoskeleton [157]. Consistent with its role in transducing extracellular signals, IQGAP1 interacts with diverse types of surface receptors in different cell types (Fig. 2.5) and impact transport of the receptors or their specific subunits. Many of these interactions and their consequent functions illustrate the duality of IQGAP1's functions as a molecular rheostat utilized by cells in a context-dependent manner. IQGAP1 likely serve as a platform on which distinct receptor-mediated modules pre-assemble at distinct membrane site, awaiting

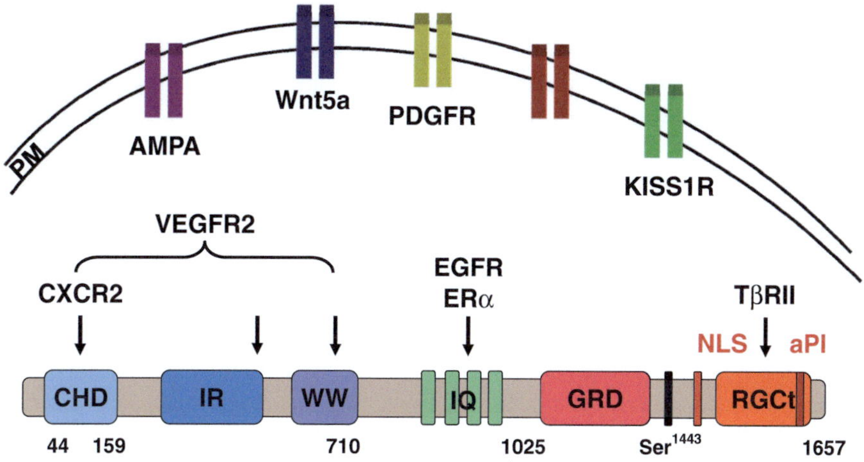

Fig. 2.5 IQGAP1 relays signals from several surface receptors. In several cell lines, IQGAP1 binds several different types of receptors, including receptor tyrosine kinases (RTK) and G-protein-coupled receptors. Those whose binding sites on IQGAP1 have been reported are indicated above the respective domain. Those receptors whose binding sites have not been reported are listed at the plasma membrane

response to different extracellular cues, leading to modulation of cytoskeletal-dependent functions. Indeed, super-resolution imaging suggests that in absence of ligands, surface receptors like EGFR rest along cortical F-actin in polymer forms, presumably to achieve concentration as an efficient way to detecting low level of ligand molecules and amplify their signal [158].

Consistent with this finding, several studies linked tyrosine phosphorylated IQGAP1 to EGF, PDGF [159, 160], VEGF [161–163], or FGF [83] signaling, revealing links with several receptor tyrosine kinases (RTKs) (Fig. 2.5). In endothelial cells stimulated by VEGF, tyrosine-phosphorylated IQGAP1 binds vascular endothelial growth factor receptor 2 (VEGFR2) and promotes migration in wound-healing assays [161]. In animal models, this process likely occurs through IQGAP1's modulation of VE-cadherin/β-catenin complex at adheren junctions and control of reactive oxygen species (ROS)-mediated angiogenesis [162]. However, the non-receptor tyrosine kinase c-Src appears to input into this function by interacting with IQGAP1 via its SH2 domain and promoting VEGFR-2-mediated angiogenesis, independent of IQGAP1's tyrosine phosphorylation [163]. Similarly, Platelet-derived growth factor (PDGF) stimulation of its receptor PDGFR induces IQGAP1-PDGFR complex formation on focal adhesions of vascular smooth muscle cell (VSMC) leading to increased migration [164]. Although mechanistic details of these processes is still lacking, it could involve IQGAP1 nuclear activity. Indeed, IQGAP1 appears to act through the redox-sensitive transcription factor NF-E2-related factor 2 (Nrf2) and impact migration of human coronary arterial endothelial cells in response to VEGF or insulin-like growth factor-1 (IGF-1) downstream of the MAPK pathway [165, 166].

Serine phosphorylated IQGAP1 has also been linked to EGF-mediated signaling (Fig. 2.5). IQGAP1 via its IQ domain, binds the kinase domain of the epidermal growth factor receptor (EGFR), irrespective of its ligand EGF, and modulates receptor activity [115]. The EGFR belongs to the ErbB (HER) family of receptors that comprises four members, ErbB1-4 (HER1-4). In response to ligand binding these receptors homodimerize and transphosphorylate several tyrosines in their cytoplasmic tails leading to receptor activation [167]. The exception is ErbB2, which has no known ligand (i.e. orphan receptor) and ErbB3, which lacks a functional kinase domain, thus these two are activated via hetero-oligomerization with other ErbB receptors [167]. Whether IQGAP1 mediates ErbB hetero-oligomers is an interesting and plausible hypothesis that could contribute to understanding the signal diversity of these receptors. The IQ motifs of IQGAP1 also bind the hinge region of the intracellular estrogen receptor ERα and modulate its estradiol-mediated transcriptional activity [168]. Moreover, in SK-OV-3.ipl human ovarian tumor cells, interaction of IQGAP1 with the hyaluronan receptor CD44, which is a transmembrane glycoprotein, links the ERα signaling to F-actin reorganization and ovarian tumor cell invasion [169]. Although the IQGAP1's domains that bind to many surface receptors has not been mapped, it is of note that many receptors bind to the IQ motifs of IQGAP1 in response to different signals in different cell types. Given that these IQ motifs also bind Ca/CaM and myosin/Mlc1 and to the Zn^{2+}/Ca^{2+}-binding protein S100B [170] and the Ca^{2+}-sensor S100P [171], it remains curious whether these differential interactions involve CaM/Zn^{2+}/Ca^{2+} signaling, myosin-dependent pathways or combinations of them.

IQGAP1 residues 1503-1657 mediate binding to the transforming growth factor TGF-β receptor (TβRII) in liver pericytes (stellate cells) [172]. This interaction appears to suppress TβRII receptor internalization that interferes with differentiation of pericytes into myofibroblasts. Stromal myofibroblasts are required for supporting metastatic colorectal tumors in the liver. Consequently, in stromal myofibroblasts associated with human colorectal liver metastases IQGAP1 is downregulated to maintain differentiation and tumor growth [173]. Although the mechanisms underlying this function remain unknown, it would appear that IQGAP1-C-mediated oncogenic signal, which prevents cell differentiation and polarization [72], must be silenced selectively in stromal cells by metastatic tumors for supporting growth in their new environment. One way to achieve this is through autoinhibition of IQGAP1 in these cells (Figs 2.3 and 2.4).

IQGAP1 has also been linked to signaling through G-protein coupled receptors (GPCRs). GPCRs are the largest and most diverse super-family of membrane proteins. They mediate much of the vertebrate physiology by transducing signals from hormones, growth factors, neurotransmitters, but also ions, lipids, cholesterol, and water [174, 175]. They appear to gain functional and cell type specificity via a diversity of interacting proteins [176]. Vertebrate GPCRs are commonly classified into five families on the basis of sequence and structural similarity; rhodopsin, secretin, glutamate, adhesion and Frizzled/Taste2 [177]. IQGAP1 interaction with the canonical Wnt/E-Cadherin/β-catenin complex downstream of the Frizzled receptor has been described earlier in this chapter. Proteomic analysis suggests that IQGAP1 bridges Wnt5a-receptor-actin-myosin-polarity (WRAMP) structure with multivesicular bodies (MVBs) in the rear of motile cells to facilitate membrane retraction and motility [178]. Through interactions with two different glutamate receptors, IQGAP1 appears to play a role in brain development and higher brain functions such as memory and cognitive activities [179, 180]. Earlier in vitro studies have implicated IQGAP1 in promoting neurite outgrowth in N1E-115 neuroblastoma cells [181]. A potential mechanism for this is that IQGAP1 binds the GluR4 subunit of the α-amino-3-hydroxy-5-methyl-4-isoxazole propionic acid (AMPA) receptor and regulates its polarized transport to the surface [179]. Similarly, it binds the NR2 subunits of the NMDAR glutamate receptor and the postsynaptic density scaffold PSD-95 and controls dendritic spine numbers in the hippocampus and lateral amygdala [180]. This effect also arises, at least in part, from IQGAP1's regulation of NR2A subunit trafficking and leads to regulation of cognitive behavior in mice model [180].

In neutrophils, IQGAP1's N-terminal 1-160 residues associates with the interleukin 8, chemokine receptor CXCR2 and promotes chemotaxis in a dynamic manner [182] that is consistent with its interactive negative and positive roles in cell migration and invasion [72]. This may further implicate IQGAP1 in the response to inflammation signals and will require further analyses. In human mammary cell lines, IQGAP1 binds the Kisspeptin receptor, KISSR1 (GPR54) and regulates EGFR transactivation, leading to cell invasion in a process modulated by ERα [183]. Because IQGAP1 also binds the ERα and modulates its nuclear activity [168], this evidence demonstrates further the ability of IQGAP1 to regulate functional spatiotemporal crosstalk between different types of receptors, including surface and nuclear receptors.

In addition to direct binding to surface receptors, IQGAP1 appears to gain functional specificity and diversity by transmitting receptors' signals through binding to adaptor proteins. The Src Homology (SH2/SH3) domain-containing adapter proteins integrates non-receptor and receptor tyrosine kinase (RTK) signaling and remodeling of the actin cytoskeleton by recruiting proline-rich effector proteins [184]. These adaptors are evolutionarily conserved and several of them relay specific functional signal from RTKs to downstream effectors, leading to cell proliferation, migration or axonal guidance. Downstream of activated ErbB receptors, IQGAP1-N, through residues 401-533, interacts with the phosphotyrosine binding (PTB) domain of the adaptor protein ShcA independent of tyrosine phosphorylation and colocalizes to lamellipodia [185]. Thus, this type of interaction may influence distinct cell functions like lamellipodia formation and cell migration mainly through Rac1 as opposed to serine-phosphorylated IQGAP1, which influences cell proliferation mainly through Cdc42 [72].

In addition, the identity of Rho regulator pairing with certain adaptors such as Nck (non-catalytic region of tyrosine kinase adaptor) may lend functional specificity to IQGAP1 in modulating actin dynamics. Recent evidence implicates the Cdc42 GEF, Intersectin, as a regulator of a Cdc42-Nck-N-WASP-dependent actin polymerization [186]. The extent of IQGAP1 signal versatility leading to functional specificity in different cell types has yet to be realized through much needed mechanistic analyses. In part, the richness of this signal versatility has been revealed from analyses of IQGAP1 function in cell motility in different cell types as discussed below.

A Master Regulator of Cell Motility

The actin cytoskeleton plays a major role in cell motility by mediating lamellipodia formation at the leading edge, which is critical for directional migration [187]. Several N-WASP-independent pathways are known to activate Arp2/3 complex mediated cell motility [188]. Also instead of N-WASP [189], WAVE2 [190, 191], is believed to be the activator of Arp2/3 complex in lamellipodia formation. Regardless of these views, in several cell types, IQGAP1 accumulates at the leading edge of migrating cells (Fig. 2.6) and promotes cell motility, and RNAi-mediated knock-down of IQGAP1 potently reduced cell motility and inhibited protrusive actin meshwork at the leading edge [71, 72, 83, 87, 104, 161, 192, 193]. Given that an interaction between IQGAP1 and WAVE was not detected in certain cell types [83], this effect underscores that IQGAP1 plays multiple essential roles in driving cell motility not limited to structural projection formation.

In this respect, IQGAP1's signaling and capacity to interface several surface receptors and transduce signals through multiple pathways like MAPK, mTOR-Akt, Cdc42 and Rac1 points to complex and delicate interplay by which it regulates cell motility according to variables such as the cell type and the nature of the signal (Fig. 2.6). For example, activated Cdc42 and Rac1 directly bind both IQGAP1 and N-WASP [14, 62, 75, 77] and they activate N-WASP independently of IQGAP1 [194]. However, while activated Cdc42 and Rac1 each synergistically promote IQGAP1-mediated actin nucleation by N-WASP, Cdc42 enhances binding of N-WASP to IQGAP1, whereas Rac1 appears to reduce such binding [58]. Recent evidence implicates phospholipids as essential players in IQGAP1-mediated cell motility. Type Iγ phosphatidylinositol 4-phosphate 5-kinase (PIPKIγ) binds and targets IQGAP1 to the leading edge where

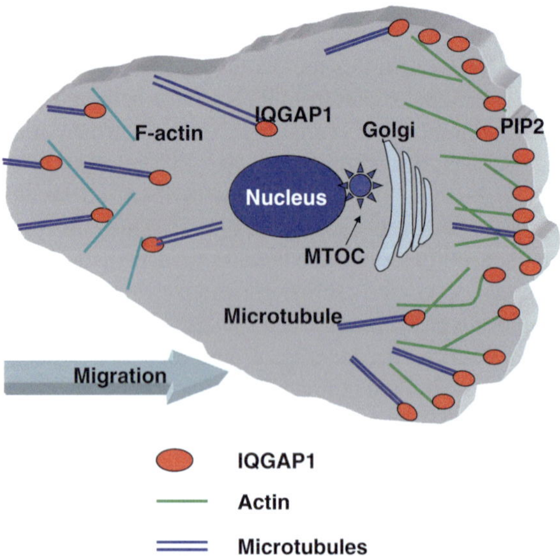

Fig. 2.6 A model, IQGAP1 coordinates cell motility in the leading and trailing edges of motile cells. The role of IQGAP1 in generating membrane ruffles in the leading edge of motile cells and regulating cell polarity has been extensively studied. Emerging evidence now demonstrates that IQGAP1 engages microtubules and focal adhesions to control the retracting end during cell motility

it binds the PIPKIγ-byproduct phosphatidylinositol 4,5 bisphosphate (PIP_2), which in turn promotes IQGAP1 actin polymerization activity and cell migration in response to integrin or growth factor receptor activation [195]. However, different signals appear to specify IQGAP1's protein–protein interactions in a context-dependent manner. In chemotactic neutrophil cells, IQGAP1, through its CHD, interacts dynamically and colocalizes at the leading edge with the chemokine receptor CXCR2 [182], analogous to its roles in FGF-mediated [83] and VEGF mediated [161] cell motility in other cell types. Additionally, the role of IQGAP1 in regulating exocytosis (see below) through the exocyst complex [71] may be required for migration [72] and membrane expansion during lamellipodia formation. Much work will be required to understanding the role of the exocytic pathway in IQGAP1-mediated cell motility.

IQGAP1 has also been implicated in modulating cell motility by modulating microtubule dynamics at both the leading and the trailing edges of motile cells (Fig. 2.6). In PDGF-treated cells, IQGAP1 forms a complex with the A-kinase-anchoring protein 220 (AKAP220) and its substrate the cAMP-dependent protein kinase (PKA) to suppress GSK3β activity and recruit CLASP2 to the leading edge of motile cells [68]. Similarly, IQGAP1 has been implicated in microtubule-based rear-end dynamics that propels motile cells forward. Interaction of IQGAP1 with Wnt5a-mediated WRAMP integrates cortical endoplasmic reticulum, localized Ca^{2+} signal, actomyosin contraction and adhesion disassembly during the dynamic structural and signaling events that lead to membrane retraction [178]. Furthermore, the ability of IQGAP1 to both nucleates branched filaments and crosslinks them to mother filaments [63, 83] and to microtubules and other types of microfilaments, may be required for providing the mechanical force necessary for directional movement.

While mechanism of IQGAP1 in modulating cell migration during wound-healing closure in fibroblast has been demonstrated [72], to date, its exact role in collective cell migration remains obscure. Collective cell migration is important during developmental events such as embryonic morphogenesis and wound healing or in pathological states such as tumor growth and spreading. During collective migration leader cells sense and migrate up an external signal gradient while pulling along adjacent cells—referred to as follower cells—through strong cell–cell contacts [196]. Rho GTPases have been implicated in controlling the mechanical and biochemical cues that determine the identity of leader versus follower cells during collective migration of epithelial sheets [197, 198]. Given its structural and signaling roles in epithelial adhesion and polarity (see below), IQGAP1 must play a crucial role in collective migration during embryonic differentiation as well as cancer spreading and should be further investigated.

A Crucial Modulator of Epithelial Adhesion and Polarity

Epithelial cells that line the cavities (lumens) of animal organs are prototype of polarized cells. These cells form columnar shape with structurally and functionally distinct apical, lateral and basal plasma membrane domains, and they adhere to one another by their lateral membranes to form sheets. The mechanisms underlying

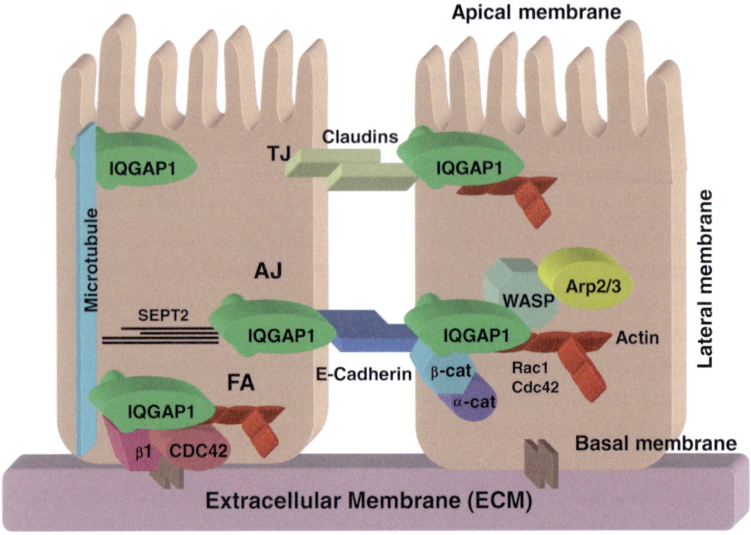

Fig. 2.7 IQGAP1 is a major regulator of epithelial cell polarity. Schematic diagram of polarized epithelia and the role of IQGAP1 in organizing tight junctions (TJ), adheren junctions (AJ) and focal adhesion (FA) complexes. TJs separate the apical and baso-lateral membrane while the AJs facilitate communications between adjacent cells and the FA complexes mediate interactions with the extracellular basement membrane (ECM). E-Cadherin is a transmembrane glycoprotein with an extracellular domain that mediates cell–cell adhesion and a cytoplasmic tails that directly binds β-catenin, which serves as a scaffold to anchor α-catenin and actin polymerization machinery (e.g. N-WASP and Arp2/3). IQGAP1 interplays with Rac1, Cdc42 and Rho A to regulate this complex and facilitate actin dynamics, thus regulating epithelial adhesion and polarity. IQGAP1 also binds microtubules (MT) and several different types of microfilaments like septins [71, 96] and claudins [205]. Deregulation of IQGAP1-signaling can ultimately lead to different types of human carcinoma

epithelial polarization are still incompletely understood. The identity of the different membranes is determined and maintained by spatial control of the cytoskeleton and polarity protein complexes, including the Rho GTPases and a host of their regulators and effectors [199–201]. Polarized epithelia maintain cell–cell contacts via adheren junctions (AJs) that provide mechanical attachments between adjacent cells. The AJs control the maturation and the maintenance of the contacts [202], and as such they are dynamic in nature and link directly to the actin cytoskeleton [203]. The main components of the AJs are the Cadherin–Catenin core complex, the regulators of tissue organization. Epithelial Cadherin (E-Cadherin) belongs to the classical transmembrane Cadherin receptors. These are glycoproteins containing extracellular domains that mediate cell–cell adhesion and cytoplasmic tails that directly bind β-catenin, α-catenin and p120 catenin in the cytoplasm [204]. Actin filaments concentrate at this site along with a number of structural and signaling regulators of F-actin dynamic that facilitate the junctional dynamics and homeostasis [203].

In addition to concentrating at lamellipodia, IQGAP1 preferentially accumulates at cell–cell junctions of polarized epithelia (Fig. 2.7) and appears to regulate the integrity of the AJs [62, 63, 71, 78, 206]. Accordingly, IQGAP1 has been widely linked to cadherin/catenin-mediated adhesion [78, 207] in different epithelial cell types, including islet pancreatic β-cells that secrete insulin [71, 208]. Consistent with its regulatory role in epithelial structure and function, IQGAP1 appears to modulate the AJs by modulating the activities of Rac1 and Cdc42 and their association and localization with E-cadherin–catenin core complex in response to different stimuli. In general, it appears to act positively, favoring cell–cell contact formation by maintaining Rac1GTP-E-Cadherin-actin meshwork formation at the contacts [104]. However, under conditions that favor cell scattering, it acts negatively, promoting cell–cell dissociation by interacting with and delocalizing α-catenin from the core complex [209]. As discussed earlier, IQGAP1 likely controls the dynamics of microfilaments at the cell–cell junctions (Fig. 2.7). However, Much work is required for deciphering the physiological mechanisms underlying IQGAP1's activities in these processes and their relation to human disease. A potential mechanism is IQGAP1's regulation of protein exocytosis at cell–cell contacts at the basolateral membranes [23]. Indeed, cell–cell contacts are considered to be major sites of regulated exocytosis [210–212], where IQGAP1 plays a major role as discussed below.

A Regulator of Exocytosis: Focus on Insulin Secretion

Trafficking is one of the major roles of the cytoskeleton, in which Rho GTPases are pivotal regulators [21]. Delivery of proteins, hormones or lipids to the plasma membrane or into the extracellular space occurs through regulated exocytosis in which vesicles accumulate near the periphery of specialized cells awaiting a signal for their fusion with the plasma membrane. Regulated exocytosis is a sequential multistep process involving budding of exocytic vesicles from the Trans Golgi Networks (TGN) and their transport along microtubule and actin tracks toward the target membrane [213]. Release of vesicles cargo into the extracellular space ensues via a cascade of protein–protein interactions, marking a series of distinct steps, including vesicle tethering, docking, priming and fusion with the plasma membrane [6, 214, 215]. While the core exocytic machinery is conserved, different cell types adopt subunit combinations suited for their specialized physiological roles [216], where dynamic F-actin remodeling plays a principal role along the different steps [217–219]. Insulin exocytosis from neuroendocrine β cells of the islets of Langherans in the pancreas follows this process albeit in a specialized manner that has been well described (see [220]). However, the molecular mechanisms remain poorly understood.

Thus, the role of actin dynamics in insulin exocytosis has been appreciated early on in studies using islets and beta cell culture [221–224]. As Rho GTPases have been implicated in a variety of exocytic pathways in several cell types [23, 47, 96, 225], their role in insulin exocytosis has been investigated as well [71, 226–232]. Similarly, posttranslational modification of RhoGTPases, such as farnesylation, and regulation by Rho GDIs have been implicated in glucose-stimulated insulin exocytosis [233, 234]. However, their specific role in pancreatic islet β cells remains elusive. For example while expression of wild type Cdc42 in beta cells has no effect [71, 228], expression of dominant active mutants appears to inhibit both mastoparan- and glucose-induced insulin exocytosis, consistent with their known growth inhibitory effect [71, 228]. On the other hand, expression of the dominant negative Cdc42 mutants induces insulin secretion [71]. Most likely, these effects reflect indirect effecting Cdc42's upstream role on actin dynamics via specialized effectors, if not a pleiotropic effect on the exocytic machinery.

Rac1 was found not only to play both negative and positive roles in insulin exocytosis [235, 236], but also it has been implicated extensively in glucose uptake by regulating trafficking of the glucose transporter GLUT4 [237–239]. This evidence likely reflects a fine balance of the crosstalk between peripheral (muscle, adipose, liver) and central (β cells), perhaps via feedback loops and involving multiple GTPases [43].

Moreover, each GTPase has distinct context- and stimulus-dependent effect on actin dynamics and studies on insulin secretion have largely relied on pharmacologic inhibitors, global genetic knockout and overexpression approaches, which result in pleiotropic effects and mask the exact fine role of the GTPase. More selective reagents such as separation-of-function mutants are required. Moreover, the exact player involved in actin remodeling in the β-cells must be analyzed. Small GTPases play distinct and separable roles along the exocytic pathway. For example, Cdc42 and Rac1 could play a more proximal role during docking, tethering and fusion or release of exocytic vesicles, in which crosstalk between Cdc42 and Rac1 occurs. Antagonistic crosstalk between Rac1 and Cdc42 plays important roles in many aspects of epithelial polarity and function [44, 240], a role that likely impacts insulin secretion as well. Studies using fluorescent biosensors revealed different pools of the same GTPase act simultaneously to influence diverse cellular functions [20], thus supporting complex spatiotemporal actions not only for different GTPases, but also for a single GTPase. The precise step in which Cdc42 or Rac1 imparts on insulin exocytosis remains unclear. While evidence for Cdc42 interaction with the t-SNARE (soluble *N*-ethylmaleimide-sensitive factor attachment protein receptor) Syntaxin 1A suggests a role in the insulin release step, the interaction appears to be indirect as the purified proteins failed to interact [228], suggesting involvement of specialized effector proteins.

Several lines of independent evidence support that IQGAP1 is the likely effector. IQGAP1 appears to impact insulin homeostasis via three interconnected mechanisms, insulin secretion, insulin synthesis and recycling of the membranes of insulin-granules after fusion we discuss here. First, its role in exocytosis is conserved; in yeast IQGAP1, Iqg1p, interacts with the exocyst and influences bud growth and septum secretion during cytokinesis [96]. In pancreatic islet β cells, it modulates glucose-stimulated insulin exocytosis through Cdc42, a role that was revealed by separation-of-function mutant analyses and protein–protein interactions [71]. The N-terminal domain of IQGAP1 interacts with mTORC1 [70] and with several subunits of the exocyst and enhances insulin exocytosis in response to glucose [71], leading to increased cell size [72]. In contrast, the C-terminal domain, which interacts with Cdc42 and Rac1, inhibited insulin exocytosis and decreased the cell size while increasing cell proliferation and transformed phenotypes [71, 72].

The IQGAP1$^{\Delta MK24}$, which disrupt the interaction with Cdc42 and the IQGAP1^{S1443A}, which abolishes phosphorylation by PKCε both were unable to support cellular transformation, supporting that Cdc42 and IQGAP1 phosphorylation at S^{1443} plays negative roles in insulin secretion. Indeed, genetic deletion of PKCε enhances insulin secretion and prevents glucose intolerance in fat-fed mice, and peptide-mediated inhibition of PKCε improves insulin availability and glucose tolerance in diabetic db/db mice [241].

Furthermore, IQGAP1 interacts with the exocyst in pancreatic β cells through a domain that enhances insulin secretion [71]. The evolutionarily conserved vesicle-tethering exocyst complex is composed of nine subunits, some of which are localized to the exocytic vesicles while other subunits like Sec3 and Exo70 are believed to be localized to the plasma membrane [242, 243]. The plasma membrane localization of Sec3 and Exo70 appears to direct the vesicles to specific sites on the membrane thereby controlling active sites of exocytosis, as in directional migration or cytokinesis [244]. In yeast, Iqg1p interacts and localizes with and controls exocytosis and the localization of the exocyst landmark subunit Sec3 during cytokines [96]. Similarly, in β cells, IQGAP1 interacts, localizes with, and controls the localization of the exocyst proximal subunit Exo70, as well as other subunits like Sec6 and Sec8 [71]. Thus, it appears that IQGAP1 has a conserved function in directing the secretory vesicles by which it regulates the latest steps of insulin exocytosis. To facilitate fusion of vesicles to the plasma membrane, the exocyst interacts with the SNAREs [245]. IQGAP1 interactions with the phospholipids PIP2 and PIP3, as well as with the plasma membrane-associated t-SNARE syntaxin1 A [71], support its essential role in the final step of insulin release. This role extends to modulating F-actin, by which it regulates this final step whereby F-actin disassembly enhances and assembly inhibits insulin granules release [43].

Second, IQGAP1 localizes to the endoplasmic reticulum and interacts with the translocon subunit Sec61β and appears to increase insulin synthesis in pancreatic β cells [71]. Much work is now required to delineate the role of iQGAP1 in insulin translational control. Third, in response to glucose, IQGAP1-Cdc42-GTP binds Rab27-GDP and the Rab27 effector Coronin 3 and promotes the recycling of insulin-granule membranes [246], a process required for more rounds of exocytosis [247]. Coronin 3 also regulates F-actin dynamic by binding to the IQGAP1-partner Arp2/3 [248]. Following insulin exocytosis from β cells, recycling of the membranes of the insulin-containing granules occurs through the process of endocytosis [247]. Endocytosis s a versatile mechanism required for various cellular functions, including nutrient uptake, but it also serve as a balance for exocytosis by recycling vesicle membranes after fusion to the plasma membrane, and thus maintaining cell homeostasis [249]. Small G proteins, F-actin and several F-actin regulators like Coronins also are important components of endocytosis [249]. Thus it appears that IQGAP1 engages several types of F-actin regulators and signaling molecules to regulate distinct aspects of insulin exocytosis. Accordingly, studies employing global actin de-polymerization with depolarizing agents cannot identify mechanisms of action required to understanding the molecular basis of normal cell function or disease.

However, while these data establish the role of IQGAP1-Cdc42 dynamic interactions in regulating insulin exocytosis, they do not exclude interplay of IQGAP1-Rac1 vs. IQGAP1-Cdc42 in coordinating insulin exocytosis with insulin-stimulated glucose uptake [43]. As discussed above, IQGAP1 has been implicated in endocytosis [246] and Rac1 has been widely implicated in glucose uptake via stimulating the transport of the glucose transporter GLUT4 [237–239]. Moreover its interaction with microtubules suggests a role in intracellular cargo transport or a novel role in coupling exocytosis-endocytosis at the cell cortex. Mechanism-based studies will be required to reveal its essential roles in insulin and glucose homeostasis, which is supported by the finding that IQGAP1, but not Cdc42, is downregulated in humans with T2D discussed below.

IQGAP1's Role in Human Disease

Mutations affecting polymerization in human α-smooth muscle *ACTA2*, α-cardiac (*ACTC*), α-skeletal (*ACTA1*), and γ-cytoplasmic actin (*ACTG1*) have been found to associate with an array of congenital and premature acquired cardiovascular, cardiomyopathies, skeletal myopathies and deafness respectively [250–255]. The mechanisms disrupted by these mutations are unknown. On the other hand, Rho GTPases, as regulators of actin dynamics and many other pivotal cellular pathways, are associated with a wide range of developmental and acquired human diseases [256–258]. Similarly, regulators of actin polymerization have been implicated in a variety of human cancers [61, 259, 260]. Thus elucidating the molecular mechanisms that control actin polymerization is essential to understanding and treating a wide range of human disease. Although IQGAP1 dysfunction can influence a myriad of human diseases, the discussion below is focused on diabetes, cancer and infection for which evidence has been accumulated.

Diabetes

T2D is a complex and multifactorial metabolic disease. In T2D pancreatic islet β cells fail to secrete sufficient insulin to overcome peripheral insulin resistance, therefore T2D has been described as a two hit-disease. Dysfunction of β cells occurs from reduced cell mass and failure to produce insulin whereas insulin resistance occurs in peripheral tissues such as the adipose, the liver and the skeletal muscles from failure of adequate glucose uptake in response to insulin. Initially, insulin resistance in obesity prompts β cells to produce more insulin, a process that apparently leads to cell demise and the manifestation of acquired T2D [261, 262]. Accordingly, a prevailing view is that dysfunction of β-cells is central to the etiology of T2D, a notion that has also been supported by genome-wide association studies (GWAS) [263, 264]. As is the case with complex diseases, mounting evidence support a polygenetic etiology for T2D, as well as associated genetic variations in a large number of cellular processes [265–267]. This complexity prompted recommendation for new approaches to interpreting GWAS data in which a set of risk loci common to the disease cohort must be analyzed instead of individual genomic loci or single nucleotide polymorphisms (SNPs) [268]. Such approach is consistent with the emerging concept implicating pathways, not single genes, in disease mechanisms.

Like the case in cancer, Rho proteins were not identified in GWAS studies in diabetes. The caveat, however, is that GWAS have limited sensitivity (resolution), which can obscure small but important changes [269], particularly in signaling proteins. Importantly, GWAS cannot reveal mechanisms of disease development [270]. Also, disease-development mechanisms can vary between populations. More high-resolution studies will require conducting GWAS among environmentally- and ethnically-distinct population. Of note, Rho proteins play important regulatory role in metabolism albeit indirectly through effector proteins. It has been well appreciated in the field of cell signaling that modest changes in signaling proteins can be amplified downstream by effector proteins and significantly impact disease development. SNPs in the rhoGEF11 (ARHGEF11) gene associated with T2D and glucose intolerance have been identified in Amish and Pima Indian groups [271]. It remains unclear whether this SNP is causal, and whether it affects insulin secretion, glucose uptake or both. Moreover, RhoGEFs are multidomain proteins and likely have Rho-independent functions [27] thus mechanistic analysis will be required to identifying their specific role in diabetes.

By contrast GWAS metadata from the DIAGRAM consortium, involving 8130 T2D cases and 38,987 controls revealed IQGAP1 SNPs associating with T2D [272]. Similarly, transcriptional meta-analyses revealed that while Cdc42 transcript in β cells remains unchanged between diabetic patients and non-diabetic control groups, IQGAP1 mRNA [273] and protein [Osman, unpublished] levels both were down-regulated in humans with T2D. This is consistent with the signaling and structural roles of IQGAP1 both upstream and downstream of Cdc42 and Rac1 and mTORC1-PI3K pathways in pancreatic β cells. Because of its regulation of cell size and proliferation [70, 72], interaction with the exocytic machinery and role in glucose-stimulated insulin secretion [71], IQGAP1 likely is the direct effector of Cdc42 and Rac1 in insulin exocytosis. Thus its dysfunction would have direct impact on the etiology of T2D both by affecting β cell function and division. Further support for IQGAP1's role in β cell biology is its association with A-kinase anchoring protein 79 (AKAP79), and cAMP-dependent protein kinase (PKA) [69].

The mechanisms underlying reduced expression of IQGAP1 in T2D patients are under investigation, and the cellular consequences of this reduction can provide avenues for future intervention with T2D etiology. A likely mechanism must entail direct effect on β cell function in insulin secretion by affecting F-actin dynamics, leading to dysregulation of docking and tethering of insulin exocytic vesicles or dysfunction in the release step. As discussed in the previous section, gain-of-function mutations of IQGAP1 in insulin exocytosis that leads to sustained, unregulated insulin secretion are expected to lead to hyperinsulinism and the development of T2D, whereas loss-of-function mutations will lead to β-cell dedifferentiation, and also T2D and/or pancreatic cancer [43]. On this basis, rebalancing intervention approaches can be developed to correct the specific defect. A crucial consideration in such approaches would be the importance of signal dynamics, as small changes upstream in the pathway translate into larger changes downstream. In this respect IQGAP1$^{IR\text{-}WW}$ expression in β cells led to small changes in the level of active S6K and resulted in greater level of Akt^{S473} activity, which manifested in increased cell proliferation and transformed phenotypes [70]. This explains why a significant but modest reduction of IQGAP1 in human β cells is associated with T2D, presumably from reduction in functional β cell mass. Indeed, reduced IQGAP1 expression diminishes its phosphorylation level that is required for promoting cell division and proliferation [72].

Infectious Diseases

Infectious viral and bacterial pathogens may not produce cytoskeletal elements, however they employ effector or mimic proteins and effective strategies to hijack the host actin polymerization machineries for invading human cells and spreading into tissues [274–277]. Among the strategies used by pathogen for invading host cells is the modulation of phagocytosis [278, 279] perturpation of epithelial tight and adheren junctions [277] in which IQGAP1 plays major organizing roles [240]. During these actions, pathogens target Rho GTPases signaling pathways at various steps of the infection process from membrane modifications to internalization and spreading [258, 279, 280]. Some bacteria use toxin to covalently modify and attenuate Rho GTPase [258]. Others utilize a Type III or Type IV secretion system to inject effector proteins that mimic Rho regulators by which they modulate Rho GTPase activity and hijack their cytoskeletal and trafficking pathways [281–287]. Once they gained entry inside the host cell, pathogens generate propelling F-actin networks known as comets for spreading from cell to cell [288]. Comet tails share similar motility and dendritic nucleation features with lamellipodia. Detailed mechanistic knowledge about the two systems is still lacking, but new technology is providing promising insights. Cryo-electron tomography and mathematical simulation suggest a mechanism in which a fishbone-like array of four actin filaments propels the tethered pathogen by fast continuous polymerization [289, 290]. This approach has potential for facilitating comparative visualization of the array of actin regulatory proteins and the proteome involved in generating the dynamic comet vs. lamellipodia structures.

IQGAP1 has been widely implicated in mediating infections by several types of human bacterial and viral pathogens and has been subject of recent reviews [61, 291]. Different pathogens either directly alter IQGAP1's expression and localization or co-opt it via interactions with effector proteins to mediate infection and spreading from cell to cell. Several studies demonstrated a crucial role for IQGAP1 in viral infection and egress. IQGAP1 co-purifies with highly purified

HIV-1 virions from human monocyte-derived macrophages [292], suggesting a role in AIDS pathobiology. It also directly interacts with the Gag protein of Moloney murine leukemia virus (MuLV) [293], the core protein of classical swine fever virus (CSFV) [294] and with the L-domain of the matrix protein VP40 of Ebola virus and appears to promote the viral budding. Together these studies suggest that IQGAP1 likely involves in multiple stages of viral infection, including early and late stages such as entry and egress. IQGAP1 interacts with the cell abscission proteins Alix and Tsg101 [128], a subunit of the endosomal sorting complex required for transport (ESCRT) also implicated in viral budding [295]. This interaction provides a potential mechanism for membrane pinching required for cytokinesis as well as for viral egress [295].

IQGAP1 has also been implicated in mediating bacterial infections. Microarray and proteomic analyses revealed that different pathogens target IQGAP1 differently, suggesting existence of versatile IQGAP1-mediated mechanisms of infection. For example, IQGAP1 binds the type III receptors Tir and Ibe of the enteropathogenic *Escherichia coli*, and the SseI effector of *Salmonella* [296–298], perhaps to facilitate infection. On the other hand, the human pathogen *Pseudomonas aeruginosa* utilizes an interaction between its quorum-sensing molecule *N*-acylhomoserine lactones and IQGAP1 as a mode of communication with host epithelia during early infection [299]. Infection by *Mycobacteria* and *Yersinia pestis* leads to downregulation of IQGAP1 mRNA and protein levels respectively [300, 301]. By contrast, *Helicobacter pylori* (*H. pylori*) infection appears to upregulates IQGAP1 mRNA and protein levels in several gastric carcinoma cell lines [302]. As infection by *H. pylori* is believed to underlie the development of gastric carcinoma, it is not surprising that IQGAP1 is highly expressed in gastric cancer tissues [61, 303]. For this reason and for the current challenges posed by the increased *H. pylori* resistance to antibiotic, the discussion in this section is focused on IQGAP1's role in mediating *H. pylori* infection. Infection with *H. pylori* has been widely implicated in the development of peptic ulcer, gastritis and gastric cancer [304–306]. While the mechanism of the infection has been extensively analyzed, the mechanism of the carcinoma development has been largely unclear. Accordingly, successful cure of *H. pylori-mediated* ulcers and gastritis has largely relied on an antibiotic regimen, but gastric cancer remains deadly. However, currently increased antibiotic resistance has rendered this approach ineffective [306] and threatens increasing the cases of gastric cancers worldwide. In addition to formulating new antibiotic-based therapies, delineating the mechanisms underlying the infection and the development of gastric cancer has become imperative to effective treatment.

A body of current research support the idea that the bacteria targets the cytoskeleton and epithelial cell polarity to generate and maintain the infection, which ultimately leads to the development of gastric cancer by altering host cell signaling [61]. In addition to increasing the expression level of IQGAP1, *H. pylori* infection leads to delocalization of IQGAP1 and E-Cadherin from adheren junctions to internal cytoplasmic vesicles [302]. This evidence supports that *H. pylori* targets adheren junctions and depolarizes gastric epithelial cells. Loss of epithelial polarity is considered the first hallmark sign of cellular transformation leading to carcinogenesis

[199, 200]. While the bacteria are considered to be un-invasive, a few have been observed intracellularly [307], but mechanisms of invasion have not been investigated, thus the significance of intracellular bacteria to gastric adenocarcinoma development remains unknown.

A prevailing view is that the *H. pylori* injection of its presumptive oncoprotein cytotoxin-associated gene A (CagA) ultimately leads to the development of the adenocarcinoma [307–309], however, this assumption has been controversial and was challenged by compelling evidence (reviewed in [61]). Entry of CagA into epithelial cells triggers a cascade of exquisitely delineated molecular and signaling events causing the host cell to assume an Epithelial-Mesenchymal Transition (EMT)-like phenotype, known as 'hummingbird' phenotype, marked by considerable actin polymerization and cellular elongation (reviewed in [308]). Although EMT is a known factor in cancer metastasis and invasive growth, the contribution of the EMT-shape to gastric carcinogenesis has been debated (reviewed in [61]). A recent study suggests that it promotes formation of gastric cancer stem cells (CSCs), capable of inducing in vitro tumorspheres as well as xenograft tumors [310]. On the other hand, contradictory evidence supports the notion that *H. pylori* prevent cell migration by subverting the dynamics of focal adhesions (FAs) [311, 312]. Intriguingly, while half of the world population is infected with *H. pylori*, gastric carcinoma arises after a long period of chronic infection, and only in about 5 % of the infected population [307]. Thus, it is unclear how the CagA-induced CSCs might account for carcinoma development in this small population. Another line of evidence suggests that in 25 % of *H. pylori*-infected mice, chronic infection accompanied by inflammation leads to accumulation of bone marrow-derived cells (BMDC) displastic lesions, which may be responsible of preneoplasia development [313]. While these studies reveal the association of gastric carcinoma with CSCs and BMDCs, the molecular mechanism underlying the etiology of the disease remain unclear.

The mechanisms involved in generating the "hummingbird" elongated shape potentially can provide clues for the eventual development of carcinoma and they have been previously discussed [61]. Additionally, new evidence shows that the cordate *Ciona intestinalis* co-opt the actomyosin cytoskeleton, to generate a contractile mechanism, analogous to what happens during cytokinesis, for producing elongated cell shape during development [314]. Given that IQGAP1 has a role in regulating this conserved machinery, it is curious whether *H. pylori* co-opt IQGAP1 to produce the hummingbird shape required for maintaining the infection. In fact, IQGAP1 involvement potentially can explain the mechanisms underlying bacterial invasion, CagA injection, bacterial spreading as well as neoplastic transformation leading to gastric cancer (reviewed in [61]). Recent evidence shows that the motile intracellular pathogen *Rickettsia parkeri*, a causative agent of human rickettsiosis (spotted fever), differentially exploits the N-WASP-Arp2/3 and the formin actin nucleation pathways at distinct phases of infection to establish a niche and then spread between cells [275].

It is curios whether the bacteria also utilizes microtubule cytoskeleton. Microtubule-mediated transport has been demonstrated for infectious viruses [118, 315], where actin–microtubule interaction plays a role, as has been demonstrated analogous to several cellular processes [213, 316]. Another line of evidence implicates the formin mDia1 in modulating microtubule dynamics at the cell cortex to facilitate viral release through RhoA [118]. Given IQGAP1's interactions with both components, by analogy, it is likely that IQGAP1 links several distinct pathways involving actin-microtubule-based functions to disease development and progression. However, in addition to its structural roles, IQGAP1 signaling appears to control these functions upstream, thus providing further mechanism(s) for cancer development that can be targeted therapeutically by allosteric inhibitors. The mechanisms described below could account for the development of gastric carcinoma following the disruption of epithelial polarity by CagA leading to the development of the hummingbird shape [61].

Role in Cancer Inception and Metastasis

Numerous studies have implicated Rho GTPases in promoting tumorigenesis and this has been subject of recent excellent reviews [260, 317–323]. Rho GTPases likely promote tumorigenesis indirectly via complex mechanisms involving crosstalk with multiple pathways like Ras and Src. Expression of dominant or constitutively activated mutants of Cdc42 leads to cellular transformation and tumor formation in animals [324], and cellular transformation by oncogenic Ras requires the activation of Cdc42 [39]. However, unlike Ras, which is mutated in a about 33 % of human cancers, mutations in Rho GTPases are rare; instead, Rho GTPase hyper-activation occurs through aberrant overexpression, loss of GAP-mediated inactivation, or inappropriate activation of RhoGEFs [325, 326]. Indeed, VAV1, which is a common GEF for Rho, Rac1 and Cdc42, largely associates with human cancers [327], essentially through activating actin polymerization and promoting tumor invasion [328, 329]. Although VAV1 has a CHD, which potentially binds actin, and conceivably can serve both as a GEF and effector of Rho GTPases, such hypothesis has yet to be proven. It is more likely that VAV1 docks to the actin polymerization machinery that includes WASP [330] to modulate F-actin dynamics by localized activation of Rho proteins. Similarly, effectors of Rho GTPases in regulating actin polymerization and cell motility like mDia, which interacts with IQGAP1, associates with human metastasis by promoting tumor cell invasion [331]. Thus, following initiation of cellular transformation, recruitment of the actin polymerization machinery may represent a distal step leading to metastasis by lesion-initiator signaling molecules like IQGAP1.

To date, only a single mutation, the M1231I in *iqgap1* coding region, has been identified in association with gastric cancer [332]. Although it remains unknown whether this variant is causal, it is presumed to constrain flexibility of a key loop that links residue 1231 region to the putative Cdc42-binding site, leading to reduced kinetics in IQGAP1-Cdc42 interactions [333]. By contrast, overexpression or mislocalization of IQGAP1 has been reported to associate with a wide range of human carcinomas (reviewed in [43, 61, 74, 334]) and the list continues to grow.

Overexpression of IQGAP1 correlates with upregulation of intermediate filaments such as vimentin and cytokeratin and is indicator of poor prognosis and tumor differentiation [131, 135]. On the other hand, lack of IQGAP1's expression has been shown to associate with favorable prognosis in certain cancers [335]. Accordingly, IQGAP1 is being sought as a therapeutic target in cancer [336, 337]. However, the mechanism(s) by which IQGAP1 initiates the different types of human tumors are unclear and may be as variable as its cellular functions and the signaling pathways it controls in different cell types. Because of its structural and signaling properties, IQGAP1 likely imparts on the initiation steps of cancer-lesion as well as progression to metastasis. Indeed, the cellular transforming capacity of Cdc42 requires IQGAP1 [72]. Expression of IQGAP1, not only increases the level of *p*IQGAP1, but also the level of active Cdc42 in cells [71, 72]. Thus regulation of IQGAP1 phosphorylation cycles may be at play and could explain why reduced expression of the phosphatase PP2A in breast cancer cell lines appears to impair targeting of IQGAP1 to F-actin-anchored, Rac1-bound E-cadherin-catenins complex, resulting in increased E-cadherin endocytosis [338]. This together with evidence from the budding yeast that Iqg1p helps targets Cdc42 and the cell polarity machineries to plasma membranes [94, 96], supports the concept that IQGAP1 acts both upstream and downstream of Cdc42 and Rac1, thus serving both as regulator and effector [72]. Furthermore, IQGAP1 interaction with several Rho GTPases likely help facilitates crosstalk and signal-dependent switch between GTPases involved in cell transformation and progression to cancer or metastasis. Support for such view can be gleaned from a switch observed during cell invasion, where phosphorylated RacGAP1 interacts with IQGAP1 at the tips of invasive pseudopodia and locally suppresses Rac while activating RhoA to promote pseudopodia growth and invasion [339].

Indeed, IQGAP1 has been implicated in tumor cell invasion in vitro and in vivo metastasis of many human cancers [72, 334, 340–342], largely through modulating actin dynamics and membrane traffic. Invading cancer cells form actin-based projections called invadopodia, which are enriched with proteases used for breaching the extracellular matrix (ECM) and progressing to metastasis. The interaction of IQGAP1 with the exocyst complex is required for invadopodia formation downstream of Cdc42 and RhoA [342]. In invasive breast carcinomas, the interaction of the exocyst complex with the endosomal WASP and Scar homolog (WASH) on late endosomes, is required for delivering the trans-membrane type 1 matrix metalloproteinase (MT1-MMP) by invadopodia [343]. Thus, IQGAP1 likely plays multiple roles in invadopodia formation, motility and function, through roles in actin-based motility and endosomal trafficking of metalloproteinase secretion required for ECM degradation. However, experimental evidence is now required.

Given its multiple roles in cells, the mechanisms of IQGAP1-induction of carcinoma will likely vary according to the cell types and the underlying signaling pathway. This is consistent with the current notion that mechanisms of cancer initiation are broader than initially thought. A core mechanism by which IQGAP1 likely initiate cancer lesions, involves the signaling pathways it controls and that have been implicated in cancer. One such evidence is that expression of the IR-WW domain, which binds MAPK and mTOR, attenuates ERK1/2 activity in response to EGF while enhancing AktSer473 level and promoting cell proliferation [70]. Indeed, Akt has been widely implicated in human cancer. However, a recent study showed that a peptide for IQGAP1 WW domain disrupted IQGAP1-ERK1/2 interactions, inhibited RAS- and RAF-driven tumorigenesis, bypassed acquired resistance to the BRAF inhibitor vemurafenib (PLX-4032) and acted as a systemically deliverable therapeutic to significantly increase the lifespan of tumor-bearing mice [337]. Similarly, disruption of IQGAP1-Erk1/2 interaction, with the flavonoid quercetin, inhibited proliferation in human myeloma cell lines [344]. However the WW domain interacts with far more proteins (Fig. 2.1), therefore more mechanistic studies are required before selective therapy can be developed. The same signaling pathway appears to be activated in the pancreatic cancer cell lines PANC1, where a recent study reported that the IQGAP1's IQ motifs interact specifically with K-Ras, irrespective of K-Ras activity and activated ERK1/2 [345]. Consistent with the fine balance known for IQGAP1 activity, this study found either overexpression or knockdown of IQGAP1 impacted the K-Ras interaction with B-Raf [345].

Finally, an important mechanism by which IQGAP1 may induce cellular transformation and potentially cancer could be through its role in stimulating DNA synthesis. In fibroblast [72] and in MDA-MB-231 [346], overexpression of IQGAP1 enhances DNA synthesis and induces transformed phenotypes through Cdc42 [72] or RhoA/C [346]. The mechanism by which IQGAP1 leads to enhanced DNA synthesis and cell cycle progression is still unclear. Interestingly, recent evidence implicates IQGAP1 in regulating actin dynamics during DNA synthesis. Nuclear actin is a major component of the nucleoskeleton and it participates in essential nuclear processes, including chromatin remodeling, gene transcription, DNA replication and double strand repair, which require actin polymerization [347, 348]. Live cell imaging studies documented transport of actin, IQGAP1 and Rac1 to the nucleus in response to hydroxyurea-induced DNA replication stress [102]. While the exact role played by IQGAP1 remains to be deciphered, it is conceivable that its dysfunction can contribute to cancer inception, progression and tumor maintenance. Thus much work lies ahead to elucidating the cell-type-, cancer-type- and cancer-stage-specific mechanisms of IQGAP1-driven cancers.

Future Directions

Despite being heavily investigated, many unknowns remain about IQGAP1's mechanisms in cell and organism physiology and pathobiology. Perhaps one of the most compelling evidence for IQGAP1 as a rheostat is that either its downregulation (knockout and Knockdown) or overexpression produces similar effect in all of its diverse cellular functions. This underscores the delicate balance of its action and also emphasizes the importance of signal dynamics where small changes in IQGAP1 level or signal result in large changes downstream such as those observed on S6K and Akt [70], and leading to manifestation of disease state. Thus, it is becoming clearer that cells use IQGAP1 as a universal rheostat (Fig. 2.8). This is particularly

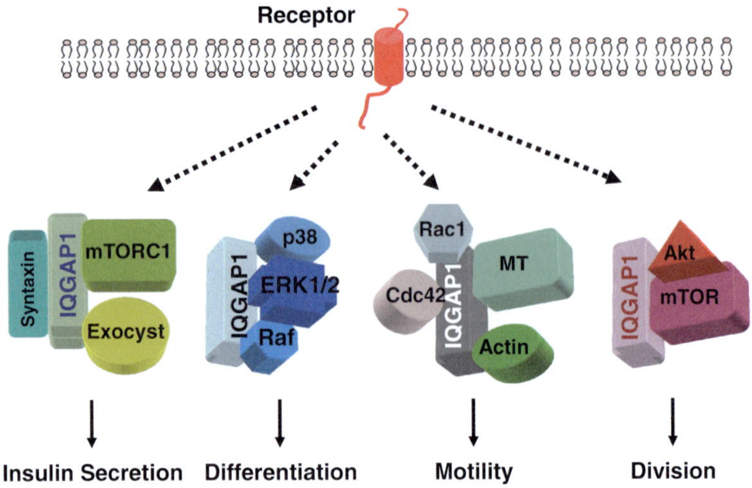

Fig. 2.8 A model of IQGAP1 as a rheostat of protein trafficking that underlie different cell functions. The preponderance of evidence from different model cell lines and organisms support a role for IQGAP1 as a scaffold signal-regulator of a variety of cell functions with an underlying mechanism in modulating protein trafficking [23, 43, 61, 70, 71]. In doing so, IQGAP1 relays a variety of extracellular signals through direct or indirect interactions with different types of receptors, including RTK and GPCR to a multitude of preformed specialized modules to effect specific functions. Dysregulation in a module can lead to specific human disease. The current challenge is to define the specific pathway for each module and elucidate the interactions in molecular detail in order to rewire that module therapeutically

evident in regulating the dynamics of actin, microtubule, exocytosis and asymmetric cell division. Future research should be directed to unraveling the specific determinants of the distinct modules.

Before proteins can be exocytosed, they must navigate the cytosol viscosity and dense meshwork of cytoskeletal filaments [349] in their journey to the plasma membrane. This process requires the action of cytoskeletal tracks and specialized motor or propeller proteins. Microtubules serve as tracks for the motor proteins Kinesin and dynein whereas actin cables serve as tracks for myosin motors to propel cargo-containing vesicles to their destinations [350]. A large body of evidence demonstrated both regulatory and structural interactions between microtubules and actin in diverse cellular processes [24, 25, 213, 316]. IQGAP1 interactions with actin, microtubules and myosin predict involvement in cytoplasmic traffic, which awaits analysis. Such studies will require mechanistic analyses with more sophisticated tools and avoidance of microtubule and actin depolymerizing agents that produce limited mechanistic insights if not confusing pleiotropic effects.

Another area of important investigation is the role of membrane microdomains potentially in anchoring and specifying the various modules organized by IQGAP1 beneath the plasma membrane or at internal membranes. For such studies, super-resolution fluorescence microscopy and/or electron microscopy might be required for visualizing the distinct modules.

An emerging area of research is the role of actin dynamics in memory control and neuronal plasticity. Translational control of actin nucleating proteins has recently been shown to underlie memory decay and forgetting required for neuron plasticity [351]. IQGAP1 plays crucial roles in actin bundling and capping [63, 85] as well as in actin nucleation [82–84], processes shown to impact memory duration [351]. This exciting area of research would pave the way for analyses of potential role of IQGAP1 in neurodegenerative diseases like Alzheimer.

A new field of study has uncovered the role of mechanical stress in modulating cell–cell contacts at adheren junctions during developmental, homeostatic and disease states. At the heart of this, myosin-dependent tension has been shown to reduce actin turnover and restrict Cadherins mobility at the junctions [352]. Given that IQGAP1 binds all these components and regulate junctional adhesion, its potential role in mechanical signaling must be analyzed. Whereas proteomic studies revealed the role of IQGAP1 in actomyosin dynamics in the rear of motile cells [178], focal adhesion turnover has been shown to be regulated by CLASP proteins-mediated localized exocytosis and ECM degradation [353]. Given the role of IQGAP1-CLASP in regulating microtubule dynamics discussed earlier, this evidence clearly implicates IQGAP1 in focal adhesion turnover-mediated cell motility and opens a new direction of investigation. Analysis of the mechanisms utilized by IQGAP1 in impacting diverse cellular functions will require more sophisticated approaches and tools, least of which are separation-of-function mutants and developing signal-sensitive biosensors. Understanding how IQGAP1 serve these multiple but highly specialized functions could provide means for targeting human disease in a specific and context-dependent manner.

Acknowledgement Work in the MO lab was supported by grants from the NIH-NCI (CA104285) and from ACS.

References

1. Ramaekers FCS, Bosman FT (2004) The cytoskeleton and disease. J Pathol 204:351–354
2. McMurray CT (2000) Neurodegeneration: diseases of the cytoskeleton? Cell Death Diff 7:861–865
3. Cortelazzo A, De Felice C, Pecorelli A, Belmonte G, Signorini C, Leoncini S, Zollo G, Capone A, Giovampaola CD, Sticozzi C, Valacchi G, Ciccoli L, Guerranti R, Hayek J (2014) Beta-actin deficiency with oxidative posttranslational modifications in Rett syndrome erythrocytes: insights into an altered cytoskeletal organization. PLoS One 9(3):e93181. doi:10.1371/journal.pone.0093181
4. Dominguez R, Holmes KC (2011) Actin structure and function. Annu Rev Biophys 40:169–186
5. Herman IM (1993) Actin isoforms. Curr Opin Cell Biol 5:48–55
6. Osman MA, Cerione RA (2006) Actin doesn't do the locomotion: secretion drives cell polarization. In: Segev N (ed) Protein trafficking: mechanisms and regulation. Landes Bioscience, Georgetown, TX
7. Pollard TD, Cooper JA (2009) Actin, a central player in cell shape and movement. Science 326(5957):1208–1212

8. Pollard TD, Blanchoin L, Mullins RD (2000) Molecular mechanisms controlling actin filament dynamics in nonmuscle cells. Annu Rev Biophys Biomol Struct 29:545–576
9. Pruyne D, Evangelista M, Yang C, Bi E, Zigmond S, Bretscher A, Boone C (2002) Role of formins in actin assembly: nucleation and barbed-end association. Science 297(5581): 612–615
10. Kim K, Yamashita A, Wear MA, Maeda Y, Cooper JA (2004) Capping protein binding to actin in yeast: biochemical mechanism and physiological relevance. J Cell Biol 164: 567–580
11. Machesky LM, Atkinson SJ, Ampe C, Vandekerckhove J, Pollard TD (1994) Purification of a cortical complex containing two unconventional actins from Acanthamoeba by affinity chromatography on profilin-agarose. J Cell Biol 127:107–115
12. Pollard TD, Borisy GG (2003) Cellular motility driven by assembly and disassembly of actin filaments. Cell 112:453–465
13. Eden S, Rohatgi R, Podtelejnikov AV, Mann M, Kirschner MW (2002) Mechanism of regulation of WAVE1-induced actin nucleation by Rac1 and Nck. Nature 418:790–793
14. Miki H, Sasaki T, Takai Y, Takenawa T (1998) Induction of filopodium formation by a WASP-related actin-depolymerizing protein N-WASP. Nature 391:93–96
15. Ismail AM, Padrick SB, Chen B, Umetani J, Rosen MK (2009) The WAVE regulatory complex is inhibited. Nat Struct Mol Biol 16(5):561–563
16. Rohatgi R, Ma L, Miki H, Lopez M, Kirchhausen T, Takenawa T, Kirschner MW (1999) The Interaction between N-WASP and the Arp2/3 complex links Cdc42-dependent signals to actin assembly. Cell 97:221–231
17. Rohatgi R, Ho HY, Kirschner MW (2000) Mechanism of N-WASP activation by CDC42 and phosphatidylinositol 4, 5-bisphosphate. J Cell Biol 150:1299–1310
18. Egile C, Loisel TP, Laurent V, Li R, Pantaloni D, Sansonetti PJ, Carlier MF (1999) Activation of the Cdc42 effector N-Wasp by the *Shigella flexneri* Icsa protein promotes actin nucleation by Arp2/3 complex and bacterial actin-based motility. J Cell Biol 146:1319–1332
19. Miki H, Miura K, Takenawa T (1996) N-WASP, a novel actin-depolymerizing protein, regulates the cortical cytoskeletal rearrangement in a PIP2-dependent manner downstream of tyrosine kinases. EMBO J 15:5326–5335
20. Pertz O (2010) Spatio-temporal Rho GTPase signaling: where are we now? J Cell Sci 123:1841–1850
21. Ridley AJ (2006) Rho GTPases and actin dynamics in membrane protrusions and vesicle trafficking. Trends Cell Biol 16:522–529
22. Etienne-Manneville S, Hall A (2002) Rho GTPases in cell biology. Nature 420:629–635
23. Osman MA (2010) An emerging role for IQGAP1 in regulating protein traffic. Sci World J 10:944–953
24. Bartolini F, Gundersen GG (2010) Formins and microtubules. Biochim Biophys Acta 1803:164–173
25. Palazzo AF, Cook TA, Alberts AS, Gundersen GG (2001) mDia mediates Rho-regulated formation and orientation of stable microtubules. Nat Cell Biol 3:723–729
26. Vetter IR, Wittinghofer A (2001) The guanine nucleotide-binding switch in three dimensions. Science 294:1299–1304
27. Cherfils J, Zeghouf M (2013) Regulation of small GTPases by GEFs, GAPs, and GDIs. Physiol Rev 93(1):269–309
28. Hancock JF, Paterson H, Marshall CJ (1990) A polybasic domain or palmitoylation is required in addition to the CAAX motif to localize $p21^{ras}$ to the plasma membrane. Cell 63:133–139
29. Adamson P, Marshall CJ, Hall A, Tilbrook PA (1992) Post-translational modifications of p21rho proteins. J Biol Chem 267:20033–20038
30. Casey PJ, Seabra MC (1996) Protein prenyltransferases. J Biol Chem 271:5289–5292
31. Leonard D, Hart MJ, Platko JV, Eva A, Henzel W, Evans T, Cerione RA (1992) The identification and characterization of a GDP-dissociation inhibitor (GDI) for the CDC42Hs protein. J Biol Chem 267:22860–22868

32. Michaelson D, Silletti J, Murphy G, D'Eustachio P, Rush M, Philips MR (2001) Differential localization of Rho GTPases in live cells regulation by hypervariable regions and RhoGDI binding. J Cell Biol 152:111–126
33. Johnson JL, Erickson JW, Cerione RA (2009) New insights into how the Rho guanine nucleotide dissociation inhibitor regulates the interaction of Cdc42 with membranes. J Biol Chem 284:23860–23871
34. Kozma R, Ahmed S, Best A, Lim L (1995) The Ras-related protein Cdc42Hs and bradykinin promote formation of peripheral actin microspikes and filopodia in Swiss 3T3 fibroblasts. Mol Cell Biol 15:1942–1952
35. Nobes CD, Hall A (1995) Rho, Rac, and Cdc42 GTPases regulate the assembly of multimolecular focal complexes associated with actin stress fibers, lamellipodia, and filopodia. Cell 81:53–62
36. Ridley AJ, Paterson HF, Johnston CL, Diekmann D, Hall A (1992) The small GTP-binding protein rac regulates growth factor-induced membrane ruffling. Cell 70:401–410
37. Ridley AJ, Hall A (1992) The small GTP-binding protein rho regulates the assembly of focal adhesions and actin stress fibers in response to growth factors. Cell 70:389–399
38. Burridge K, Wennerberg K (2004) Rho and Rac take center stage. Cell 116:167–179
39. Olson MF, Ashworth A, Hall A (1995) An essential role for Rho, Rac, and Cdc42 GTPases in cell cycle progression through G1. Science 269:1270–1272
40. Minden A, Lin A, Claret F-X, Abo A, Karin M (1995) Selective activation of the JNK signaling cascade and c-Jun transcriptional activity by the small GTPases Rac and Cdc42Hs. Cell 81:1147–1157
41. Coso OA, Chiariello M, Yu J-C, Teramoto H, Crespo P, Xu N, Miki T, Gutkind JS (1995) The small GTP-binding proteins Rac1 and Cdc42 regulate the activity of the JNK/SAPK signaling pathway. Cell 81:1137–1146
42. Lamaze C, Chuang T, Terlecky LJ, Bokoch GM, Schmid SL (1996) Regulation of receptor-mediated endocytosis by Rho and Rac. Nature 382:177–179
43. Osman MA, Sarkar FH, Rodriguez-Boulan E (2013) A molecular rheostat on the interface of cancer and diabetes. Biochim Biophys Acta 1836:166–176
44. Diebold BA, Fowler B, Lu J, Dinauer MC, Bokoch GM (2004) Antagonistic crosstalk between Rac and Cdc42 GTPases regulates generation of reactive oxygen species. J Biol Chem 279:28136–28142
45. Machacek M, Hodgson L, Welch C, Elliott H, Nalbant P, Pertz O, Abell A, Johnson GL, Hahn KM, Danuser G (2009) Coordination of Rho GTPase activities during cell protrusion. Nature 461:99–103
46. Gasman S, Chasserot-Golaz S, Popoff MR, Aunis D, Bader MF (1999) Involvement of Rho GTPases in calcium-regulated exocytosis from adrenal chromaffin cells. J Cell Sci 112:4763–4771
47. Ory S, Gasman S (2011) Rho GTPases and exocytosis: what are the molecular links? Semin Cell Dev Biol 22(1):27–32. doi:10.1016/j.semcdb.2010.12.002
48. Fukumoto Y, Kaibuchi K, Hori Y, Fujioka H, Araki S, Ueda T, Kikuchi A, Takai Y (1990) Molecular cloning and characterization of a novel type of regulatory protein (GDI) for the rho proteins, Ras p21-like small GTP-binding proteins. Oncogene 5:1321–1328
49. Ueda T, Kikuchi A, Ohga N, Yamamoto J, Takai Y (1990) Purification and characterization from bovine brain cytosol of a novel regulatory protein inhibiting the dissociation of GDP from and the subsequent binding of GTP to rhoB p20, a ras p21-like GTP-binding protein. J Biol Chem 265:9373–9380
50. Rossman KL, Der CJ, Sondek J (2005) GEF means go: turning on RHO GTPases with guanine nucleotide-exchange factors. Nat Rev Mol Cell Biol 6:167–180
51. Cote JF, Vuori K (2007) GEF what? Dock180 and related proteins help Rac to polarize cells in new ways. Trends Cell Biol 17:383–393
52. Tcherkezian J, Lamarche-Vane N (2007) Current knowledge of the large RhoGAP family of proteins. Biol Cell 99:67–86

53. Hao Y, Wong R, Feig LA (2008) RalGDS couples growth factor signaling to Akt activation. Mol Cell Biol 28:2851–2859
54. Zheng Y, Bagrodia S, Cerione RA (1994) Activation of phosphoinositide 3-kinase activity by Cdc42Hs binding to p85. J Biol Chem 269:18727–18730
55. Kozma R, Ahmed S, Best A, Lim L (1996) The GTPase-activating protein n-chimaerin cooperates with Rac1 and Cdc42Hs to induce the formation of lamellipodia and filopodia. Mol Cell Biol 16:5069–5080
56. Marshal CJ (1996) Ras effectors. Curr Opin Cell Biol 8:197–204
57. Ma L, Rohatgi R, Kirschner MW (1998) The Arp2/3 complex mediates actin polymerization induced by the small GTP-binding protein Cdc42. Proc Natl Acad Sci U S A 95:15362–15367
58. Wallrabe H, Cai Y, Sun Y, Periasamy A, Luzes R, Fang X, Kan HM, Cameron LC, Schafer DA, Bloom GS (2013) IQGAP1 interactome analysis by in vitro reconstitution and live cell 3-color FRET microscopy. Cytoskeleton 70(12):819–836
59. Burack WR, Shaw AS (2000) Signal transduction: hanging on a scaffold. Curr Opin Cell Biol 12:211–216
60. Ferrell JE Jr (2000) What do scaffold proteins really do? Science's STKE 2000(52):pe1
61. Osman MA, Bloom GS, Tagoe EA (2013) *Helicobacter pylori*-induced alteration of epithelial cell signaling and polarity: a possible mechanism of gastric carcinoma etiology and disparity. Cytoskeleton 70:349–359
62. Hart MJ, Callow MG, Souza B, Polakis P (1996) QGAP1, a calmodulin-binding protein with a rasGAP-related domain, is a potential effector for cdc42Hs. EMBO J 15:2997–3005
63. Bashour A-M, Fullerton AT, Hart MJ, Bloom GS (1997) IQGAP1, a Rac- and Cdc42-binding protein, directly binds and cross-inks microfilaments. J Cell Biol 137:1555–1566
64. Fukata M, Kuroda S, Fujii K, Nakamura T, Shoji I, Matsuura Y, Okawa K, Iwamatsu A, Kikuchi A, Kaibuchi K (1997) Regulation of cross-linking activity of actin filament by IQGAP1, a target for Cdc42. J Biol Chem 272:29579–29583
65. Brandt DT, Grosse R (2007) Get to grips: steering local actin dynamics with IQGAPs. EMBO Rep 8:1019–1023
66. Briggs MW, Sacks DB (2003) IQGAP proteins are integral components of cytoskeletal regulation. EMBO Rep 4:571–574
67. Mateer SC, Wang N, Bloom GS (2003) IQGAPS: integrators of the cytoskeleton, cell adhesion machinery and signaling networks. Cell Motil Cytoskeleton 55(3):147–155
68. Logue JS, Whiting JL, Tunquist B, Sacks DB, Langeberg LK, Wordeman L, Scott JD (2011) AKAP220 protein organizes signaling elements that impact cell migration. J Biol Chem 286:39269–39281
69. Nauert JB, Rigas JD, Lester LB (2003) Identification of an IQGAP1/AKAP79 complex in beta-cells. J Cell Biochem 90:97–108
70. Tekletsadik YK, Sonn R, Osman MA (2012) A conserved role for IQGAP1 in regulating TOR complex 1. J Cell Sci 125:2041–2052
71. Rittmeyer EN, Daniel S, Hsu SC, Osman MA (2008) A dual role for IQGAP1 in regulating exocytosis. J Cell Sci 121:391–408
72. Wang JB, Sonn R, Tekletsadik YK, Samorodnitsky D, Osman MA (2009) IQGAP1 regulates cell proliferation through a novel CDC42-mTOR pathway. J Cell Sci 122:2024–2033
73. Roy M, Li Z, Sacks DB (2004) IQGAP1 binds ERK2 and modulates its activity. J Biol Chem 279:17329–17337
74. White CD, Brown MD, Sacks DB (2009) IQGAPs in cancer: a family of scaffold proteins underlying tumorigenesis. FEBS Lett 583:1817–1824
75. McCallum SJ, Wu WJ, Cerione RA (1996) Identification of a putative effector of Cdc42Hs with high sequence similarity to the RasGAP-related protein IQGAP1 and a Cdc42Hs binding partner with similarity to IQGAP2. J Biol Chem 271:21732–21737
76. Joyal JL, Annan RS, Ho Y-D, Huddleston ME, Carr SA, Hart MJ, Sacks DB (1997) Calmodulin modulates the interaction between IQGAP1 and Cdc42. J Biol Chem 272:15419–15425

77. Kuroda S, Fukata M, Kobayashi K, Nakafuku M, Nomura N, Iwamatsu A, Kaibuchi K (1996) Identification of IQGAP as a putative target for the small GTPases, Cdc42 and Rac1. J Biol Chem 271:23363–23367
78. Kuroda S, Fukata M, Nakagawa M, Fujii K, Nakamura T, Ookubo T, Izawa I, Nagase T, Nomura N, Tani H, Shoji I, Matsuura Y, Yonehara S, Kaibuchi K (1998) Role of IQGAP1, a target of the small GTPases Cdc42 and Rac1, in regulation of E-cadherin-mediated cell-cell adhesion. Science 281:832–835
79. Weissbach L, Settleman J, Kalady MF, Snijders AJ, Murthy AE, Yan Y-X, Bernards A (1994) Identification of a human rasGAP-related protein containing calmodulin-binding domains. J Biol Chem 269:20517–20521
80. Kurella VB, Richard JM, Parke CL, Lecour LF Jr, Bellamy HD, Worthylake DK (2009) Crystal structure of the GAP-related domain from IQGAP1. J Biol Chem 284:14857–14865
81. Grohmanova K, Schlaepfer D, Hess D, Gutierrez P, Beck M, Kroschewski R (2004) Phosphorylation of IQGAP1 modulates its binding to Cdc42, revealing a new type of rho-GTPase regulator. J Biol Chem 279:48495–48504
82. Brandt DT, Marion S, Griffiths G, Watanabe T, Kaibuchi K, Grosse R (2007) Dia1 and IQGAP1 interact in cell migration and phagocytic cup formation. J Cell Biol 178:193–200
83. Benseñor LB, Kan HM, Wang N, Wallrabe H, Davidson LA, Cai Y, Schafer DA, Bloom GS (2007) IQGAP1 regulates cell motility by linking growth factor signaling to actin assembly. J Cell Sci 120:658–669
84. Le Clainche C, Schlaepfer D, Ferrari A, Klingauf M, Grohmanova K, Veligodskiy A, Didry D, Le D, Egile C, Carlier MF, Kroschewski R (2007) IQGAP1 stimulates actin assembly through the N-WASP-Arp2/3 pathway. J Biol Chem 282:426–435
85. Pelikan-Conchaudron A, Le Clainche C, Didry D, Carlier MF (2011) The IQGAP1 protein is a calmodulin-regulated barbed end capper of actin filaments: possible implications in its function in cell migration. J Biol Chem 286:35119–35128
86. Fukata M, Watanabe T, Noritake J, Nakagawa M, Yamaga M, Kuroda S, Matsuura Y, Iwamatsu A, Perez F, Kaibuchi K (2002) Rac1 and Cdc42 capture microtubules through IQGAP1 and CLIP-170. Cell 109:873–885
87. Watanabe T, Wang S, Noritake J, Sato K, Fukata M, Takefuji M, Nakagawa M, Izumi N, Akiyama T, Kaibuchi K (2004) Interaction with IQGAP1 links APC to Rac1, Cdc42, and actin filaments during cell polarization and migration. Dev Cell 7:871–883
88. Watanabe T, Noritake J, Kakeno M, Matsui T, Harada T, Wang S, Itoh N, Sato K, Matsuzawa K, Iwamatsu A et al (2009) Phosphorylation of CLASP2 by GSK-3beta regulates its interaction with IQGAP1, EB1 and microtubules. J Cell Sci 122:2969–2979
89. Johnson M, Sharma M, Brocardo MG, Henderson BR (2011) IQGAP1 translocates to the nucleus in early S-phase and contributes to cell cycle progression after DNA replication arrest. Int J Biochem Cell Biol 43:65–73
90. Dixon MJ, Gray A, Schenning M, Agacan M, Tempel W, Tong Y, Nedyalkova L, Park H-W, Leslie NR, van Aalten DMF, Downes CP, Batty IH (2012) IQGAP proteins reveal an atypical phosphoinositide (aPI) binding domain with a pseudo C2 domain fold. J Biol Chem 287:22483–22496
91. Wang DS, Shaw R, Winkelmann JC, Shaw G (1994) Binding of PH domains of β-adrenergic-receptor kinase and β-spectrin to WD40/β-transducin repeat containing regions of the β-subunit of trimeric G-proteins. Biochem Biophys Res Commun 203:29–35
92. Yao L, Kawakami Y, Kawakami T (1994) The pleckstrin homology domain of Bruton tyrosine kinase interacts with protein kinase C. Proc Natl Acad Sci 91:9175
93. Erickson JW, Cerione RA, Hart MJ (1997) Identification of an actin cytoskeletal complex that includes IQGAP and the Cdc42 GTPase. J Biol Chem 272:24443–24447
94. Osman MA, Cerione RA (1998) Iqg1p, a yeast homologue of the mammalian IQGAPs, mediates cdc42p effects on the actin cytoskeleton. J Cell Biol 142:443–455
95. Epp AJ, Chant J (1997) An IQGAP-related protein controls actin-ring formation and cytokinesis in yeast. Curr Biol 7:921–929

96. Osman MA, Konopka JB, Cerione RA (2002) Iqg1p links spatial and secretion landmarks to polarity and cytokinesis. J Cell Biol 159:601–611
97. Mateer SC, McDaniel AE, Nicolas V, Habermacher GM, Lin M-JS, Cromer DA, King ME, Bloom GS (2002) The mechanism for regulation of the F-actin binding activity of IQGAP1 by calcium/calmodulin. J Biol Chem 277:12324–12333
98. Mateer ST, Morris LE, Cromer DA, Benseñor LB, Bloom GS (2004) Actin filament binding by a monomeric IQGAP1 fragment with a single calponin homology domain. Cell Motil Cytoskeleton 58:231–241
99. Li R, Debreceni B, Jia B, Gao Y, Tigyi G, Zheng Y (1999) Localization of the PAK1-, WASP-, and IQGAP1-specifying regions of Cdc42. J Biol Chem 274:29648–29654
100. Breitsprecher D, Goode BL (2013) Formins at a glance. J Cell Sci 126:1–7
101. Baarlink C, Wang H, Grosse R (2013) Nuclear actin network assembly by formins regulates the SRF coactivator MAL. Science 340:864–867
102. Johnson MA, Sharma M, Mok MT, Henderson BR (2013) Stimulation of in vivo nuclear transport dynamics of actin and its co-factors IQGAP1 and Rac1 in response to DNA replication stress. Biochim Biophys Acta 1833:2334–2347
103. Daou P, Hasan S, Breitsprecher D, Baudelet E, Camoin L, Audebert S, Goode BL, Badache A (2014) Essential and non-redundant roles for Diaphanous formins in cortical microtubule capture and directed cell migration. Mol Biol Cell 25(5):658–668
104. Noritake J, Fukata M, Sato K, Nakagawa M, Watanabe T, Izumi N, Wang S, Fukata Y, Kaibuchi K (2004) Positive role of IQGAP1, an effector of Rac1, in actin-meshwork formation at sites of cell-cell contact. Mol Biol Cell 15:1065–1076
105. Arnal I, Heichette C, Diamantopoulos GS, Chrétien D (2004) CLIP-170/tubulin-curved oligomers coassemble at microtubule ends and promote rescues. Curr Biol 14:2086–2095
106. Mimori-Kiyosue Y, Grigoriev I, Lansbergen G, Sasaki H, Matsui C, Severin F, Galjart N, Grosveld F, Vorobjev I, Tsukita S et al (2005) CLASP1 and CLASP2 bind to EB1 and regulate microtubule plus-end dynamics at the cell cortex. J Cell Biol 168:141–153
107. Choi JH, Bertram PG, Drenan R, Carvalho J, Zhou HH, Zheng XFS (2002) The FKBP12-rapamycin-associated protein (FRAP) is a CLIP-170 kinase. EMBO Rep 3:988–994
108. Bañón-Rodríguez I, Gálvez-Santisteban M, Vergarajauregui S, Bosch M, Borreguero-Pascual A, Martín-Belmonte F (2014) EGFR controls IQGAP basolateral membrane localization and mitotic spindle orientation during epithelial morphogenesis. EMBO J 33:129–145
109. Gervais L, Casanova J (2010) In vivo coupling of cell elongation and lumen formation in a single cell. Curr Biol 20:359–366
110. Saotome I, Curto M, McClatchey AI (2004) Ezrin is essential for epithelial organization and villus morphogenesis in the developing intestine. Dev Cell 6:855–864
111. Jayanandanan N, Mathew R, Leptin M (2014) Guidance of subcellular tubulogenesis by actin under the control of a synaptotagmin-like protein and Moesin. Nat Commun 5:3036. doi:10.1038/ncomms4036
112. Gauthier BR, Wollheim CB (2008) Synaptotagmins bind calcium to release insulin. Am J Physiol Endocrinol Metab 295:E1279–E1286
113. Liu J, Guidry JJ, Worthylake DK (2013) Conserved sequence repeats of IQGAP1 mediate binding to Ezrin. J Proteome Res. doi:10.1021/pr400787p
114. Kanwar N, Wilkins JA (2011) IQGAP1 involvement in MTOC and granule polarization in NK cell cytotoxicity. Eur J Immunol 41:2763–2773
115. McNulty DE, Li Z, White CD, Sacks DB, Annan RS (2011) MAPK Scaffold IQGAP1 binds the EGF receptor and modulates its activation. J Biol Chem 286:15010–15021
116. Cotton CU, Hobert ME, Ryan S, Carlin CR (2013) Basolateral EGF receptor sorting regulated by functionally distinct mechanisms in renal epithelial cells. Traffic 14:337–354
117. Sun QY, Schatten H (2006) Role of NuMA in vertebrate cells: review of an intriguing multifunctional protein. Front Biosci 11:1137–1146
118. Arakawa Y, Cordeiro JV, Way M (2007) The release of vaccinia virus from infected cells requires RhoA-mDia modulation of cortical actin. Cell Host Microbe 1:227–240

119. Stinchcombe JC, Majorovits E, Bossi G, Fuller S, Griffiths GM (2006) Centrosome polarization delivers secretory granules to the immunological synapse. Nature 443:462–465
120. Gorman JA, Babich A, Dick CJ, Schoon RA, Koenig A, Gomez TS, Burkhardt JK, Billadeau DD (2012) The cytoskeletal adaptor protein IQGAP1 regulates TCR-mediated signaling and filamentous actin dynamics. J Immunol 188:6135–6144
121. Barral Y (2010) Cell biology. Septins at the nexus. Science 329:1289–1290
122. Saarikangas J, Barral Y (2011) The emerging functions of septins in metazoans. EMBO Rep 12:1118–1126
123. Spiliotis ET, Gladfelter AS (2012) Spatial guidance of cell asymmetry: septin GTPases show the way. Traffic 13:195–203
124. Founounou N, Loyer N, Le Borgne R (2013) Septins regulate the contractility of the actomyosin ring to enable adherens junction remodelling during cytokinesis of epithelial cells. Dev Cell 24:242–255
125. Mostowy S, Janel S, Forestier C, Roduit C, Kasas S, Pizarro-Cerdá J, Cossart P, Lafont F (2011) A role for septins in the interaction between the Listeria monocytogenes INVASION PROTEIN InlB and the Met receptor. Biophys J 100:1949–1959
126. Gilden JK, Peck S, Chen YC, Krummel MF (2012) The septin cytoskeleton facilitates membrane retraction during motility and blebbing. J Cell Biol 196:103–114
127. Baumann S, König J, Koepke J, Feldbrügge M (2014) Endosomal transport of septin mRNA and protein indicates local translation on endosomes and is required for correct septin filamentation. EMBO Rep 15:94–102
128. Morita E, Sandrin V, Chung HY, Morham SG, Gygi SP, Rodesch CK, Sundquist WI (2007) Human ESCRT and ALIX proteins interact with proteins of the midbody and function in cytokinesis. EMBO J 26:4215–4227
129. Kinoshita M, Field CM, Coughlin ML, Straight AF, Mitchison TJ (2002) Self- and actin-templated assembly of mammalian septins. Dev Cell 3:791–802
130. Mavrakis M, Azou-Gros Y, Tsai FC, Alvarado J, Bertin A, Iv F, Kress A, Brasselet S, Koenderink GH, Lecuit T (2014) Septins promote F-actin ring formation by crosslinking actin filaments into curved bundles. Nat Cell Biol 16:322–334
131. Tsubota A, Matsumoto K, Mogushi K, Nariai K, Namiki Y, Hoshina S, Hano H, Tanaka H, Saito H, Tada N (2010) IQGAP1 and vimentin are key regulator genes in naturally occurring hepatotumorigenesis induced by oxidative stress. Carcinogenesis 31:504–511
132. Thiery JP (2002) Epithelial-mesenchymal transitions in tumour progression. Nat Rev Cancer 2:442–454
133. Huber MA, Kraut N, Beug H (2005) Molecular requirements for epithelial-mesenchymal transition during tumor progression. Curr Opin Cell Biol 17:548–558
134. Satelli A, Li S (2011) Vimentin in cancer and its potential as a molecular target for cancer therapy. Cell Mol Life Sci 68:3033–3046
135. Govaere O, Komuta M, Berkers J, Spee B, Janssen C, de Luca F, Katoonizadeh A, Wouters J, van Kempen LC, Durnez A, Verslype C, De Kock J, Rogiers V, van Grunsven LA, Topal B, Pirenne J, Vankelecom H, Nevens F, van den Oord J, Pinzani M, Roskams T (2014) Keratin 19: a key role player in the invasion of human hepatocellular carcinomas. Gut 63:674–685
136. Pufall MA, Graves BJ (2002) Autoinhibitory domains: modular effectors of cellular regulation. Annu Rev Cell Dev Biol 18:421–462
137. Hall BE, Yang SS, Bar-Sagi D (2002) Autoinhibition of Sos by intramolecular interactions. Front Biosci 7:d288–d294
138. Freedman TS, Sondermann H, Friedland GD, Kortemme T, Bar-Sagi D, Marqusee S, Kuriyan J (2006) A Ras-induced conformational switch in the Ras activator Son of sevenless. Proc Natl Acad Sci U S A 103:16692–16697
139. Aghazadeh B, Lowry WE, Huang XY, Rosen MK (2000) Structural basis for relief of autoinhibition of the Dbl homology domain of proto-oncogene Vav by tyrosine phosphorylation. Cell 102:625–633

140. Lorenz M, Yamaguchi H, Wang Y, Singer RH, Condeelis J (2004) Imaging Sites of N-WASP Activity in Lamellipodia and Invadopodia of Carcinoma Cells. Curr Biol 14:697–703
141. Kim AS, Kakalis LT, Abdul-Manan N, Liu GA, Rosen MK (2000) Autoinhibition and activation mechanisms of the Wiskott–Aldrich syndrome protein. Nature 404:151–158
142. Li F, Higgs HN (2003) The mouse formin mDia1 is a potent actin nucleation factor regulated by autoinhibition. Curr Biol 13(15):1335–1340
143. Hagan GN, Lin Y, Magnuson MA, Avruch J, Czech MP (2008) A Rictor-Myo1c complex participates in dynamic cortical actin events in 3T3-1 adipocytes. Mol Cell Biol 28:4215–4226
144. Laplante M, Sabatini DN (2009) mTOR signaling at a glance. J Cell Sci 122:3589–3594
145. Sarbassov DD, Guertin DA, Ali SM, Sabatini DM (2005) Phosphorylation and regulation of Akt/PKB by the rictor–mTOR complex. Science 307:1098–1101
146. Hemmings BA, Restuccia DF (2012) PI3K-PKB/Akt Pathway. Cold Spring Harb Perspect Biol 4:a011189
147. Dibble CC, Asara JM, Manning BD (2009) Manning, characterization of rictor phosphorylation sites reveals direct regulation of mTOR complex 2 by S6K1. Mol Cell Biol 29:5657–5670
148. Boulbes D, Chen CH, Shaikenov T, Agarwal NK, Peterson TR, Addona TA, Keshishian H, Carr SA, Magnuson MA, Sabatini DM, Sarbassov DD (2010) Rictor phosphorylation on the Thr-1135 site does not require mammalian target of rapamycin complex 2. Mol Cancer Res 8:896–906
149. Treins C, Warne PH, Magnuson MA, Pende M, Downward J (2010) Rictor is a novel target of p70 S6 kinase-1. Oncogene 29:1003–1016
150. Julien LA, Carriere A, Moreau J, Roux PP (2010) mTORC1-activated S6K1 phosphorylates Rictor on threonine 1135 and regulates mTORC2 signaling. Mol Cell Biol 30:908–921
151. Fruman DA, Rommel C (2014) PI3K and cancer: lessons, challenges and opportunities. Nat Rev Drug Discov 13:140–156
152. Mendoza MC, Er EE, Blenis J (2011) The Ras-ERK and PI3K-mTOR pathways: crosstalk and compensation. Trends Biochem Sci 36:320–328
153. Sbroggiò M, Carnevaleb D, Berteroa A, Cifelli G, Blasioa ED, Mascio G, Hirsch E, Bahou WF, Turco E, Silengo L, Brancaccio M, Lembo G, Tarone G (2011) IQGAP1 regulates ERK1/2 and AKT signaling in the heart and sustains functional remodeling upon pressure overload. Cardiovasc Res 91:456–464
154. Brown MD, Sacks DB (2009) Protein scaffolds in MAP kinase signaling. Cell Signal 21:462–469
155. Mendoza MC, Er EE, Zhang W, Ballif BA, Elliott HL, Danuser G, Blenis J (2011) ERK-MAPK drives lamellipodia protrusion by activating the WAVE2 regulatory complex. Mol Cell 41:661–671
156. Chen B, Brinkmann K, Chen Z, Pak CW, Liao Y, Shi S, Henry L, Grishin NV, Bogdan S, Rosen MK (2014) The WAVE regulatory complex links diverse receptors to the actin cytoskeleton. Cell 156:195–207
157. Li W, Fan J, Woodley DT (2001) Nck/Dock: an adapter between cell surface receptors and the actin cytoskeleton. Oncogene 20:6403–6417
158. Needham SR, Hirsch M, Rolfe DJ, Clarke DT, Zanetti-Domingues LC, Wareham R, Martin-Fernandez ML (2013) Measuring EGFR separations on cells with approximately 10 nm resolution via fluorophore localization imaging with photobleaching. PLoS One 8:e62331
159. Blagoev B, Ong SE, Kratchmarova I, Mann M (2004) Temporal analysis of phosphotyrosine-dependent signaling networks by quantitative proteomics. Nat Biotechnol 22:1139–1145
160. Kratchmarova I, Blagoev B, Haack-Sorensen M, Kassem M, Mann M (2005) Mechanism of divergent growth factor effects in mesenchymal stem cell differentiation. Science 308:1472–1477
161. Yamaoka-Tojo M, Ushio-Fukai M, Hilenski L, Dikalov SI, Chen YE, Tojo T, Fukai T, Fujimoto M, Patrushev NA, Wang N et al (2004) IQGAP1, a novel VEGF receptor binding protein involved in reactive oxygen species-dependent endothelial migration and proliferation. Circ Res 95:276–283

162. Yamaoka-Tojo M, Tojo T, Kim HW, Hilenski L, Patrushev NA, Zhang L, Fukai T, Ushio-Fukai M (2006) IQGAP1 mediates VE-cadherin-based cell-cell contacts and VEGF signaling at adherence junctions linked to angiogenesis. Arterioscler Thromb Vasc Biol 26:1991–1997
163. Meyer RD, Sacks DB, Rahimi N (2008) IQGAP1-dependent signaling pathway regulates endothelial cell proliferation and angiogenesis. PLoS One 3(12):e3848. doi:10.1371/journal.pone.0003848
164. Kohno T, Urao N, Ashino T, Sudhahar V, Inomata H, Yamaoka-Tojo M, McKinney RD, Fukai T, Ushio-Fukai M (2013) IQGAP1 links PDGF receptor-{beta} signal to focal adhesions involved in vascular smooth muscle cell migration: role in neointimal formation after vascular injury. Am J Physiol Cell Physiol 305:C591–C600
165. Valcarcel-Ares MN, Gautam T, Warrington JP, Bailey-Downs L, Sosnowska D, de Cabo R, Losonczy G, Sonntag WE, Ungvari Z, Csiszar A (2012) Disruption of Nrf2 signaling impairs angiogenic capacity of endothelial cells: implications for microvascular aging. J Gerontol A Biol Sci Med Sci 67:821–829
166. Cheung KL, Lee JH, Shu L, Kim L-H, Sacks DB, Kong A-NT (2013) The Ras GTPase-activating-like protein IQGAP1 mediates Nrf2 protein activation via the mitogen-activated protein kinase/extracellular signal-regulated kinase (ERK) Kinase (MEK)-ERK pathway. J Biol Chem 288:22378–22386
167. Lemmon MA, Schlessinger J (2010) Cell signaling by receptor tyrosine kinases. Cell 141:117–1134
168. Erdemir HH, Li Z, Sacks DB (2014) IQGAP1 binds to estrogen receptor-α and modulates its function. J Biol Chem 289(13):9100–9112
169. Bourguignon LWY, Gilad E, Rothman K, Peyrollier K (2005) Hyaluronan-CD44 interaction with IQGAP1 promotes Cdc42 and ERK signaling, leading to actin binding, Elk-1/estrogen receptor transcriptional activation, and ovarian cancer progression. J Biol Chem 280: 11961–11972
170. Mbele GO, Deloulme JC, Gentil BJ, Delphin C, Ferro M, Garin J, Takahashi M, Baudier J (2002) The zinc- and calcium-binding S100B interacts and co-localizes with IQGAP1 during dynamic rearrangement of cell membranes. J Biol Chem 277:49998–50007
171. Heil A, Nazmi AR, Koltzscher M, Poeter M, Austermann J, Assard N, Baudier J, Kaibuchi K, Gerke V (2011) S100P is a novel interaction partner and regulator of IQGAP1. J Biol Chem 286:7227–7238
172. Liu C, Billadeau DD, Abdelhakim H, Leof E, Kaibuchi K, Bernabeu C, Bloom GS, Yang L, Boardman L, Shah VH, Kang N (2013) IQGAP1 suppresses TβRII-mediated myofibroblastic activation and metastatic growth in liver. J Clin Invest 123(3):1138–1156
173. Lu J, Qu Y, Liu Y, Jambusaria R, Han Z, Ruthel G, Freedman BD, Harty RN (2013) Host IQGAP1 and Ebola virus VP40 interactions facilitate virus-like particle egress. J Virol 87:7777–7780
174. Oldhami WM, Hamm HE (2008) Heterotrimeric G protein activation by G-protein-coupled receptors. Nat Rev Mol Cell Biol 9:60–71
175. Katritch V, Cherezov V, Stevens RC (2013) Structure-function of the G protein–coupled receptor superfamily. Annu Rev Pharmacol Toxicol 53:531–556
176. Ritter SL, Hall RA (2008) Fine-tuning of GPCR activity by receptor-interacting proteins. Nat Rev Mol Cell Biol 10:819–830
177. Malin C, Schi HB (2008) Structural diversity of G protein-coupled receptors and significance for drug discovery. Nat Rev Drug Discov 7:339–357
178. Witze ES, Connacher MK, Houel S, Schwartz MP, Morphew MK, Reid L, Sacks DB, Anseth KS, Ahn NG (2013) Wnt5a directs polarized calcium gradients by recruiting cortical endoplasmic reticulum to the cell trailing edge. Dev Cell 26:645–657
179. Nuriya M, Oh S, Huganir RL (2005) Phosphorylation-dependent interactions of alpha-Actinin-1/IQGAP1 with the AMPA receptor subunit GluR4. J Neurochem 95:544–552
180. Gao C, Frausto SF, Guedea AL, Tronson NC, Jovasevic V, Leaderbrand K, Corcoran KA, Guzmán YF, Swanson GT, Radulovic J (2011) IQGAP1 regulates NR2A signaling, spine density, and cognitive processes. J Neurosci 31:8533–8542

181. Li Z, McNulty DE, Marler KJ, Lim L, Hall C, Annan RS, Sacks DB (2005) IQGAP1 promotes neurite outgrowth in a phosphorylation-dependent manner. J Biol Chem 280:13871–13878
182. Neel NF, Sai J, Ham A-JL, Sobolik-Delmaire T, Mernaugh RL, Richmond A (2011) IQGAP1 is a novel CXCR2-interacting protein and essential component of the “Chemosynapse”. PLoS One 6(8):e23813. doi:10.1371/journal.pone.0023813
183. Cvetkovic D, Dragan M, Leith SJ, Mir ZM, Leong HS, Pampillo M, Lewis JD, Babwah AV, Bhattacharya M (2013) KISS1R induces invasiveness of estrogen receptor-negative human mammary epithelial and breast cancer cells. Endocrinology 154:1999–2014
184. Buday L, Wunderlich L, Tamás P (2002) The Nck family of adapter proteins: regulators of actin cytoskeleton. Cell Signal 14:723–731
185. Smith MJ, Hardy WR, Li GY, Goudreault M, Hersch S, Metalnikov P, Starostine A, Pawson T, Ikura M (2010) The PTB domain of ShcA couples receptor activation to the cytoskeletal regulator IQGAP1. EMBO J 29(5):884–896. doi:10.1038/emboj.2009.399, Epub 2010 Jan 14
186. Humphries AC, Donnelly SK, Way M (2014) Cdc42 and the RhoGEF intersectin-1 collaborate with Nck to promote N-WASP-dependent actin polymerisation. J Cell Sci 127:673–685
187. Ridley AJ (2011) Life at the leading edge. Cell 145:1012–1022
188. Miki H, Takenawa T (2003) Regulation of actin dynamics by WASP family proteins. J Biochem 134:309–313
189. Benesch S, Polo S, Lai FP, Anderson KI, Stradal TE, Wehland J, Rottner K (2005) N-WASP deficiency impairs EGF internalization and actin assembly at clathrin-coated pits. J Cell Sci 118:3103–3115
190. Suetsugu S, Yamazaki D, Kurisu S, Takenawa T (2003) Differential roles of WAVE1 and WAVE2 in dorsal and peripheral ruffle formation for fibroblast cell migration. Dev Cell 5:595–609
191. Yamazaki D, Suetsugu S, Miki H, Kataoka Y, Nishikawa S, Fujiwara T, Yoshida N, Takenawa T (2003) WAVE2 is required for directed cell migration and cardiovascular development. Nature 424:452–456
192. Mataraza JM, Briggs MW, Li Z, Entwistle A, Ridley AJ, Sacks DB (2003) IQGAP1 promotes cell motility and invasion. J Biol Chem 278:41237–41245
193. Dong PX, Jia N, Xu ZJ, Liu YT, Li DJ, Feng YJ (2008) Silencing of IQGAP1 by shRNA inhibits the invasion of ovarian carcinoma HO-8910PM cells *in vitro*. J Exp Clin Cancer Res 27:77
194. Tomasevic N, Jia Z, Russell A, Fujii T, Hartman JJ, Clancy S, Wang M, Beraud C, Wood KW, Sakowicz R (2007) Differential regulation of WASP and N-WASP by Cdc42, Rac1, Nck, and PI(4,5)P2. Biochemistry 46:3494–3502
195. Choi S, Thapa N, Hedman AC, Li Z, Sacks DB, Anderson RA (2013) IQGAP1 is a novel phosphatidylinositol 4,5 bisphosphate effector in regulation of directional cell migration. EMBO J 32:2617–2630
196. Weijer CJ (2009) Collective cell migration in development. J Cell Sci 122:3215–3223
197. Friedl P, Wolf K, Zegers MM (2014) Rho-directed forces in collective migration. Nat Cell Biol 16:208–210
198. Reffay M, Parrini MC, Cochet-Escartin O, Ladoux B, Buguin A, Coscoy S, Amblard F, Camonis J, Silberzan P (2014) Interplay of RhoA and mechanical forces in collective cell migration driven by leader cells. Nat Cell Biol 16:217–223
199. Tanos B, Rodriguez-Boulan E (2008) The epithelial polarity program: machineries involved and their hijacking by cancer. Oncogene 27:6939–6957
200. Nelson JW (2009) Remodeling epithelial cell organization: transitions between front–rear and apical–basal polarity. Cold Spring Harb Perspect Biol 1(1):a000513. doi:10.1101/cshperspect.a000513
201. Roignot J, Peng X, Mostov K (2013) Polarity in mammalian epithelial morphogenesis. Cold Spring Harb Perspect Biol 5:a013789
202. Hartsock A, Nelson WJ (2008) Adherens and tight junctions: structure, function and connections to the actin cytoskeleton. Biochim Biophys Acta 1778:660–669

203. Ratheesh A, Yap AS (2012) A bigger picture: classical Cadherins and the dynamic actin cytoskeleton. Nat Rev Mol Cell Biol 13:673–679
204. Niessen C, Leckband D, Yap AS (2011) Tissue organization by classical cadherin adhesion molecules: dynamic molecular and cellular mechanisms of morphogenetic regulation. Physiol Rev 91:691–731
205. Tanos BE, Perez Bay AE, Salvarezza S, Vivanco I, Mellinghoff I, Osman M, Sacks DB, Rodriguez-Boulan E (2015) J Cell Sci 128(5):853–862. doi:10.1242/jcs.118703
206. Yamashiro S, Noguchi T, Mabuchi I (2003) Localization of two IQGAPs in cultured cells and early embryos of *Xenopus laevis*. Cell Motil Cytoskeleton 55:36–50
207. Kuroda S, Fukata M, Nakagawa M, Kaibuchi K (1999) Cdc42, Rac1, and their effector IQGAP1 as molecular switches for cadherin-mediated cell-cell adhesion. Biochem Biophys Res Commun 262:1–6
208. Yan J, Yang Y, Zhang H, King C, Kan HM, Cai Y, Yuan CX, Bloom GS, Hua X (2009) Menin interacts with IQGAP1 to enhance intercellular adhesion of beta-cells. Oncogene 28:973–982
209. Fukata M, Nakagawa M, Itoh N, Kawajiri A, Yamaga M, Kuroda S, Kaibuchi K (2001) Involvement of IQGAP1, an effector of Rac1 and Cdc42 GTPases, in cell-cell dissociation during cell scattering. Mol Cell Biol 21:2165–2183
210. Spiliotis ET, Nelson JW (2003) Spatial control of exocytosis. Curr Opin Cell Biol 15:430–437
211. Kreitzer G, Schmoranzer J, Low SH, Li X, Gan Y, Weimbs T, Simon SM, Rodriguez-Boulan E (2003) Three-dimensional analysis of post-Golgi carrier exocytosis in epithelial cells. Nat Cell Biol 5:126–136
212. Nelson JW, Rodriguez-Boulan E (2004) Unravelling protein sorting. Nat Cell Biol 6:282–284
213. Schroeder HW III, Mitchell C, Shuman H, Holzbauer ELF, Goldman YE (2010) Motor number controls cargo switching at actin-microtubule intersections in vitro. Curr Biol 20:687–696
214. Gerber SH, Südhof TC (2002) Molecular determinants of regulated exocytosis. Diabetes 51(Suppl 1):S3–S11
215. Shandala T, Kakavanos-Plew R, Ng YS, Bader C, Sorvina A, Parkinson-Lawrence EJ, Brooks RD, Borlace GN, Prodoehl MJ, Brooks DA (2012). Molecular machinery regulating exocytosis. In: Weigert R (ed) crosstalk and integration of membrane trafficking pathways. InTech, ISBN: 978-953-51-0515-2. doi:10.5772/39110. Available from: http://www.intechopen.com/books/crosstalk-and-integration-of-membrane-trafficking-pathways/molecular-machinery-regulating-exocytosis
216. Burgoyne RD, Morgan A (2003) Secretory granule exocytosis. Physiol Rev 83:581–632
217. Gutierrez LM (2012) New insights into the role of the cortical cytoskeleton in exocytosis from neuroendocrine cells. Int Rev Cell Mol Biol 295:109–137
218. Nightingale TD, Cutler DF, Cramer LP (2012) Actin coats and rings promote regulated exocytosis. Trends Cell Biol 6:329–337
219. Porat-Shliom N, Milberg O, Masedunskas A, Weigert R (2012) Multiple roles for the actin cytoskeleton during regulated exocytosis. Cell Mol Life Sci. doi:10.1007/s00018-012-1156-5
220. Hou JC, Min L, Pessin JE (2009) Insulin granule biogenesis, trafficking and exocytosis. Vitam Horm 80:473–506
221. Wilson JR, Ludowyke RI, Biden TJ (2001) A redistribution of actin and myosin IIA accompanies Ca(2?)-dependent insulin secretion. FEBS Lett 492:101–106
222. Wang JL, Easom RA, Hughes JH, McDaniel ML (1990) Evidence for a role of microfilaments in insulin release from purified beta-cells. Biochem Biophys Res Commun 171:424–430
223. Swanston-Flatt SK, Carlsson L, Gylfe E (1980) Actin filament formation in pancreatic beta-cells during glucose stimulation of insulin secretion. FEBS Lett 117:299–302
224. Howell SL, Tyhurst M (1980) Regulation of actin polymerization in rat islets of Langerhans. Biochem J 192:381–383
225. Hong-Geller E, Cerione RA (2000) Cdc42. Rac stimulate exocytosis of secretory granules by activating the IP(3)/calcium pathway in RBL-2H3 mast cells. J Cell Biol 148:481–494

226. Kowluru A, Li G, Rabaglia ME, Segu VB, Hofmann F, Aktories K, Metz SA (1997) Evidence for differential roles of the Rho subfamily of GTP-binding proteins in glucose- and calcium-induced insulin secretion from pancreatic beta cells. Biochem Pharmacol 54:1097–1108
227. Kowluru A (2010) Small G proteins in islet beta-cell function. Endocr Rev 31:52–78
228. Daniel S, Noda M, Cerione RA, Sharp GW (2002) A link between Cdc42 and syntaxin is involved in mastoparan-stimulated insulin release. Biochemistry 41:9663–9671
229. Amin RH, Chen HQ, Veluthakal R, Silver RB, Li J, Li G et al (2003) Mastoparan-induced insulin secretion from insulin-secreting betaTC3 and INS-1 cells: evidence for its regulation by Rho subfamily of G proteins. Endocrinology 144:4508–4518
230. Hammar E, Tomas A, Bosco D, Halban PA (2009) Role of the Rho-ROCK (Rho associated kinase) signaling pathway in the regulation of pancreatic beta cell function. Endocrinology 150:2072–2079
231. Jayaram B, Syed I, Kyathanahalli CN, Rhodes CJ, Kowluru A (2011) Arf nucleotide binding site opener [ARNO] promotes sequential activation of Arf6, Cdc42 and Rac1 and insulin secretion in INS 832/13 beta-cells and rat islets. Biochem Pharmacol 81:1016–1027
232. Asahara S, Shibutani Y, Teruyama K, Inoue HY, Kawada Y, Etoh H et al (2013) Ras-related C3 botulinum toxin substrate 1 (RAC1) regulates glucose stimulated insulin secretion via modulation of F-actin. Diabetologia 56:1088–1097
233. Kowluru A, Veluthakal R (2005) Rho guanosine diphosphate-dissociation inhibitor plays a negative modulatory role in glucose-stimulated insulin secretion. Diabetes 54:3523–3529
234. Kowluru A, Veluthakal R, Rhodes CJ, Kamath V, Syed I, Koch BJ (2010) Protein farnesylation-dependent Raf/extracellular signal-related kinase signaling links to cytoskeletal remodeling to facilitate glucose induced insulin secretion in pancreatic beta-cells. Diabetes 59:967–977
235. Li J, Luo R, Kowluru A, Li G (2004) Novel regulation by Rac1 of glucose- and forskolin-induced insulin secretion in INS-1 beta-cells. Am J Physiol Endocrinol Metab 286:E818–E827
236. Kowluru A (2011) Friendly, and not so friendly, roles of Rac1 in islet beta-cell function: lessons learnt from pharmacological and molecular biological approaches. Biochem Pharmacol 81:965–975
237. Ueda S, Kitazawa S, Ishida K, Nishikawa Y, Matsui M, Matsumuto H, Aoki T, Nozaki S, Takeda T, Tamori Y, Aiba A, Kahn CR, Kataoka T, Satoh T (2010) Crucial role of the small GTPase Rac1 in insulin-stimulated translocation of glucose transporter 4 to the mouse skeletal muscle sarcolemma. FASEB J 24:2254–2261
238. Chiu TT, Jensen TE, Sylow L, Richter EA, Klip A (2011) Rac1 signalling towards GLUT4/glucose uptake in skeletal muscle. Cell Signal 23:1546–1554
239. Balamatsias D, Kong AM, Waters JE, Sriratana A, Gurung R, Bailey CG, Rasko JE, Tiganis T, Macaulay SL, Mitchell CA (2011) Identification of P-Rex1 as a novel Rac1-guanine nucleotide exchange factor (GEF) that promotes actin remodeling and GLUT4 protein trafficking in adipocytes. J Biol Chem 286:43229–43240
240. Noritake U, Watanabe T, Sato K, Wang S, Kaibuchi K (2005) IQGAP1: a key regulator of adhesion and migration. J Cell Sci 118:2085–2092
241. Schmitz-Peiffer C, Laybutt DR, Burchfield JG, Gurisik E, Narasimhan S, Mitchell CJ, Pedersen DJ, Braun U, Cooney GJ, Leitges M, Biden TJ (2007) Inhibition of PKCε improves glucose-stimulated insulin secretion and reduces insulin clearance. Cell Metab 6:320–328
242. He B, Xi F, Zhang X, Zhang J, Guo W (2007) Exo70 interacts with phospholipids and mediates the targeting of the exocyst to the plasma membrane. EMBO J 26:4053–4065
243. He B, Guo W (2009) The exocyst complex in polarized exocytosis. Curr Opin Cell Biol 21:537–542
244. Liu J, and Guo W. (2012). The exocyst complex in exocytosis and cell migration. Protoplasma. 249(3):587–97. doi: 10.1007/s00709-011-0330-1. Epub 2011 Oct 14
245. Morgera F, Sallah MR, Dubuke ML, Gandhi P, Brewer DN, Carr CM, Munson M (2012) Regulation of exocytosis by the exocyst subunit Sec6 and the SM protein Sec1. Mol Biol Cell 23:337–346, 1939–4586

246. Kimura T, Yamaoka M, Taniguchi S, Okamoto M, Takei M, Ando T, Iwamatsu A, Watanabe T, Kaibuchi K, Ishizaki T, Niki I (2013) Activated Cdc42-bound IQGAP1 determines the cellular endocytic site. Mol Cell Biol 33:4834–4843
247. Vo YP, Hutton JC, Angleson JK (2004) Recycling of the dense-core vesicle membrane protein phogrin in Min6 beta-cells. Biochem Biophys Res Commun 324:1004–1010
248. Rosentreter A, Hofmann A, Xavier CP, Stumpf M, Noegel AA, Clemen CS (2007) Coronin 3 involvement in F-actin-dependent processes at the cell cortex. Exp Cell Res 313:878–895
249. Doherty GJ, McMahon HT (2009) Mechanisms of endocytosis. Annu Rev Biochem 78:857–902
250. Guo DC, Pannu H, Tran-Fadulu V, Papke CL, Yu RK, Avidan N, Bourgeois S, Estrera AL, Safi HJ, Sparks E, Amor D, Ades L, McConnell V, Willoughby CE, Abuelo D, Willing M, Lewis RA, Kim DH, Scherer S, Tung PP, Ahn C, Buja LM, Raman CS, Shete SS, Milewicz DM (2007) Mutations in smooth muscle alphaactin (ACTA2) lead to thoracic aortic aneurysms and dissections. Nat Genet 39:1488–1493
251. Guo DC, Papke CL, Tran-Fadulu V, Regalado ES, Avidan N, Johnson RJ, Kim DH, Pannu H, Willing MC, Sparks E, Pyeritz RE, Singh MN, Dalman RL, Grotta JC, Marian AJ, Boerwinkle EA, Frazier LQ, LeMaire SA, Coselli JS, Estrera AL, Safi HJ, Veeraraghavan S, Muzny DM, Wheeler DA, Willerson JT, Yu RK, Shete SS, Scherer SE, Raman CS, Buja LM, Milewicz DM (2009) Mutations in smooth muscle alpha-actin (ACTA2) cause coronary artery disease, stroke, and Moyamoya disease, along with thoracic aortic disease. Am J Hum Genet 84:617–627
252. Morisaki H, Akutsu K, Ogino H, Kondo N, Yamanaka I, Tsutsumi Y, Yoshimuta T, Okajima T, Matsuda H, Minatoya K, Sasaki H, Tanaka H, Ishibashi-Ueda H, Morisaki T (2009) Mutation of ACTA2 gene as an important cause of familial and nonfamilial nonsyndromatic thoracic aortic aneurysm and/or dissection (TAAD). Hum Mutat 30:1406–1411
253. Sparrow JC, Nowak KJ, Durling HJ, Beggs AH, Wallgren-Pettersson C, Romero N, Nonaka I, Laing NG (2003) Muscle disease caused by mutations in the skeletal muscle alpha-actin gene (ACTA1). Neuromuscul Disord 13:519–531
254. Pearson GD, Devereux R, Loeys B, Maslen C, Milewicz D, Pyeritz R, Ramirez F, Rifkin D, Sakai L, Svensson L, Wessels A, Van Eyk J, Dietz HC (2008) Report of the National Heart, Lung, and Blood Institute and National Marfan Foundation Working Group on research in Marfan syndrome and related disorders. Circulation 118(7):785–791
255. Milewicz DM, Guo DC, Tran-Fadulu V, Lafont AL, Papke CL, Inamoto S, Kwartler CS, Pannu H (2008) Genetic basis of thoracic aortic aneurysms and dissections: focus on smooth muscle cell contractile dysfunction. Annu Rev Genomics Hum Genet 9:283–302
256. Newey SE, Velamoor V, Govek EE, Van Aelst L (2005) Rho GTPases, dendritic structure, and mental retardation. J Neurobiol 64:58–74
257. Loirand G, Pacaud P (2010) The role of Rho protein signaling in hypertension. Nat Rev Cardiol 7:637–647
258. Aktories K (2011) Bacterial protein toxins that modify host regulatory GTPases. Nat Rev Microbiol 9:487–498
259. Kanaan Z, Qadan M, Eichenberger MR, Galandiuk S (2010) The actin-cytoskeleton pathway and its potential role in inflammatory bowel disease-associated human colorectal cancer. Genet Test Mol Biomarkers 14:347–353
260. Vega FM, Ridley AJ (2008) Rho GTPases in cancer cell biology. FEBS Lett 582:2093–2101
261. Weir GC, Marselli L, Marchetti P, Katsuta H, Jung MH, Bonner-Weir S (2009) Towards better understanding of the contributions of overwork and glucotoxicity to the β-cell inadequacy of type 2 diabetes. Diabetes Obes Metab 11(Suppl 4):82–90
262. Bonner-Weir S, Li L, O-Yahalom W-C, Guo L, Weir GC, Sharma A (2010) β-cell growth and regeneration: replication is only part of the story. Diabetes 59:2340–2348
263. Mathews AEW, Mathews CE (2012) Inherited b-cell dysfunction in lean individuals with type 2 diabetes. Diabetes 61:1659–1660
264. Bonnycastle LL, Chines PS, Hara T, Huyghe JR, Swift AJ, Heikinheimo P, Mahadevan J, Peltonen S, Huopio H, Nuutila P, Narisu N, Goldfeder RL, Stitze MLl LS, Boehnke M, Urano

F, Collins FS, Laakso M (2013) Autosomal dominant diabetes arising from a wolfram syndrome 1 mutation. Diabetes 62:3943–3950
265. McCarthy MI, Zeggini E (2006) Genetics of type 2 diabetes. Curr Diab Rep 6:147–154
266. Zeggini E, Scott L, Saxena R, Voight B (2008) Meta-analysis of genome-wide association data and large-scale replication identifies additional susceptibility loci for type 2 diabetes. Nat Genet 40:638–645
267. Morris A, Voight B, Teslovich T, Ferreira T, Segre A, Steinthorsdottir V et al (2012) Large-scale association analysis provides insights into the genetic architecture and pathophysiology of type 2 diabetes. Nat Genet 44:981–990
268. Sirovich L (2014) Genomic data and disease forecasting: application to type 2 diabetes (T2D). PLoS One 9(1):e85684. doi:10.1371/journal.pone.0085684
269. McCarthy MI, Hirschhorn JN (2008) Genome-wide association studies: potential next steps on a genetic journey. Hum Mol Genet 17(R2):R156–R165
270. Manolio T (2010) Genome-wide association studies and assessment of the risk of disease. N Engl J Med 363:166–176
271. Fu M, Sabra MM, Damcott C, Pollin T, Ma L, Ott S et al (2007) Evidence that Rho guanine nucleotide exchange factor 11 (ARHGEF11) on 1q21 is a Type 2 diabetes susceptibility gene in the old order Amish. Diabetes 56:1363–1368
272. Mercader JM, Puiggros M, Segrè AV, Planet E, Sorianello E, Sebastian D et al (2012) Identification of novel type 2 diabetes candidate genes involved in the crosstalk between the mitochondrial and the insulin signaling systems. PLoS Genet 8(12):e1003046. doi:10.1371/journal.pgen.1003046
273. Marselli L, Thorne J, Ahn YB, Omer A, Sgroi DC, Libermann TH, Out H, Sharma A, Bonner-Weir S, Weir GC (2008) Gene expression of purified beta-cell tissue obtained from human pancreas with laser capture microdissection. J Clin Endocrinol Metab 93:1046–1053
274. da Silva CV, Cruz L, Araújo Nda S, Angeloni MB, Fonseca BB, Gomes Ade O, Carvalho Fdos R, Gonçalves AL, Barbosa Bde F (2012) A glance at Listeria and Salmonella cell invasion: different strategies to promote host actin polymerization. Int J Med Microbiol 302:19–32
275. Reed SC, Lamason RL, Risca VI, Abernathy E, Welch MD (2014) Rickettsia actin-based motility occurs in distinct phases mediated by different actin nucleators. Curr Biol 24:98–103
276. Welch MD, Way M (2013) Arp2/3-mediated actin-based motility: a tail of pathogen abuse. Cell Host Microbe 14:242–255
277. Ireton K (2013) Molecular mechanisms of cell-cell spread of intracellular bacterial pathogens. Open Biol 3(7):130079. doi:10.1098/rsob.130079
278. Dong N, Liu L, Shao F (2010) A bacterial effector targets host DH-PH domain RhoGEFs and antagonizes macrophage phagocytosis. EMBO J 29:1363–1376
279. Hayward RD, Koronakis V (2002) Direct modulation of the host cell cytoskeleton by Salmonella actin-binding proteins. Trends Cell Biol 12:15–20
280. Mounier J, Laurent V, Hall A, Fort P, Carlier M-F, Sansonetti PJ, Egile C (1999) Rho family GTPases control entry of Shigella flexneri into epithelial cells but not intracellular motility. J Cell Sci 112:2069–2080
281. Hardt WD, Chen LM, Schuebel KE, Bustelo XR, Galan JE (1998) *S. typhimurium* encodes an activator of Rho GTPases that induces membrane ruffling and nuclear responses in host cells. Cell 93:815–826
282. Fu Y, Galan JE (1999) A *Salmonella* protein antagonizes Rac-1 and Cdc42 to mediate host-cell recovery after bacterial invasion. Nature 401:293–297
283. Stebbins CE, Galan JE (2000) Modulation of host signaling by a bacterial mimic: structure of the *Salmonella* effector SptP bound to Rac1. Mol Cell 6:1449–1460
284. Prehna G, Ivanov MI, Bliska JB, Stebbins CE (2006) *Yersinia* virulence depends on mimicry of host rho-family nucleotide dissociation inhibitors. Cell 126:869–880
285. Alto NM, Shao F, Lazar CS, Brost RL, Chua G, Mattoo S, McMahon SA, Ghosh P, Hughes TR, Boone C, Dixon JE (2006) Identification of a bacterial type III effector family with G protein mimicry functions. Cell 124:133–145

286. Klink BU, Barden S, Heidler TV, Borchers C, Ladwein M, Stradal TE, Rottner K, Heinz DW (2010) Structure of Shigella IpgB2 in complex with human RhoA: implications for the mechanism of bacterial guanine nucleotide exchange factor mimicry. J Biol Chem 285:17197–17208
287. Cossart P, Roy CR (2010) Manipulation of host membrane machinery by bacterial pathogens. Curr Opin Cell Biol 22:547–554
288. Cameron LA, Svitkina TM, Vignjevic D, Theriot JA, Borisy GG (2001) Dendritic organization of actin comet tails. Curr Biol 11:130–135
289. Mueller J, Pfanzelter J, Winkler C, Narita A, Le Clainche C, Nemethova M, Carlier M-F, Maeda Y, Welch MD, Ohkawa T, Schmeiser C, Resch GP, Small JV (2014) Electron tomography and simulation of baculovirus actin comet tails support a tethered filament model of pathogen propulsion. PLoS Biol 12(1):e1001765. doi:10.1371/journal.pbio.1001765
290. Robinson R (2014) Actin comets: traversing the cellular universe. PLoS Biol 12(1):e1001766. doi:10.1371/journal.pbio.1001766
291. Kim H, White CD, Sacks DB (2011) IQGAP1 in microbial pathogenesis: targeting the actin cytoskeleton. FEBS Lett 585:723–729
292. Chertova E, Chertov O, Coren LV, Roser JD, Trubey CM, Bess JW Jr, Sowder RC II, Barsov E, Hood BL, Fisher RJ, Nagashima K, Conrads TP, Veenstra TD, Lifson JD, Ott DE (2006) Proteomic and biochemical analysis of purified human immunodeficiency virus type 1 produced from infected monocyte-derived macrophages. J Virol 80:9039–9052
293. Leung J, Yueh A, Appah FS Jr, Yuan B, de los Santos K, Goff SP (2006) Interaction of Moloney murine leukemia virus matrix protein with IQGAP. EMBO J 25:2155–2166
294. Gladue DP, Holinka LG, Fernandez-Sainz IJ, Prarat MV, O'Donnell V, Vepkhvadze NG, Lu Z, Risatti GR, Borca MV (2011) Interaction between core protein of classical swine fever virus with cellular IQGAP1 protein appears essential for virulence in swine. Virology 412:68–74
295. Carlton JG, Martin-Serrano J (2007) Parallels between cytokinesis and retroviral budding: a role for the ESCRT machinery. Science 316:1908–1912
296. Brown MD, Bry L, Li Z, Sacks DB (2008) Actin pedestal formation by enteropathogenic *Escherichia coli* is regulated by IQGAP1, calcium, and calmodulin. J Biol Chem 283:35212–35222
297. Buss C, Muller D, Ruter C, Heusipp G, Schmidt MA (2009) Identification and characterization of Ibe, a novel type III effector protein of A/E pathogens targeting human IQGAP1. Cell Microbiol 11:661–677
298. McLaughlin LM, Govoni GR, Gerke C, Gopinath S, Peng K et al (2009) The *Salmonella* SPI2 effector SseI mediates long-term systemic infection by modulating host cell migration. PLoS Pathog 5(11):e1000671. doi:10.1371/journal.ppat.1000671
299. Karlsson T, Turkina MV, Yakymenko O, Magnusson K-E, Vikström E (2012) The *Pseudomonas aeruginosa N*-Acylhomoserine lactone quorum sensing molecules target IQGAP1 and modulate epithelial cell migration. PLoS Pathog 8(10):e1002953. doi:10.1371/journal.ppat.1002953
300. McGarvey JA, Wagner D, Bermudez LE (2004) Differential gene expression in mononuclear phagocytes infected with pathogenic and non-pathogenic mycobacteria. Clin Exp Immunol 136:490–500
301. Rogers JV, Choi YW, Giannunzio LF, Sabourin PJ, Bornman DM, Blosser EG, Sabourin CL (2007) Transcriptional responses in spleens from mice exposed to *Yersinia pestis CO92*. Microb Pathog 43:67–77
302. Conlin VS, Curtis SB, Zhao Y, Moore EDW, Smith VC, Meloche RM, Finlay BB, Buchan AMJ (2004) *Helicobacter pylori* infection targets adherens junction regulatory proteins and results in increased rates of migration in human gastric epithelial cells. Infect Immun 72:5181–5192
303. Wu Y, Chen YC, Sang JR, Xu WR (2011) RhoC protein stimulates migration of gastric cancer cells through interaction with scaffold protein IQGAP1. Mol Med Rep 4:697–703
304. Blaser MJ, Perez-Perez GI, Kleanthous H, Cover TL, Peek RM, Chyou PH, Stemmermann GN, Nomura A (1995) Infection with *Helicobacter pylori* strains possessing *cagA* is associ-

ated with an increased risk of developing adenocarcinoma of the stomach. Cancer Res 55:2111–2115
305. Hatakeyama M, Higashi H (2005) *Helicobacter pylori* CagA: a new paradigm for bacterial carcinogenesis. Cancer Sci 96:835–843
306. Mégraud F (2013) Current recommendations for Helicobacter pylori therapies in a world of evolving resistance. Gut Microbes 4(6):541–548
307. Fox GJ, Wang CT, Parsonnet J (2006) Helicobacter, chronic inflammation, and cancer. In: Hausen ZH (ed) Infections causing human cancer. Wiley-VCH, Germany, pp 386–467
308. Bauer B, Meyer TF (2011) The human gastric pathogen helicobacter pylori and its association with gastric cancer and ulcer disease. Ulcers:23. doi:10.1155/2011/340157. Article ID 340157
309. Wessler S, Gimona M, Rieder G (2011) Regulation of the actin cytoskeleton in Helicobacter pylori-induced migration and invasive growth of gastric epithelial cells. Cell Commun Signal 9:27. http://www.biosignaling.com/content/9/1/27
310. Bessède E, Staedel C, Acuña Amador LA, Nguyen PH, Chambonnier L, Hatakeyama M, Belleannée G, Mégraud F, Varon C (2013) Helicobacter pylori generates cells with cancer stem cell properties via epithelial-mesenchymal transition-like changes. Oncogene. doi:10.1038/onc.2013.380
311. Tsutsumi R, Takahashi A, Azuma T, Higashi H, Hatakeyama M (2006) Focal adhesion kinase is a substrate and downstream effector of SHP-2 complexed with *Helicobacter pylori* CagA. Mol Cell Biol 26:1–276
312. Schneider S, Weydig C, Wessler S (2008) Targeting focal adhesions: *Helicobacter pylori-host* communication in cell migration. Cell Commun Signal 6:2. doi:10.1186/1478-811X-6-2. http://www.biosignaling.com/content/6/1/2
313. Varon C, Dubus P, Mazurier F, Asencio C, Chambonnier L, Ferrand J, Giese A, Senant-Dugot N, Carlotti M, Mégraud F (2012) Helicobacter pylori infection recruits bone marrow-derived cells that participate in gastric preneoplasia in mice. Gastroenterology 142:281–291
314. Sehring IM, Dong B, Denker E, Bhattachan P, Deng W, Mathiese BT, Jiang D (2014) An equatorial contractile mechanism drives cell elongation but not cell division. PLoS Biol 12(2):e1001781. doi:10.1371/journal.pbio.1001781
315. Greber UF, Way M (2006) A superhighway to virus infection. Cell 124:741–754
316. Rodriguez OC, Schaefer AW, Mandato CA, Forscher P, William M, Bement WM, Waterman-Storer CM (2003) Conserved microtubule–actin interactions in cell movement and morphogenesis. Nat Cell Biol 5:599–609
317. Ellenbroek SI, Collard JG (2007) Rho GTPases: functions and association with cancer. Clin Exp Metastasis 24:657–672
318. Tang Y, Olufemi L, Wang MT, Nie D (2008) Role of Rho GTPases in breast cancer. Front Biosci 13:759–776
319. Karlsson R, Pedersen ED, Wang Z, Brakebusch C (2009) Rho GTPase function in tumorigenesis. Biochim Biophys Acta 1796:91–98
320. Parri M, Chiarugi P (2010) Rac and Rho GTPases in cancer cell motility control. Cell Commun Signal 8:23
321. Mulloy JC, Cancelas JA, Filippi MD, Kalfa TA, Guo F, Zheng Y (2010) Rho GTPases in hematopoiesis and hemopathies. Blood 115:936–947
322. Pai S, Kim C, Williams DA (2010) Rac GTPases in human diseases. Dis Markers 29:177–187
323. Mack NA, Whalley HJ, Castillo-Lluva S, Malliri A (2011) The diverse roles of Rac signaling in tumorigenesis. Cell Cycle 10:1571–1581
324. Lin R, Cerione RA, Manor D (1999) Specific contributions of the small GTPases Rho, Rac, and Cdc42 to Dbl transformation. J Biol Chem 274:23633–23641
325. Schnelzer A, Prechtel D, Knaus U, Dehne K, Gerhard M, Graeff H, Harbeck N, Schmitt M, Lengyel E (2000) Rac1 in human breast cancer. Overexpression, mutation analysis, and characterization of a new isoform, Rac1b. Oncogene 19:3013–3020
326. Vigil D, Cherfils J, Rossman KL, Der CJ (2010) Ras superfamily GEFs and GAPs: validated and tractable targets for cancer therapy? Nat Rev Cancer 10:842–857

327. Palmby TR, Abe K, Karnoub AE, Der CJ (2004) Vav transformation requires activation of multiple GTPases and regulation of gene expression. Mol Cancer Res 2:702–711
328. Montalvo-Ortiz BL, Castillo-Pichardo L, Hernández E, Humphries-Bickley T, De la Mota-Peynado A, Cubano LA, Vlaar CP, Dharmawardhane S (2012) Characterization of EHop-016, novel small molecule inhibitor of Rac GTPase. J Biol Chem 287:13228–13238
329. Razidlo GL, Schroeder B, Chen J, Billadeau DD, Mark A, McNiven MA (2014) Vav1 as a central regulator of invadopodia assembly. Curr Biol 24:86–93
330. Coppolino MG, Krause M, Hagendorff P, Monner DA, Trimble W, Grinstein S, Wehland J, Sechi AS (2001) Evidence for a molecular complex consisting of Fyb/SLAP, SLP-76, Nck, VASP and WASP that links the actin cytoskeleton to Fcgamma receptor signalling during phagocytosis. J Cell Sci 114:4307–4318
331. Narumiya S, Tanji M, Ishizaki T (2009) Rho signaling, ROCK and mDia1, in transformation, metastasis and invasion. Cancer Metastasis Rev 28:65–76
332. Morris LE, Bloom GS, Frierson HF, Powell SM (2005) Nucleotide variants within the IQGAP1 gene in diffuse-type gastric cancers. Genes Chromosomes Cancer 42:280–286
333. Elliott SF, Allen G, Timson DJ (2012) Biochemical analysis of the interactions of IQGAP1 C-terminal domain with CDC42. World J Biol Chem 3:53–60
334. Johnson M, Sharma M, Henderson BR (2009) IQGAP1 regulation and roles in cancer. Cell Signal 21:1471–1478
335. Walch A, Seidl S, Hermannstädter C, Rauser S, Deplazes J, Langer R, von Weyhern CH, Sarbia M, Busch R, Feith M, Gillen S, Höfler H, Luber B (2008) Combined analysis of Rac1, IQGAP1, Tiam1 and E-cadherin expression in gastric cancer. Mod Pathol 21:544–552
336. Sanchez-Laorden B, Viros A, Marais R (2013) Mind the IQGAP. Cancer Cell 23:715–717
337. Jameson KL, Mazur PK, Zehnder AM, Zhang J, Zarnegar B, Sage J, Khavari PA (2013) IQGAP1 scaffold-kinase interaction blockade selectively targets RAS-MAP kinase–driven tumors. Nat Med 19:626–630
338. Suzuki K, Takahashi K (2006) Induction of E-cadherin endocytosis by loss of protein phosphatase 2A expression in human breast cancers. Biochem Biophys Res Commun 349:255–260
339. Jacquemet G, Green DM, Bridgewater RE, von Kriegsheim A, Humphries MJ, Norman JC, Caswell PT (2013) RCP-driven α5β1 recycling suppresses Rac and promotes RhoA activity via the RacGAP1-IQGAP1 complex. J Cell Biol 202:917–935
340. Clark EA, Golub TR, Lander ES, Hynes RO (2000) Genomic analysis of metastasis reveals an essential role for RhoC. Nature 406:532–535
341. Jadeski L, Mataraza JM, Jeong HW, Li Z, Sacks DB (2008) IQGAP1 stimulates proliferation and enhances tumorigenesis of human breast epithelial cells. J Biol Chem 283:1008–1017
342. Sakurai-Yageta M, Recchi C, Le Dez G, Sibarita J-B, Daviet L, Camonis J, D'Souza-Schorey C, Chavrier P (2008) The interaction of IQGAP1 with the exocyst complex is required for tumor cell invasion downstream of Cdc42 and RhoA. J Cell Biol 181:985–998
343. Monteiro P, Rossé C, Castro-Castro A, Irondelle M, Lagoutte E, Paul-Gilloteaux P, Desnos C, Formstecher E, Darchen F, Perrais D, Gautreau A, Hertzog M, Chavrier P (2013) Endosomal WASH and exocyst complexes control exocytosis of MT1-MMP at invadopodia. J Cell Biol 203:1063–1079
344. Ma Y, Jin Z, Huang J, Zhou S, Ye H, Jiang S, Yu K (2014) Quercetin suppresses the proliferation of multiple myeloma cells by downregulating IQGAP1 expression and ERK activation. Leuk Lymphoma 55(11):2597–2604
345. Matsunaga H, Kubota K, Inoue T, Isono F, Ando O (2014) IQGAP1 selectively interacts with K-Ras but not with H-Ras and modulates K-Ras function. Biochem Biophys Res Commun 44:360–364
346. Casteel DE, Turner S, Schwappacher R, Rangaswami H, Su-Yuo J, Zhuang S, Boss GR, Pilz RB (2012) Rho isoform-specific interaction with IQGAP1 promotes breast cancer cell proliferation and migration. J Biol Chem 287:38367–38378
347. Shumaker DK, Kuczmarski ER, Goldman RD (2003) The nucleoskeleton: lamins and actin are major players in essential nuclear functions. Curr Opin Cell Biol 2003(15):358–366

348. Andrin C, McDonald D, Attwood KM, Rodrigue A, Ghosh S, Mirzayans R, Masson JY, Dellaire G, Hendzel MJ (2012) A requirement for polymerized actin in DNA double-strand break repair. Nucleus 3:384–395
349. Luby-Phelps K (2000) Cytoarchitecture and physical properties of cytoplasm: volume, viscosity, diffusion, intracellular surface area. Int Rev Cytol 192:189–221
350. Karcher RL, Deacon SW, Gelfand VI (2002) Motor-cargo interactions: the key to transport specificity. Trends Cell Biol 12:21–27
351. Hadziselimovic N, Vukojevic V, Peter F, Milnik A, Fastenrath M, Fenyves BG, Hieber P, Demougin P, Vogler C, Quervain DJ-F, Papassotiropoulos A, Stetak A (2014) Forgetting is regulated via musashi-mediated translational control of the Arp2/3 complex. Cell 156:1153–1166
352. Engl W, Arasi B, Yap LL, Thiery JP, Viasnoff V (2014) Actin dynamics modulate mechanosensitive immobilization of E-cadherin at adheren junctions. Nat Cell Biol 16:587–594
353. Stehbens SJ, Paszek M, Pemble H, Ettinger A, Gierke S, Wittmann T (2014) CLASPs link focal-adhesion-associated microtubule capture to localized exocytosis and adhesion site turnover. Nat Cell Biol 16:561–573

Chapter 3
Regulation of the Cytoskeleton by the Rho Family of GTPases in Hematopoietic Stem Cells in Health and Disease

Ramesh C. Nayak, Kyung-Hee Chang, and Jose A. Cancelas

Introduction

Rho-family GTPases comprise a main branch of the Ras superfamily of small GTPases. Twenty-two human members of the Rho family have been identified, and can be subdivided into ten groups on the basis of their identity to Cdc42, Rac1, RhoA, RhoD, Rif/RhoF, Rnd3/RhoE, TTF/RhoH, Chp/RhoV, mitochondrial Rho (Miro1/RhoT1), or Rho-related BTB-domain-containing protein (RhoBTB). They are low molecular weight guanine nucleotide binding proteins and function as binary molecular switches by cycling between an active GTP-bound state and an inactive GDP-bound state [1]. Rho GTPases are involved in most, if not all, actin-dependent processes such as those involved in migration, adhesion, morphogenesis, axon guidance, and phagocytosis. In classic fibroblast studies, activation of RhoA, Rac1, and Cdc42 led to the reorganization of the actin cytoskeleton into distinct structures: stress fibers and focal adhesions, veil-like lamellipodia, and filopodial microspikes, respectively [2, 3]. In addition to their regulation of the actin cytoskeleton, Rho GTPases have been shown to play vital roles in a number of cellular processes such as the regulation of enzymatic activities, cell adhesion, intracellular signaling cascades, endocytosis, vesicle trafficking, G1 cell cycle progression, oncogenesis, gene transcription, microtubule dynamics, cell polarity, and asymmetric cell division (ACD) [1, 4–14].

R.C. Nayak
Division of Experimental Hematology and Cancer Biology,
Cincinnati Children's Hospital Medical Center, Cincinnati, OH, USA

K.-H. Chang • J.A. Cancelas (✉)
Division of Experimental Hematology and Cancer Biology,
Cincinnati Children's Hospital Medical Center, Cincinnati, OH, USA

Hoxworth Blood Center, University of Cincinnati College of Medicine,
Cincinnati, OH 45267, USA
e-mail: jose.cancelas@uc.edu; jose.cancelas@cchmc.org

H. Schatten (ed.), *The Cytoskeleton in Health and Disease*,
DOI 10.1007/978-1-4939-2904-7_3

In an earlier review, we described the regulation of the actin and microtubule cytoskeleton by Rho GTPases [15]. The actin and microtubule cytoskeleton mediates a variety of essential biological functions in all eukaryotic cells. In addition to providing a structural framework around which cell shape and polarity are defined, its dynamic properties provide the driving force for cells to move and to divide. Understanding the biochemical mechanisms by which Rho GTPases control the cell cytoskeleton is thus a major goal of cell biology and has implications for human health and disease.

Regulation of Rho GTPase Activity

Rho GTPases are molecular switches and cycle between guanosine triphosphate (GTP)-bound (on) and guanosine diphosphate (GDP)-bound (off) state. Cycling between GTP- and GDP-bound states is controlled primarily by three classes of regulatory molecules: GTPase-activating proteins (GAPs) enhance the relatively slow intrinsic GTPase activity of Rho proteins; guanine nucleotide-exchange factors (GEFs) catalyze the exchange of GDP for GTP; and Rho GTPase guanine nucleotide dissociation inhibitors (Rho GDIs) interact and stabilize the GDP-bound form. GAPs suppress Rho activity, whereas GEFs promote it. In cells, GTP is preferentially loaded into Rho GTPases during nucleotide exchange, because GTP is found at substantially higher concentrations than GDP. Rho GDIs bind to the prenylated forms of Rho-GDP and regulate their distribution between cytosol and the membrane [15]. The number of GEFs (>60) and GAPs (>80) identified so far in the human genome is far more than the number of Rho GTPases (currently 22), suggesting a tight and spatio-temporal regulation of Rho GTPase activity [16–18]. In addition to GEF/GAP regulation, post translational modifications, such as ubiquitination, phosphorylation, isoprenylation, transglutamination, AMPylation, SUMOylation, and oxidation of conserved cysteine residues, critically control the expression and activity of the Rho GTPases [13, 19–22]. A summary of the regulation of Rho GTPases activity is outlined in Fig. 3.1. Rho GTPase expression is also regulated by various micro RNAs (MiRNAs) at the mRNA level [20]. The constitutively active GTP-bound RhoH and Rnd isoforms are regulated by tissue-specific differential expression [23, 24].

The first mammalian GEF, Dbl, isolated in 1985 as an oncogene in an NIH 3T3 focus formation assay using DNA from a human diffuse B-cell lymphoma [25], was found to contain a region of ~180 amino acids that showed significant sequence similarity to CDC24, a protein identified genetically as an upstream activator of CDC42 in yeast [26, 27]. Dbl and Cdc24 represented the initial members of a new family of GEFs. The region of homology among the Dbl family members contains an ~200 residue Dbl homology (DH) domain and an adjacent, C-terminal, ~100 residue pleckstrin homology (PH) domain. The GEFs with the same substrate specificity often have ~20 % sequence identity. The breakpoint cluster region (BCR) was identified as the first RhoGAP isoform, and contains a 150 amino acid GAP domain which is required to induce the GTPase activity of the Rho GTPases [28, 29].

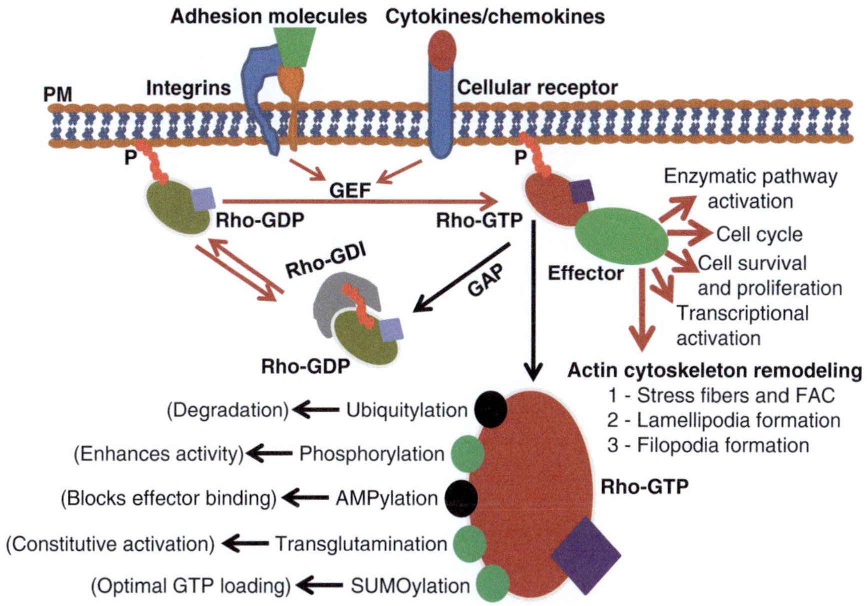

Fig. 3.1 Regulation of Rho GTPase activity by guanine nucleotide-exchange factor/GTPase-activating protein (GEF/GAP) cycling and post-translational covalent modifications. The interaction of cytokines/growth factors and adhesion molecules in the extracellular matrix with their cellular receptors expressed on hematopoietic stem cells (HSCs) triggers the activation of GEFs. GEFs catalyze the exchange of GDP with GTP resulting in Rho GTPase activation. The activated Rho GTPases interact with effector proteins and modulate cytoskeletal rearrangement, leading to the formation of actin stress fibers and focal adhesion contact (FAC), lemellipodia, and filopodia. The activated Rho GTPases also modulate various enzymatic pathways, cell cycle progression, and transcriptional activation, and therefore, regulate a number of cellular processes, such as survival, proliferation, adhesion, migration, differentiation etc. Rho GTPase activating proteins (Rho-GAPs) inactivate Rho GTPase by inducing intrinsic GTPase activity which hydrolyzes bound GTP. Rho GDIs (GDP dissociation inhibitors) bind to the iso-prenylated and GDP-bound form of Rho GTPases and regulate membrane versus cytosolic distribution. Rho GDI also prevents spontaneous activation as well as degradation of GDP-bound Rho GTPases. The post-translational and covalent modifications such as ubiquitination, AMPylation, transglutamination, phosphorylation, and SUMOylation constitutively activate or inactivate the RhoGTPases. *P* = iso-prenylation

Rho GTPase Regulation of HSC Activities

Hematopoietic stem cells (HSCs) are the most actively dividing somatic cells and produce billions of blood cells every day. The two critical features; indefinite self-renewability and multi-lineage differentiation potential, enable the HSCs to maintain their own levels while generating mature blood cells indefinitely. The tight regulation and balance between self-renewal and differentiation potential is critical for steady-state hematopoiesis. During mammalian embryonic development, HSCs first originate in the extra embryonic yolk sac and the para aortic splanchnopleura/aorta gonad mesonephros (AGM) of the embryo proper. During the prenatal and perinatal period, HSCs migrate to the fetal liver and spleen, and finally to the bone marrow (BM), where they reside and produce all hematopoietic cells in adequate quantity throughout life [30–35]. Various cell types in the BM microenvironment play critical roles in regulating HSC activities during both steady-state and stress hematopoiesis, for example, osteoblasts, endothelial cells, perivascular CXCL12 producing reticular cells, nestin$^+$ mesenchymal stem cells, and leptin receptor+ perivascular stromal cells [36–43]. The stromal derived factor (SDF-1α) and stem cell factor (SCF) produced and secreted by BM niche cells, and the fibronectin in the extracellular matrix (ECM), interact with the cellular receptors CXCR4, c-Kit, and integrins, respectively, which are expressed on HSCs, and provide signals for the maintenance of HSC activities [44–51]. Rho GTPases integrate these niche signals into the various cellular processes of HSCs by modulating cytoskeletal rearrangement, gene expression, and the activity of enzymes involved in proliferation and survival pathways. At any given time, a small fraction of hematopoietic stem cells and progenitors (HSC/P) leave the BM, migrate towards peripheral blood (PB), and then migrate back to the BM niche. The migration of HSC/P from the BM niche to the PB is called mobilization, and the migration from the PB towards the BM microenvironment is called homing [52, 53]. Stem cell mobilization and homing are exploited therapeutically for HSC/P transplantation in BM failure and leukemic patients [53–56]. The migration of HSC/P during embryonic development, steady state hematopoiesis, and after stem cell transplantation is critically regulated by Rho GTPases. The multidimensional roles of different Rho GTPases in the regulation of HSC/P activities are depicted in Fig. 3.2.

RhoA Controls the Multilineage Commitment Potential of Hematopoietic Progenitor Cells

RhoA, a member of the Rho family of GTPases, is well characterized in most cell types. Until recently, its role in the activities of HSCs was controversial. Inhibition of RhoA through the retroviral mediated expression of the dominant negative (DN) mutant RhoA N19 has been found to enhance the self-renewal and engraftment potential of HSCs [57]. Another study showed that elevated RhoA activity after deletion of p190-RhoGAP, a negative regulator of RhoA, was advantageous for the self-renewal and long-term engraftment potential of HSC/Ps [58]. The interpretation of these data is affected by the possible off-target effect of the over-expression of the dominant negative mutant form of RhoA in HSC/P. Recently, Zhou et al. [59] conclusively described the functions of RhoA in HSC/P activities using a conditional gene-targeted murine model. The results of this study showed that: RhoA deficiency causes a multilineage hematopoietic failure which is associated with a blockage of hematopoiesis at the multipotent progenitor stage; RhoA$^{-/-}$ HSCs retain long-term engraftment potential, but fail to produce multi potent progenitors and

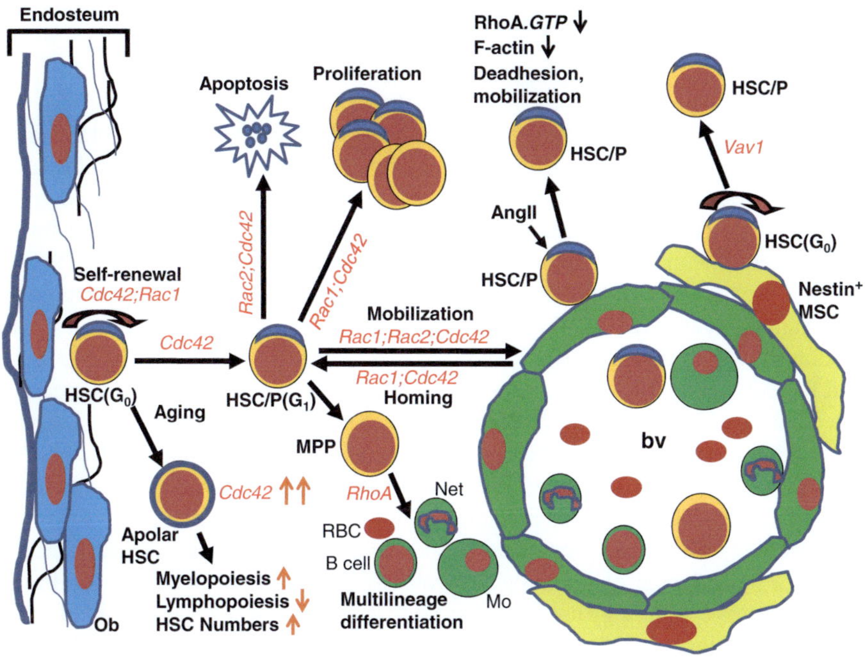

Fig. 3.2 Rho GTPase regulation of hematopoietic stem cell and progenitors (HSC/P) activities. Rho GTPases such as RhoA, Rac (Rac1 and Rac2), Cdc42, and the guanine nucleotide-exchange factor (GEF) Vav, regulate HSC/P functions. Rac1 and Cdc42 control the self-renewal, and Cdc42 regulates the entry of quiescent HSCs into the cell cycle. Rac1 and Cdc42 control HSC/P proliferation, whereas Rac2 and Cdc42 regulate their survival. Rac1 and Cdc42 are required for HSC homing to the bone marrow microenvironment. Rac1, Rac2, and CDC42 play critical roles in the retention of HSC in the bone marrow microenvironment. The elevated level of Cdc42 seen in aged HSC affects HSC polarity, and apolar aged HSCs show a myeloid biased phenotype. RhoA controls the cytokinesis and multilineage differentiation of multi potent progenitors (MPPs) leading to formation of mature blood cells. Vav1, a hematopoietic specific GEF, regulates the localization of HSCs near nestin+ mesenchymal stem cells in the perivascular bone marrow niche. An increased level of angiotensin II (Ang II) induces deadhesion of HSC/P from endothelial cells and therefore, enhances HSC/P mobilization to the peripheral blood. Ang II treatment reduces the level of activated RhoA, resulting in a decreased F-actin content in the cortical region of HSC/Ps that eventually leads to deadhesion and mobilization. *bv* blood vessel, *Net* neutrophil, *Mo* monocyte, *HSC* hematopoietic stem cell, *Ob* osteoblast

mature blood cells; RhoA null HSC/Ps show reduced actomyosin-signaling and defective cytokinesis; Defective cytokinesis and programmed necrosis of the hematopoietic progenitors result in a bloodless phenotype; and the loss of RhoA results in a mitotic failure of progenitors manifested by an accumulation of multinucleated cells due to failed cytokinesis and abscission of the cleavage furrow post-telophase, and increased programmed necrosis. Recently, in a hyperangiotensinamia murine model and in sickle cell patients, we have demonstrated that treatment with angiotensin reduces the RhoA activity and actomyosin signaling that leads to de-adhesion of the HSC/P from endothelial cells and mobilization to the PB [146].

Rac GTPases Regulate HSC/P Homing, Retention, and Migration

Rac GTPases modulate actin cytoskeleton rearrangement and F-actin branching, and regulate membrane protrusion, lemellipodia formation, and directional migration in fibroblasts and many other cell types [1]. The SDF-1α, SCF, and epidermal growth factor (EGF) produced and secreted by niche cells, and the fibronectin in ECM interact with the receptors CXCR4, c-Kit, EGFR and integrin, respectively, which are expressed on HSC/P and trigger the activation of Rho GTPases through upstream regulator GEFs. The chemotactic migration of HSC/P towards SDF-1α is mediated by both Rac1 and Rac2, and regulated by the Rac GEF Tiam1 [60]. Rac1 and Rac2 integrate the signals derived from SCF/c-Kit and fibronectin/b1-integrins interactions with various cellular processes, such as homing, migration, and interaction with the microenvironment, and the long-term engraftment potential of HSC/P [61–63]. Initial studies employed Rho GTPase inhibitors (bacterial toxins) to understand the role of Rho GTPases in the regulation of HSC/P activities [64]. However, the specificity of the inhibitors towards individual Rho GTPases was questionable due to a high degree of homology between different isoforms of Rho GTPases (92 % sequence homology between Rac 1 and Rac2). The Rac GTPase subfamily is comprised of three highly homologous isoforms: Rac1, Rac2, and Rac3. Rac1 is ubiquitously expressed, Rac2 is hematopoietic specific, and Rac3 is highly expressed in the central nervous system [65, 66]. Therefore, the non-specific and off-target effects of GTPase inhibitors are highly predictable. Our group has conclusively identified the specific and overlapping functions of the individual Rho GTPase isoforms using gene-targeted mouse models deficient in Rac1, Rac2, Cdc42, RhoH, and specific GAPs and GEFs. Yang et al. studied the role of hematopoietic specific Rac2 using a gene-targeted mouse model. The Rac2$^{-/-}$ mice showed increased numbers of circulating HSC/Ps in the PB due to defective adhesion to the BM microenvironment. HSC/Ps lacking Rac2 were defective in actin cytoskeleton remodeling and α4β1-mediated adhesion to fibronectin or vascular cell adhesion protein 1 (VCAM 1) [67]. Interestingly, the Rac2 null HSC/Ps showed increased migration towards a CXCL-12 (SDF-1α) gradient, possibly through the compensatory upregulation of Rac1 and Cdc42 activities. This study, for the first time, suggested a critical role for Rac2 in the retention of HSC/Ps in the BM microenvironment, and indicated the involvement of cross-talk between Rac and Cdc42. Later, Gu et al. [63] studied and dissected the specific and overlapping roles of Rac1 and Rac2 in HSC/Ps. Deletion of both Rac1 and Rac2 led to massive egress of HSC/Ps into the PB from the BM. They also found that Rac1$^{-/-}$, but not Rac2$^{-/-}$, HSC/Ps failed to engraft in lethally irradiated recipient mice. The Rac 2 deficient HSC/Ps showed normal short-term engraftment, however, Rac2 null HSC/Ps showed reduced adhesion to fibronectin in vitro, which explained the impaired BM retention and increased mobilization seen in the Rac2$^{-/-}$ HSC/Ps. HSC/Ps lacking Rac2 showed increased apoptosis due to defective activation of the PI3K/Akt survival pathways. The Rac1 and Rac2 mutant phenotypes, such as defective proliferation, reduced adhesion to fibronectin, impaired migration towards a SDF-1α gradient, and increased mobilization, were more severe in HSC/Ps lacking both Rac1 and Rac2, suggesting that Rac1 and Rac2 play overlapping roles. However, when analyzed carefully, it was found that deletion of *Rac1* after engraftment did not impair steady-state hematopoiesis [68]. The engraftment failure of Rac1$^{-/-}$ HSC/P was due to impaired spatial localization with respect to the BM endosteal niche. However, Rac1$^{-/-}$ HSC/Ps showed normal homing to the BM medullary cavity. Although Rac2$^{-/-}$ HSCs showed near normal short-term engraftment, they were impaired in their long-term hematopoietic reconstitution due to an abnormal interaction with the BM microenvironment [69]. Sanchez-Aguilera et al. [70] in a gene-targeted mouse model demonstrated that Vav1, an upstream activator of Rac GTPases, controls the retention of quiescent HSCs near nestin^{+} mesenchymal stem cells. Transplanted Vav1$^{-/-}$ HSP/Cs showed impaired early localization near nestin^{+} perivascular mesenchymal stem cells. The engraftment defect seen in the Vav1$^{-/-}$ HSP/Cs was due to impaired responses to SDF-1α, decreased circadian- and pharmacologically-induced mobilization in vivo, and dysregulated Rac/Cdc42 activation.

Cdc42 Controls HSC Polarity and Asymmetric Cell Division

Asymmetric cell division (ACD) is a process in which cytokinesis of a mother cell generates two daughter cells of unequal fate. This process is a critical regulator of embryonic development and adult tissue stem cell homeostasis [71, 72]. In most cases one of the daughter cells behaves as a mother cell and replenishes the stem cell pool, and the second acquires a matured phenotype for tissue regeneration and repair. The molecular and cellular mechanisms of ACD have been extensively studied in the development of invertebrate embryos [72, 73]. In higher mammals, ACD is well characterized in mature epithelial cells and neural stem/progenitor cells [74, 75]. The basic mechanisms and key regulators of ACD between invertebrate and higher mammalian cells are highly conserved [76, 77]. ACD is accomplished in three sequential steps; (1) establishment of the axis of polarity during late interphase or early prophase; (2) orientation of the mitotic spindles along the axis of polarity; and (3) cell division leading to the formation of two unequal daughter cells harboring different polarity proteins and cell fate determinants [71]. During the initial stages, the cell polarity proteins and cell fate determinants asymmetrically localize to the opposite poles of the cells, and thereby establish an axis of polarity [78]. The process of ACD begins with the polarized localization of apical polarity complex proteins PAR3 (Bazooka in *Drosophila*), PAR6 and atypical protein kinase C (aPKC) [79–82]. The apical polarity complex PAR3-PAR6-aPKC recruits other polarity regulators such as inscuteable (Ins) [79, 83], partner of inscuteable (Pin) [84, 85], and guanine nucleotide binding protein-αi (Gαi) [86] to the apical side of the cells. This is followed by basal recruitment of cell fate determinants such as Numb, Brat, Miranda, Prospero, and partner of Numb (PON) at the pro-metaphase stage of the cytokinesis [87–92]. The daughter cell harboring Numb, an endocytic protein and Notch signaling inhibitor, is differentiated to the mature phenotype. Several studies support the hypothesis that aPKC dependent phosphorylation is critical for the asymmetric localization of Numb [71, 93, 94]. Following asymmetric distribution of polarity proteins and cell fate determinants, the next step is the correct orientation of the mitotic spindle along the axis of polarity. Pin recruits the microtubule motor proteins, dynein and kinesin, through its interaction with the dynein-binding protein Mud (Drosophila homologue of mammalian NuMA) and the Discs-large (Dlg)-kinesin heavy chain (Khc-73) protein complex, respectively, and thereby anchors the astral microtubule of the mitotic spindle to the apical cell cortex [95, 96].

The continuous rearrangement (assembly and disassembly) of actin and the microtubule cytoskeleton plays a decisive role in establishing the polarity of migrating fibroblasts, stem cells during embryonic development, neural stem/progenitor cells, and mature epithelial cells [1, 97]. The Rho family of small GTP-binding proteins, mainly Rac1, Rac2 and Cdc42, regulate the cell polarity by modulating the rearrangement of actin and the microtubule cytoskeleton through their interactions with actin and tubulin-binding protein [1]. The role of Rho GTPase Cdc42 in the establishment of the polarity axis during ACD is well documented in embryonic development and stem cell biology. Cdc42 activates the apical polarity complex PAR3-PAR6-aPKC by directly interacting with adaptor protein PAR6 through its Cdc42- and Rac-interactive binding (CRIB) domain and plays a critical role in establishing the axis of polarity in the Drosophila neuroblast, C. elegans zygote, and mammalian neural stem/progenitor cells [98, 99]. The interaction of Cdc42 and PAR6 activates the aPKC, a Ca^{++} and DAG (diacyl glycerol) independent serine/threonine kinase. The aPKC phosphorylates and directs the localization of the cell fate-determinant Numb toward the opposite pole of the cells [99, 100].

Unlike in invertebrate embryogenesis and mammalian neurogenesis, in HSCs, the mechanisms of ACD and the roles of polarity proteins and cell fate determinants are poorly understood. The development of the in vitro single cell culture and the paired daughter cells assay offered, for the first time, clues about the ACD of primitive HSCs [101, 102]. The daughter cells derived from one primitive (Lineage^{-}c-Kit^{+}Sca1^{+}CD34$^{-/lo}$) HSC possessed a differential ability for proliferation in vitro and hematopoietic reconstitution in vivo, suggesting the occurrence of ACD. Using the paired daughter cells assay, it was found that a majority of the primitive HSCs underwent ACD, whereas, the hematopoietic progenitors predominantly divided in a symmetric commitment manner. However, these studies did not provide direct evidence of the differential distribution of polarity proteins and subsequent ACD in HSCs. Using a Notch driven green fluorescent reporter system in a real-time microscopic imaging method, Wu et al. demonstrated that HSCs undergo both asymmetric and symmetric cell divisions [103]. In the case of ACD, the Notch inhibitor Numb was asymmetrically segregated into one of the daughter cells, and the daughter cell harboring Numb eventually differentiated and lost the Notch-driven expression of green fluorescent protein. Pro differentiation and pro self-renewal environments modulate the balance between the asymmetric and symmetric cell division of HSCs. Also, the presence of an oncogene has been shown to skew the symmetry of cell division. Mushashi2 (Msi2), a cell polarity RNA binding protein, has been shown to regulate normal HSCs and leukemic stem cell activities [104, 105]. In neural progenitor cells, Msi1 directly binds Numb mRNA in the 3′ untranslated region, inhibits its translation and results in the down-regulation of the level of Numb protein, and the subsequent up-regulation of Notch activity [106]. In chronic myelogenous leukemia the progression of the disease from the chronic phase to the aggressive blast crisis phase is correlated with an increase in Msi2 and a decrease in Numb protein levels. An overexpression of Numb induces the differentiation and slows the disease progression [105, 107]. Together, these results suggest that ACD, and the expression and distribution pattern of the cell fate determinant Numb, regulates the activity of normal and malignant HSCs. However, contradictory reports exist about the roles of Numb, Numb-like protein, and the apical polarity complex PAR3-PAR6-aPKC, in the regulation of HSC activities. Wilson et al. [108] showed that the conditional and combined deletion of Numb and Numb-like protein did not affect the self-renewal or multi-lineage differentiation potential of HSCs. Also, the role of Msi2 has been further analyzed using a conditional gene-targeted mouse model [109]. Deletion of Msi2 resulted in depletion of quiescent HSCs due to increased differentiation, and the Msi2 loss-of-function phenotype is independent of Numb. In a conditional gene-targeted mouse model, Sengupta et al. [110] have shown that the constitutive or inducible deletion of the apical polarity complex proteins aPKC ζ and/or aPKC λ does not affect steady-state or stress-induced hematopoiesis. The percentage of polarized primitive HSC population remained unaffected in the absence of both homologues of aPKC. However, the role of these polarity proteins and cell-fate determinants in the regulation of malignant or leukemic stem cell cannot be ruled out.

Recently, the role of the Rho GTPase Cdc42 and the planar cell polarity proteins Frizzled (Fz) and Flamingo (Fmi) in the polarity and ACD of HSCs has been studied in detail. A summary of the role of Cdc42, and the core planar cell polarity proteins Fz and Fmi, in the regulation of ACD of HSCs is given in Fig. 3.3. In 2013, Florian et al. [111], first demonstrated the role of Cdc42 in the regulation of HSC polarity and ACD. In this study, the polarized distribution of tubulin and Cdc42 was used as a surrogate marker for the quantitation of HSC polarity. The level of GTP-bound activated Cdc42 critically regulates HSC polarity. Aged HSCs with elevated Cdc42 activity lost their polarity and long-term repopulation potential in vivo. Casin, a Cdc42 specific Rho GTPse inhibitor, induced HSC polarity and rejuvenated the aged HSCs, with an in vivo long-term reconstitution potential similar to young HSC. The pharmacological inhibition of Cdc42 activity in aged HSCs restored the level and spatial distribution of histone H4 lysine 16, suggesting a role for Cdc42 in the epigenetic regulation of HSCs. The canonical Wnt/β-catenin signaling pathway regulates the self-renewal of HSCs [112]. The non-canonical Wnt signaling pathway, mediated through the planar cell polarity proteins Fmi, Fz and disheveled (Dsh) has been shown to maintain the quiescent long-term primitive HSCs [113]. Fz, a seven-span transmembrane protein, and Fmi, a cadherin family transmembrane protein, are highly expressed and distributed in a polarized manner in primitive long-term HSC. Fz mediated non-canonical Wnt signaling suppresses the Ca^{2+}-NFAT-IFNg

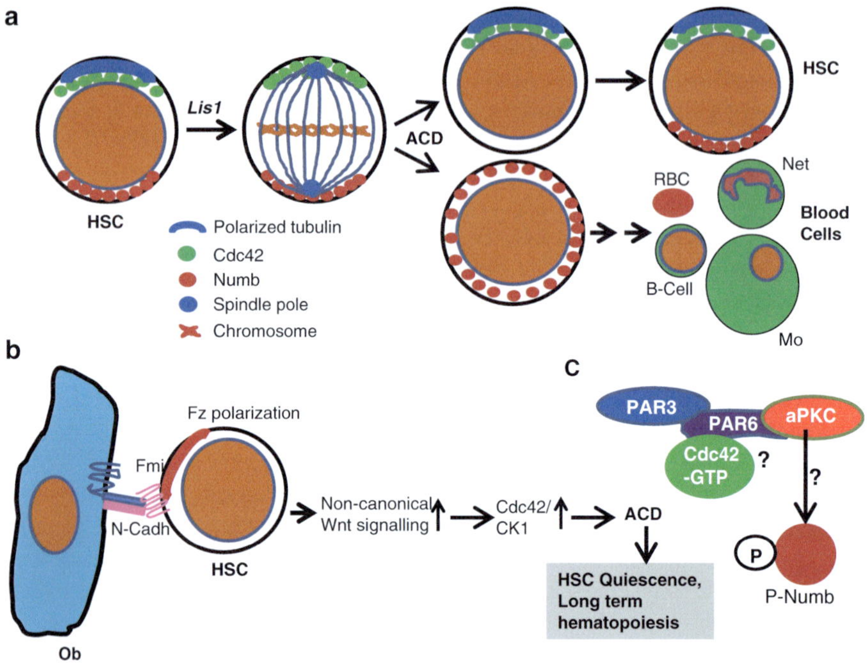

Fig. 3.3 Cdc42 regulates the polarity and asymmetric cell division of hematopoietic stem cells (HSCs). (**a**) The polarized nature of HSCs is determined by analyzing the asymmetric cell distribution (ACD) of Cdc42, Numb, and tubulin. The polarized distribution of Cdc42 and tubulin establishes the axis of polarity, with Numb and Cdc42 at the opposite poles. However, the cellular distribution of tubulin and Cdc42 with respect to Numb has yet to be analyzed in the same HSC. Lis1, a dynein binding protein, regulates the spindle orientation and ensures the occurrence of both symmetric and asymmetric division, and thereby, the self-renewal versus differentiation potential of HSCs. The committed progenitors generate mature red blood cells (RBC). (**b**) The core planar cell polarity (PCP) proteins, Flamingo (Fmi) and Frizzled (Fz), regulate HSC quiescence through the activation of non-canonical Wnt signaling in the endosteal niche. The interaction of Fmi, a cadherin-like transmembrane receptor expressed in the HSC, with N–cadherin (N-Cadh) expressed in the osteoblast (Ob) guides the polarized distribution of seven-span transmembrane receptor Fz at the Ob/ HSC interface, and activates non-canonical Wnt signaling mediated through the Cdc42/casein kinase-1 (CK1) pathway. The activated Cdc42 regulates HSC ACD and quiescence, and long term hematopoiesis. (**c**) The sub-cellular distribution of the apical polarity protein complex PAR3-PAR6-aPKC has not yet been analyzed in HSCs in detail. The interaction of GTP-bound Cdc42 with PAR6 and direct evidence of aPKC mediated Numb phosphorylation and asymmetric distribution of p-Numb is yet to be established in HSCs. *Mo* monocyte, *Net* neutrophil, *P* phosphorylation, *CK1* casein kinase-1

pathway through the Cdc42-PAK1-CK1 (casein kinase1) complex, and antagonizes the canonical Wnt signaling pathway. Another intriguing report describes how the switching from the canonical to non-canonical Wnt pathway takes place in aged HSCs [114]. In aged long-term HSCs a dramatic increase in the expression of Wnt5B was shown to activate the non-canonical Wnt signaling pathway, while inhibiting the canonical Wnt/β-catenin pathway. Treatment of young HSCs with Wnt5B resulted in the activation of Cdc42 and the loss of HSC tubulin polarity, and with Wnt5B resulted in the aged HSC phenotype in vivo. These studies offer concrete evidence for the role of Cdc42 in the establishment of HSC polarity and subsequent ACD. However, it has yet to be demonstrated in HSC whether PAR6 is the direct target of activated Cdc42 for the activation of the apical polarity complex PAR3-PAR6-aPKC and the subsequent aPKC-mediated phosphorylation of Numb resulting in its asymmetric segregation to one of the daughter cells. It is also possible that activated Cdc42, through its interaction with the WASP/Arp2/3 complex, induces actin polymerization in a polarized manner, and that the crosstalk between actin and the microtubule cytoskeleton regulates the tubulin polarization of HSCs.

Two recent reports demonstrating the roles of the dynein binding protein Lis1, and the acto-myosin complex protein myosin II (MIIB), in the ACD of HSCs, further indicate that actin and the microtubule cytoskeleton regulate HSC polarity and asymmetric division. Lis1 facilitates the anchoring of the mitotic spindle to the cell cortex through its interaction with dynein and the dynactin complex, and ensures the proper orientation of the mitotic spindle during cell division [115]. The deletion of *Lis1* results in the depletion of HSC pools, a bloodless phenotype and embryonic lethality. The incorrect positioning of the mitotic spindle and a defective inheritance of the cell fate determinant Numb in the absence of Lis1 led to accelerated differentiation of HSCs and exhaustion of the stem cell pool. Recently, MIIB has been shown to be distributed in a polarized manner in HSCs and to regulate the ACD. The RhoA effector protein, Rho kinase (ROCK), induced the phosphorylation and activation of the myosin light-chain that in turn induced the acto-myosin contractile force required for the formation of actin stress fibers and focal adhesion contact. The regulation of ACD by MIIB tempted us to speculate on a direct role for RhoA GTPase in this process [116].

Rho GTPases and Human HSC/P Disease

The importance of Rho GTPase regulated signaling pathways in human biology is highlighted by the identification of genetic alterations in all classes of protein that interact with the GEFs, GAPs, GDIs, receptors, and downstream targets. Rho GEF genes rearrangements and deletions, have been identified in developmental and neuro-degenerative disorders, and in cancer.

Rho GTPases and Leukemia and Lymphoma

Many GEFs, including Dbl, Lbc, Lfc, Vav, Net1, Ect2, and Tiam, were originally isolated as oncogenes using an in vitro NIH 3T3 fibroblast transformation assay with DNA derived from various human tumors [25, 117–120]. Further studies revealed that constitutively active Rho, Rac, or Cdc42 also induced transformation, strongly suggesting that the oncogenic activity of GEFs is mediated through dysregulated activation of Rho GTPases [121].

In lymphohematopoiesis, Rho GTPase activity has been implicated in the transformation of HSCs or hematopoietic progenitors acquiring self-renewal ability. For instance, the Rho family of GTPases has been implicated in the pathogenesis of leukemias and lymphomas.

BCR-ABL leukemias result from the reciprocal translocation between the BCR gene on Chromosome 22 and the Abelson leukemia (ABL) gene on Chromosome 9. Several types of fusion protein can be generated including the forms p210-BCR-ABL and p190-BCR-ABL. The wild-type BCR fragment contains a DH-PH GEF domain located centrally, and a GAP domain in its C-terminus. In BCR-ABL leukemias, the reciprocal, 9:22 chromosome translocation results in a fusion of the N-terminal sequences derived from BCR, with the non-receptor tyrosine kinase ABL. The two most common BCR-ABL fusion proteins are a 210 kDa protein, which contains the DH-PH domains, but not the GAP domain and is associated with chronic myeloid leukemia (CML), and a 190 kDa protein which lacks the DH-PH and GAP domains of BCR and is associated with acute lymphoblastic leukemia (ALL).

CML is a hematological malignancy that is characterized by an uncontrolled expansion of immature myeloid cells and their premature release into the circulation. CML is caused by the expression of the fusion oncoprotein p210-BCR-ABL, a constitutively active tyrosine kinase which regulates a variety of signaling cascades, including Ras, extracellular-signal regulated kinase (ERK), Akt, c-Jun activated kinase (JNK), p38, CrkL, signal transducer and activator of transcription 5 (STAT5), and nuclear factor-κB (NF-κB) [122]. Expression of p210-BCR-ABL, confers a proliferative advantage to cells, and induces abnormal adhesion and migration of hematopoietic progenitor cells [123, 124]. These effects can be suppressed by the tyrosine kinase inhibitor, imatinib mesylate. Imatinib is a highly effective drug in CML treatment, however, tyrosine kinase inhibition resistance due to mutations in p210-BCR-ABL and other causes of HSC resistance to drug treatment have increased the interest in better defining the signaling pathways activated by p210-BCR-ABL [125].

p210-BCR-ABL contains additional functional domains of interest. In particular a DH domain with GEF that can activate Rho GTPases, and a Src-homology3 (SH3) domain which can recruit other proteins with GEF activity as well as Rac proteins, which have been shown to activate a variety of signaling molecules that coincide with known downstream targets of p210-BCR-ABL. Hyperactivation of Rac1 and Rac2 and, to a lesser extent, Rac3 in HSC/P isolated from chronic phase CML patients has been demonstrated [126]. In an experimental model of CML in mice, Rac GTPases were also shown to be hyperactivated in primary HSC/P expressing p210-BCR-ABL after retrovirus-mediated gene transfer [126]. While a Rac1 deficiency did not modify the median survival of p210-BCR-ABL-expressing mice, the median survival of p210-BCR-ABL-expressing Rac2-deficient mice was significantly increased (from 21 to 43 days), and the median survival of p210-BCR-ABL-expressing Rac1/Rac2-deficient mice was even more strikingly increased (to 92 days). Expression of p210-BCR-ABL in Rac1/Rac2-deficient HSC/Ps also led to an altered disease phenotype, with mice showing oligoclonal leukemias of myeloid, lymphoid, or bi-lineage immunophenotypes. In this murine model, ERK, JNK, p38, Akt, STAT5, and CrkL signaling were all attenuated in splenocytes harvested from p210-BCR-ABL-expressing Rac2-deficient leukemic recipients, and almost completely abrogated in the Rac1/Rac2-deficient cells. Mechanistic studies in a binary transgenic animal model of inducible CML-like disease further unveiled the specific role of Rac2. These studies demonstrated that Rac2 is required for leukemogenesis and is a potent therapeutic target for CML expressing p210-BCR-ABL, where it is required for the proliferation and survival of leukemic stem cells and progenitors [127].

These alterations in signaling correlated with the overall survival seen in animals from each of these genotypes. The decreased activation of downstream pathways was not due to decreased ABL tyrosine kinase activity, as autophosphorylation of p210-BCR-ABL was still noted in these cells [126]. STAT5 phosphorylation was also detectable in leukemic cells regardless of the presence or absence of Rac1 and Rac2 GTPase activity. Activation of CrkL, which has been suggested as an effector that binds directly to p210-BCR-ABL [128], was decreased in Rac2-deficient and practically abrogated in Rac1/Rac2-deficient leukemias, suggesting that CrkL activation is dependent on other proteins as well. These data suggested that Rac1 and especially Rac2 were critical for p210-BCR-ABL transformation and myeloproliferative disorder (MPD) development in vivo. Interestingly, Rac3 appeared hyperactivated in splenocytes derived after long latency in Rac1/Rac2-deficient animals. Rac3 was originally discovered by screening the p210-BCR-ABL-expressing erythroid blastic-phase CML cell line K562 [129]. Rac3 activation has been demonstrated in p190-BCR-ABL-expressing malignant precursor B-lineage lymphoblasts [130] suggesting that Rac3 hyperactivation could play a specific role in cancer development and invasiveness. These data, along with the observed differences in survival mediated by Rac1- versus Rac2-deficient HSCs, support the hypothesis that individual Rac GTPases play unique roles in the development of p210-BCR-ABL-mediated disease.

Based on these genetic data, the effect of the Rac inhibitor NSC23766 on p210-BCR-ABL induced transformation was examined. NSC23766, a first-generation, Rac-specific small molecule inhibitor [131], was developed using computer-assisted virtual screening based on the GEF-Rac1 GTPase complex. NSC23766 was found to fit into a shallow surface groove of Rac1 that has been shown to be critical for GEF-specification. It was shown to effectively inhibit Rac protein binding and activation by the Rac-specific GEFs TrioN or Tiam1 in a dose-dependent manner. In contrast, NSC23766 did not interfere with the binding or activation of Cdc42 or RhoA by their respective GEFs. NSC23766 induced decreased Rac-dependent p21-activated kinase (PAK) activation and mobilization of normal HSCs in mouse studies [132]. In either the murine model of p210-BCR-ABL-induced MPD or in a human xenogeneic transplantation assay, NSC23766 inhibited Rac GTPase activity and impaired leukemogenesis [131]. These studies defined Rac as an attractive molecular target in p210-BCR-ABL transformed HSCs.

Interestingly, p190-BCR-ABL which lacks a GEF-domain, also induces activation of the Rho family of GTPases, specifically Rac and Rho, and induces pre-B lymphoblastic leukemias which are initiated by the transformation of B-cell progenitors. p190-BCR-ABL recruits other GEFs to activate the Rho family of GTPases. Vav3, but not its homologues Vav1 or Vav2, was found to be required for p190-BCR-ABL lymphoblastic leukemogenesis, proliferation, and especially leukemic progenitor survival. Forced expression of Vav3 restored leukemogenesis, and a deficiency of Rac2 phenocopied the effect on leukemogenesis impairment which had been induced by the Vav3 deficiency. Vav3 mediated activation of Rac and Rho seems to repress the expression and/or activation of pro-apoptotic BH3-only molecules, included Bax, Bak, Bad, Bim or Bik [133].

The mixed-lineage leukemia gene (MLL) on chromosome 11q23 is rearranged in both acute myelogenous leukemia (AML) and acute lymphoblastic leukemias (ALL) and constitutes a group of leukemias associated with poor prognosis. These MLL gene rearrangements consist of reciprocal translocations that fuse the amino terminus of *MLL* to a diverse group of partner genes [134, 135]. In a fashion similar to p210-BCR-ABL CML, the leukemia initiating cell is a HSC or progenitor with HSC-like characteristics [136], and MLL rearrangements can be found in up to 7–10 % of acute leukemias. In AML, the most common partner gene for *MLL* is *AF9* on chromosome 9p22. Recently, Somervaille and Cleary [137] showed that the activity of the small Rho GTPase proteins Rac1 and Cdc42 are increased in murine cells expressing MLL-AF9. In a xenograft model of MLL-AF9 leukemia, Wei et al. [138] targeted the Rac1 signaling pathway pharmacologically or by gene-silencing, which resulted in rapid apoptosis of *MLL-AF9*-expressing cells. Confirming these data, in a panel of AML cell lines, Muller et al. demonstrated that the MLL gene-rearranged cell lines ML-2 and THP-1 displayed a profound dependence on Rac signaling, and treatment with NSC23766 inhibited the growth of these cells in vitro, and in an xenograft model in vivo [139].

T-cell lymphomas are a heterogeneous and poorly understood group of non-Hodgkin lymphomas. Angioimmunoblastic T-cell lymphomas (AITL) are characterized by skin rash, generalized lymphadenopathy, splenomegaly, pleural effusion, ascites, anemia and thrombocytopenia, and an increase of circulating large granular lymphocytes with CD3(−) and CD16(+), CD56(+) with T-cell receptor γ-chain gene rearrangement. Although these lymphomas seem to arise from more differentiated T-cells, the cellular origin of these lymphomas is unclear and a cellular origin in a T-cell progenitor/precursor cannot be ruled out. In AITL, highly prevalent RHOA mutations encoding a p.Gly17Val alteration are present in approximately two-thirds of cases. RHOA Gly17Val protein seems to interfere with RHOA signaling in the downstream activation of Rho effectors in biochemical and cellular assays, an effect potentially mediated by the sequestration of activated GEF proteins [140, 141]. The mechanism behind how this loss-of-function mutation induces lymphomagenesis is unknown.

Rho GTPase Activity in HSC Activity in Vasculopathies and Sickle Cell Disease

Unlike their intrinsic activity in leukemias and lymphomas, Rho GTPase activity has also been recognized as a major feature of complications in non-malignant disease. Patients in organ failure of vasculo-endothelial origin have an increased circulating pool of HSC/P [142, 143] which may represent a homeostatic stress response that contributes to vascular damage repair [144]. An example of systemic vasculo-endothelial disease is sickle cell disease (SCD). SCD results from the substitution of a single nucleotide, valine to glutamic acid, at the sixth amino acid of the β-globin chain of hemoglobin A. SCD is characterized by globin polymerization that results in red cell dehydration, hemolysis and subsequent stress erythropoiesis. The common features of SCD are the activation of multiple signaling pathways associated with endothelial damage [142], and an increased pool of primitive hematopoietic progenitors is found in circulation [143]. Vascular pathology is a common feature of SCD patients [145]. Recent data have associated hyperangiotensinemia in SCD with increased circulation of HSC/P, including primitive progenitors at an unmatched magnitude with increased levels of circulating erythroid committed progenitors. The mechanism postulated is that hyperangiotensinemia results in HSC/P de-adhesion from BM endothelial cells through changes in the balance of activated Rho family GTPases, Rho and Rac, and cytoskeletal rearrangements in BM endothelial cells (BMEC) and HSC/P [146].

Concluding Remarks

After more than 20 years of elucidation of the crucial roles of the Rho family of GTPases in the regulation of basic cytoskeletal activities, we are still unveiling some of their functions. Changes in activity through induction, repression, activation, or inhibition, through a complex array of signaling pathways control the migration, adhesion and transformation of HSC/P. This chapter is far from complete. It is expected that in the near future, more information will become available about the multiple mechanisms regulated by the Rho family of GTPases which will be crucial to our understanding of the basic mechanisms of cell division, migration, and the transformation of hematopoietic cells, and enable us to develop novel targeted therapies in hematopoietic disease.

Acknowledgements The authors want to thank Ms. Margaret O'Leary for her editorial assistance. We also thank the Mouse and Research Flow Cytometry Core Facilities of Cincinnati Children's Hospital Medical Center, both supported by the National Institutes of Health/Centers of Excellence for Molecular Hematology Grant 1P30DK090971-01. We also want to thank the Hematology Repository for human specimens from sickle cell anemia patients and controls. This chapter was partly funded by the Heimlich Institute of Cincinnati, U.S. Department of Defense 10580355, the National Institutes of Health/NHLBI R01-HL087159 and HL087159S1 (J.A.C.);

the Center for Clinical Translational Science and Training Awards and the National Blood Foundation (K.H.C.) and funds from the Hoxworth Blood Center and the Cincinnati Children's Hospital Medical Center.

References

1. Jaffe AB, Hall A (2005) Rho GTPases: biochemistry and biology. Annu Rev Cell Dev Biol 21:247–269
2. Ridley AJ, Hall A (1992) The small GTP-binding protein rho regulates the assembly of focal adhesions and actin stress fibers in response to growth factors. Cell 70:389–399
3. Ridley AJ, Paterson HF, Johnston CL, Diekmann D, Hall A (1992) The small GTP-binding protein rac regulates growth factor-induced membrane ruffling. Cell 70:401–410
4. Etienne-Manneville S, Hall A (2002) Rho GTPases in cell biology. Nature 420:629–635
5. Olson MF, Ashworth A, Hall A (1995) An essential role for Rho, Rac, and Cdc42 GTPases in cell cycle progression through G1. Science 269:1270–1272
6. Etienne-Manneville S (2004) Cdc42–the centre of polarity. J Cell Sci 117:1291–1300
7. Joyce D, Bouzahzah B, Fu M, Albanese C, D'Amico M, Steer J, Klein JU, Lee RJ, Segall JE, Westwick JK et al (1999) Integration of Rac-dependent regulation of cyclin D1 transcription through a nuclear factor-kappaB-dependent pathway. J Biol Chem 274:25245–25249
8. Welsh CF, Roovers K, Villanueva J, Liu Y, Schwartz MA, Assoian RK (2001) Timing of cyclin D1 expression within G1 phase is controlled by Rho. Nat Cell Biol 3:950–957
9. Coso OA, Chiariello M, Yu JC, Teramoto H, Crespo P, Xu N, Miki T, Gutkind JS (1995) The small GTP-binding proteins Rac1 and Cdc42 regulate the activity of the JNK/SAPK signaling pathway. Cell 81:1137–1146
10. Miralles F, Posern G, Zaromytidou AI, Treisman R (2003) Actin dynamics control SRF activity by regulation of its coactivator MAL. Cell 113:329–342
11. Diekmann D, Abo A, Johnston C, Segal AW, Hall A (1994) Interaction of Rac with p67phox and regulation of phagocytic NADPH oxidase activity. Science 265:531–533
12. Takeya R, Sumimoto H (2003) Molecular mechanism for activation of superoxide-producing NADPH oxidases. Mol Cells 16:271–277
13. Wang HR, Zhang Y, Ozdamar B, Ogunjimi AA, Alexandrova E, Thomsen GH, Wrana JL (2003) Regulation of cell polarity and protrusion formation by targeting RhoA for degradation. Science 302:1775–1779
14. Cassimeris L (2002) The oncoprotein 18/stathmin family of microtubule destabilizers. Curr Opin Cell Biol 14:18–24
15. DerMardirossian C, Bokoch GM (2005) GDIs: central regulatory molecules in Rho GTPase activation. Trends Cell Biol 15:356–363
16. Schmidt A, Hall A (2002) Guanine nucleotide exchange factors for Rho GTPases: turning on the switch. Genes Dev 16:1587–1609
17. Moon SY, Zheng Y (2003) Rho GTPase-activating proteins in cell regulation. Trends Cell Biol 13:13–22
18. Tcherkezian J, Lamarche-Vane N (2007) Current knowledge of the large RhoGAP family of proteins. Biol Cell 99:67–86
19. Boulter E, Estrach S, Garcia-Mata R, Feral CC (2012) Off the beaten paths: alternative and crosstalk regulation of Rho GTPases. FASEB J 26:469–479
20. Liu M, Bi F, Zhou X, Zheng Y (2012) Rho GTPase regulation by miRNAs and covalent modifications. Trends Cell Biol 22:365–373
21. Lang P, Gesbert F, Delespine-Carmagnat M, Stancou R, Pouchelet M, Bertoglio J (1996) Protein kinase A phosphorylation of RhoA mediates the morphological and functional effects of cyclic AMP in cytotoxic lymphocytes. EMBO J 15:510–519

22. Castillo-Lluva S, Tatham MH, Jones RC, Jaffray EG, Edmondson RD, Hay RT, Malliri A (2010) SUMOylation of the GTPase Rac1 is required for optimal cell migration. Nat Cell Biol 12:1078–1085
23. Hansen SH, Zegers MM, Woodrow M, Rodriguez-Viciana P, Chardin P, Mostov KE, McMahon M (2000) Induced expression of Rnd3 is associated with transformation of polarized epithelial cells by the Raf-MEK-extracellular signal-regulated kinase pathway. Mol Cell Biol 20:9364–9375
24. Riento K, Guasch RM, Garg R, Jin B, Ridley AJ (2003) RhoE binds to ROCK I and inhibits downstream signaling. Mol Cell Biol 23:4219–4229
25. Eva A, Aaronson SA (1985) Isolation of a new human oncogene from a diffuse B-cell lymphoma. Nature 316:273–275
26. Bender A, Pringle JR (1989) Multicopy suppression of the cdc24 budding defect in yeast by CDC42 and three newly identified genes including the ras-related gene RSR1. Proc Natl Acad Sci U S A 86:9976–9980
27. Ron D, Zannini M, Lewis M, Wickner RB, Hunt LT, Graziani G, Tronick SR, Aaronson SA, Eva A (1991) A region of proto-dbl essential for its transforming activity shows sequence similarity to a yeast cell cycle gene, CDC24, and the human breakpoint cluster gene, BCR. New Biol 3:372–379
28. Diekmann D, Brill S, Garrett MD, Totty N, Hsuan J, Monfries C, Hall C, Lim L, Hall A (1991) BCR encodes a GTPase-activating protein for p21rac. Nature 351:400–402
29. Bax B (1998) Domains of rasGAP and rhoGAP are related. Nature 392:447–448
30. Moore MA, Metcalf D (1970) Ontogeny of the haemopoietic system: yolk sac origin of in vivo and in vitro colony forming cells in the developing mouse embryo. Br J Haematol 18:279–296
31. Dieterlen-Lievre F, Martin C (1981) Diffuse intraembryonic hemopoiesis in normal and chimeric avian development. Dev Biol 88:180–191
32. Muller AM, Medvinsky A, Strouboulis J, Grosveld F, Dzierzak E (1994) Development of hematopoietic stem cell activity in the mouse embryo. Immunity 1:291–301
33. Yoder MC, Hiatt K (1997) Engraftment of embryonic hematopoietic cells in conditioned newborn recipients. Blood 89:2176–2183
34. Kumaravelu P, Hook L, Morrison AM, Ure J, Zhao S, Zuyev S, Ansell J, Medvinsky A (2002) Quantitative developmental anatomy of definitive haematopoietic stem cells/long-term repopulating units (HSC/RUs): role of the aorta-gonad-mesonephros (AGM) region and the yolk sac in colonisation of the mouse embryonic liver. Development 129:4891–4899
35. Mikkola HK, Orkin SH (2006) The journey of developing hematopoietic stem cells. Development 133:3733–3744
36. Gong JK (1978) Endosteal marrow: a rich source of hematopoietic stem cells. Science 199:1443–1445
37. Adams GB, Scadden DT (2006) The hematopoietic stem cell in its place. Nat Immunol 7:333–337
38. Li L, Xie T (2005) Stem cell niche: structure and function. Annu Rev Cell Dev Biol 21:605–631
39. Morrison SJ, Spradling AC (2008) Stem cells and niches: mechanisms that promote stem cell maintenance throughout life. Cell 132:598–611
40. Williams DA, Cancelas JA (2006) Leukaemia: niche retreats for stem cells. Nature 444:827–828
41. Calvi LM, Adams GB, Weibrecht KW, Weber JM, Olson DP, Knight MC, Martin RP, Schipani E, Divieti P, Bringhurst FR et al (2003) Osteoblastic cells regulate the haematopoietic stem cell niche. Nature 425:841–846
42. Ding L, Saunders TL, Enikolopov G, Morrison SJ (2012) Endothelial and perivascular cells maintain haematopoietic stem cells. Nature 481:457–462
43. Mendez-Ferrer S, Michurina TV, Ferraro F, Mazloom AR, Macarthur BD, Lira SA, Scadden DT, Ma'ayan A, Enikolopov GN, Frenette PS (2010) Mesenchymal and haematopoietic stem cells form a unique bone marrow niche. Nature 466:829–834

44. Peled A, Grabovsky V, Habler L, Sandbank J, Arenzana-Seisdedos F, Petit I, Ben-Hur H, Lapidot T, Alon R (1999) The chemokine SDF-1 stimulates integrin-mediated arrest of CD34(+) cells on vascular endothelium under shear flow. J Clin Invest 104:1199–1211
45. Peled A, Petit I, Kollet O, Magid M, Ponomaryov T, Byk T, Nagler A, Ben-Hur H, Many A, Shultz L et al (1999) Dependence of human stem cell engraftment and repopulation of NOD/SCID mice on CXCR4. Science 283:845–848
46. Peled A, Kollet O, Ponomaryov T, Petit I, Franitza S, Grabovsky V, Slav MM, Nagler A, Lider O, Alon R et al (2000) The chemokine SDF-1 activates the integrins LFA-1, VLA-4, and VLA-5 on immature human CD34(+) cells: role in transendothelial/stromal migration and engraftment of NOD/SCID mice. Blood 95:3289–3296
47. Driessen RL, Johnston HM, Nilsson SK (2003) Membrane-bound stem cell factor is a key regulator in the initial lodgment of stem cells within the endosteal marrow region. Exp Hematol 31:1284–1291
48. Miyazawa K, Williams DA, Gotoh A, Nishimaki J, Broxmeyer HE, Toyama K (1995) Membrane-bound steel factor induces more persistent tyrosine kinase activation and longer life span of c-kit gene-encoded protein than its soluble form. Blood 85:641–649
49. Williams DA, Rios M, Stephens C, Patel VP (1991) Fibronectin and VLA-4 in haematopoietic stem cell-microenvironment interactions. Nature 352:438–441
50. Papayannopoulou T, Craddock C, Nakamoto B, Priestley GV, Wolf NS (1995) The VLA4/VCAM-1 adhesion pathway defines contrasting mechanisms of lodgement of transplanted murine hemopoietic progenitors between bone marrow and spleen. Proc Natl Acad Sci U S A 92:9647–9651
51. Scott LM, Priestley GV, Papayannopoulou T (2003) Deletion of alpha4 integrins from adult hematopoietic cells reveals roles in homeostasis, regeneration, and homing. Mol Cell Biol 23:9349–9360
52. Kessinger A, Sharp JG (2003) The whys and hows of hematopoietic progenitor and stem cell mobilization. Bone Marrow Transplant 31:319–329
53. Cancelas JA, Hernandez-Jodra M, Zamora C, Perez-Oteyza J, Brieva JA, Roldan E, Navas G, Garcia-Larana J, Lopez J, Odriozola J (1994) Circulating stem cell collection in lymphoma and myeloma after mobilization with cyclophosphamide and granulocyte colony-stimulating factor for autologous transplantation. Vox Sang 67:362–367
54. Briddell RA, Hartley CA, Smith KA, McNiece IK (1993) Recombinant rat stem cell factor synergizes with recombinant human granulocyte colony-stimulating factor in vivo in mice to mobilize peripheral blood progenitor cells that have enhanced repopulating potential. Blood 82:1720–1723
55. Devine SM, Flomenberg N, Vesole DH, Liesveld J, Weisdorf D, Badel K, Calandra G, DiPersio JF (2004) Rapid mobilization of CD34+ cells following administration of the CXCR4 antagonist AMD3100 to patients with multiple myeloma and non-Hodgkin's lymphoma. J Clin Oncol 22:1095–1102
56. Ryan MA, Nattamai KJ, Xing E, Schleimer D, Daria D, Sengupta A, Kohler A, Liu W, Gunzer M, Jansen M et al (2010) Pharmacological inhibition of EGFR signaling enhances G-CSF-induced hematopoietic stem cell mobilization. Nat Med 16:1141–1146
57. Xu H, Eleswarapu S, Geiger H, Szczur K, Daria D, Zheng Y, Settleman J, Srour EF, Williams DA, Filippi MD (2009) Loss of the Rho GTPase activating protein p190-B enhances hematopoietic stem cell engraftment potential. Blood 114:3557–3566
58. Ghiaur G, Lee A, Bailey J, Cancelas JA, Zheng Y, Williams DA (2006) Inhibition of RhoA GTPase activity enhances hematopoietic stem and progenitor cell proliferation and engraftment. Blood 108:2087–2094
59. Zhou X, Florian MC, Arumugam P, Chen X, Cancelas JA, Lang R, Malik P, Geiger H, Zheng Y (2013) RhoA GTPase controls cytokinesis and programmed necrosis of hematopoietic progenitors. J Exp Med 210:2371–2385
60. del Pozo MA, Vicente-Manzanares M, Tejedor R, Serrador JM, Sanchez-Madrid F (1999) Rho GTPases control migration and polarization of adhesion molecules and cytoskeletal ERM components in T lymphocytes. Eur J Immunol 29:3609–3620

61. Taylor ML, Metcalfe DD (2000) Kit signal transduction. Hematol Oncol Clin North Am 14:517–535
62. Alai M, Mui AL, Cutler RL, Bustelo XR, Barbacid M, Krystal G (1992) Steel factor stimulates the tyrosine phosphorylation of the proto-oncogene product, p95vav, in human hemopoietic cells. J Biol Chem 267:18021–18025
63. Gu Y, Filippi MD, Cancelas JA, Siefring JE, Williams EP, Jasti AC, Harris CE, Lee AW, Prabhakar R, Atkinson SJ et al (2003) Hematopoietic cell regulation by Rac1 and Rac2 guanosine triphosphatases. Science 302:445–449
64. Whetton AD, Lu Y, Pierce A, Carney L, Spooncer E (2003) Lysophospholipids synergistically promote primitive hematopoietic cell chemotaxis via a mechanism involving Vav 1. Blood 102:2798–2802
65. Reibel L, Dorseuil O, Stancou R, Bertoglio J, Gacon G (1991) A hemopoietic specific gene encoding a small GTP binding protein is overexpressed during T cell activation. Biochem Biophys Res Commun 175:451–458
66. Bolis A, Corbetta S, Cioce A, de Curtis I (2003) Differential distribution of Rac1 and Rac3 GTPases in the developing mouse brain: implications for a role of Rac3 in Purkinje cell differentiation. Eur J Neurosci 18:2417–2424
67. Yang FC, Atkinson SJ, Gu Y, Borneo JB, Roberts AW, Zheng Y, Pennington J, Williams DA (2001) Rac and Cdc42 GTPases control hematopoietic stem cell shape, adhesion, migration, and mobilization. Proc Natl Acad Sci U S A 98:5614–5618
68. Cancelas JA, Lee AW, Prabhakar R, Stringer KF, Zheng Y, Williams DA (2005) Rac GTPases differentially integrate signals regulating hematopoietic stem cell localization. Nat Med 11:886–891
69. Jansen M, Yang FC, Cancelas JA, Bailey JR, Williams DA (2005) Rac2-deficient hematopoietic stem cells show defective interaction with the hematopoietic microenvironment and long-term engraftment failure. Stem Cells 23:335–346
70. Sanchez-Aguilera A, Lee YJ, Lo Celso C, Ferraro F, Brumme K, Mondal S, Kim C, Dorrance A, Luo HR, Scadden DT et al (2011) Guanine nucleotide exchange factor Vav1 regulates perivascular homing and bone marrow retention of hematopoietic stem and progenitor cells. Proc Natl Acad Sci U S A 108:9607–9612
71. Knoblich JA (2010) Asymmetric cell division: recent developments and their implications for tumour biology. Nat Rev Mol Cell Biol 11:849–860
72. Knoblich JA (2008) Mechanisms of asymmetric stem cell division. Cell 132:583–597
73. Gonczy P (2008) Mechanisms of asymmetric cell division: flies and worms pave the way. Nat Rev Mol Cell Biol 9:355–366
74. Gotz M, Huttner WB (2005) The cell biology of neurogenesis. Nat Rev Mol Cell Biol 6:777–788
75. Lin D, Edwards AS, Fawcett JP, Mbamalu G, Scott JD, Pawson T (2000) A mammalian PAR-3-PAR-6 complex implicated in Cdc42/Rac1 and aPKC signalling and cell polarity. Nat Cell Biol 2:540–547
76. Betschinger J, Knoblich JA (2004) Dare to be different: asymmetric cell division in Drosophila, C. elegans and vertebrates. Curr Biol 14:R674–R685
77. Rolls MM, Doe CQ (2003) Cell polarity: from embryo to axon. Nature 421:905–906
78. Schaefer M, Knoblich JA (2001) Protein localization during asymmetric cell division. Exp Cell Res 271:66–74
79. Wodarz A, Ramrath A, Kuchinke U, Knust E (1999) Bazooka provides an apical cue for inscuteable localization in Drosophila neuroblasts. Nature 402:544–547
80. Petronczki M, Knoblich JA (2001) DmPAR-6 directs epithelial polarity and asymmetric cell division of neuroblasts in Drosophila. Nat Cell Biol 3:43–49
81. Wodarz A, Ramrath A, Grimm A, Knust E (2000) Drosophila atypical protein kinase C associates with Bazooka and controls polarity of epithelia and neuroblasts. J Cell Biol 150:1361–1374
82. Rolls MM, Albertson R, Shih HP, Lee CY, Doe CQ (2003) Drosophila aPKC regulates cell polarity and cell proliferation in neuroblasts and epithelia. J Cell Biol 163:1089–1098

83. Schober M, Schaefer M, Knoblich JA (1999) Bazooka recruits inscuteable to orient asymmetric cell divisions in Drosophila neuroblasts. Nature 402:548–551
84. Yu F, Morin X, Cai Y, Yang X, Chia W (2000) Analysis of partner of inscuteable, a novel player of Drosophila asymmetric divisions, reveals two distinct steps in inscuteable apical localization. Cell 100:399–409
85. Schaefer M, Shevchenko A, Shevchenko A, Knoblich JA (2000) A protein complex containing inscuteable and the Galpha-binding protein pins orients asymmetric cell divisions in Drosophila. Curr Biol 10:353–362
86. Schaefer M, Petronczki M, Dorner D, Forte M, Knoblich JA (2001) Heterotrimeric G proteins direct two modes of asymmetric cell division in the Drosophila nervous system. Cell 107:183–194
87. Knoblich JA, Jan LY, Jan YN (1995) Asymmetric segregation of Numb and Prospero during cell division. Nature 377:624–627
88. Spana EP, Doe CQ (1995) The prospero transcription factor is asymmetrically localized to the cell cortex during neuroblast mitosis in Drosophila. Development 121:3187–3195
89. Shen CP, Jan LY, Jan YN (1997) Miranda is required for the asymmetric localization of Prospero during mitosis in Drosophila. Cell 90:449–458
90. Lee CY, Wilkinson BD, Siegrist SE, Wharton RP, Doe CQ (2006) Brat is a Miranda cargo protein that promotes neuronal differentiation and inhibits neuroblast self-renewal. Dev Cell 10:441–449
91. Betschinger J, Mechtler K, Knoblich JA (2006) Asymmetric segregation of the tumor suppressor brat regulates self-renewal in Drosophila neural stem cells. Cell 124:1241–1253
92. Lu B, Rothenberg M, Jan LY, Jan YN (1998) Partner of Numb colocalizes with Numb during mitosis and directs Numb asymmetric localization in Drosophila neural and muscle progenitors. Cell 95:225–235
93. Smith CA, Lau KM, Rahmani Z, Dho SE, Brothers G, She YM, Berry DM, Bonneil E, Thibault P, Schweisguth F et al (2007) aPKC-mediated phosphorylation regulates asymmetric membrane localization of the cell fate determinant Numb. EMBO J 26:468–480
94. Atwood SX, Prehoda KE (2009) aPKC phosphorylates Miranda to polarize fate determinants during neuroblast asymmetric cell division. Curr Biol 19:723–729
95. Siller KH, Cabernard C, Doe CQ (2006) The NuMA-related Mud protein binds Pins and regulates spindle orientation in Drosophila neuroblasts. Nat Cell Biol 8:594–600
96. Bowman SK, Neumuller RA, Novatchkova M, Du Q, Knoblich JA (2006) The Drosophila NuMA Homolog Mud regulates spindle orientation in asymmetric cell division. Dev Cell 10:731–742
97. Hall A (2005) Rho GTPases and the control of cell behaviour. Biochem Soc Trans 33:891–895
98. Johansson A, Driessens M, Aspenstrom P (2000) The mammalian homologue of the Caenorhabditis elegans polarity protein PAR-6 is a binding partner for the Rho GTPases Cdc42 and Rac1. J Cell Sci 113(Pt 18):3267–3275
99. Hutterer A, Betschinger J, Petronczki M, Knoblich JA (2004) Sequential roles of Cdc42, Par-6, aPKC, and Lgl in the establishment of epithelial polarity during Drosophila embryogenesis. Dev Cell 6:845–854
100. Etienne-Manneville S, Hall A (2003) Cell polarity: par6, aPKC and cytoskeletal crosstalk. Curr Opin Cell Biol 15:67–72
101. Ema H, Takano H, Sudo K, Nakauchi H (2000) In vitro self-renewal division of hematopoietic stem cells. J Exp Med 192:1281–1288
102. Takano H, Ema H, Sudo K, Nakauchi H (2004) Asymmetric division and lineage commitment at the level of hematopoietic stem cells: inference from differentiation in daughter cell and granddaughter cell pairs. J Exp Med 199:295–302
103. Wu M, Kwon HY, Rattis F, Blum J, Zhao C, Ashkenazi R, Jackson TL, Gaiano N, Oliver T, Reya T (2007) Imaging hematopoietic precursor division in real time. Cell Stem Cell 1:541–554

104. Kharas MG, Lengner CJ, Al-Shahrour F, Bullinger L, Ball B, Zaidi S, Morgan K, Tam W, Paktinat M, Okabe R et al (2010) Musashi-2 regulates normal hematopoiesis and promotes aggressive myeloid leukemia. Nat Med 16:903–908
105. Ito T, Kwon HY, Zimdahl B, Congdon KL, Blum J, Lento WE, Zhao C, Lagoo A, Gerrard G, Foroni L et al (2010) Regulation of myeloid leukaemia by the cell-fate determinant Musashi. Nature 466:765–768
106. Imai T, Tokunaga A, Yoshida T, Hashimoto M, Mikoshiba K, Weinmaster G, Nakafuku M, Okano H (2001) The neural RNA-binding protein Musashi1 translationally regulates mammalian numb gene expression by interacting with its mRNA. Mol Cell Biol 21:3888–3900
107. Zhao C, Chen A, Jamieson CH, Fereshteh M, Abrahamsson A, Blum J, Kwon HY, Kim J, Chute JP, Rizzieri D et al (2009) Hedgehog signalling is essential for maintenance of cancer stem cells in myeloid leukaemia. Nature 458:776–779
108. Wilson A, Ardiet DL, Saner C, Vilain N, Beermann F, Aguet M, Macdonald HR, Zilian O (2007) Normal hemopoiesis and lymphopoiesis in the combined absence of numb and numblike. J Immunol 178:6746–6751
109. Park SM, Deering RP, Lu Y, Tivnan P, Lianoglou S, Al-Shahrour F, Ebert BL, Hacohen N, Leslie C, Daley GQ et al (2014) Musashi-2 controls cell fate, lineage bias, and TGF-beta signaling in HSCs. J Exp Med 211:71–87
110. Sengupta A, Duran A, Ishikawa E, Florian MC, Dunn SK, Ficker AM, Leitges M, Geiger H, Diaz-Meco M, Moscat J et al (2011) Atypical protein kinase C (aPKCzeta and aPKClambda) is dispensable for mammalian hematopoietic stem cell activity and blood formation. Proc Natl Acad Sci U S A 108:9957–9962
111. Florian MC, Dorr K, Niebel A, Daria D, Schrezenmeier H, Rojewski M, Filippi MD, Hasenberg A, Gunzer M, Scharffetter-Kochanek K et al (2012) Cdc42 activity regulates hematopoietic stem cell aging and rejuvenation. Cell Stem Cell 10:520–530
112. Reya T, Duncan AW, Ailles L, Domen J, Scherer DC, Willert K, Hintz L, Nusse R, Weissman IL (2003) A role for Wnt signalling in self-renewal of haematopoietic stem cells. Nature 423:409–414
113. Sugimura R, He XC, Venkatraman A, Arai F, Box A, Semerad C, Haug JS, Peng L, Zhong XB, Suda T et al (2012) Noncanonical Wnt signaling maintains hematopoietic stem cells in the niche. Cell 150:351–365
114. Florian MC, Nattamai KJ, Dorr K, Marka G, Uberle B, Vas V, Eckl C, Andra I, Schiemann M, Oostendorp RA et al (2013) A canonical to non-canonical Wnt signalling switch in haematopoietic stem-cell ageing. Nature 503:392–396
115. Zimdahl B, Ito T, Blevins A, Bajaj J, Konuma T, Weeks J, Koechlein CS, Kwon HY, Arami O, Rizzieri D et al (2014) Lis1 regulates asymmetric division in hematopoietic stem cells and in leukemia. Nat Genet 46:245–252
116. Shin JW, Buxboim A, Spinler KR, Swift J, Christian DA, Hunter CA, Leon C, Gachet C, Dingal PC, Ivanovska IL et al (2014) Contractile forces sustain and polarize hematopoiesis from stem and progenitor cells. Cell Stem Cell 14:81–93
117. Katzav S, Martin-Zanca D, Barbacid M (1989) Vav, a novel human oncogene derived from a locus ubiquitously expressed in hematopoietic cells. EMBO J 8:2283–2290
118. Miki T, Smith CL, Long JE, Eva A, Fleming TP (1993) Oncogene ect2 is related to regulators of small GTP-binding proteins. Nature 362:462–465
119. Chan AM, Takai S, Yamada K, Miki T (1996) Isolation of a novel oncogene, NET1, from neuroepithelioma cells by expression cDNA cloning. Oncogene 12:1259–1266
120. Whitehead I, Kirk H, Tognon C, Trigo-Gonzalez G, Kay R (1995) Expression cloning of lfc, a novel oncogene with structural similarities to guanine nucleotide exchange factors and to the regulatory region of protein kinase C. J Biol Chem 270:18388–18395
121. Jaffe AB, Hall A (2002) Rho GTPases in transformation and metastasis. Adv Cancer Res 84:57–80
122. Ren R (2005) Mechanisms of BCR-ABL in the pathogenesis of chronic myelogenous leukaemia. Nat Rev Cancer 5:172–183

123. Ramaraj P, Singh H, Niu N, Chu S, Holtz M, Yee JK, Bhatia R (2004) Effect of mutational inactivation of tyrosine kinase activity on BCR/ABL-induced abnormalities in cell growth and adhesion in human hematopoietic progenitors. Cancer Res 64:5322–5331
124. Zhao RC, Jiang Y, Verfaillie CM (2001) A model of human p210(BCR/ABL)-mediated chronic myelogenous leukemia by transduction of primary normal human CD34(+) cells with a BCR/ABL-containing retroviral vector. Blood 97:2406–2412
125. Deininger M (2008) Resistance and relapse with imatinib in CML: causes and consequences. J Natl Compr Canc Netw 6(Suppl 2):S11–S21
126. Thomas EK, Cancelas JA, Chae HD, Cox AD, Keller PJ, Perrotti D, Neviani P, Druker BJ, Setchell KD, Zheng Y et al (2007) Rac guanosine triphosphatases represent integrating molecular therapeutic targets for BCR-ABL-induced myeloproliferative disease. Cancer Cell 12:467–478
127. Thomas EK, Cancelas JA, Zheng Y, Williams DA (2008) Rac GTPases as key regulators of p210-BCR-ABL-dependent leukemogenesis. Leukemia 22:898–904
128. Oda T, Heaney C, Hagopian JR, Okuda K, Griffin JD, Druker BJ (1994) Crkl is the major tyrosine-phosphorylated protein in neutrophils from patients with chronic myelogenous leukemia. J Biol Chem 269:22925–22928
129. Haataja L, Groffen J, Heisterkamp N (1997) Characterization of RAC3, a novel member of the Rho family. J Biol Chem 272:20384–20388
130. Cho YJ, Zhang B, Kaartinen V, Haataja L, de Curtis I, Groffen J, Heisterkamp N (2005) Generation of rac3 null mutant mice: role of Rac3 in BCR/ABL-caused lymphoblastic leukemia. Mol Cell Biol 25:5777–5785
131. Gao Y, Dickerson JB, Guo F, Zheng J, Zheng Y (2004) Rational design and characterization of a Rac GTPase-specific small molecule inhibitor. Proc Natl Acad Sci U S A 101:7618–7623
132. Cancelas JA, Lee AW, Prabhakar R, Stringer KF, Zheng Y, Williams DA (2005) Rac GTPases differentially integrate signals regulating hematopoietic stem cell localization. Nat Med 11:886–891
133. Chang KH, Sanchez-Aguilera A, Shen S, Sengupta A, Madhu MN, Ficker AM, Dunn SK, Kuenzi AM, Arnett JL, Santho RA et al (2012) Vav3 collaborates with p190-BCR-ABL in lymphoid progenitor leukemogenesis, proliferation, and survival. Blood 120:800–811
134. De Braekeleer M, Morel F, Le Bris MJ, Herry A, Douet-Guilbert N (2005) The MLL gene and translocations involving chromosomal band 11q23 in acute leukemia. Anticancer Res 25:1931–1944
135. Dimartino JF, Cleary ML (1999) Mll rearrangements in haematological malignancies: lessons from clinical and biological studies. Br J Haematol 106:614–626
136. Krivtsov AV, Twomey D, Feng Z, Stubbs MC, Wang Y, Faber J, Levine JE, Wang J, Hahn WC, Gilliland DG et al (2006) Transformation from committed progenitor to leukaemia stem cell initiated by MLL-AF9. Nature 442:818–822
137. Somervaille TC, Cleary ML (2006) Identification and characterization of leukemia stem cells in murine MLL-AF9 acute myeloid leukemia. Cancer Cell 10:257–268
138. Wei J, Wunderlich M, Fox C, Alvarez S, Cigudosa JC, Wilhelm JS, Zheng Y, Cancelas JA, Gu Y, Jansen M et al (2008) Microenvironment determines lineage fate in a human model of MLL-AF9 leukemia. Cancer Cell 13:483–495
139. Muller LU, Schore RJ, Zheng Y, Thomas EK, Kim MO, Cancelas JA, Gu Y, Williams DA (2008) Rac guanosine triphosphatases represent a potential target in AML. Leukemia 22:1803–1806
140. Palomero T, Couronne L, Khiabanian H, Kim MY, Ambesi-Impiombato A, Perez-Garcia A, Carpenter Z, Abate F, Allegretta M, Haydu JE et al (2014) Recurrent mutations in epigenetic regulators, RHOA and FYN kinase in peripheral T cell lymphomas. Nat Genet 46:166–170
141. Yoo HY, Sung MK, Lee SH, Kim S, Lee H, Park S, Kim SC, Lee B, Rho K, Lee JE et al (2014) A recurrent inactivating mutation in RHOA GTPase in angioimmunoblastic T cell lymphoma. Nat Genet 46:371–375
142. Telen MJ (2007) Role of adhesion molecules and vascular endothelium in the pathogenesis of sickle cell disease. Hematology/the Education Program of the American Society of Hematology. American Society of Hematology. Education Program:84–90

143. Croizat H, Ponchio L, Nicolini FE, Nagel RL, Eaves CJ (2000) Primitive haematopoietic progenitors in the blood of patients with sickle cell disease appear to be endogenously mobilized. Br J Haematol 111:491–497
144. Korbling M, Estrov Z, Champlin R (2003) Adult stem cells and tissue repair. Bone Marrow Transplant 32(Suppl 1):S23–S24
145. Hebbel RP, Osarogiagbon R, Kaul D (2004) The endothelial biology of sickle cell disease: inflammation and a chronic vasculopathy. Microcirculation 11:129–151
146. Chang KH, Nayak RC, Roy S, Perumbeti A, Wellendorf AM, Bezold KY, Pirman M, Starnes J, Loberg A, Zhou X, Inagami T, Zheng Y, Malik P, Cancelas JA (2014) Vasculopathy-associated hyperangiotensinemia mobilizes haematopoietic stem cells/progenitors through endothelial AT2R and cytoskeletal dysregulation. Nat Commun. doi:10.1038/ncomms6914

Chapter 4
The Role of the Cytoskeleton in Cell Migration, Its Influence on Stem Cells and the Special Role of GFAP in Glial Functions

Juliana de Mattos Coelho-Aguiar, Felipe Andreiuolo, Henrike Gebhardt, Luiz Henrique Geraldo, Bruno Pontes, Diana Isabel Lourenço Matias, Joana Balça-Silva, Diego Pinheiro Aguiar, Anália do Carmo, Maria Celeste Lopes, Rolf Mentlein, and Vivaldo Moura-Neto

J.M. Coelho-Aguiar • D.I.L. Matias
Instituto Estadual do Cérebro Paulo Niemeyer, Secretaria do Estado de Saúde do Rio de Janeiro, Rio de Janeiro, Brazil

Instituto de Ciências Biomédicas, Universidade Federal do Rio de Janeiro, Rio de Janeiro, Brazil
e-mail: jumcoelho@gmail.com; dimtias@gmail.com

F. Andreiuolo
Department of Pathology, Centre Hospitalier Sainte-Anne, Université Paris Descartes, Paris, France
e-mail: f.andreiuolo@ch-sainte-anne.fr

H. Gebhardt • R. Mentlein
Department of Anatomy, University of Kiel, Kiel, Germany
e-mail: h.gebhardt@anat.uni-kiel.de; rment@anat.uni-kiel.de

L.H. Geraldo • B. Pontes
Instituto de Ciências Biomédicas, Universidade Federal do Rio de Janeiro, Rio de Janeiro, Brazil
e-mail: lh_geraldo@icb.ufrj.br; brunoaccpontes@gmail.com

J. Balça-Silva
Center for Neuroscience and Cell Biology, Institute for Biomedical Imaging and Life Sciences (CNC.IBILI), University of Coimbra, Coimbra, Portugal

Faculty of Medicine, University of Coimbra (FMUC), Coimbra, Portugal
e-mail: joanabalca.silva@gmail.com

H. Schatten (ed.), *The Cytoskeleton in Health and Disease*,
DOI 10.1007/978-1-4939-2904-7_4

D.P. Aguiar
Instituto de Ciências Biomédicas, Universidade Federal do Rio de Janeiro,
Rio de Janeiro, Brazil

Instituto Nacional de Traumatologia e Ortopedia, Divisão de Pesquisa, Rio de Janeiro, Brazil
e-mail: diegopaguiar@gmail.com

A. do Carmo
Center for Neuroscience and Cell Biology, Institute for Biomedical Imaging and Life Sciences (CNC.IBILI), University of Coimbra, Coimbra, Portugal
e-mail: analiacarmo@gmail.com

M.C. Lopes
Center for Neuroscience and Cell Biology, Institute for Biomedical Imaging and Life Sciences (CNC.IBILI), University of Coimbra, Coimbra, Portugal

Faculty of Pharmacy, University of Coimbra (FFUC), Coimbra, Portugal
e-mail: celeste@ff.uc.pt

V. Moura-Neto (✉)
Instituto de Ciências Biomédicas, Universidade Federal do Rio de Janeiro, Rio de Janeiro, Brazil

Instituto Estadual do Cérebro Paulo Niemeyer, Secretaria do Estado de Saúde do Rio de Janeiro, Rio de Janeiro, Brazil

Universidade do Grande Rio (UNIGRANRIO), Brazil
e-mail: vivaldomouraneto@globo.com

Abbreviations

AD	Alzheimer's disease
Arp2/3	Actin-related proteins 2/3
bFGF	Basic fibroblast growth factor
CCL	C-C motif chemokine
CCR7	C-C chemokine receptor type 7
CDC42	Cell division control protein 42 homolog
CNS	Central nervous system
CNTF	Ciliary neurotrophic factor
CSF	Cerebrospinal fluid
CX3CR1	CX3C chemokine receptor 1
CXCR	C-X-C chemokine receptor
DOCK180	Dedicator of cytokinesis protein
ECM	Extracellular matrix
ELMO1	Engulfment and cell-motility protein 1
ERK	Extracellular signal-regulated protein kinase
ERM	Ezrin/Radixin/Moesin
FAK	Focal adhesion kinase
G proteins	guanine nucleotide-binding proteins
GABA	Gamma amino butyric acid
GEF	Guanine nucleotide exchange factor
GFAP	Glial fibrillary acidic protein
GPCRs	G-protein-coupled receptors

HD1	*Hemidesm*osomal 1 protein
hMSC	Human mesenchymal stem cells
HSPC	Hematopoietic stem/progenitor cell
IF	Intermediate filament
JAK	Janus kinase
MAPKs	Mitogen-activated protein kinases
mDia1	Diaphanous homolog 1 of Drosophila
MFs	Microfilaments
MLC	Myosin light chain
MSCs	Mesenchymal stem cells
MTs	Microtubules
NSCs	Neural stem cells
PBSF	Pre-B cell growth-stimulating factor = SDF1, stromal cell-derived factor 1
PKB	Protein kinase B
PNS	Peripheral nervous system
RAC1	Ras-related C3 botulinum toxin substrate 1
RDC-1	G protein-coupled receptor = CXCL12
RNAi	RNA interference
ROCK	Rho-associated coiled-coil forming protein kinases
SDF-1	Stromal cell-derived factor-1= CXCL12
STAT3	Signal transducer and activator of transcription 3
TGFβ1	Transforming growth factor beta 1
TNF	Tumor necrosis factor
VASP	Vasodilator-stimulated phosphoprotein
XCR1	C sub-family of chemokine receptors 1

Introduction

The cytoskeleton is a dynamic network of proteins organized in fibrillar or globular filaments in the cell cytoplasm [1]. Three types of filaments are common to many eukaryotic cells: (1) intermediate filaments provide mechanical strength and resistance to shear stress; (2) microtubules determine the positions of membrane-enclosed organelles and direct intracellular transport; and (3) actin filaments determine the shape of the cell's surface and are necessary for cell migration. Actin filaments also interact with accessory proteins that link them to other cellular components, as well as to each other [1].

This chapter initially addresses the role of the cytoskeleton in one of its main functions, cell migration. The dynamics and the mechanical aspects of the actin filaments are essential to this process, as are the signaling pathways induced by the chemokines and their receptors. We also give special attention to the cytoskeletal proteins of the stem cells, in the origin of cell functions. Finally, the chapter discusses the participation of the cytoskeleton in functions performed by different types of glial cells, focusing on the role of a particular intermediate filament, the glial fibrillary acidic protein (GFAP), in the health and disease.

An Overview of Cell Migration

Cell migration has fascinated cell biologists, biochemists, and recently also physicists and mathematicians. This is not surprising, since it is an essential process that occurs during different stages and at different times, ranging from organism development to normal adult life and also during disease states [2–4].

Cells in multicellular organisms can move in different directions, through the extracellular matrix, over each other, or even between each other. Cells move in three basic steps: (1) extending the plasma membrane forward at the front, or leading edge, of the cell in a protrusion; (2) moving the cell body; and (3) retracting the rear part of the cell [5]. These steps involve two main cytoskeleton filaments, microtubules and actin filaments; the microtubules are required for polarization [6, 7] while the actin filaments are the main players during migration and protrusion formation [8].

Cells are able to extend four different types of protrusions at the leading edge, lamellipodia, filopodia, blebs, and invadopodia. All these structures have their own functions and contribute to cell migration in specific ways. Lamellipodia are able to extend long distances through the extracellular matrix, pulling cells through the tissues [2]. Filopodia explore the cell's surroundings [9, 10]. Membrane blebs help in cell migration during development [11], and invadopodia are protrusions that allow degradation of the extracellular matrix, and help cells to pass through tissues [12].

In the following parts of this section we discuss the mechanical aspects of the formation of a protrusive migratory structure called the lamellipodium, and how the plasma membrane regulates the behavior of this structure, as well as its influence during cell migration. We also discuss some chemokines that induce migratory processes followed by cytoskeletal changes.

Lamellipodium

The thin protrusive region at the leading edge of migrating fibroblasts in culture was termed the "lamellipodium" by Abercrombie et al. [13]. Abercrombie et al. [14] showed that these structures contain actin filaments arranged in a branched structure, but not microtubules. First described in fibroblasts, lamellipodia have also been observed in many other cell types such as precursor cells, epithelial cells, and neural crest cells [2, 15].

For many years, a group of proteins called the Actin-Related Proteins 2/3 (Arp2/3) complex was thought to be the primary mediator of actin polymerization in lamellipodia. First described as a nucleator of actin polymerization [16], the Arp2/3 complex binds to actin filaments and induces the formation of branched actin networks [17]. Branched actin networks were also observed in electron-microscopy images of lamellipodia [18]. However, it is now known that the extent of actin filament branching can vary depending on the cell type and conditions, as a recent report found only a few branches in the leading edge of cells [19]. Not only branching but also the balance of other known actin-binding proteins can contribute to the extension of the lamellipodium. For example, more capping protein activity reduces actin length and increases nucleation by Arp2/3 [20]. On the other hand, an increase in the expression of vasodilator-stimulated phosphoprotein (VASP) (a protein known to promote filament elongation) was reported to generate longer filaments [21, 22]. More recently, other actin nucleators were found to contribute to lamellipodial protrusion, including several members of the Formin family of proteins. Formins were described as protecting actin filaments from capping and also as promoting filament elongation without branching. One of these proteins, diaphanous homolog 1 of Drosophila (mDia1), was first reported to localize at the lamellipodia of migrating cells [23].

The final essential factor in cell movement is the plasma membrane. The lamellipodial protrusion will encounter the physical barrier imposed by the membrane, and this barrier will also restrict cell migration [24].

Membrane Mechanical Properties Orchestrates Cell Migration

The mechanical characteristics of the plasma membrane, particularly its plasma membrane tension and bending modulus, play central roles in cell motility and cytoskeleton remodeling [25–28].

A direct way to assess these responses to forces is by measuring two elastic parameters of the cell membrane: its bending modulus [28] and its membrane tension [29], using a technique based on extracting the membrane tether from the cell by pulling on it with an attached microsphere trapped in an optical tweezers [30]. The experimental procedure is illustrated in Fig. 4.1. Analysis of the force-extension curve, together with measurement of the tether radius, yield these two elastic parameters and also information regarding the membrane–cytoskeleton interaction [31]. Tether pulling with optically trapped beads is the only known direct method for these measurements [26].

The mechanical load exerted by the membrane at the leading edge of cells can locally influence the dynamic growth and organization of the actin network [25, 32–34]. The high membrane tension in the lamellipodia of motile cells directly influences the protrusion [35–37]. Simultaneously at the rear of the cell, the same membrane load can exert a pulling force that induces retraction [38, 39]. However, this mechanical load imposed by the membrane was also reported to be influenced by forces generated from the actin cytoskeletal protrusion itself [27, 31, 32, 40, 41].

One possibility is that the membrane mechanical parameters are determined primarily through a balance of forces between the cytoskeleton and the hydrostatic pressure acting on the membrane [29, 42]. Another possibility is that these parameters are controlled mainly by the interaction between the membrane and the cytoskeleton [31, 43]. Regardless of the exact mechanism, membrane mechanical properties have emerged as important regulators that coordinate local dynamics over cellular scales.

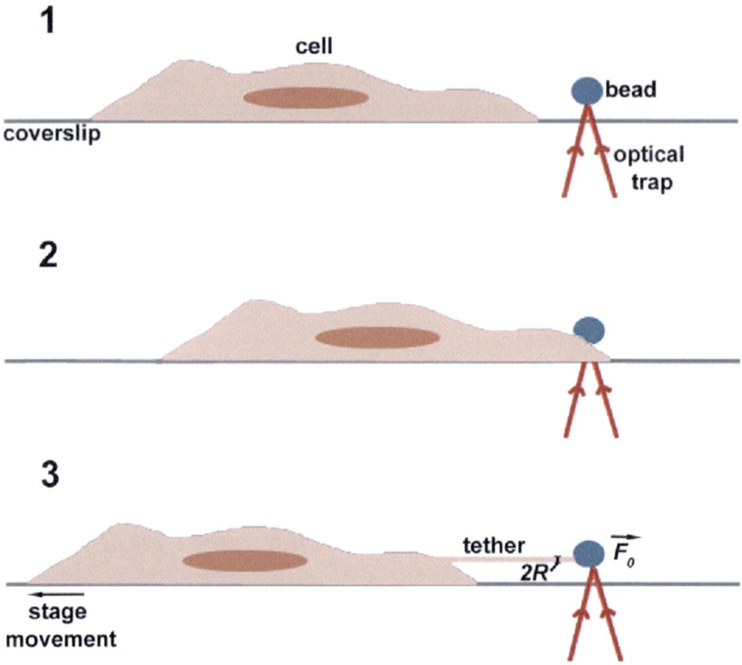

Fig. 4.1 Schematic representation of a tether extraction experiment. Situation (*1*): a bead is trapped in an optical trap. Situation (*2*): a bead is placed against the cell surface. Situation (*3*): by moving the microscope stage, a membrane tether, with radius *R* and force F_0, is formed

Apart from these dynamic aspects, recent data also suggest that cell specialization and/or differentiation can account for the differences in the mechanical properties of the membrane, and that these differences are reflected in their specialized functions [44]. A question that remains unanswered is how the threshold value of these membrane parameters are set. The answer is still not clear, and may vary with different cell types. Pontes et al. [43] began to test this hypothesis by measuring the membrane tension and bending modulus for a variety of cells. These authors observed that the elastic parameters for neurons are close to those obtained for an isolated cell membrane (a membrane disconnected from the cytoskeleton), suggesting a weaker interaction between the membrane and the adjacent F-actin cortex in this cell type. They also observed that the parameters did not change within the different neuronal cell regions, i.e., the cell body, neurite and growth cone. They found very similar membrane mechanical parameters for astrocytes and glioblastoma cells, supporting the idea that these two cell types have the same origin and also share similar functions, for example giving support to neurons in the brain [45]. Macrophages and microglial cells have substantially higher values for the membrane mechanical parameters. When activated, these two phagocytic cells decrease their bending modulus by a factor of 3. This reduction can be interpreted as an easier way to bend the cell surface, which is advantageous during phagocytosis.

Taken together, these observations are striking examples which demonstrate that different cells performing different functions show different mechanical parameters. These new findings suggest the possibility of characterizing cells based not only on morphological and biochemical analyses, but now on their mechanical properties as well.

Chemokines Induce Migratory Processes

Chemokines are *chemo*tactic cyto*kines*, comprising a large superfamily of small peptides (approximately 8–17 kDa) that currently number 47 in humans [46, 47]. They can be classified according to their amino-acid structure into four groups, based on the variations of a conserved cysteine motif in the mature sequence of the proteins [48, 49]. Chemokines bind to the chemokine receptor subfamily of class A G-protein-coupled receptors (GPCRs), which comprises ten CCR family members, seven C-X-C chemokine receptors (CXCR) family members, the "C" sub-family of chemokine receptors 1 (XCR1) and CX3C chemokine receptor 1 (CX3CR1) [46, 50]. These GPCRs signal through heterotrimeric G-proteins, and regulate a diversity of signal transduction pathways involved in chemotaxis and cell survival.

Chemokines were first described for their role in chemotaxis and migration of leukocytes to lymphoid tissues and sites of injury, and the signaling pathways activated by their receptors lead to changes and reorganization within the cytoskeleton proteins. They also proved to be important in the development and homeostasis of the immune system and various other organs, and in pathophysiological processes associated with osteoporosis [51], obesity and insulin resistance [52], viral infections [53, 54], immune responses [55, 56], mobilization of progenitors to the bone marrow [57] and autoimmune encephalomyelitis [58]. More recently, chemokines emerged as key mediators of cancer progression, by interfering with the homing of cancer cells to metastatic sites and the recruitment of a number of different cell types to the tumor microenvironment, such as tumor-associated macrophages, tumor-associated neutrophils, lymphocytes, cancer-associated fibroblasts, myeloid-derived suppressor cells and endothelial cells [48, 49, 59, 60, 61].

Processing the chemokine gradients into migratory or adhesive responses occurs in multiple dynamic steps that regulate changes in the cytoskeleton and cellular adhesion [46, 62–65]. Binding of the chemokines to their G-coupled receptors can lead to downstream activation of different signaling pathways (Fig. 4.2), such as protein kinase B (PKB/Akt) and mitogen-activated protein kinases (MAPKs) [66, 67]. Another activated pathway may be the Janus kinase family (JAK), activated in a Gαi-independent fashion [68, 69]. The Rho family of GTPases and their downstream effectors were also implicated in chemokine-elicited migration. One of the important groups of Rho effectors is the Rho-associated coiled-coil forming protein kinases (ROCK) I and II, which enhance myosin light chain (MLC) phosphorylation by both inhibiting MLC phosphatase and phosphorylating MLC, thereby regulating actin–myosin contraction [66, 70]. ROCK isoforms also regulate lymphocyte polarity

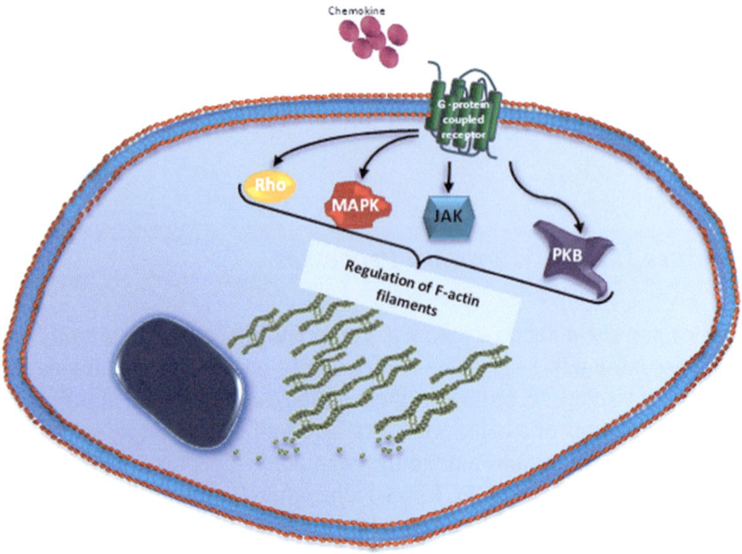

Fig. 4.2 Signaling pathways involved in cytoskeleton regulation. Chemokines bind to G-protein-coupled receptors, and consequently can activate different signaling pathways, such as protein kinase B (PKB), mitogen-activated protein kinases (MAPKs), the Janus kinases family (JAK) and the Rho family GTPases. All these signaling pathways are involved in modulating the cytoskeleton proteins, leading to their reorganization, more specifically of the F-actin filaments

and migration through members of the Ezrin/Radixin/Moesin (ERM) family of proteins [66, 71]. Rho GTPases also control cytoskeletal remodeling through effector proteins from the mDia family of formins, which, as mentioned above, are actin-nucleating proteins favoring the formation of long straight actin filaments. Lack of mDia1 expression significantly reduces T cell homing to secondary lymphoid organs [72].

Little is known about the correlation between chemokines and the cytoskeleton. At least C-C motif chemokine 19 (CCL19)/CCL21-CCR7 and CXCL12-CXCR4 constitute an exception.

The Influence of the Chemokines CCL21/CCL19 and Their Receptor CCR7 on the Cytoskeleton

From a physiological perspective, all these intracellular events that occur in the lymphocyte homing process depend on a combination of interactions between different chemokines and their receptors, according to the cell type involved: T-cell homing and traffic of lymphocytes into and within secondary lymphoid tissues rely largely on CCR7 and its ligands CCL21/CCL19 [73, 74], as well as a minor contribution from CXCL12-CXCR4 interactions [75, 76]. Bardi et al. [77] reported that both the CCR7-mediated polarization and chemotaxis are dependent on the Rho kinases, but not on MAPK extracellular signal-regulated protein kinase (ERK)-2, as previously described [78–80]. The C-C chemokine receptor type 7 (CCR7) and other chemokine receptors such as CXCR4 also activate leukocyte integrins, which are important for the endothelial adhesion and arrest of rolling lymphocytes [81–83], possibly through downstream activation of RhoA [84–86].

Although B cell integrin activation is also primarily induced by CCR7 and CXCR4, their homing also requires the activation of CXCR5, whose expression is restricted to B cells and a subset of CD4+ T cells [76].

The expression of CCR7 and CCL21 has been described in many cancers (especially melanoma, breast cancer, and head and neck cancers), and was correlated with actin polymerization and lamellipodium formation, which contribute to increased tumor-cell migration, invasion and metastatic potential [87–91].

The Influence of the Chemokine CXCL12 and Its Receptor CXCR4 on the Cytoskeleton

CXCL12, better known as stromal cell-derived factor-1 (SDF-1), was first described as pre-B cell growth-stimulating factor (PBSF) [92], and activates integrins in B-cells as mentioned above. The chemokine CXCL12 and its receptor CXCR4 are well known for their role in the metastasis of breast cancer [93, 94]. However, CXCL12 is constitutively expressed in a broad range of tissues, e.g. in bone marrow, spleen, liver, lung and brain, as well as in most types of tumors [92, 95]. This chemokine is the only known ligand for CXCR4, also known as Fusin/LESTR/CD184 [96–98]. CXCR4 is a G-protein-coupled seven transmembrane receptor and is widely expressed by many different cell types including hematopoietic cells, leukocytes, endothelial cells, central nervous system (CNS) cells, and cells of the gastrointestinal tract.

Physiologically, CXCL12 is important for the homing of CXCR4-expressing hematopoietic cells to the bone marrow [99] and for guiding CXCR4-positive cells from different tissues to their niche [100]. CXCL12/CXCR4 knockout is lethal and leads to several impairments in CNS development and hematopoiesis in mice [101]. Furthermore, the CXCL12-CXCR4 axis plays a role in angiogenesis and inflammation (e.g. recruitment of lymphocytes). The chemokine CXCL12 can also bind to another chemokine receptor, CXCR7/G Protein-Coupled Receptor (RDC-1) [102, 103]. However, the connection between this pathway and the cytoskeleton is poorly understood.

Li et al. [104] showed that after CXCR4 stimulation, a signaling pathway that leads to the reorganization of the actin-cytoskeleton becomes activated. After binding of CXCL12 to CXCR4, the heterodimeric G-protein dissociates into the Gαi- and Gβγ-subunits. The Gαi2-subunit interacts with the N-terminus of the engulfment and cell-motility protein 1 (ELMO1) which forms a complex with the DOCK180 (Dedicator of cytokinesis) protein. The ELMO1/DOCK180 complex serves as a guanine nucleotide exchange factor (GEF), activating the small GTP-binding protein (G protein) Rac1 [104].

It is well known that small GTPases such as RhoA, Rac and CDC42 control the dynamics of the cytoskeleton [105]. Rac, which is activated by ELMO1/DOCK180, can remove the capping proteins and activate the Arp2/3 complex, which induces the growth of actin filaments and the formation of new actin branches from existing ones [106, 107].

Another cytoskeleton modulation by CXCL12 is the activation of Focal Adhesion Kinase (FAK) and Paxillin. After CXCL12 binds to CXCR4, the Janus kinase 2 (JAK2) and the MAP-kinase ERK1/2 pathways become activated by phosphorylation. Activated JAK2 phosphorylates Signal transducer and activator of transcription 3 (STAT3), and pSTAT3 and pERK are able to phosphorylate FAK and Paxillin, activating these proteins, which leads to actin cytoskeleton reorganization [108]. In conclusion, CXCL12 in known to influence the cytoskeleton reorganization in two different ways, through the Gαi2-ELMO1/Dock180-Rac1 activation and the JAK2-pSTAT3/pERK-pFAK/paxillin activation.

Apart from the chemokines previously described in detail, the CXCL9/Mig chemokine is also known to activate the small GTPases Rac1 and RhoA via its receptor CXCR3 on human melanoma cells, also leading to cytoskeletal changes [109].

Role of Cytoskeleton in Maintenance of Stem-Cell Properties

Nowadays, the application of stem cells in regenerative medicine is one of the major fields in biomedical research. Because of their ability to self-renew and differentiate into specific lineages, stem cells play an important role in the development of cell-based therapies [110–112]. The implementation of these new therapies made it necessary to investigate the cellular and molecular mechanisms involved in the regulation of stem-cell differentiation, growth, and phenotypic expression. The most recent studies have indicated that the regulation of stem-cell growth and fate is also dependent on the crosstalk between the extracellular matrix (ECM) ligands and the stem-cell surface receptors [113, 114]. Therefore, during their differentiation into specific lineages, stem cells are subjected to extracellular stimuli that determine a number of morphological alterations associated with the expression of cytoskeletal proteins, actin filaments, microtubules, intermediate filaments, and their downstream effectors [115–117]. These alterations are also crucial in establishing the migration phenotype observed in different types of stem cells, as was previously documented in mesenchymal stem cells (MSCs) and in the embryonic neural stem cells (NSCs) [113, 115–118]. A brief summary of the most important conclusions that explain the role of the cytoskeleton in stemness maintenance and its contribution to stem-cell migration and differentiation is given below.

Increasing evidence shows that a diverse array of environmental factors contributes to the control of stem-cell activity, differentiation and migration. According to the tissue microenvironment factors, stem cells are believed to modulate their cytoskeleton, to migrate and move away from their niche, and then to differentiate [119]. Studies on myocardial development and on capillary endothelial cells have

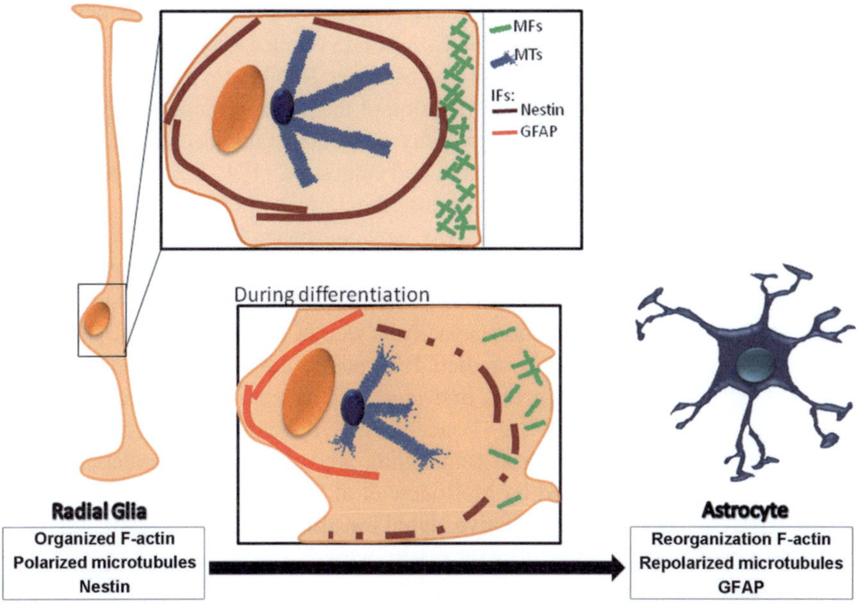

Fig. 4.3 Changes of the cytoskeleton in a neural stem cell during differentiation. During the differentiation process, the neural stem cell (represented by radial glia) undergoes cytoskeletal changes, in the microfilaments (MFs), microtubules (MTs) and intermediate filaments (IFs). Alterations in the stem-cell cytoskeleton may involve the disorganization of MFs, which become more dispersed; destabilization of the MTs; and also modulation of the expression of IF proteins such as Nestin and GFAP, leading to the differentiation of stem cells into astrocytes, for example

demonstrated that alterations in cell shape might regulate cellular differentiation [113, 115, 116, 118, 120]. These results show that the cytoskeleton is a key player in the differentiation and migration of stem cells [121] (Fig. 4.3).

Recent studies showed that during osteogenic differentiation, the actin cytoskeleton of MSCs becomes more dispersed, similarly to that of osteoblasts, and the disruption of the actin cytoskeleton decreases osteogenesis in favor of adipogenesis [121]. These results have been crucial in the area of tissue engineering of bone and cartilage, which attempts to develop new therapeutic strategies for the treatment of musculoskeletal trauma and diseases [43, 122, 123]. A number of studies have shown that the stem-cell fate and the adhesive interactions between the stem cells and the substrate can be influenced through the control of their shape by artificial extracellular matrices.

Alterations in the cytoskeleton are also dependent on microtubules that contribute to migration and to stem-cell polarization. The regulation of microtubules is usually dependent on the Rho GTPases, in particular RhoA, Rac1 and Cdc42. Previous studies have reported that migrating hematopoietic stem and progenitor cells growing on MSCs display a polarized morphology, with the formation of an uropod at the rear pole and a leading edge at the front, which is involved in microtubule destabilization. The uropod formation seems to be dependent on the activity of RhoA and its downstream effector Rho-associated coiled-coil containing protein kinase (ROCK I). When RhoA is inhibited using the Rho kinase inhibitor (Y-27652) or RNA interference (RNAi), the polarization of the hematopoietic stem/progenitor cells (HSPCs) and their migration capability are considerably decreased, indicating the crucial role of microtubules in stem-cell migration [124, 125]. Vertelov et al. [126] showed that in hypoxic conditions, the human mesenchymal stem cells (hMSCs) showed increased RhoA activity, and consequently it may contribute not only to increasing migration, but also to preserving MSCs in an undifferentiated state, as compared to normoxic conditions. Moreover, the microtubules seem to be important to maintain the migration capacity as well as the polarity of NSCs [127].

Two of the most thoroughly studied IF proteins are nestin and GFAP [128] (Fig. 4.3). In the 1990s, nestin was first identified as a marker of neuroepithelial stem/progenitor cells in the CNS by Lendahl and collaborators [129]. Nowadays, it is considered to be a marker for distinguishing precursor from differentiated cells [130–132]. A study performed by Mellodew et al. [133] also showed that loss of nestin expression could be a predictive signal for differentiation of NSCs. GFAP is classically known as a marker of mature astrocytes. However, several studies have been conducted in order to evaluate its contribution to the maintenance of stem-cell features. Previous studies suggested that primary astrocyte cultures from the post-natal and adult mouse brain could contain GFAP-expressing cells that may act as multipotent NSCs when transferred to neurogenic conditions [134, 135]. GFAP functions are addressed in detail in section "GFAP Expression and Its Functions in Astrocytes".

Cytoskeleton Alterations During Disease Progression: The Role of Stem Cells in Cancer

The cancer stem-cell theory predicts that not all cancer cells in a tumor exhibit the same tumor-growing ability, and that only a small population of cells with stem-cell properties drives tumor growth. The proliferation, survival and migration of tumor stem cells seem to be dependent on the local microenvironment. Although highly controlled during embryonic development, the ECM is commonly deregulated in cancer [136, 137] and seems to contribute to the development of chemo- and radio-resistance of tumor cells. Under normal conditions, the ECM receptors allow stem cells to anchor to the local microenvironment where their properties can be maintained [136, 137]. This anchorage physically constrains stem cells to make direct contacts with the microenvironment cells, which produce paracrine-signaling molecules that are essential for maintaining stem-cell properties [136, 137].

Considering that tumor cells possess an increased proliferation and migration ability, we could hypothesize that due to the occurrence of genetic mutations and microenvironment alterations, the characteristics of the stem-cell cytoskeleton become distorted. The most recent studies have shown that cancer cells express the same cytoskeleton markers as benign stem cells [130]. Therefore, the problem seems to be associated with the degree of expression of the cytoskeleton markers and with the signaling pathways that become activated.

One of the first lines of evidence that the cytoskeleton is involved in the tumor phenotype was the experiment conducted by Vasioukhin et al. [138], who set up conditional gene targeting to knockout genes in the stem cells and basal epidermal layer of mouse skin. They started to knock out the α-catenin, and observed that mouse skin rapidly took on the appearance of squamous cell carcinoma in situ. More recently, Rampazzo et al. [139] demonstrated that the treatment of tumor stem cells isolated from glioblastoma samples, with Wnt ligands, or the induction of β-catenin overexpression mediates neuronal differentiation and halts proliferation in primary glioblastoma cells.

In prostate cancer, deregulation of the non-canonical Wnt/Ca^{2+} pathway leads to F-actin filament rearrangements and consequently to the reduction of cancer progression [140].

Understanding the complexities of the stem cell cytoskeleton in cell homeostasis and in tumor development is a challenging exercise, not only to understand the physiology of many diseases but also to implement new therapeutic strategies.

Role of Intermediate Filaments in Glial Cells: Example

Glial Fibrillary Acidic Protein (GFAP) in Health and Disease

The intermediate filaments are components of the cytoskeleton that are specific to each cell type. These filaments confer mechanical force and resistance on the cells, and are regulated developmentally and tissue-specifically. GFAP is the main intermediate filament (IF) protein in astrocytes, although other intermediate filament proteins such as nestin, vimentin and synemin can also be found in these cells [141]. A combination of vimentin and nestin is observed in immature astrocytes, while vimentin and GFAP are found in mature astrocytes [37, 142, 143] (Fig. 4.3). Only GFAP seems to be capable of forming homodimers [143].

GFAP was discovered in the brains of patients with multiple sclerosis. It was initially termed ‘plaque protein’ and was first isolated, purified and the amino-acid content determined over 40 years ago. Immunostaining for GFAP has been performed since 1975 [144, 145].

Although GFAP was originally thought to be an astrocyte-specific IF [145], curiously, several investigators have shown that it is present in different amounts in various types of tissues, such as enteric glia [146], Schwann cells [147], and even outside the CNS, in chondrocytes, fibroblasts and myoepithelial cells [147, 148], lymphocytes [149] and stellate cells from the liver, kidney, pancreas, lungs, and testes [150, 151].

Astrocytes are heterogeneous, and their biology varies according to the particular physiological state and time frame, and also to their location in the CNS [151, 152].

It is a challenging task to determine the specific role of GFAP in these multiple environments and in different physiological and pathological conditions.

The human GFAP polypeptide consists of 432 amino acids and has a molecular mass of 55 kDa. The gene contains 9 exons and spans over 10 Kb in chromosome 17p21 [153]. GFAP belongs to the family of type 3 intermediate filaments.

The constitutional transcription of the human GFAP gene is controlled by a TATA-like sequence CATAAA, located 29 base pairs downstream of the RNA start site [154]. Multiple sites seem to be involved in the regulation of GFAP expression, with important roles for phosphorylation and DNA methylation in GFAP transcription [155]. The demethylation of the GFAP promoter activates GFAP transcription [156]. GFAP expression during development is also controlled by acetylation in neural stem cells, and has been shown to be significantly reduced by acetylation in mature astrocytes [157, 158].

The expression of GFAP has multiple regulatory factors, including various hormones, cytokines and growth factors, including interleukins 1 and 2, tumor necrosis factor (TNF), ciliary neurotrophic factor (CNTF), basic fibroblast growth factor (bFGF), transforming growth factor beta 1 (TGFβ1) and glutamate. Interestingly, this regulation seems to be partially controlled by the interaction between astrocytes and cortical neurons, mainly through TGFβ1; and varies in different regions of the brain [159–163].

GFAP Alternative Splicing-Isoforms

At least nine isoforms of GFAP mRNA exist, and are generated by alternative mRNA splicing and polyadenylation signal selection [155, 164–169]. Please refer to Middeldorp and Hol for an excellent review of GFAP biology and GFAP isoforms [155]. Seven of these isoforms are present in humans.

GFAPα was the first to be identified, and is also the most abundant and the most studied.

GFAPβ has a transcription site located 169 nucleotides above the site for GFAPα, which corresponds to a 5′ region that is not transcribed in the main isoform [170]. GFAPβ is the main isoform in Schwann cells from the rat peripheral nervous system (PNS), but its mRNA comprises only 5–10 % of the total GFAP mRNA in the CNS. It has been found in normal hamster brain and in a case of human glioma [171].

GFAPγ was first described in spleen and bone marrow from mice. GFAPγ lacks exon 1 and includes the last 126 bp of intron 1–2, comprises around 5 % of all GFAP mRNA in the CNS in mice, and is also present in a small proportion in humans [167].

The splice variants GFAPΔEx6, GFAPΔ164, and GFAPΔ135 skip sequences in exon 6/7 and have been observed in tissue from patients with Alzheimer's disease, focal lesions in chronic epilepsy, and a specific astrocyte subtype. These out-of-frame splice forms completely lack the tail domain [166].

GFAPκ is the result of an alternative splicing at the 3′ end of the GFAP pre-mRNA, and the consequent inclusion of an alternative exon termed 7B [164].

The C-terminal domain of GFAPκ is therefore different from those present in other isoforms.

GFAPζ was described in mice, originating from the initial report of a transcript including the last 284 bp of intron 8–9 [167, 172].

GFAPδ is the most often studied isoform, after the canonical form GFAPα. It was initially described in rats in 1999, from an alternative splicing, resulting in substitution of the last two exons by an alternative exon called 7+ [165]. Nielsen and coworkers [168] described the corresponding RNA in humans, terming it 7a, and naming the isoform GFAP epsilon; and more recently have come to a consensus on the name GFAPδ [173]. The exon 7a has its own polyadenylation signal inhibiting the expression of exons 8 and 9. It has been isolated only in mammals, and seems to be subject to a different evolutionary pressure than the other exons [174]. Among the higher primates, the exons 7+ are 100 % identical, with the exception of an alanine on codon 426, conserved in only 9 % of human alleles, and replaced by a valine in 21 % and by a threonine in 70 %. The potential phosphorylation of this threonine residue could explain a positive selection for this change [175].

Incorporation of exon 7a results in a substitution of the 42 C-terminal amino acids by a new c-terminal domain of 41 amino acids. Among the different isoforms, only GFAPδ and GFAPκ show a modification in the C-terminal portion. As a consequence, GFAPδ and GFAPκ are incapable of forming homodimers, but are able to form heterodimers with GFAPα and vimentin [164, 175]. GFAPκ mRNA comprises around 5–10 % of the total GFAP mRNA in humans [164, 175, 176].

GFAPδ is expressed in proliferating neurogenic astrocytes during development, in the adult human brain as well as in radial glia cells [169, 177–179]. It can be detected by immunohistochemistry, particularly in astrocytes of the *glia limitans* and in different forms of gliosis [172, 180], in contrast to earlier studies on post-mortem material [169]. Moreover, its expression seems to parallel that of vimentin in normal and reactive astrocytes, but not in glial tumors [180]. Interestingly, differential GFAP isoform expression in mice does not seem to be linked with aging or reactive gliosis [172]. The specific functions of the different GFAP isoforms have not been well established. These isoforms often differ from GFAPα in the C-tail domain, the region responsible for interaction with other cell components. The interaction of GFAPα was specifically shown with presenilins 1 and 2 [168] and αβcristallin [176]. Therefore an assembly-compromised role of GFAPδ as a modulator of the GFAP filament surface has been postulated [155].

GFAP Expression and Its Functions in Astrocytes

GFAP is phylogenetically ancient. The human GFAP polypeptide shows a 90 % homology with its murine and porcine counterparts, and about 85 % homology with goldfish GFAP [174].

During development, GFAP is expressed in radial glia, bipolar cells which express vimentin and nestin and which have been shown to be neural precursors [181, 182] (Fig. 4.3). Studies differ with respect to the exact moment when GFAP expression can first be detected in these cells, varying from gestational week 6–12;

these differences are probably due to the location in the brain or to the detection techniques used [155, 183, 184]. Nervous-system neural precursor cells show a progressive shift in intermediate filament expression, from vimentin to GFAP. In the normal adult brain, only certain subpopulations of astrocytes seem to co-express vimentin and GFAP such as Bergmann Glia, subpopulations of corpus callosum, hippocampus, subpial, and rare white-matter astrocytes [185, 186].

The functions of GFAP are not yet completely elucidated, and include a role in the long-term maintenance of the brain parenchyma structure, the proper functioning of the blood–brain barrier [187], myelination [188], astrocyte proliferation [189, 190], and astrocytic modulation of some neuronal functions, such as the formation and protection of synapses [191, 192]. GFAP is involved in other important and fundamental cellular processes, and is probably implicated in astrocyte motility [193, 194] and exocytosis of astrocytic gliotransmitters [195, 196]. GFAP is also important in the regulation and maintenance of glutamate transporters in the astrocyte plasma membrane, a key mechanism for glutamate uptake and its metabolism and for the formation of GABA [197].

GFAP, Pathological States and Disease

GFAP Knockouts

Knockout mice for intermediate filaments (GFAP−/−, Vimentin−/−, GFAP−/−, and vimentin−/−) do not show major changes in their development, adult life, and reproduction [198–200]. These authors found no major differences in brain architecture and cellularity in comparison to wild-type animals. However, another group of researchers working on GFAP-null rodents reported contrasting results, showing myelination defects in the spinal cord, optic nerve and corpus callosum, and hydrocephalus in half GFAP-null mice after 18 months [201]. Astrocytes lacking intermediate filaments exhibit normal morphology, but lack the ability to form normal glial scars [202], have restricted motility in vitro [203], and are highly sensitive to ischemia and trauma [204]. GFAP−/− mice are also more sensitive to neurodegeneration induced by kainic acid or mechanical trauma, which is not observed in wild-type animals [205].

GFAP and Gliosis

GFAP expression rises as a consequence of inflammation and various CNS diseases such as trauma, ischemia, genetic diseases, toxic lesions, and degenerative diseases [206]. In all these situations, astrocytes react to injury, in a process usually called astrogliosis or simply gliosis. Despite having general morphological features in common, astrocytes may vary morphologically and chronologically in their responses according to the nature, intensity and localization of the lesion. The kinetics of this response is usually rapid, and can be detected 1 h after the insult, with a maximum intensity at between 3 and 7 days [207].

Astrocytic gliosis has classically been described morphologically by the hypertrophy of the cell soma and processes, which is roughly proportional to the severity of the insult and the proximity of the astrocyte to it. More recently it has been well established that there is an increase in the GFAP cellular content, and, depending on the severity of the reaction, also an increase in the number of astrocytes [208–210]. Constitutional GFAP expression is heterogeneous among different astrocyte populations, and in the normal state not all astrocytes express detectable levels of GFAP. With increasing intensity of gliosis, most astrocytes will express GFAP, and in severe gliosis one also observes astrocyte proliferation, with subsequent overlap and disruption of individual astrocyte domains [151]. Therefore, GFAP has been generally used as a marker of gliosis [206], even though astrocyte reactivity and GFAP upregulation due to different stimuli may be associated with different changes in transcriptome profiles and cell function [211]. In other words, GFAP levels can be generally increased in various CNS pathological states such as trauma, ischemia, infections, and neurodegenerative diseases.

Interestingly, low levels of GFAP can be detected in the cerebrospinal fluid (CSF) in healthy individuals. The reason has not been well established, but it has been postulated that the presence of GFAP might be related to some degree of astrocyte death and release of the protein into the extracellular space, as normal astrocytes do not secrete GFAP [212]. GFAP levels in the CSF can be elevated in association with several conditions, including traumatic, vascular, developmental, inflammatory, neoplastic, and degenerative diseases. We briefly describe certain diseases in which GFAP seems to have a particular role.

Alzheimer's Disease

Alzheimer's disease (AD) is characterized mainly by two neuropathological alterations, the formation of neurofibrillary tangles, and amyloid deposits in the brain [213]. Reactive astrogliosis has been well described in AD, although its role in this disease is not yet completely understood. Reactive astrogliosis in AD is usually focal, and reactive astrocytes are intimately associated with amyloid plaques or diffuse amyloid deposits, surrounding them and forming miniature scars all around [151]. The intensity of reactive astrogliosis, as determined by GFAP levels, has been reported to increase in parallel with increasing disease morphological burden [214, 215], and in some studies can be correlated with cognitive impairment [216].

Alexander's Disease

A gradual increase in astrocyte GFAP content is usually observed during adult life in mice, primates, and humans [217, 218]. Mice induced to overexpress GFAP die for reasons that are currently unknown, and their astrocytes exhibit accumulation of the protein in numerous cytoplasmic Rosenthal fibers [219]. Likewise, Alexander's disease, a rare human disease, is caused in 95 % of cases by mutations of GFAP, leading to its accumulation in the cytoplasm, associated with proteins such as small heat shock, ubiquitin and β cristallin in the form of Rosenthal fibers [220, 221]. There are two clinical forms, one with early onset, which progresses with childhood leukodystrophy with striking clinical signs of megalencephaly, seizures and psychomotor delay; and a later-onset form, often revealed by difficulty swallowing and speaking, autonomic dysfunction and ataxia. Several different mutations have been reported in the human GFAP gene in Alexander's disease, mostly heterozygous missense changes predicting the production of full-length mutant and wild-type proteins, and subsequently alterations such as small in-frame insertions and deletions, and in-frame skipping of an entire exon or frameshifts at the extreme C-terminal end [222]. The pathophysiology is explained by the increase in the toxic function of abnormal tissue deposits [223]. For the list of GFAP mutations in this disease please refer to the Waisman Center of the University of Wisconsin-Madison (http://www.waisman.wisc.edu/alexander/mutations.html).

Interestingly, Rosenthal fibers are also present in certain pathological conditions, namely tumors such as pilocytic astrocytomas and gangliogliomas, and even some types of chronic gliosis, for example, those observed in the periphery of craniopharyngiomas or hemangioblastomas.

GFAP and Gliomas

The development of immunohistochemistry for GFAP was an important advance in surgical neuropathology, including the diagnosis of brain tumors. The expression of GFAP in primary glial tumors has been extensively studied since the beginning of the 1980s, and is now widely used in diagnostic neuropathology [224–226]. Independently of the histological grade, every tumor with astrocytic differentiation is expected to show at least some positivity for GFAP [227]. GFAP-positive cells can also be observed in some other glioma types, such as oligodendroglioma, where often small cells named "minigemistocytes" express GFAP [227]; and in ependymomas [226, 228]. A significant proportion of choroid-plexus tumors can also express GFAP, even focally [226].

Some studies have shown in vitro a negative correlation between GFAP expression levels and the malignant transformation of astrocytes [229–231]. However in experimental astrocytoma murine models, GFAP expression does not seem to affect tumor progression [232].

GFAP serum levels seem to be significantly elevated in patients with glioblastoma multiforme [233]. This implies damage to the blood–brain barrier, since this protein is not usually detectable in the serum or is present in very low levels. Some other conditions that can lead to its rise in the serum are stroke, hemorrhage, trauma and multiple sclerosis [212].

Enteric Glia Cytoskeleton

GFAP is also the main marker of enteric glia, a peripheral glial-cell type derived from the neural crest cells that has close morphological, molecular and functional similarities to astrocytes (reviewed by Coelho-Aguiar et al. [234]). These glial cells are crucial for the proper functioning of the gut. Their disruption in mice by the targeting of GFAP-positive cells leads to increased permeability of the mucosal epithelium, followed by an inflammation process and disruption of the ileum and jejunum structure [235–237].

Interestingly, as in the CNS, GFAP is also expressed in glia-like progenitors in the gut. Some of these GFAP-positive cells can generate multilineage colonies in vitro and also give rise to glial and neuronal cells in vivo in graft studies, in injury conditions or in experiments of myenteric plexus ablation with benzalkonium chloride detergent [238–245].

Similarly to reactive gliosis in the CNS, there is an increase of GFAP levels in enteric glia of inflamed gut regions, notably in inflammatory bowel diseases [246]. An increase in GFAP expression was also observed in colon biopsies of patients with *Clostridium difficile* infections [246]. Patients with Parkinson's disease show α synuclein aggregates (the pathological trait of Parkinson's disease) also in the ENS, and also an increased expression and reduced phosphorylation of GFAP, which is also observed in neurodegenerative processes in the CNS [247, 248]. These informations suggest an association between enteric inflammation and glial dysregulation.

Furthermore, specific roles are known for other components of the cytoskeleton in enteric glia, the F-actin filaments and microtubules. Little is known about this, but Ca^{2+} dependent responses have been identified in enteric glia, related to their functions in neurotransmission. Calcium enters enteric glial cells through a mechanism termed capacitative calcium entry, which is responsible for the maintenance of calcium storage in different cell types. It has been shown that in cultured myenteric glia, the disruption of the actin filaments or microtubules can decrease and even completely inhibit calcium entry [249]. These experiments confirmed the importance of the cytoskeleton for the physical interactions between the calcium storage organelle and the plasma membrane as a capacitative calcium-entry mechanism, which replenishes the cell after depletion of the intracellular calcium store. It is probable that the same events occur in astrocytes and other cell types to maintain their intracellular calcium store, but this has not yet been investigated.

Schwann Cell Cytoskeleton

Another neural crest-derived GFAP-expressing type of glia is the Schwann cells, which are responsible for the myelinization of the entire peripheral nervous system (PNS). Schwann cells can be found in myelinating or non-myelinating forms. This cell type is capable of great migratory and differentiation capacities during development and even in adults, to repair and replace the myelin of injured axons. The physical interaction of these cells with the surrounding microenvironment involves changes in the cytoskeleton.

These cells constitute the other example of peripheral glia that express GFAP. However, it is clear that they show a different GFAP isoform expression from that of astrocytes and enteric glia. Moreover, they express this IF at lower levels than astrocytes, and even more reduced levels in relation to enteric glia [250].

GFAP expression is a marker of non-myelinating Schwann cells. GFAP appears after differentiation of Schwann-cell precursors into immature Schwann cells [251]. Its expression depends on contact with non-myelinated axons, and is downregulated in myelinating Schwann cells. On the other hand, vimentin is expressed in myelin-forming Schwann cells. In experiments with sciatic-nerve transection, a reduction of GFAP mRNA levels was observed. Moreover, the immunodetection of GFAP also decreased, while vimentin expression increased [252].

Interactions among Schwann cells, the extracellular matrix and axons are mediated by surface receptors and are transduced by the cytoskeleton proteins. These interactions are essential to the recovery of neuronal transmission after axonal nerve injury. The axonal regeneration process follows well-established steps. After disruption of myelin sheets, Schwann cells dedifferentiate and proliferate. There is an increase in the expression of adhesion molecules and cytoskeleton proteins such as GFAP and vimentin. Then, these glial cells form bands of Büngner, which serve as a substrate for axonal regrowth. After that, they enwrap the axons and form myelin [253]. All these stages require continuous reorganization of the Schwann cell cytoskeleton.

In GFAP-null mice, the development of peripheral axons and their myelin is normal, as is their functioning. However, lack of GFAP leads to retardation in nerve regeneration after injury, probably because of a problem in Schwann cell proliferation. This study, developed by Triolo and coworkers [253], revealed that GFAP interacts with integrin $\alpha v\beta 8$, which interacts with fibrin, thus acting in the early steps of Schwann-cell proliferation. These investigators also showed that vimentin interacts with integrin $\alpha 5\beta 1$, which connects to fibronectin, acting in the subsequent steps of Schwann-cell proliferation and nerve regeneration [253].

Another cytoskeleton protein associated with integrins in Schwann cells is merlin, also known as schwannomin. Merlin is a perinuclear protein that translocates to cytoplasm during differentiation and becomes associated with integrin $\beta 1$ in myelinating Schwann cells. This study suggests that merlin links the integrin to the microfilaments, supporting myelination [254]. Other proteins associated with intermediate filaments interact with integrins, such as hemidesmosomal 1 protein (HD1), plectin and dystonin [255]. Targeted disruption of dystonin, for example, results in problems in the interaction with the axon basement membrane and in PNS dysmyelination [256].

Conclusion

As set out in this chapter, the first step in cell locomotion is the formation of cytoskeleton protrusions such as lamellipodia at the periphery of a cell. This protrusion is composed of F-actin filaments. Study of the mechanical characteristics of cytoskeleton proteins and plasma membrane tension in lamellipodia was advanced by the use of optical tweezers, which enable measurement of the bending modulus and membrane tension. Important inductors of the migratory phenotype are the chemokines, small peptides that elicit cytoskeleton responses through signaling by their G protein-coupled receptors. CCL19/CCL21 and their receptor CCR7, and CXCL12 and its receptor CXCR4 have physiological roles in guiding different cell types to their niches, as well as acting in tumor-cell migration by eliciting reorganization of the actin cytoskeleton.

We next discussed the role of the cytoskeleton in the performance of stem cells. Actin microfilaments and microtubules are reorganized during differentiation, and are important for stem-cell fate and polarization. Intermediate filaments, specific to each cell type, may also be involved. Neural stem and precursor cells, for example, express nestin, while GFAP is typical for mature astrocytes but is also found in neural precursors. In the last section we explored the specific characteristics of the intermediate filament GFAP in its multiple isoforms. GFAP is essential for astrocyte functions, and its disruption is implicated in several disorders and diseases. Peripheral glial cells that also express GFAP are the enteric glial cells, which are the counterparts of astrocytes in the gut; and the Schwann cells, which are responsible for the myelination of the peripheral nerves.

Acknowledgments This work was supported by National Institute for Translational Neuroscience (INNT) from Ministry of Science and Technology; Brazilian Federal Agency for the Support and Evaluation of Graduate Education (CAPES) from the Ministry of Education; National Council for Technological and Scientific Development (CNPq); Rio de Janeiro State Research Foundation (FAPERJ); PhD program on Morphological Sciences (PCM) from the Federal University of Rio de Janeiro (UFRJ); Ary Frauzino Foundation to Cancer Research; and Pró-Saúde Hospital Administration - Beneficent Association for Social and Hospital Assistance.

References

1. Alberts B (2002) Molecular biology of the cell, 4th edn, vol xxxiv. Garland Science, New York, 1464 p
2. Friedl P, Gilmour D (2009) Collective cell migration in morphogenesis, regeneration and cancer. Nat Rev Mol Cell Biol 10(7):445–457
3. Mogilner A (2009) Mathematics of cell motility: have we got its number? J Math Biol 58(1-2):105–134
4. Petrie RJ, Doyle AD, Yamada KM (2009) Random versus directionally persistent cell migration. Nat Rev Mol Cell Biol 10(8):538–549
5. Ridley AJ et al (2003) Cell migration: integrating signals from front to back. Science 302(5651):1704–1709
6. Etienne-Manneville S, Hall A (2003) Cdc42 regulates GSK-3beta and adenomatous polyposis coli to control cell polarity. Nature 421(6924):753–756
7. Pletjushkina OJ et al (1994) Taxol-treated fibroblasts acquire an epithelioid shape and a circular pattern of actin bundles. Exp Cell Res 212(2):201–208
8. Euteneuer U, Schliwa M (1984) Persistent, directional motility of cells and cytoplasmic fragments in the absence of microtubules. Nature 310(5972):58–61
9. Eilken HM, Adams RH (2010) Dynamics of endothelial cell behavior in sprouting angiogenesis. Curr Opin Cell Biol 22(5):617–625

10. Gupton SL, Gertler FB (2007) Filopodia: the fingers that do the walking. Sci STKE 2007(400):p. re5
11. Charras G, Paluch E (2008) Blebs lead the way: how to migrate without lamellipodia. Nat Rev Mol Cell Biol 9(9):730–736
12. Buccione R, Caldieri G, Ayala I (2009) Invadopodia: specialized tumor cell structures for the focal degradation of the extracellular matrix. Cancer Metastasis Rev 28(1–2):137–149
13. Abercrombie M, Heaysman JE, Pegrum SM (1970) The locomotion of fibroblasts in culture. I. Movements of the leading edge. Exp Cell Res 59(3):393–398
14. Abercrombie M, Heaysman JE, Pegrum SM (1971) The locomotion of fibroblasts in culture. IV. Electron microscopy of the leading lamella. Exp Cell Res 67(2):359–367
15. Weijer CJ (2009) Collective cell migration in development. J Cell Sci 122(Pt 18):3215–3223
16. Mullins RD, Heuser JA, Pollard TD (1998) The interaction of Arp2/3 complex with actin: nucleation, high affinity pointed end capping, and formation of branching networks of filaments. Proc Natl Acad Sci U S A 95(11):6181–6186
17. Campellone KG, Welch MD (2010) A nucleator arms race: cellular control of actin assembly. Nat Rev Mol Cell Biol 11(4):237–251
18. Svitkina TM, Borisy GG (1999) Arp2/3 complex and actin depolymerizing factor/cofilin in dendritic organization and treadmilling of actin filament array in lamellipodia. J Cell Biol 145(5):1009–1026
19. Urban E et al (2010) Electron tomography reveals unbranched networks of actin filaments in lamellipodia. Nat Cell Biol 12(5):429–435
20. Akin O, Mullins RD (2008) Capping protein increases the rate of actin-based motility by promoting filament nucleation by the Arp2/3 complex. Cell 133(5):841–851
21. Bear JE, Gertler FB (2009) Ena/VASP: towards resolving a pointed controversy at the barbed end. J Cell Sci 122(Pt 12):1947–1953
22. Breitsprecher D et al (2011) Molecular mechanism of Ena/VASP-mediated actin-filament elongation. EMBO J 30(3):456–467
23. Chesarone MA, DuPage AG, Goode BL (2010) Unleashing formins to remodel the actin and microtubule cytoskeletons. Nat Rev Mol Cell Biol 11(1):62–74
24. Keren K (2011) Cell motility: the integrating role of the plasma membrane. Eur Biophys J 40(9):1013–1027
25. Houk AR et al (2012) Membrane tension maintains cell polarity by confining signals to the leading edge during neutrophil migration. Cell 148(1–2):175–188
26. Gauthier NC, Masters TA, Sheetz MP (2012) Mechanical feedback between membrane tension and dynamics. Trends Cell Biol 22(10):527–535
27. Masters TA et al (2013) Plasma membrane tension orchestrates membrane trafficking, cytoskeletal remodeling, and biochemical signaling during phagocytosis. Proc Natl Acad Sci U S A 110(29):11875–11880
28. Gittes F et al (1993) Flexural rigidity of microtubules and actin filaments measured from thermal fluctuations in shape. J Cell Biol 120(4):923–934
29. Diz-Munoz A, Fletcher DA, Weiner OD (2013) Use the force: membrane tension as an organizer of cell shape and motility. Trends Cell Biol 23(2):47–53
30. Bo L, Waugh RE (1989) Determination of bilayer membrane bending stiffness by tether formation from giant, thin-walled vesicles. Biophys J 55(3):509–517
31. Pontes B et al (2011) Cell cytoskeleton and tether extraction. Biophys J 101(1):43–52
32. Gauthier NC et al (2011) Temporary increase in plasma membrane tension coordinates the activation of exocytosis and contraction during cell spreading. Proc Natl Acad Sci U S A 108(35):14467–14472
33. Keren K et al (2008) Mechanism of shape determination in motile cells. Nature 453(7194):475–480
34. Risca VI et al (2012) Actin filament curvature biases branching direction. Proc Natl Acad Sci U S A 109(8):2913–2918
35. Sheetz MP, Dai J (1996) Modulation of membrane dynamics and cell motility by membrane tension. Trends Cell Biol 6(3):85–89

36. Morris CE, Homann U (2001) Cell surface area regulation and membrane tension. J Membr Biol 179(2):79–102
37. Apodaca G (2002) Modulation of membrane traffic by mechanical stimuli. Am J Physiol Renal Physiol 282(2):F179–F190
38. Ofer N, Mogilner A, Keren K (2011) Actin disassembly clock determines shape and speed of lamellipodial fragments. Proc Natl Acad Sci U S A 108(51):20394–20399
39. Cai Y et al (2010) Cytoskeletal coherence requires myosin-IIA contractility. J Cell Sci 123(Pt 3):413–423
40. Hissa B et al (2013) Membrane cholesterol removal changes mechanical properties of cells and induces secretion of a specific pool of lysosomes. PLoS One 8(12):e82988
41. Lieber AD et al (2013) Membrane tension in rapidly moving cells is determined by cytoskeletal forces. Curr Biol 23(15):1409–1417
42. Abu Shah E, Keren K (2013) Mechanical forces and feedbacks in cell motility. Curr Opin Cell Biol 25(5):550–557
43. Pontes B et al (2013) Membrane elastic properties and cell function. PLoS One 8(7):e67708
44. Janmey PA, McCulloch CA (2007) Cell mechanics: integrating cell responses to mechanical stimuli. Annu Rev Biomed Eng 9:1–34
45. Faria J et al (2006) Interactive properties of human glioblastoma cells with brain neurons in culture and neuronal modulation of glial laminin organization. Differentiation 74(9–10):562–572
46. Zlotnik A, Yoshie O (2000) Chemokines: a new classification system and their role in immunity. Immunity 12(2):121–127
47. Lazennec G, Richmond A (2010) Chemokines and chemokine receptors: new insights into cancer-related inflammation. Trends Mol Med 16(3):133–144
48. Vindrieux D, Escobar P, Lazennec G (2009) Emerging roles of chemokines in prostate cancer. Endocr Relat Cancer 16(3):663–673
49. Ali S, Lazennec G (2007) Chemokines: novel targets for breast cancer metastasis. Cancer Metastasis Rev 26(3–4):401–420
50. Mantovani A, Bonecchi R, Locati M (2006) Tuning inflammation and immunity by chemokine sequestration: decoys and more. Nat Rev Immunol 6(12):907–918
51. Binder NB et al (2009) Estrogen-dependent and C-C chemokine receptor-2-dependent pathways determine osteoclast behavior in osteoporosis. Nat Med 15(4):417–424
52. Chavey C et al (2009) CXC ligand 5 is an adipose-tissue derived factor that links obesity to insulin resistance. Cell Metab 9(4):339–349
53. Kohlmeier JE et al (2008) The chemokine receptor CCR5 plays a key role in the early memory CD8+ T cell response to respiratory virus infections. Immunity 29(1):101–113
54. Yoder A et al (2008) HIV envelope-CXCR4 signaling activates cofilin to overcome cortical actin restriction in resting CD4 T cells. Cell 134(5):782–792
55. Tiwari S et al (2009) Targeting of the GTPase Irgm1 to the phagosomal membrane via PtdIns(3,4)P(2) and PtdIns(3,4,5)P(3) promotes immunity to mycobacteria. Nat Immunol 10(8):907–917
56. Molon B et al (2005) T cell costimulation by chemokine receptors. Nat Immunol 6(5):465–471
57. Pitchford SC et al (2009) Differential mobilization of subsets of progenitor cells from the bone marrow. Cell Stem Cell 4(1):62–72
58. Reboldi A et al (2009) C-C chemokine receptor 6-regulated entry of TH-17 cells into the CNS through the choroid plexus is required for the initiation of EAE. Nat Immunol 10(5):514–523
59. Balkwill F, Charles KA, Mantovani A (2005) Smoldering and polarized inflammation in the initiation and promotion of malignant disease. Cancer Cell 7(3):211–217
60. Mantovani A (2009) Cancer: inflaming metastasis. Nature 457(7225):36–37
61. Mentlein R, Hattermann K, Held-Feindt J (2013) Migration, metastasis, and more: the role of chemokines in the proliferation, spreading, and metastasis of tumors. In: Resende RR, Ulrich H (eds) Trends in the cell proliferation and cancer research. Springer, Netherlands, pp 339–358

62. Moser B, Loetscher P (2001) Lymphocyte traffic control by chemokines. Nat Immunol 2(2):123–128
63. von Andrian UH, Mackay CR (2000) T-cell function and migration. Two sides of the same coin. N Engl J Med 343(14):1020–1034
64. Sallusto F, Mackay CR, Lanzavecchia A (2000) The role of chemokine receptors in primary, effector, and memory immune responses. Annu Rev Immunol 18:593–620
65. Sanchez-Madrid F, del Pozo MA (1999) Leukocyte polarization in cell migration and immune interactions. EMBO J 18(3):501–511
66. Ridley AJ (2001) Rho GTPases and cell migration. J Cell Sci 114(Pt 15):2713–2722
67. Thelen M (2001) Dancing to the tune of chemokines. Nat Immunol 2(2):129–134
68. Mellado M et al (1998) The chemokine monocyte chemotactic protein 1 triggers Janus kinase 2 activation and tyrosine phosphorylation of the CCR2B receptor. J Immunol 161(2):805–813
69. Mellado M et al (2001) Chemokine signaling and functional responses: the role of receptor dimerization and TK pathway activation. Annu Rev Immunol 19:397–421
70. Amano M, Fukata Y, Kaibuchi K (2000) Regulation and functions of Rho-associated kinase. Exp Cell Res 261(1):44–51
71. Lee JH et al (2004) Roles of p-ERM and Rho-ROCK signaling in lymphocyte polarity and uropod formation. J Cell Biol 167(2):327–337
72. Sakata D et al (2007) Impaired T lymphocyte trafficking in mice deficient in an actin-nucleating protein, mDia1. J Exp Med 204(9):2031–2038
73. Gunn MD et al (1998) A chemokine expressed in lymphoid high endothelial venules promotes the adhesion and chemotaxis of naive T lymphocytes. Proc Natl Acad Sci U S A 95(1):258–263
74. Stein JV et al (2000) The CC chemokine thymus-derived chemotactic agent 4 (TCA-4, secondary lymphoid tissue chemokine, 6Ckine, exodus-2) triggers lymphocyte function-associated antigen 1-mediated arrest of rolling T lymphocytes in peripheral lymph node high endothelial venules. J Exp Med 191(1):61–76
75. Campbell JJ et al (1998) Chemokines and the arrest of lymphocytes rolling under flow conditions. Science 279(5349):381–384
76. Okada T et al (2002) Chemokine requirements for B cell entry to lymph nodes and Peyer's patches. J Exp Med 196(1):65–75
77. Bardi G, Niggli V, Loetscher P (2003) Rho kinase is required for CCR7-mediated polarization and chemotaxis of T lymphocytes. FEBS Lett 542(1–3):79–83
78. Boehme SA et al (1999) Activation of mitogen-activated protein kinase regulates eotaxin-induced eosinophil migration. J Immunol 163(3):1611–1618
79. Bonacchi A et al (2001) Signal transduction by the chemokine receptor CXCR3: activation of Ras/ERK, Src, and phosphatidylinositol 3-kinase/Akt controls cell migration and proliferation in human vascular pericytes. J Biol Chem 276(13):9945–9954
80. Kampen GT et al (2000) Eotaxin induces degranulation and chemotaxis of eosinophils through the activation of ERK2 and p38 mitogen-activated protein kinases. Blood 95(6):1911–1917
81. Berlin-Rufenach C et al (1999) Lymphocyte migration in lymphocyte function-associated antigen (LFA)-1-deficient mice. J Exp Med 189(9):1467–1478
82. Laudanna C, Alon R (2006) Right on the spot. Chemokine triggering of integrin-mediated arrest of rolling leukocytes. Thromb Haemost 95(1):5–11
83. Warnock RA et al (1998) Molecular mechanisms of lymphocyte homing to peripheral lymph nodes. J Exp Med 187(2):205–216
84. Giagulli C et al (2004) RhoA and zeta PKC control distinct modalities of LFA-1 activation by chemokines: critical role of LFA-1 affinity triggering in lymphocyte in vivo homing. Immunity 20(1):25–35
85. Laudanna C, Campbell JJ, Butcher EC (1996) Role of Rho in chemoattractant-activated leukocyte adhesion through integrins. Science 271(5251):981–983
86. Pasvolsky R et al (2008) RhoA is involved in LFA-1 extension triggered by CXCL12 but not in a novel outside-in LFA-1 activation facilitated by CXCL9. J Immunol 180(5):2815–2823

87. Cunningham HD et al (2010) Expression of the C-C chemokine receptor 7 mediates metastasis of breast cancer to the lymph nodes in mice. Transl Oncol 3(6):354–361
88. Mburu YK et al (2012) Chemokine receptor 7 (CCR7) gene expression is regulated by NF-kappaB and activator protein 1 (AP1) in metastatic squamous cell carcinoma of head and neck (SCCHN). J Biol Chem 287(5):3581–3590
89. Jung JI et al (2015) High-fat diet-induced obesity increases lymphangiogenesis and lymph node metastasis in the B16F10 melanoma allograft model: roles of adipocytes and M2-macrophages. Int J Cancer 136(2):258–270
90. Shields JD et al (2007) Chemokine-mediated migration of melanoma cells towards lymphatics–a mechanism contributing to metastasis. Oncogene 26(21):2997–3005
91. Shields JD et al (2007) Autologous chemotaxis as a mechanism of tumor cell homing to lymphatics via interstitial flow and autocrine CCR7 signaling. Cancer Cell 11(6):526–538
92. Nagasawa T, Kikutani H, Kishimoto T (1994) Molecular cloning and structure of a pre-B-cell growth-stimulating factor. Proc Natl Acad Sci U S A 91(6):2305–2309
93. Muller A et al (2001) Involvement of chemokine receptors in breast cancer metastasis. Nature 410(6824):50–56
94. Liu Y et al (2010) Correlation effect of EGFR and CXCR4 and CCR7 chemokine receptors in predicting breast cancer metastasis and prognosis. J Exp Clin Cancer Res 29:16
95. Tashiro K et al (1993) Signal sequence trap: a cloning strategy for secreted proteins and type I membrane proteins. Science 261(5121):600–603
96. Loetscher M et al (1994) Cloning of a human seven-transmembrane domain receptor, LESTR, that is highly expressed in leukocytes. J Biol Chem 269(1):232–237
97. Feng Y et al (1996) HIV-1 entry cofactor: functional cDNA cloning of a seven-transmembrane, G protein-coupled receptor. Science 272(5263):872–877
98. Bleul CC et al (1996) The lymphocyte chemoattractant SDF-1 is a ligand for LESTR/fusin and blocks HIV-1 entry. Nature 382(6594):829–833
99. Aiuti A et al (1997) The chemokine SDF-1 is a chemoattractant for human CD34+ hematopoietic progenitor cells and provides a new mechanism to explain the mobilization of CD34+ progenitors to peripheral blood. J Exp Med 185(1):111–120
100. Reiss K et al (2002) Stromal cell-derived factor 1 is secreted by meningeal cells and acts as chemotactic factor on neuronal stem cells of the cerebellar external granular layer. Neuroscience 115(1):295–305
101. Ma Q et al (1998) Impaired B-lymphopoiesis, myelopoiesis, and derailed cerebellar neuron migration in CXCR4- and SDF-1-deficient mice. Proc Natl Acad Sci U S A 95(16):9448–9453
102. Balabanian K et al (2005) The chemokine SDF-1/CXCL12 binds to and signals through the orphan receptor RDC1 in T lymphocytes. J Biol Chem 280(42):35760–35766
103. Odemis V et al (2012) The presumed atypical chemokine receptor CXCR7 signals through G(i/o) proteins in primary rodent astrocytes and human glioma cells. Glia 60(3):372–381
104. Li H et al (2013) Association between Galphai2 and ELMO1/Dock180 connects chemokine signalling with Rac activation and metastasis. Nat Commun 4:1706
105. Takabayashi T et al (2009) Lipopolysaccharides increase the amount of CXCR4, and modulate the morphology and invasive activity of oral cancer cells in a CXCL12-dependent manner. Oral Oncol 45(11):968–973
106. Dong X et al (2005) P-Rex1 is a primary Rac2 guanine nucleotide exchange factor in mouse neutrophils. Curr Biol 15(20):1874–1879
107. Sun CX, Magalhaes MA, Glogauer M (2007) Rac1 and Rac2 differentially regulate actin free barbed end formation downstream of the fMLP receptor. J Cell Biol 179(2):239–245
108. Gao H et al (2009) Activation of signal transducers and activators of transcription 3 and focal adhesion kinase by stromal cell-derived factor 1 is required for migration of human mesenchymal stem cells in response to tumor cell-conditioned medium. Stem Cells 27(4):857–865
109. Robledo MM et al (2001) Expression of functional chemokine receptors CXCR3 and CXCR4 on human melanoma cells. J Biol Chem 276(48):45098–45105

110. Lima FR et al (2012) Glioblastoma: therapeutic challenges, what lies ahead. Biochim Biophys Acta 1826(2):338–349
111. Singh A, Settleman J (2010) EMT, cancer stem cells and drug resistance: an emerging axis of evil in the war on cancer. Oncogene 29(34):4741–4751
112. Thomson JA et al (1998) Embryonic stem cell lines derived from human blastocysts. Science 282(5391):1145–1147
113. Guilak F et al (2009) Control of stem cell fate by physical interactions with the extracellular matrix. Cell Stem Cell 5(1):17–26
114. Lutolf MP, Gilbert PM, Blau HM (2009) Designing materials to direct stem-cell fate. Nature 462(7272):433–441
115. Gage FH (2000) Mammalian neural stem cells. Science 287(5457):1433–1438
116. Mamsen LS et al (2012) The migration and loss of human primordial germ stem cells from the hind gut epithelium towards the gonadal ridge. Int J Dev Biol 56(10–12):771–778
117. Spradling A, Drummond-Barbosa D, Kai T (2001) Stem cells find their niche. Nature 414(6859):98–104
118. Treiser MD et al (2010) Cytoskeleton-based forecasting of stem cell lineage fates. Proc Natl Acad Sci U S A 107(2):610–615
119. Engler AJ et al (2006) Matrix elasticity directs stem cell lineage specification. Cell 126(4):677–689
120. Kinney MA, Saeed R, McDevitt TC (2014) Mesenchymal morphogenesis of embryonic stem cells dynamically modulates the biophysical microtissue niche. Sci Rep 4:4290
121. McBeath R et al (2004) Cell shape, cytoskeletal tension, and RhoA regulate stem cell lineage commitment. Dev Cell 6(4):483–495
122. Collart-Dutilleul PY et al (2014) Initial stem cell adhesion on porous silicon surface: molecular architecture of actin cytoskeleton and filopodial growth. Nanoscale Res Lett 9(1):564
123. Galkin VE, Orlova A, Egelman EH (2012) Actin filaments as tension sensors. Curr Biol 22(3):R96–R101
124. Fonseca AV, Corbeil D (2011) The hematopoietic stem cell polarization and migration: a dynamic link between RhoA signaling pathway, microtubule network and ganglioside-based membrane microdomains. Commun Integr Biol 4(2):201–204
125. Fonseca AV et al (2010) Polarization and migration of hematopoietic stem and progenitor cells rely on the RhoA/ROCK I pathway and an active reorganization of the microtubule network. J Biol Chem 285(41):31661–31671
126. Vertelov G et al (2013) High targeted migration of human mesenchymal stem cells grown in hypoxia is associated with enhanced activation of RhoA. Stem Cell Res Ther 4(1):5
127. Titushkin I, Cho M (2007) Modulation of cellular mechanics during osteogenic differentiation of human mesenchymal stem cells. Biophys J 93(10):3693–3702
128. Hyder CL et al (2011) Insights into intermediate filament regulation from development to ageing. J Cell Sci 124(Pt 9):1363–1372
129. Lendahl U, Zimmerman LB, McKay RD (1990) CNS stem cells express a new class of intermediate filament protein. Cell 60(4):585–595
130. Kleeberger W et al (2007) Roles for the stem cell associated intermediate filament Nestin in prostate cancer migration and metastasis. Cancer Res 67(19):9199–9206
131. Suzuki S et al (2010) The neural stem/progenitor cell marker nestin is expressed in proliferative endothelial cells, but not in mature vasculature. J Histochem Cytochem 58(8):721–730
132. Wiese C et al (2004) Nestin expression–a property of multi-lineage progenitor cells? Cell Mol Life Sci 61(19–20):2510–2522
133. Mellodew K et al (2004) Nestin expression is lost in a neural stem cell line through a mechanism involving the proteasome and Notch signalling. Brain Res Dev Brain Res 151(1–2):13–23
134. Hitoshi S et al (2004) Primitive neural stem cells from the mammalian epiblast differentiate to definitive neural stem cells under the control of Notch signaling. Genes Dev 18(15):1806–1811
135. Sachewsky N et al (2014) Primitive neural stem cells in the adult mammalian brain give rise to GFAP-expressing neural stem cells. Stem Cell Rep 2(6):810–824

136. Jinka R et al (2012) Alterations in cell-extracellular matrix interactions during progression of cancers. Int J Cell Biol 2012:219196
137. Moura-Neto V et al (2014) Glioblastomas and the special role of adhesion molecules in their invasion. In: Sedo A, Mentlein R (eds) Glioma cell biology. Springer, Vienna, pp 293–315
138. Vasioukhin V et al (2001) Hyperproliferation and defects in epithelial polarity upon conditional ablation of alpha-catenin in skin. Cell 104(4):605–617
139. Rampazzo E et al (2013) Wnt activation promotes neuronal differentiation of glioblastoma. Cell Death Dis 4:e500
140. Wang Q et al (2010) A novel role for Wnt/Ca^{2+} signaling in actin cytoskeleton remodeling and cell motility in prostate cancer. PLoS One 5(5):e10456
141. Jing R et al (2007) Synemin is expressed in reactive astrocytes in neurotrauma and interacts differentially with vimentin and GFAP intermediate filament networks. J Cell Sci 120(Pt 7):1267–1277
142. Sancho-Tello M et al (1995) Developmental pattern of GFAP and vimentin gene expression in rat brain and in radial glial cultures. Glia 15(2):157–166
143. Eliasson C et al (1999) Intermediate filament protein partnership in astrocytes. J Biol Chem 274(34):23996–24006
144. Eng LF, Ghirnikar RS, Lee YL (2000) Glial fibrillary acidic protein: GFAP-thirty-one years (1969–2000). Neurochem Res 25(9–10):1439–1451
145. Eng LF et al (1971) An acidic protein isolated from fibrous astrocytes. Brain Res 28(2):351–354
146. Kato H et al (1990) Immunocytochemical characterization of supporting cells in the enteric nervous system in Hirschsprung's disease. J Pediatr Surg 25(5):514–519
147. Hainfellner JA et al (2001) Fibroblasts can express glial fibrillary acidic protein (GFAP) in vivo. J Neuropathol Exp Neurol 60(5):449–461
148. Viale G et al (1991) Glial fibrillary acidic protein immunoreactivity in normal and diseased human breast. Virchows Arch A Pathol Anat Histopathol 418(4):339–348
149. Riol H et al (1997) Detection of the peripheral nervous system (PNS)-type glial fibrillary acidic protein (GFAP) and its mRNA in human lymphocytes. J Neurosci Res 48(1):53–62
150. Carotti S et al (2008) Glial fibrillary acidic protein as an early marker of hepatic stellate cell activation in chronic and posttransplant recurrent hepatitis C. Liver Transpl 14(6):806–814
151. Sofroniew MV, Vinters HV (2010) Astrocytes: biology and pathology. Acta Neuropathol 119(1):7–35
152. Brenner M (2014) Role of GFAP in CNS injuries. Neurosci Lett 565:7–13
153. Reeves SA et al (1989) Molecular cloning and primary structure of human glial fibrillary acidic protein. Proc Natl Acad Sci U S A 86(13):5178–5182
154. Nakatani Y, Brenner M, Freese E (1990) An RNA polymerase II promoter containing sequences upstream and downstream from the RNA startpoint that direct initiation of transcription from the same site. Proc Natl Acad Sci U S A 87(11):4289–4293
155. Middeldorp J, Hol EM (2011) GFAP in health and disease. Prog Neurobiol 93(3):421–443
156. Namihira M et al (2009) Committed neuronal precursors confer astrocytic potential on residual neural precursor cells. Dev Cell 16(2):245–255
157. Kanski R et al (2014) Histone acetylation in astrocytes suppresses GFAP and stimulates a reorganization of the intermediate filament network. J Cell Sci 127(Pt 20):4368–4380
158. Zhou Q et al (2011) Histone deacetylase inhibitors SAHA and sodium butyrate block G1-to-S cell cycle progression in neurosphere formation by adult subventricular cells. BMC Neurosci 12:50
159. de Sampaio e Spohr TC et al (2002) Neuro-glia interaction effects on GFAP gene: a novel role for transforming growth factor-beta1. Eur J Neurosci 16(11):2059–2069
160. Gomes FC et al (1999) Neurons induce GFAP gene promoter of cultured astrocytes from transgenic mice. Glia 26(2):97–108
161. Romao LF et al (2008) Glutamate activates GFAP gene promoter from cultured astrocytes through TGF-beta1 pathways. J Neurochem 106(2):746–756
162. Sousa Vde O et al (2004) Glial fibrillary acidic protein gene promoter is differently modulated by transforming growth factor-beta 1 in astrocytes from distinct brain regions. Eur J Neurosci 19(7):1721–1730

163. Rutka JT et al (1997) Role of glial filaments in cells and tumors of glial origin: a review. J Neurosurg 87(3):420–430
164. Blechingberg J et al (2007) Identification and characterization of GFAPkappa, a novel glial fibrillary acidic protein isoform. Glia 55(5):497–507
165. Condorelli DF et al (1999) Structural features of the rat GFAP gene and identification of a novel alternative transcript. J Neurosci Res 56(3):219–228
166. Hol EM et al (2003) Neuronal expression of GFAP in patients with Alzheimer pathology and identification of novel GFAP splice forms. Mol Psychiatry 8(9):786–796
167. Zelenika D et al (1995) A novel glial fibrillary acidic protein mRNA lacking exon 1. Brain Res Mol Brain Res 30(2):251–258
168. Nielsen AL et al (2002) A new splice variant of glial fibrillary acidic protein, GFAP epsilon, interacts with the presenilin proteins. J Biol Chem 277(33):29983–29991
169. Roelofs RF et al (2005) Adult human subventricular, subgranular, and subpial zones contain astrocytes with a specialized intermediate filament cytoskeleton. Glia 52(4):289–300
170. Feinstein DL, Weinmaster GA, Milner RJ (1992) Isolation of cDNA clones encoding rat glial fibrillary acidic protein: expression in astrocytes and in Schwann cells. J Neurosci Res 32(1):1–14
171. Galea E, Dupouey P, Feinstein DL (1995) Glial fibrillary acidic protein mRNA isotypes: expression in vitro and in vivo. J Neurosci Res 41(4):452–461
172. Kamphuis W et al (2012) GFAP isoforms in adult mouse brain with a focus on neurogenic astrocytes and reactive astrogliosis in mouse models of Alzheimer disease. PLoS One 7(8):e42823
173. Thomsen R et al (2013) Alternative mRNA splicing from the glial fibrillary acidic protein (GFAP) gene generates isoforms with distinct subcellular mRNA localization patterns in astrocytes. PLoS One 8(8):e72110
174. Singh R et al (2003) Genetic polymorphism and sequence evolution of an alternatively spliced exon of the glial fibrillary acidic protein gene, GFAP. Genomics 82(2):185–193
175. Nielsen AL, Jorgensen AL (2004) Self-assembly of the cytoskeletal glial fibrillary acidic protein is inhibited by an isoform-specific C terminus. J Biol Chem 279(40):41537–41545
176. Perng MD et al (2008) Glial fibrillary acidic protein filaments can tolerate the incorporation of assembly-compromised GFAP-delta, but with consequences for filament organization and alphaB-crystallin association. Mol Biol Cell 19(10):4521–4533
177. Leonard BW et al (2009) Subventricular zone neural progenitors from rapid brain autopsies of elderly subjects with and without neurodegenerative disease. J Comp Neurol 515(3): 269–294
178. Middeldorp J et al (2010) GFAPdelta in radial glia and subventricular zone progenitors in the developing human cortex. Development 137(2):313–321
179. van den Berge SA et al (2010) Longterm quiescent cells in the aged human subventricular neurogenic system specifically express GFAP-delta. Aging Cell 9(3):313–326
180. Andreiuolo F et al (2009) GFAPdelta immunostaining improves visualization of normal and pathologic astrocytic heterogeneity. Neuropathology 29(1):31–39
181. Gotz M, Hartfuss E, Malatesta P (2002) Radial glial cells as neuronal precursors: a new perspective on the correlation of morphology and lineage restriction in the developing cerebral cortex of mice. Brain Res Bull 57(6):777–788
182. Kriegstein A, Alvarez-Buylla A (2009) The glial nature of embryonic and adult neural stem cells. Annu Rev Neurosci 32:149–184
183. deAzevedo LC et al (2003) Cortical radial glial cells in human fetuses: depth-correlated transformation into astrocytes. J Neurobiol 55(3):288–298
184. Sarnat HB (1992) Regional differentiation of the human fetal ependyma: immunocytochemical markers. J Neuropathol Exp Neurol 51(1):58–75
185. Lazarides E (1982) Intermediate filaments: a chemically heterogeneous, developmentally regulated class of proteins. Annu Rev Biochem 51:219–250
186. Yamada T et al (1992) Vimentin immunoreactivity in normal and pathological human brain tissue. Acta Neuropathol 84(2):157–162

187. Pekny M et al (1998) Impaired induction of blood-brain barrier properties in aortic endothelial cells by astrocytes from GFAP-deficient mice. Glia 22(4):390–400
188. Gimenez YRM et al (2000) Comparative anatomy of the cerebellar cortex in mice lacking vimentin, GFAP, and both vimentin and GFAP. Glia 31(1):69–83
189. Messing A et al (1998) Fatal encephalopathy with astrocyte inclusions in GFAP transgenic mice. Am J Pathol 152(2):391–398
190. Rutka JT et al (1994) Effects of antisense glial fibrillary acidic protein complementary DNA on the growth, invasion, and adhesion of human astrocytoma cells. Cancer Res 54(12):3267–3272
191. Emirandetti A et al (2006) Astrocyte reactivity influences the number of presynaptic terminals apposed to spinal motoneurons after axotomy. Brain Res 1095(1):35–42
192. Tyzack GE et al (2014) Astrocyte response to motor neuron injury promotes structural synaptic plasticity via STAT3-regulated TSP-1 expression. Nat Commun 5:4294
193. Elobeid A et al (2000) Effects of inducible glial fibrillary acidic protein on glioma cell motility and proliferation. J Neurosci Res 60(2):245–256
194. Yoshida T et al (2007) The functional alteration of mutant GFAP depends on the location of the domain: morphological and functional studies using astrocytoma-derived cells. J Hum Genet 52(4):362–369
195. Berg A et al (2013) Axonal regeneration after sciatic nerve lesion is delayed but complete in GFAP- and vimentin-deficient mice. PLoS One 8(11):e79395
196. Potokar M et al (2010) Intermediate filaments attenuate stimulation-dependent mobility of endosomes/lysosomes in astrocytes. Glia 58(10):1208–1219
197. Sullivan SM et al (2007) Cytoskeletal anchoring of GLAST determines susceptibility to brain damage: an identified role for GFAP. J Biol Chem 282(40):29414–29423
198. Gomi H et al (1995) Mice devoid of the glial fibrillary acidic protein develop normally and are susceptible to scrapie prions. Neuron 14(1):29–41
199. McCall MA et al (1996) Targeted deletion in astrocyte intermediate filament (GFAP) alters neuronal physiology. Proc Natl Acad Sci U S A 93(13):6361–6366
200. Pekny M et al (1995) Mice lacking glial fibrillary acidic protein display astrocytes devoid of intermediate filaments but develop and reproduce normally. EMBO J 14(8):1590–1598
201. Liedtke W et al (1996) GFAP is necessary for the integrity of CNS white matter architecture and long-term maintenance of myelination. Neuron 17(4):607–615
202. Pekny M et al (1999) Abnormal reaction to central nervous system injury in mice lacking glial fibrillary acidic protein and vimentin. J Cell Biol 145(3):503–514
203. Lepekhin EA et al (2001) Intermediate filaments regulate astrocyte motility. J Neurochem 79(3):617–625
204. Nawashiro H et al (1998) Mice lacking GFAP are hypersensitive to traumatic cerebrospinal injury. Neuroreport 9(8):1691–1696
205. Otani N et al (2006) Enhanced hippocampal neurodegeneration after traumatic or kainate excitotoxicity in GFAP-null mice. J Clin Neurosci 13(9):934–938
206. Eng LF, Ghirnikar RS (1994) GFAP and astrogliosis. Brain Pathol 4(3):229–237
207. Alonso G, Privat A (1993) Reactive astrocytes involved in the formation of lesional scars differ in the mediobasal hypothalamus and in other forebrain regions. J Neurosci Res 34(5):523–538
208. Faulkner JR et al (2004) Reactive astrocytes protect tissue and preserve function after spinal cord injury. J Neurosci 24(9):2143–2155
209. Hatten ME et al (1991) Astroglia in CNS injury. Glia 4(2):233–243
210. Wanner IB et al (2013) Glial scar borders are formed by newly proliferated, elongated astrocytes that interact to corral inflammatory and fibrotic cells via STAT3-dependent mechanisms after spinal cord injury. J Neurosci 33(31):12870–12886
211. Anderson MA, Ao Y, Sofroniew MV (2014) Heterogeneity of reactive astrocytes. Neurosci Lett 565:23–29
212. Liem RK, Messing A (2009) Dysfunctions of neuronal and glial intermediate filaments in disease. J Clin Invest 119(7):1814–1824
213. Hardy J (2006) A hundred years of Alzheimer's disease research. Neuron 52(1):3–13

214. Muramori F, Kobayashi K, Nakamura I (1998) A quantitative study of neurofibrillary tangles, senile plaques and astrocytes in the hippocampal subdivisions and entorhinal cortex in Alzheimer's disease, normal controls and non-Alzheimer neuropsychiatric diseases. Psychiatry Clin Neurosci 52(6):593–599
215. Simpson JE et al (2010) Astrocyte phenotype in relation to Alzheimer-type pathology in the ageing brain. Neurobiol Aging 31(4):578–590
216. Kashon ML et al (2004) Associations of cortical astrogliosis with cognitive performance and dementia status. J Alzheimers Dis 6(6):595–604, discussion 673–681
217. Porchet R et al (2003) Analysis of glial acidic fibrillary protein in the human entorhinal cortex during aging and in Alzheimer's disease. Proteomics 3(8):1476–1485
218. Sloane JA et al (2000) Astrocytic hypertrophy and altered GFAP degradation with age in subcortical white matter of the rhesus monkey. Brain Res 862(1–2):1–10
219. Messing A, Brenner M (2003) GFAP: functional implications gleaned from studies of genetically engineered mice. Glia 43(1):87–90
220. Alexander WS (1949) Progressive fibrinoid degeneration of fibrillary astrocytes associated with mental retardation in a hydrocephalic infant. Brain 72(3):373–381, 3 pl
221. Brenner M et al (2001) Mutations in GFAP, encoding glial fibrillary acidic protein, are associated with Alexander disease. Nat Genet 27(1):117–120
222. Flint D et al (2012) Splice site, frameshift, and chimeric GFAP mutations in Alexander disease. Hum Mutat 33(7):1141–1148
223. Brenner M, Messing A (2015) A new mutation in GFAP widens the spectrum of Alexander disease. Eur J Hum Genet 23(1):1–2
224. Tascos NA, Parr J, Gonatas NK (1982) Immunocytochemical study of the glial fibrillary acidic protein in human neoplasms of the central nervous system. Hum Pathol 13(5):454–458
225. de Armond SJ, Eng LF, Rubinstein LJ (1980) The application of glial fibrillary acidic (GFA) protein immunohistochemistry in neurooncology. A progress report. Pathol Res Pract 168(4):374–394
226. Louis DN et al (2007) The 2007 WHO classification of tumours of the central nervous system. Acta Neuropathol 114(2):97–109
227. Herpers MJ, Budka H (1984) Glial fibrillary acidic protein (GFAP) in oligodendroglial tumors: gliofibrillary oligodendroglioma and transitional oligoastrocytoma as subtypes of oligodendroglioma. Acta Neuropathol 64(4):265–272
228. Cruz-Sanchez FF et al (1988) An immunohistological study of 66 ependymomas. Histopathology 13(4):443–454
229. Rutka JT, Smith SL (1993) Transfection of human astrocytoma cells with glial fibrillary acidic protein complementary DNA: analysis of expression, proliferation, and tumorigenicity. Cancer Res 53(15):3624–3631
230. Toda M et al (1999) Suppression of glial tumor growth by expression of glial fibrillary acidic protein. Neurochem Res 24(2):339–343
231. Weinstein DE, Shelanski ML, Liem RK (1991) Suppression by antisense mRNA demonstrates a requirement for the glial fibrillary acidic protein in the formation of stable astrocytic processes in response to neurons. J Cell Biol 112(6):1205–1213
232. Wilhelmsson U et al (2003) Loss of GFAP expression in high-grade astrocytomas does not contribute to tumor development or progression. Oncogene 22(22):3407–3411
233. Gallego Perez-Larraya J et al (2014) Diagnostic and prognostic value of preoperative combined GFAP, IGFBP-2, and YKL-40 plasma levels in patients with glioblastoma. Cancer 120(24):3972–3980
234. Coelho-Aguiar JM et al (2015) The enteric glia: identity and functions. Glia 63(6):921–935
235. Aube AC et al (2006) Changes in enteric neurone phenotype and intestinal functions in a transgenic mouse model of enteric glia disruption. Gut 55(5):630–637
236. Bush TG et al (1998) Fulminant jejuno-ileitis following ablation of enteric glia in adult transgenic mice. Cell 93(2):189–201
237. Cornet A et al (2001) Enterocolitis induced by autoimmune targeting of enteric glial cells: a possible mechanism in Crohn's disease? Proc Natl Acad Sci U S A 98(23):13306–13311

238. Almond S et al (2007) Characterisation and transplantation of enteric nervous system progenitor cells. Gut 56(4):489–496
239. Bondurand N et al (2003) Neuron and glia generating progenitors of the mammalian enteric nervous system isolated from foetal and postnatal gut cultures. Development 130(25): 6387–6400
240. Hetz S et al (2014) In vivo transplantation of neurosphere-like bodies derived from the human postnatal and adult enteric nervous system: a pilot study. PLoS One 9(4):e93605
241. Joseph NM et al (2011) Enteric glia are multipotent in culture but primarily form glia in the adult rodent gut. J Clin Invest 121(9):3398–3411
242. Kruger GM et al (2002) Neural crest stem cells persist in the adult gut but undergo changes in self-renewal, neuronal subtype potential, and factor responsiveness. Neuron 35(4):657–669
243. Laranjeira C et al (2011) Glial cells in the mouse enteric nervous system can undergo neurogenesis in response to injury. J Clin Invest 121(9):3412–3424
244. Liu MT et al (2009) 5-HT4 receptor-mediated neuroprotection and neurogenesis in the enteric nervous system of adult mice. J Neurosci 29(31):9683–9699
245. Rauch U et al (2006) Isolation and cultivation of neuronal precursor cells from the developing human enteric nervous system as a tool for cell therapy in dysganglionosis. Int J Colorectal Dis 21(6):554–559
246. von Boyen GB et al (2011) Distribution of enteric glia and GDNF during gut inflammation. BMC Gastroenterol 11:3
247. Clairembault T et al (2014) Enteric GFAP expression and phosphorylation in Parkinson's disease. J Neurochem 130(6):805–815
248. Devos D et al (2013) Colonic inflammation in Parkinson's disease. Neurobiol Dis 50:42–48
249. Lin T et al (2003) The role of the cytoskeleton in capacitative calcium entry in myenteric glia. Neurogastroenterol Motil 15(3):277–287
250. Jessen KR, Mirsky R (1985) Glial fibrillary acidic polypeptides in peripheral glia. Molecular weight, heterogeneity and distribution. J Neuroimmunol 8(4–6):377–393
251. Jessen KR, Mirsky R (2005) The origin and development of glial cells in peripheral nerves. Nat Rev Neurosci 6(9):671–682
252. Bianchini D et al (1992) GFAP expression of human Schwann cells in tissue culture. Brain Res 570(1–2):209–217
253. Triolo D et al (2006) Loss of glial fibrillary acidic protein (GFAP) impairs Schwann cell proliferation and delays nerve regeneration after damage. J Cell Sci 119(Pt 19):3981–3993
254. Obremski VJ, Hall AM, Fernandez-Valle C (1998) Merlin, the neurofibromatosis type 2 gene product, and beta1 integrin associate in isolated and differentiating Schwann cells. J Neurobiol 37(4):487–501
255. Rezniczek GA et al (1998) Linking integrin alpha6beta4-based cell adhesion to the intermediate filament cytoskeleton: direct interaction between the beta4 subunit and plectin at multiple molecular sites. J Cell Biol 141(1):209–225
256. Bernier G et al (1998) Dystonin is an essential component of the Schwann cell cytoskeleton at the time of myelination. Development 125(11):2135–2148

Chapter 5
Centrosome–Microtubule Interactions in Health, Disease, and Disorders

Heide Schatten and Qing-Yuan Sun

Introduction

The centrosome, known as the primary microtubule organizing center (MTOC) of eukaryotic cells, is a multifunctional organelle of ca. 1 μm in diameter without membrane boundaries. It plays essential roles in all cell cycle stages and undergoes remodeling during the transition from interphase to mitosis (G2/M) to become division-competent and assemble a functional mitotic apparatus that is able to separate chromosomes accurately to the dividing daughter cells (reviewed in [1]). Centrosome dysfunctions have been implicated in numerous diseases and disorders including Alstrom syndrome, Bardet–Biedl syndrome, retinitis pigmentosa, deafness, obesity, diabetes mellitus, lissencephaly, mental and behavioral disorders, malformations, juvenile autosomal recessive Parkinson disease, rheumatoid arthritis, polycystic kidney disease, ciliopathies, cancer and others (reviewed by Badano et al. [2], Nigg and Raff [3], Gerdes et al. [4], Bettencourt-Dias et al. [5], Schatten [6]). The centrosome is a versatile organelle that functions as major hub for signal transduction molecules and orchestrates signal transduction through its microtubule network. The centrosome further is involved in Golgi functions, in directing organelle movement such as mitochondria and cargo such as enzyme-containing vesicles and therefore, centrosome dysfunctions have wide-ranging impacts on cellular metabolism and cellular health. As detailed in Chap. 10 by Li and Hu, the centrosome also plays a role in primary cilia formation, thereby interacting with the

H. Schatten (✉)
Department of Veterinary Pathobiology, University of Missouri,
1600 E Rollins Street, Columbia, MO 65211, USA
e-mail: SchattenH@missouri.edu

Q.-Y. Sun
State Key Laboratory of Reproductive Biology, Institute of Zoology,
Chinese Academy of Sciences, Beijing 100080, China

H. Schatten (ed.), *The Cytoskeleton in Health and Disease*,
DOI 10.1007/978-1-4939-2904-7_5

extra- and intracellular microenvironment. Detailed data are available for primary cilia dysfunctions in polycystic kidney disease in which the proteins polycystin-1, polycystin-2, polaris, and cystin are affected [7].

The structure and functional components of the centrosome and the centrosome duplication cycle have been reviewed previously [1, 8, 9–12, 187] and will only be briefly addressed here while this chapter is focused on centrosome–microtubule interactions and their dysfunctions in disease and disorders. As indicated above, numerous diseases have directly been linked to centrosome dysfunctions (reviewed by Badano et al. [2], Nigg and Raff [3], Gerdes et al. [4], Bettencourt-Dias et al. [5], Schatten [6]) and others have been linked to centrosome-related signaling dysfunctions. Some of the major diseases and disorders linked to centrosomes will be addressed here while others will be mentioned and references will be provided. In this chapter we will primarily focus on (1) Centrosome–microtubule dynamics; (2) Centrosome dysfunctions in aging cells; (3) Centrosome dysfunctions in cancer cells.

Section 1: Centrosome–Microtubule Dynamics

As indicated above, the centrosome is a multifunctional cellular organelle without membrane boundaries that relies on precise regulation to nucleate microtubules for specific and varying functions throughout the cell cycle. Several regulators of centrosome functions have been determined in which kinases and phosphatases as well as post-translational modifications play major roles. Such modifications allow centrosomes to nucleate and organize different microtubule formations. Whereas the interphase centrosome nucleates the large interphase aster that radiates from the nucleus-associated interphase centrosome the mitotic centrosomes organize polar asters and the central mitotic spindle that is critical for chromosome alignment and separation. As will be detailed below significant remodeling of the centrosome takes place at the transition from G2/M to modify the interphase centrosome into division-competent mitotic centrosomes, a process termed centrosome maturation. During this process centrosomes become enabled to nucleate an increased number of mitotic microtubules as a result of increased amounts of γ-tubulin that associate with the mitotic centrosome.

As mentioned above, the centrosome holds key roles in cell cycle regulation and in several other complex cellular functions that directly or indirectly affect cell cycle progression and cellular metabolism. Numerous kinases are involved in the transition from G2 to mitosis [9–12]. These kinases primarily play a role in centrosome protein phosphorylation while dephosphorylation takes place when cells exit mitosis.

In a typical mammalian somatic cell, a perpendicularly oriented centriole pair is embedded in a centrosomal matrix (Fig. 5.1) that oftentimes is also referred to as pericentriolar material (PCM). The centrosomal matrix (PCM) is composed of a lattice of coiled-coil proteins and contains specific centrosomal proteins including the γ-tubulin ring complexes (γ-TuRCs), pericentrin, centrin, and calcium-sensitive

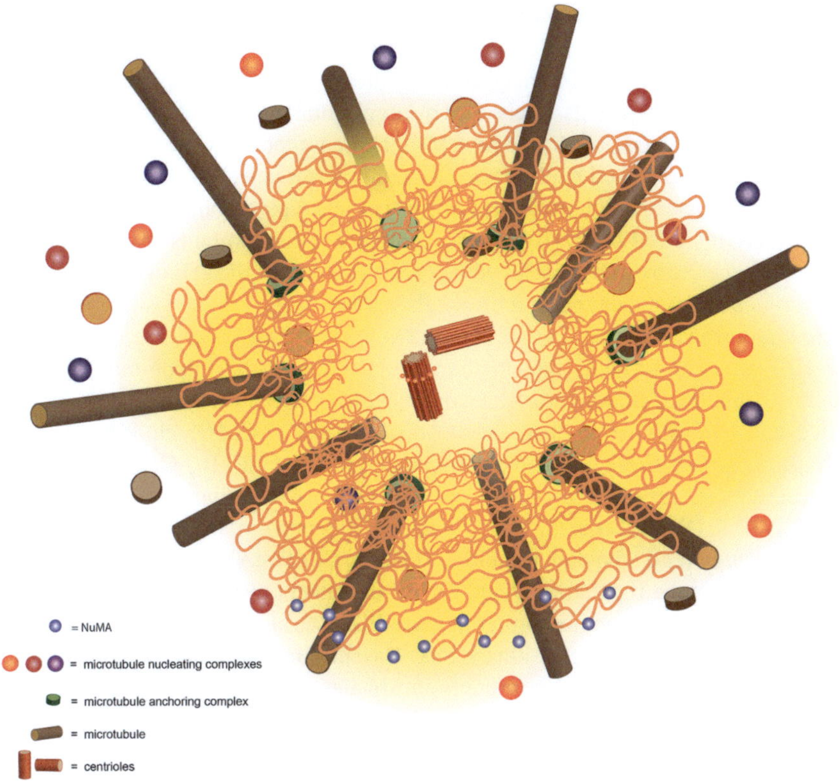

Fig. 5.1 A typical centrosome in somatic cells is composed of centrosomal material, also referred to as pericentriolar material (PCM), surrounding two perpendicularly oriented centrioles. Embedded in this matrix are centrosomal proteins such as γ-tubulin and the γ-tubulin ring complexes that nucleate microtubules along with numerous associated proteins as described in the text. Other components within the centrosomal matrix include the microtubule anchoring complexes. Modified from Schatten and Sun [13]

components ([14]; reviewed in [1]). The composition of proteins embedded within the centrosomal matrix varies in different cell cycle stages while centrioles, on the other hand, do not significantly change in their molecular composition throughout the cell cycle. However, centriole duplication takes place in a precisely regulated process to assure that centrioles are duplicated only once during the cell cycle. In mammalian cells, centrioles display a typical composition of nine outer triplet microtubules forming a barrel-shaped small tube without containing central microtubules. Centrioles duplicate through a semiconservative duplication process during which a younger (daughter) centriole forms perpendicular to the older (mother) centriole. The mother centriole is distinguished from the daughter centriole by appendages as a characteristic feature which indicates structural and functional differences. In mammalian cells, centrioles are involved in the assembly of specific centrosome proteins and in the duplication of centrosomal material [15].

The exact composition of centrosomal material is not yet clear, as it is difficult to generate precise data with our currently available methods but it has been reported that as many as 500 proteins may be associated with the interphase centrosome structure [16] although it is likely that a large number of these proteins may be centrosome-associated proteins or proteins that are temporarily localized to centrosomes during specific cell cycle stages. A more conservative estimate may include about 60–100 centrosomal proteins to be present in a typical somatic cell interphase centrosome (reviewed in [17]). Of these, centrosome core proteins are tightly associated with the centrosome matrix while others are part of the cell cycle-dependent structural centrosomal changes in most cell systems.

So far, purified centrosomes have been analyzed by mass spectrometry, revealing several classes of proteins that include *structural proteins* (alpha-tubulin, beta-tubulin, γ-tubulin, γ-tubulin complex components 1–6, centrin 2 and 3, AKAP450, pericentrin/kendrin, ninein, pericentriolar material 1 (PCM1), ch-TOG protein, C-Nap1, Cep250, Cep2, centriole-associated protein CEP110, Cep1, centriolin, centrosomal P4.1-associated protein (CPAP), CLIP-associating proteins CLASP1 and CLASP 2, ODF2, cenexin, Lis1, Nudel, EB1, centractin, myomegalin); *regulatory molecules* (cell division protein 2 (Cdc2), Cdk1, cAMP-dependent protein kinase type II-alpha regulatory chain, cAMP-dependent protein kinase-alpha catalytic subunit, serine/threonine protein kinase Plk1, serine/threonine protein kinase Nek2, serine/threonine protein kinase Sak, Casein kinase I, delta and epsilon isoforms, protein phosphatase 2A, protein phosphatase 1 alpha isoform, 14-3-3 proteins, epsilon and gamma isoforms); *motor and motor-related proteins* (dynein heavy chain, dynein intermediate chain, dynein light chain, dynactin 1, p150 Glued, dynactin 2, p50, dynactin 3); and *heat shock proteins* (heat shock protein Hsp90, TCP subunits, and heat shock protein Hsp73).

Gamma-tubulin is an essential centrosomal protein that is primarily found in the centrosome matrix structure, but it can also serve as nucleating sites in areas other than the centrosome and it can be associated with microtubule walls. It also is important for microtubule nucleation from the Golgi and it associates with the plasma membrane during cellular polarization. The major nucleating complex for microtubules from centrosomes is the ca. 2.2-MDa γ-TuRC that is present in all cells studied so far [18] and consists of 12 or 14 γ-tubulin molecules. Hundreds of γ-TuRCs may be embedded in the centrosome matrix, dependent on the requirements for microtubule nucleation which differs in different cell cycle stages. The large γ-TuRC contains 5-7 small complexes, the γTuSCs (around 280 kDa) that each comprises two molecules of γ-tubulin and one molecule each of GCP (γ-tubulin complex protein) 2 and 3 [19]. The γ-TuSCs associate with the γ-TuRC by condensation and association with proteins GCP4, GCP5, GCP6, and GCP-WD/NEDD1. Two functional genes for γ-tubulin have been identified for mammalian cells (TUBG1, TUBG2). Posttranslational modifications of γ-tubulin have been reported which includes phosphorylation and mono-ubiquitination. Complexes of γ-tubulin with protein tyrosine kinases of the Src family, polo-like kinase, microtubule affinity-regulating kinase 4 (MARK 4) or phosphoinositide 3-kinase have also been documented. Various proteins and protein complexes are needed to anchor the γ-TuRC to the centrosome matrix including the large coiled-coil A-kinase anchoring proteins [19–22, 23–26], Cep135 [27], ninein, augmin, Cep192/SPD2, AKAP450/CG-NAP, pericentrin/kendrin, and CDK5RAP2/centrosomin. Dynactin plays a major role in microtubule anchorage at centrosomes as well as at non-centrosomal anchorage sites. It is preferentially localized to the mother centriole [28–30]. Several of the microtubule minus-end binding proteins including those of the γ-TuRC are accumulated at the proximal ends of centrioles. Tubulin polyglutamylation of the centriole walls modulates interaction between tubulin- and microtubule-associated proteins. Much interest has been focused on how the activity of the γ-TuRC is regulated and significant new data have been produced that identified new components that interact with or regulate the γ-TuRC such as NME7 [31] and TACC3 Protein [32]. More specific information on γ-TuRC regulation is available in several recent original and review papers [31–34]. New methods including subdiffraction-resolution fluorescence microscopy combined with site-specific antibody analyses have generated new insights into high-order spatial organization of the centrosome structure [35–38].

As mentioned above, the number of microtubules nucleated by the γ-TuRC varies in different cell cycle stages. In interphase, fewer but longer microtubules are nucleated, while in mitosis, γ-TuRCs become increased to nucleate more microtubules. Mitotic microtubules are shorter, larger in number, and highly dynamic. The regulation of microtubule nucleation includes cell cycle-specific proteins that participate in the centrosome maturation process and include the small GTPase Ran, Aurora A kinase, polo-like kinases, and others (reviewed in [6]).

Pericentrin forms a ca. 3-MDa complex with γ-tubulin and depends on dynein for assembly onto centrosomes [39]. Pericentrin is part of the pericentrin/AKAP450 centrosomal targeting (PACT) domain [40] involved in recruiting γ-tubulin to centrosomes [20, 21]. Mutation of the pericentrin gene results in loss of recruitment of several other centrosomal proteins which becomes manifested in diseases or disorders (reviewed in [2]). **Centrins** are primarily associated with centrioles, but are also components of centrosomes with an essential role in centrosome duplication ([15, 41–43]; reviewed in [44]).

NuMA (**Nu**clear **M**itotic **A**pparatus protein): One of the critical proteins enabling mitotic and meiotic centrosome functions is NuMA. NuMA becomes a significant centrosome-associated protein during mitosis as well as meiosis when it forms an insoluble crescent around the centrosome area facing toward the central mitotic or meiotic spindle. NuMA is important for cross-linking spindle microtubules and for tethering microtubules precisely into the bipolar mitotic or meiotic apparatus [45]. NuMA is a multifunctional protein (reviewed in [46, 47] that serves as nuclear matrix protein in the nucleus during interphase but it is not associated with interphase centrosomes. NuMA becomes dispersed into the cytoplasm during nuclear envelope breakdown and associates with microtubules for translocation to the centrosomal area. Cdk1/cyclin B-dependent phosphorylation is important for this process to take place [48]; the association with microtubules and translocation to centrosomes requires dynein–dynactin-mediation; failure of this mediation will result in meiotic and mitotic dysfunctions [185].

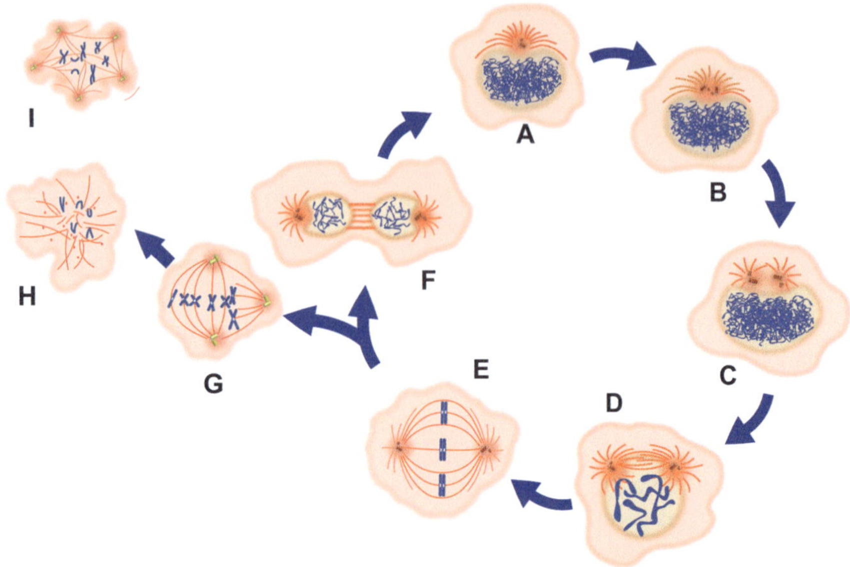

Fig. 5.2 Centrosome cycle in somatic cells. (*A*) The interphase centrosome is located next to the nucleus and organizes the interphase microtubule array. (*B*) Centriole and centrosome duplication is synchronized with DNA duplication during the S phase. (*C*) Separation of the duplicated centrosome complex to the forming mitotic poles takes place during early prophase. (*D*) Establishment of the bipolar mitotic apparatus with centrosomes localized at the center of the mitotic poles. (*E*) Compacted metaphase centrosomes organize a well-focused metaphase spindle. (*F*) Reorganized centrosomal material closely associated with the reforming daughter nuclei of dividing cells. (*G–I*) Centrosomal abnormalities resulting in tripolar (*G*) or multipolar (*H–I*) spindle formations. Modified from Schatten [6]

Regulation of the Centrosome Complex The regulation of the centrosome complex is critically important for accurate functions throughout the cell cycle and for coordination with several cell cycle events. It includes accurate duplication of the centrioles as well as the centrosomal material for precise coordination of centrosome and chromosome dynamics. In mitosis as well as meiosis, centrosomes organize microtubules that attach to kinetochores as part of a complex molecular machinery that assures accurate separation and equal distribution of chromosomes to the dividing cells. Centrosomes and chromosomes undergo coordinated duplication cycles in parallel pathways to assure accurate cell divisions. Cell cycle abnormalities occur when these events become misregulated and desynchronized, as will be addressed in sections 2 and 3.

Centrosome Duplication It is important that centrosomes are duplicated only once during the cell cycle in a process that is well synchronized with the DNA cycle to assure precise chromosome partitioning to the dividing daughter cells (Fig. 5.2). Centrosome duplication and DNA replication both require hyperphosphorylation of the retinoblastoma (RB) protein and activation of Cdk2. The program of duplication and block for reduplication has been reviewed (reviewed in [1, 9–12]) and it has been shown that centrosome duplication starts to take place shortly before the G2 cell cycle stage in a precisely orchestrated duplication program. The process begins with disorientation of the pair of centrioles, centriole duplication, centriole disjunction, and separation of sister centrioles (reviewed in [49, 50]). While this process has been well studied in somatic cells and quite detailed knowledge has been accumulated on centriole duplication, we still do not yet fully understand duplication of centrosomal material and centrosome dynamics throughout the cell cycle. Centrosome duplication and separation is frequently correlated to the better understood centriole dynamics and centriole duplication cycle. Our knowledge about centrosome duplication primarily relates to phosphorylation (reviewed in [9–12]). We know that the initiation of centrosome duplication is under cytoplasmic control and driven by cyclin-dependent kinase 2 (Cdk2) complexed with cyclin E or cyclin A that rises during the late G1 stage (reviewed in [51]). The initiation of centrosome duplication further requires calcium/calmodulin-dependent kinase II (CaMKII) [52] that localizes to spindle poles [53] and phosphorylates centrosome proteins in vitro [54]. Specific centrosome proteins depend on multiple signaling to allow the transition from G2 to mitosis; the G2/M cell cycle transition is critical for centrosome phosphorylation to become division competent and allow cell proliferation (reviewed in [1]). Polo-like kinases are required for multiple stages of mitotic progression and they are further involved in centrosome separation [55–57].

The block to centriole reduplication may involve ubiquitin-mediated proteolysis of centrosomal proteins, as several components of the SCF (Skp1/cullin/Fbox; ubiquitin ligase complex) proteolysis pathway as well as the 26S proteasome are localized to centrosomes in human cells [9–11, 58–61]. The centrosomal protein centrin plays a role in centrosome disjunction at the G2/prophase transition, and it has also been shown that the Nek2 kinase is involved in this process ([62]; reviewed in [63]). Centrosome separation is in part driven by plus- and minus end-directed microtubule motor proteins which takes place in interphase around the nucleus and during mitosis at the mitotic poles.

As indicated above, the G2/M transition represents a critical phase during which centrosomes mature to become division-competent in a process requiring Cdk1/cyclin B as well as Cdk1/cyclin A (reviewed in [64]). Cdk1/cyclin B activation is detected in centrosomes during prophase [65]. During the G2/M phase several important centrosomal proteins are acquired including polo-like kinase 1 (Plk1) [66], and NuMA [67]. On the other hand, interphase centrosome proteins such as C-Nap1 [68] or Nlp [69] are removed. γ-TuRC recruitment to the centrosome increases prior to mitosis to nucleate increased numbers of microtubules for spindle formation. Polo and Aurora A kinases [70] and cdc2/cyclin B kinase [65] are precisely regulated during G2/M and during mitosis. Misregulation is associated with diseases and disorders as will be addressed in sections 2 and 3. Exit from mitosis requires degradation of cyclin B which is achieved by proteins binding to the anaphase-promoting complex/cyclosome (APC/C); the activated APC/CCdc20 degrades cyclin B and securin to allow cell cycle exit from mitosis [71–74]. Microtubule motor proteins are important for the assembly of a functional mitotic centrosome which includes shuttling of the proteins pericentrin, centrin, ninein, and NuMA along microtubules toward the centrosomal area.

As of now we do not yet fully understand how centrosome proteins are associated with the centrosomal matrix but it is possible that the centrosomal matrix may play a role in clustering of centrosome proteins during mitosis ([75]; reviewed in [6]). Invertebrate models have provided some information on the centrosomal matrix structure which revealed that this material may contain fibrous cytoskeleton-like material. In the Spisula model, material left after high-salt extraction of centrosomal proteins contained a fibrous component [76, 77], perhaps composed of filament-like material that may have intermediate filament-like characteristics. Our own studies on sea urchin centrosomes revealed filament-like proteins that could be detected with Ah6, a monoclonal antibody to intermediate filament-like proteins [78]. In addition to these proteins microtubule-associated proteins (MAPs) in the centrosomal matrix may be involved in centrosome clustering. Other proteins that may play a role in functions of the centrosomal matrix includes HSET (kinesin-related protein), as HSET depletion blocks centrosome clustering and promotes multipolar divisions [79].

The interactions of microtubules with centrosomes and their interdependent regulation is complex and we do not yet fully understand the factors that affect centrosomes and microtubules and their influence on each other which includes signal transduction pathways that play a role in communication between microtubules and centrosomes. Signal transductions that have been explored allowed close insights into this important relationship in somatic as well as in reproductive cells and determined critically important molecular mechanisms that effectively regulate centrosome and microtubule dynamics, their interactions with each other, and communication with other cellular components. Many of the signaling molecules that have been identified to colocalize with centrosomes use centrosomes as their central docking station for cellular communications in which microtubules provide the distribution network. As centrosomes are able to modify the microtubule network they can facilitate changes to accommodate cell cycle-dependent signaling requirements. Key signaling molecules that associate with centrosomes include the mitogen activated protein kinase (MAPK) that plays a critical role in centrosome and microtubule regulation during meiosis, mitosis, and cell division ([80]; reviewed in [1]). Polo and Aurora A kinases [70], and cdc2/cyclin B kinase [65] are other important meiotic and mitotic cell cycle regulators that are concentrated at the centrosome (reviewed in [1]). These kinases are critical for centrosome regulation, as abnormalities have been linked to centrosome pathologies affecting cellular health. In mouse meiotic maturation p38α MAPK, a centrosome-associated protein, has been shown to regulate spindle assembly, spindle length and chromosome segregation [81]. Depletion of p38α affects other proteins and results in spindle pole defects and aneuploidy.

Posttranslational Tubulin Modifications The effects of posttranslational modifications on microtubule dynamics and functions have been reviewed [82–84] and their importance for modulations of cytoskeletal and cellular functions have been highlighted. PTMs are chemical modifications that regulate microtubule activity and interactions with other cellular molecules and components by creating marks on microtubules for specialized interactions and function-specific activities [83]. PTM-related microtubule dysfunctions have been linked to diseases such as cancer, diabetes, heart diseases, neurodegenerative diseases, and various others (reviewed in [82, 84]).

Several tubulin PTMs have been studied in somatic and reproductive cells and include acetylation and detyrosination/tyrosination. These PTMs have been implicated in microtubule stability, in networking with other proteins, and in targeted associations with the microtubule motor proteins dynein and/or kinesin that accommodate transport of cargo molecules along microtubules. The association of dynein with microtubules is important for transport of centrosome proteins such as pericentrin and centrosome-associated proteins such as NuMA to remodel centrosomes throughout the cell cycle (reviewed in [1, 82]). These PTMs are also important for localized function-specific stabilization of microtubules [83, 85, 86] which is an important aspect for stabilizing labile microtubules at the minus ends facing the centrosomal area.

Acetylation of microtubules is a reversible PTM that is mediated by acetyltransferase [87] while tubulin deacetylation is mediated by two known enzymes, the histone deacetylase 6 (HDAC6) [88, 89] to reverse acetylation of Lys40, and sirtuin 2 (SIRT2) [90]. HDAC6 functions can be inhibited by trichostatin A (TSA) [91, 92] and tubastatin A [93]. The reversible detyrosination/tyrosination cycle plays a role in the recruitment of microtubule-binding proteins and specific molecular motors. Detyrosination is achieved by the removal of a Tyr functional group from tubulin, whereas tyrosination is achieved by re-addition of Tyr that returns tubulin to its nascent state [94].

The Primary Cilia-Centrosome Cycle Primary cilia are tightly correlated with the centrosome cycle and the regulatory relationship between primary cilia functions and the cell cycle has clearly been established (reviewed in [95, 186]). The primary cilium–centriole–centrosome cycle starts during G1 when the distal end of the mother centriole becomes associated with a membrane vesicle (reviewed by Pan and Snell [95]) followed by growth into an axoneme that is surrounded by the enlarging ciliary vesicle that fuses with the plasma membrane. During the subsequent S phase centrioles duplicate and lengthen; the mature length of the primary cilium is achieved during G2. As detailed in Chap. 10 by Li and Hu, the primary cilium is a non-motile single cilium composed of 9 outer microtubule doublets with no central microtubule pair (“9 + 0”); it is covered by a specialized receptor-rich plasma membrane. The primary cilium protrudes from almost all cells in our body [96–98] and it communicates signals from the external cellular environment to the cell body. The molecular aspects of this cilium and functions are addressed in Chap. 10 by Li and Hu of this book. Numerous diseases have been associated with primary cilia dysfunctions that have been well elaborated for polycystic kidney syndrome and include many diseases and disorders grouped under the umbrella of ciliopathies. Several studies have revealed details of signal transduction cascades between primary cilia and the centrosome that are essential for accurate cell cycle progression [1, 97–103]. These are reviewed in Chap. 10 by Li and Hu.

Section 2: Centrosome Dysfunctions in Aging Cells

By now centrosome dysfunctions in diseases and disorders have been studied by a number of different investigators (reviewed by Badano et al. [2], Nigg and Raff [3], Gerdes et al. [4], Bettencourt-Dias et al. [5], Schatten [6]) but centrosome dysfunctions in aging cells are still largely unexplored although it has clearly been shown that centrosomes and the cytoskeleton are affected by aging in reproductive cells [13], in stem cells [104, 105], and in various cultured cells [106, 107]. The best examples of centrosome and cytoskeletal changes during aging come from oocyte cells in which aging occurs rapidly when fertilization does not take place within a certain time frame (reviewed in [108–110]). Furthermore, oocyte aging is well known to be associated with aneuploidy primarily in women past 35 years of age which results in low fertilization rates, affecting many women in advanced ages who desire to have children and seek treatment in in vitro fertilization (IVF) clinics.

Oocytes of most mammalian species are arrested at the meiosis II (MII) stage and remain arrested until fertilization takes place which typically occurs soon after ovulation. The window for optimal fertilization varies in different species. If fertilization does not take place within a certain time frame unfertilized oocytes remain in the oviduct or in culture and will undergo time-dependent quality changes in a process termed oocyte aging. In unfertilized MII stage oocytes several aging effects take place in humans and all mammalian animal models studied so far. In humans, oocyte aging occurs within a 24 h time frame and includes changes in calcium metabolism, decrease in enzyme activity, decrease in essential organelle functions such as mitochondria resulting in decrease in ATP production, destabilization of the microfilament and microtubule cytoskeleton, loss of centrosome integrity associated with loss of spindle integrity, and loss of cohesion between sister chromatids resulting in chromosome mis-segregation and aneuploidy. Loss of spindle integrity includes dispersion of centrosomal proteins including γ-tubulin and NuMA from the centrosome core structure. The mechanisms underlying these changes and consequences or causes are not well understood but several pieces of information have been accumulated in recent years. It has been shown in human oocytes that microtubules become destabilized in aging oocytes which prevents accurate motor-driven transport of centrosomal proteins along microtubules to form and maintain a functional centrosome [111, 185]. Because of the high demand to overcome the effects of oocyte aging to allow fertilization, embryo development and the birth of healthy babies this area of research has progressed more rapidly compared to research on aging in somatic cells and studies on oocyte aging have become important for procedures that can be applied in IVF clinics to overcome the aging effects. One of the important goals to overcome the effects of aging is to understand and target the mechanisms underlying loss of spindle integrity and prevent aneuploidy [13].

In fresh oocytes, centrosome dynamics are precisely regulated and include active maintenance of centrosomes until fertilization takes place. As mentioned above, the MII spindle in mammalian reproductive systems is highly dynamic and becomes unstable if regulation by a complex set of kinases and other regulatory proteins fails or if fertilization does not take place within a certain period of time (reviewed in [108, 110, 112, 113]). The kinases involved in the process of meiosis includes CDK1/cyclin B and other kinases such as PKA, AKT, MAPK, Aurora A, CaMKII, the phosphatases CDC5, CDK14s and others that participate in the meiotic process.

As mentioned above, aneuploidy resulting from oocyte aging is associated with disintegration of centrosomal proteins such as NuMA and γ-tubulin from the centrosomal core structure of the MII spindle which coincides with the formation of numerous small centrosomal aggregates in the ooplasm. It is not yet clear whether the centrosomal core structure itself is affected by oocyte aging or whether microtubule destabilization results in loss of microtubule motor activities in which transport of centrosomal proteins such as NuMA and pericentrin is impaired, thereby affecting proper maintenance of a functional centrosome. It is possible that loss of microtubule stability is the result of loss of microtubule acetylation [85] that prevents accurate association of the motor proteins dynein and kinesins with microtubules (reviewed in [82]). Other factors that affect microtubule and centrosome dynamics and stability include signal transductions that may be misguided in aging oocytes (reviewed in [1, 13]), resulting in an inability to maintain spindle integrity.

Dispersion of centrosomal components including NuMA and γ-tubulin from the meiotic spindle poles in aging oocytes has been reported for several non-rodent mammalian animal models as well as for humans ([114, 115]; reviewed in [108, 111, 116]). Oocyte aging affects meiotic regulation which not only can lead to aneuploidy affecting fertilization but also to subsequent cell and developmental abnormalities resulting in abortion, disease, or developmental defects (reviewed in [108, 110, 112, 113, 116]). Diseases that may be associated with meiotic aneuploidies to become manifested later in life include childhood cancer with characteristic centrosome dysfunctions that may originate from aberrant oocyte centrosomes.

Centrosomes are primarily located at the two meiotic spindle poles in mature MII oocytes. It is important to emphasize that, while these centrosomes serve functions as known for mitotic centrosomes, there are important differences between meiotic centrosomes and mitotic centrosomes. Whereas a typical somatic cell centrosome contains a pair of centrioles (reviewed in [1]) centrally located within the centrosome, centrosomes of the oocyte's meiotic spindle are acentriolar. These acentriolar centrosomes contain centrosomal proteins that are known for mitotic centrosomes embedded in the centrosome matrix including γ-tubulin, pericentrin, centrin, and the nuclear mitotic apparatus protein, NuMA, but the central centriole pair is absent (reviewed in [1]). As the quantity and specific composition of centrosomal proteins may differ in different cell systems it can also be different in oocytes of different animal species.

As mentioned above, in aging oocytes one of the most noticeable features of aging is the deterioration of the meiotic spindle with disintegration of centrosomal proteins from the centrosomal core structure that is correlated with loss of microtubule stability (reviewed in [13]). As we do not yet fully understand the underlying reasons for centrosome and microtubule instability in aging MII spindles it is possible that the absence of centrioles plays a role in the rapid loss of spindle integrity. In somatic cells centrioles are intimately involved in centrosome dynamics, potentially contributing to stability that may be absent in acentriolar centrosomes.

In a variety of other cell systems centrosome dysfunctions have also been implicated in aging; for example, supernumerary centrosome abnormalities have been observed in senescing cells [106] which may be the result of cell cycle abnormalities in which signal transductions are altered. As indicated above, centrosomes undergo remodeling at the transition from G2/M which is the stage during which many of the cell cycle regulators are downregulated in aging cells and affect centrosome functions. We know that specific kinases and phosphatases are important for cytoskeletal regulation in meiotic and mitotic spindles (reviewed in [82, 117]); in aging cells, studies have shown that centrosomes have lower activity in centrosome-associated protein kinases [118, 119] which includes Plk that serves an important role in centrosome functions [107]. Furthermore, mis-orientation of centrosomes has been shown in aging stem cells which results in decrease in cell divisions [104], contributing to declines in spermatogenesis during aging.

Microtubule associated proteins (MAPs) and posttranslational modifications play important roles in the regulation of microtubule stability. Such stabilizing factors may be lost in aging cells contributing to microtubule instability. Other factors for microtubule stabilization include interactions with centrosomes and cell membranes (reviewed in [82]). Individual microtubules if not regulated by specific kinases and phosphatases undergo individual microtubule aging which becomes important when considering aging of microtubules in the meiotic spindle of aging oocytes (reviewed in [82, 117]). Changes in posttranslational tubulin modifications have been associated with loss of microtubule stability with strong effects on microtubule functions in several cell systems. PTM changes resulting in microtubule instability during aging have also been determined in neuronal cells in which microtubule PTM dysfunctions have been linked to diseases such as Alzheimer's and Parkinson's.

Our previous experiments have shown that MII spindle microtubules are acetylated at the microtubule-centrosome interface area [85] which was correlated with microtubule stabilization. In aging oocytes loss of spindle integrity is first seen in this specific area which suggests that loss of acetylation plays a role in spindle instability and is in part a causative factor for aneuploidy. Restoring stability of these microtubules may be possible by treatment of aging oocytes with deacetylation inhibitors such as inhibitors of HDA6 that have been used in cancer cell therapy [120, 121]. This treatment strategy may allow microtubule stabilization in the meiotic spindle to prevent de-acetylation-related microtubule instability.

Our previous studies have also shown that restoring signal transduction that may have been impacted during aging will halt or reverse the aging process (reviewed in [13]). For example, the use of caffeine to delay or prevent oocyte aging has been proposed by Kikuchi et al. [122, 123] who found that controlling the activity of MPF can reverse oocyte aging (reviewed in [108]). These investigators also showed that both MPF and MAPK are critical for maintaining oocyte spindle integrity, and that MPF and MAPK activities gradually decrease during oocyte aging [124–128]. Continuous treatment with 10 mM caffeine could prevent the decline in MPF and

MAPK activity in aging bovine oocytes ([129]; reviewed in [108]) and continuous treatment with caffeine could restore spindle integrity in aging porcine oocytes [112] with chromosomes, microtubules and the centrosomal proteins γ-tubulin and NuMA displaying normal appearance as known for fresh oocytes (reviewed in [13]).

Section 3: Centrosome Dysfunctions in Cancer Cells

The incidence of cancer development and progression increases with aging and multiple factors play a role in changes that take place during this process. It is well known that the mutation rate increases in cells that have reached replicative senescence. Misguided signal transductions have been implicated in cancer development and progression which may in part be similar to those seen during physiological aging. One hallmark characteristic common to all cancer cells is abnormal cell division which is strongly related to abnormal centrosome functions. It has been recognized as early as 1914 [130] that centrosomes are affected in cancer cells which adversely affects chromosome segregation and cell division, two hallmark features that are clearly seen in cancer cells and tissue ([131–133]; reviewed in [1]).

Theodor Boveri's classic remarkable discoveries and brilliant data interpretations ([130]; translated into English in [134]) ignited a new era in modern cancer research when the significance of centrosomes was again recognized and when it was possible to apply new technologies to explore changes in centrosomes as important aspect for abnormal cancer cell proliferation (reviewed in [1]). While well-regulated centrosomes form the bipolar mitotic spindle in mitosis cancer cell centrosomes frequently form multipolar spindles with consequences for aneuploidy and genomic instability. Abnormal multipolar mitoses resulting from supernumerary centrosomes have been well documented in numerous cancers and have been well analyzed in HPV-associated lesions in cervical cancers in which centrosome abnormalities are already detected in early stages of tumor development [135]. While environmental insult and in some cases viral infections are known to be cancer-inducing factors we do not yet fully understand the underlying mechanisms leading to centrosomal abnormalities which in many cases may have multifactorial components. As cause and effect studies are still being explored in attempts to determine when centrosomes become dysfunctional during the cascade of events leading to the observed abnormalities (reviewed in [136]), it has clearly been determined that cancer cell centrosomes are significantly different from noncancer cell centrosomes which includes their state of abnormal phosphorylation which had first been recognized when examining breast adenocarcinoma cells [137]. Increased phosphorylation of cancer cell centrosomes [137] is associated with increases in microtubule nucleation and abnormal organization leading to aberrant attachment of microtubules to chromosomes (reviewed in [6]). Increased γ-tubulin expression has been shown in breast carcinoma cells [138, 139] and gliomas.

Centrosomes have been recognized as major microtubule organizing centers (MTOCs) and their role as major hub for signal transduction molecules that participate in signal transduction cascades through the microtubule network. The effects of misguided signal transduction on the formation of abnormal centrosomes and in turn the effects of abnormal centrosomes on signal transduction has been recognized. This interdependence may be part of a vicious cycle in which regulation of centrosomes and regulation of signal transduction by centrosomes are both affected in cancer cells.

Several proteins in cancer cell centrosomes are overexpressed or display abnormalities and include the centrosome-associated protein NuMA. The NuMA region on chromosome 11q13 has been associated with breast cancer susceptibility [140]. NuMA misregulation may further contribute to abnormalities in cancer cells, as NuMA requires specific signaling for its centrosome-associated functions in which signaling of cyclin B is important (reviewed in [46]); cyclin B signals may be affected in cancer cells. For NuMA's relocation into the nucleus following exit from mitosis it has to become dissociated from the mitotic spindle poles. This process requires cdc1/cyclin B activity [141]. Destruction of cyclin B allows exit from mitosis. If NuMA does not become relocated properly into the nucleus for its interphase functions NuMA can form cytoplasmic focal points in the cytoplasm that organize abnormal microtubule asters [141] that can contribute to mitotic abnormalities. Such abnormal mitotic formations in cancer cells have also been observed to originate from basal bodies of dislodged primary cilia that become located in the cancer cell cytoplasm and form supernumerary nucleation sites for microtubule-based asters that participate in aberrant chromosome segregation [131–133, 142].

Dysfunctions of structural, regulatory, and motor-related proteins may be other contributors to centrosome abnormalities. In cancer cells, the cell cycle coordination between chromosomes and centrosomes are lost, resulting in asynchronous misregulation and misguided duplication cycles. Multipolar centrosomes that separate chromosomes unequally to the dividing cells will contribute to imbalanced distribution of chromosomes resulting in cells that may lack tumor suppressor genes while others may have increases in tumor promoting genes.

Dissociation of centrosome cycles from DNA cycles have been reported after irradiation. Loss of proteins that are important for critical cell functions may lead to loss of cell polarity and increased cancer cell formations. For example, loss of Plk3 function will result in loss of cell shape [57], affecting microtubule functions underneath the plasma membrane resulting in loss of cellular polarity in cancer cells and tissue. Cascades of events may follow and may include loss of signal transduction processes, imbalanced or disrupted transport of centrosome proteins resulting in additional centrosomal pathologies related to centrosome and microtubule functions with consequences for failures in organelle and vesicle distribution. Secondary pathologies can develop as a result of disruption in transport leading to interconnected communication failures for which cause and effects are difficult to establish. Signal transduction events have been well studied in breast cancer in which numerous centrosomal abnormalities have been reported (reviewed by Kais and Parvin [58, 187], Fisk [9], Fukasawa [11], Korzeniewski and Duensing [143], Saladino et al. [144], Yan and Chng [145].

The causes for changes leading to aberrant centrosomes in cancer cells are known in some cases but not in others. We know that environmental stress can result in the formation of aggresomes, aggregates that are thought to be the result of misfolded proteins [146–150]; they are oftentimes located in close proximity to centrosomes and some of the aggresomes contain γ-tubulin. Aggresomes are associated with disease or disorders including Parkinson's and dementia [151].

Other factors that play a role in cancer initiation and progression include epigenetic modifications which includes aberrant hypermethylation that has been implicated in inactivation of checkpoint genes that may influence cell cycle-dependent centrosome abnormalities. Such abnormalities have been reported for pancreatic cancer [152, 153].

Among the best-studied changes in cancer cell centrosomes is overexpression of specific centrosome proteins that results in abnormal centrosome configurations and aneuploidy [133, 154]. These studies strongly correlate abnormal centrosomes with cancer development and progression in which increased centrosome number and volume, supernumerary centrioles, accumulation of increased PCM, and abnormal phosphorylation of centrosomes are characteristic for cancer cells in which cell polarity is lost [2, 137]. Centrosome misregulation is associated with abnormal microtubule nucleation, abnormal spindle formation, and chromosomal mis-segregation. As mentioned above, loss of tumor suppressor genes are among the factors that affect accurate centrosome functions.

Changes in Aurora A have been implicated in centrosome amplification in breast cancer and other cancers. In animal models, overexpression of Aurora A kinase (AURKA), an important centrosome-associated serine/threonine kinase, was strongly associated with tumor development [155, 156]. These experiments showed that Aurora A localizes to centrosomes and overexpression of Aurora A causes multipolar mitotic spindles that play a role in early development of mammary tumors. Further studies [156] showed that the pro-survival AKT pathway is activated, preventing cell death while promoting abnormal cell proliferation in which tetraploid cells with accumulated centrosomes were generated.

Genes implicated in centrosome amplification are in part responsible for deregulation of centrosome duplication and subsequent reactions that lead to cascades of cell cycle-related abnormalities. Critical cell cycle regulators are lost in cancer cells which includes loss of the tumor suppressor p53, resulting in multiple cycles of centrosome duplication in one S phase in which centrosome numbers become increased [157]. Viral oncoproteins can inactivate p53 resulting in cells with supernumerary centrosomes which has clearly been shown for the papillomavirus (reviewed in [135]). Loss of p53 following genotoxic stress or mitogenic stimulation has been documented for breast cancer cells in which changes in the CDK2/cyclin-dependent pathway has been implicated [158, 159].

Other tumor suppressor genes have directly or indirectly been implicated in changes of centrosome functions and include the breast- and ovary-specific tumor suppressor gene BRCA1 that has been shown to play a role in deregulation of centrosome duplication. BRCA1 is involved in G2/M checkpoint functions and it plays a role in preventing centrosome overduplication. While we do not yet fully understand the precise mechanisms by which BRCA1 affects centrosome regulation a model for the regulation of the centrosome by BRCA1 has been presented by Kais and Parvin [58]. The model suggests that in regulated cell cycles BRCA1 ubiquitinates the already duplicated centrosomes to inhibit reduplication. Supernumerary centrosome are the result of loss of BRCA1 during the S phase. The model suggests that overexpression of AURKA mimics the effects of BRCA1 loss. Furthermore, overexpressed AURKA overrides the spindle checkpoint and may thereby contribute to abnormal mitosis.

In regulated cell cycles BRCA1 forms a complex with the BRCA1-associated RING domain 1 (BARD1) functioning as E3 ubiquitin ligase. The BRCA1–BARD1 complex plays a role in maintaining centrosome homeostasis by ubiquitinating γ-tubulin, preventing abnormal duplication and abnormal microtubule nucleation by γ-tubulin. Centrosome abnormalities have also been reported in transgenic mice in which the BRCA1-associated centrosomal ninein-like protein (Nlp) is overexpressed, causing spontaneous breast tumorigenesis perhaps as a result of Nlp mimicking BRCA1 loss [160]. BRCA2 also plays a role in centrosome functions, as in regulated cell cycles the BRCA2-associated protein NPM forms a complex with ROCK2 to maintain numerical centrosome integrity; centrosome overduplication and fragmentation may be the result of aberrant regulation of this protein [113].

The centrosomal kinase Nek2 is important for centrosome regulation and centrosome accumulation has been reported in breast epithelial cells in which Nek2 was misregulated [161]. Several other centrosomal components are involved in cancer initiation and progression [162–165] that have been detailed in recent reviews [6, 58, 9, 11, 143–145, 187]. Other factors also play a role in centrosome abnormalities and may include structural defects of the centrosome matrix (reviewed in [6]). One line of research has focused more recently on centrosome clustering as an important factor in centrosome abnormalities in cancer cells. Centrosome clustering is important for mitosis to accumulate centrosomal material equally at the two mitotic poles. Studies have shown that centrosome amplification, cell cycle control dysfunctions, and aggregation of centrosomal material at the mitotic spindle poles are associated with centrosome clustering abnormalities (reviewed in [75]). While multipolar mitotic cells are easy to identify as abnormalities, amplified cancer cell centrosomes can cluster into abnormal bipolar spindles. These abnormalities are not easily discernible from regular bipolar spindles with non-amplified centrosomes but they will nucleate and organize abnormal microtubule formations resulting in chromosome mis-segregations and aneuploidy as shown in *Drosophila* cells [166] as well as in cancer cells [75, 144]. We do not yet fully understand the mechanisms underlying centrosome clustering in normal cells and dysfunctions in cancer cells; more research is needed to analyze the regulation of centrosome clustering. Studies by Kwon et al. [79] have determined that the actin cytoskeleton plays a role in centrosome clustering. Our previous experiments using the invertebrate sea urchin as experimental system revealed that microtubules and microfilaments are required for centrosome dynamics (Schatten et al. [167]) which may also be important for centrosome clustering. We do not yet know whether or not the filamentous components identified in sea urchin centrosomes [78] and in Spisula extracts [76, 77] or perhaps other cytoskeletal elements (reviewed in [6]) play a role in the mechanisms underlying centrosome clustering. Microtubule motor proteins have also been implicated in the clustering process which has been discussed in detail by Krämer et al. [75]. It has been proposed that preventing centrosome clustering into an abnormal bipolar mitotic apparatus may provide new targets for cancer therapy. Centrosomes that are not able to cluster form multiple microtubule asters followed by fragmented cell divisions and cell death with fewer chances for cancer cell viability [168].

The Role of Primary Cilia in Cancer Development and Progression Because of the close relationships between centrosomes and primary cilia centrosome abnormalities may also affect primary cilia formation and functions which may further contribute to cancer development and progression. During progressive stages of cancer development the basal body of the primary cilium becomes dislodged and locates in the cytoplasm [131, 132, 142] where it may form small asters and participate in the mitotic process during subsequent cell divisions. As the oncogenic Aurora A kinase (Aurora A) is localized to the basal body of primary cilia [169] it may further play a role in centrosome amplification and contribute to primary cilia-cell cycle dysfunctions.

Centrosomes as Target for Cancer Therapy and Prevention Cancer is a complex heterogeneous disease that can have different causes and different centrosome abnormalities which presents complexities for the design of effective treatment strategies (reviewed in [6]) requiring multiple targeted treatment approaches to eradicate different subpopulations of cells in cancer tissue. Centrosomes are increasingly being discussed as new targets for cancer treatment, as centrosomes are central to cell division and may be a major driver for abnormal cell divisions.

Targeting cancer cell centrosomes may include targeting misguided signaling pathways, overexpressed centrosome proteins, abnormal centrosome clustering, abnormal primary cilia dynamics, overexpressed phosphorylation such as Aurora A that is implicated in centrosome hyperphosphorylation or other components in the phosphorylation cascade; it further includes different molecules that play a role in centrosome function such as the aryl hydrocarbon receptor (AhR) and cyclin E, as reported by Korzeniewski et al. [164], and several others (reviewed in [6]). Such approaches to target multiple centrosome abnormalities are possible either through the development of new pharmaceuticals or through plant derivatives or dietary ingredients that have been shown to affect centrosome–microtubule interactions during mitosis and cell division.

Plant derivatives that have been developed into a cancer-targeting pharmaceuticals includes paclitaxel (or taxol), originally isolated from the plant *Taxus brevifolia*. Taxol primarily inhibits microtubule depolymerization, thereby preventing progression of mitosis and cell division [170, 171]. It has been shown that taxol interacts with microtubules at the centrosome–microtubule nucleation sites [172, 173], and it had been proposed that centrosomes in taxol-treated cells may lose their capacity to nucleate microtubules [172]. Other drugs that had been explored as anti-cancer drugs include colcemid and nocodazole that prevent and disrupt microtubule polymerization.

Currently, curcumin, a natural polyphenol found in the rhizomes of *Curcuma longa* (turmeric) is being investigated for its anti-cancer activities which led to new promising strategies including the use of theragnostic curcumin-encapsulated nanoparticles that will increase bioavailability and allow more potent clinical applications [174]. Our recent preliminary experiments showed an effect of curcumin on the microtubule cytoskeleton in several cancer cell lines (Schatten et al. unpublished).

Other studies have focused on the antimitotic drug griseofulvin that arrests cells at the G2/M transition stage in a concentration-dependent manner [175–178]. Studies have shown that griseofulvin affects the NFκB pathway and that the NFκB pathway and centrosome dynamics are connected [179]. This finding is intriguing, as griseofulvin has recently been shown to specifically inhibit supernumerary centrosome clustering in cancer cells which indicates its potential as drug that interferes with centrosome dynamics (reviewed in [75]). This line of potential drug development has not yet been explored in detail but it is well worth pursuing, as prevention of centrosome clustering may preferentially affect cancer cell centrosomes leading to cellular fragmentation followed by cell death. Griseofulvin has already been approved as an effective orally administered antifungal drug that interferes with microtubule dynamics in vivo and in vitro [176, 180–184] and it induces multipolar mitoses in tumor cells [175, 177, 178, 181]. It may interfere with microtubule minus ends at the centrosome–microtubule interacting sites.

Other drugs that are currently considered to interfere with the microfilament system in cancer cells are discussed in Chap. 16 of this book by Brayford et al.

Conclusion and Future Directions

In recent years, significant progress has been made in our understanding of centrosome dynamics and centrosome interactions with microtubules in several cell systems. Centrosome dysfunctions have been identified and characterized in aging cells and in cancer cells which allowed targeting of centrosomes for therapeutic interventions. New imaging methods have been applied to analyze centrosome structure in more detail than previously possible and new technological advances have allowed close insights into the composition and regulation of the centrosome organelle and its interactions with other cellular components. Studies in reproductive and somatic cells have determined centrosome abnormalities in aging cells in which centrosomal proteins disperse from the centrosomal matrix leading to centrosome disintegration and microtubule instability. These studies have also determined that the aging process is reversible to a certain extent by experimentally manipulating specific signal transduction processes. New information on cancer cell centrosomes has allowed an analysis of detailed signal transductions that are misguided in cancer cells and lead to centrosome hyperphosphorylation with centrosomes being phosphorylated throughout the cell cycle without undergoing dephosphorylation which takes place during exit from mitosis in regulated cell cycles. Centrosome amplification and multipolar centrosomes have further been analyzed in cancer cells and these abnormalities suggested new target sites for cancer therapies which includes inhibiting misguided signaling pathways, or inhibiting centrosome clustering to induce excessive centrosome fragmentation resulting in cancer cell fragmentation followed by cell death.

Many questions remain to be answered and include questions on the nature of the centrosomal matrix and on how centrosomal proteins associate with matrix components. We also do not yet fully understand the mechanisms of centrosome duplication, especially in reproductive cell systems in which centrosome abnormalities have been implicated in developmental disorders and/or embryo loss. Understanding the mechanisms that play a role in a regulated centrosome cycle will allow to determine molecular abnormalities that may be corrected in centrosome-impaired diseases or disorders.

Acknowledgments It is a pleasure to gratefully acknowledge Donald Connor's professional help with the illustrations and support from MU's Research Council for studies on centrosome dysfunctions.

References

1. Schatten H (2008) The mammalian centrosome and its functional significance. Histochem Cell Biol 129:667–686
2. Badano JL, Teslovich TM, Katsanis N (2005) The centrosome in human genetic disease. Nat Rev Genet 6:194–205
3. Nigg EZ, Raff JW (2009) Centrioles, centrosomes, and cilia in health and disease. Cell 139:663–678
4. Gerdes JM, Davis EE, Katsanis N (2009) The vertebrate primary cilium in development, homeostasis, and disease. Cell 137:32–45
5. Bettencourt-Dias M, Hildebrandt F, Pellman D, Woods G, Godinho SA (2011) Centrosomes and cilia in human disease. Trends Genet 27(8):307–315
6. Schatten H (2013) The impact of centrosome abnormalities on breast cancer development and progression with a focus on targeting centrosomes for breast cancer therapy. In: Schatten H (ed) Cell and molecular biology of breast cancer. Springer Science and Business Media/LLC, New York
7. Yoder BK et al (2002) The polycystic kidney disease proteins, polycystin-1, polycystin-2, polaris, and cystin, are co-localized in renal cilia. J Am Soc Nephrol 13:2508–2516
8. Schatten H, Sun Q-Y (2010) The role of centrosomes in fertilization, cell division and establishment of asymmetry during embryo development. Semin Cell Dev Biol 21:174–184
9. Fisk HA (2012) Many pathways to destruction: the centrosome and its control by and role in regulated proteolysis (Chapter 8). In: Schatten H (ed) The centrosome. Springer Science and Business Media, New York
10. Prosser SL, Fry AM (2012) Regulation of the centrosome cycle by protein degradation (Chapter 9). In: Schatten H (ed) The centrosome. Springer Science and Business Media, New York
11. Fukasawa K (2012) Molecular links between centrosome duplication and other cell cycle associated events (Chapter 10). In: Schatten H (ed) The centrosome. Springer Science and Business Media, New York
12. Boutros R (2012) Regulation of centrosomes by cyclin-dependent kinases (Chapter 11). In: Schatten H (ed) The centrosome. Springer Science and Business Media, New York

13. Schatten H, Sun QY (2015) Centrosome and microtubule functions and dysfunctions in meiosis: implications for age-related infertility and developmental disorders. Reprod Fertil Dev. doi:10.1071/RD14493. [Epub ahead of print] PMID:25903261
14. Salisbury JL (2004) Centrosomes: Sfi1p and centrin unravel a structural riddle. Curr Biol 14:R27–R29
15. Salisbury JL, Suino KM, Busby R, Springett M (2002) Centrin-2 is required for centriole duplication in mammalian cells. Curr Biol 12:1287–1292
16. Andersen JS, Wilkinson CJ, Mayor T, Mortensen P, Nigg EA, Mann M (2003) Proteomic characterization of the human centrosome by protein correlation profiling. Nature 426: 570–574
17. Wilkinson CJ, Andersen JS, Mann M, Nigg EA (2004) A proteomic approach to the inventory of the human centrosome. In: Nigg E (ed) Centrosomes in development and disease. Wiley-VCA, Weinheim, pp 125–142
18. Hannak E, Oegema K, Kirkham M, Gonczy P, Habermann B, Hyman AA (2002) The kinetically dominant assembly pathway for centrosomal asters in *Caenorhabditis elegans* is γ-tubulin dependent. J Cell Biol 157:591–602
19. Murphy S, Urbani L, Stearns T (1998) The mammalian gamma-tubulin complex contains homologues of yeast spindle pole body component sspc97p and spc98p. J Cell Biol 141:663–674
20. Dictenberg J, Zimmerman W, Sparks C, Young A, Vidair C, Zheng Y, Carrington W, Fay F, Doxsey SJ (1998) Pericentrin and gamma tubulin form a protein complex and are organized into a novel lattice at the centrosome. J Cell Biol 141:163–174
21. Doxsey SJ, Stein P, Evans L, Calarco P, Kirschner M (1994) Pericentrin, a highly conserved protein of centrosomes involved in microtubule organization. Cell 76:639–650
22. Flory MR, Davis TN (2003) The centrosomal proteins pericentrin and kendrin are encoded by alternatively spliced products of one gene. Genomics 82:401–405
23. Kawaguchi S, Zheng Y (2004) Characterization of a Drosophila centrosome protein CP309 that shares homology with Kendrin and CG-NAP. Mol Biol Cell 15:37–45
24. Keryer G, Di Fiore B, Celati C, Lechtreck KF, Mogensen M, Delouvee A, Lavia P, Bornens M, Tassin AM (2003) Part of Ran is associated with AKAP450 at the centrosome: involvement in microtubule-organizing activity. Mol Biol Cell 14:4260–4271
25. Steadman BT, Schmidt PH, Shanks RA, Lapierre LA, Goldenring JR (2002) Transforming acidic coiled-coil-containing protein 4 interacts with centrosomal AKAP350 and the mitotic spindle apparatus. J Biol Chem 277(33):30165–30176
26. Takahashi M, Yamagiwa A, Nishimura T, Mukai H, Ono Y (2002) Centrosomal proteins CG-NAP and kendrin provide microtubule nucleation sites by anchoring gamma-tubulin ring complex. Mol Biol Cell 13:3235–3245
27. Ohta T, Essner R, Ryu JH, Palazzo RE, Uetake Y, Kuriyama R (2002) Characterization of Cep135, a novel coiled-coil centrosomal protein involved in microtubule organization in mammalian cells. J Cell Biol 156:87–99
28. Quintyne NJ, Gill SR, Eckley DM, Crego CL, Compton DA, Schroer TA (1999) Dynactin is required for microtubule anchoring at fibroblast centrosomes. J Cell Biol 147:321–334
29. Schroer TA (2001) Microtubules don and doff their caps: dynamic attachments at plus and minus ends. Curr Opin Cell Biol 13:92–96
30. Quintyne NJ, Schroer TA (2002) Distinct cell cycle-dependent roles for dynactin and dynein at centrosomes. J Cell Biol 159:245–254
31. Liu P, Choi Y-K, Qi RZ (2014) NME7 is a functional component of the γ-tubulin ring complex. Mol Biol Cell 25:2017–2025
32. Singh P, Thomas GE, Gireesh KK, Manna TK (2014) TACC3 protein regulates microtubule nucleation by affecting γ-tubulin ring complexes. J Biol Chem 289(46):31719–31735
33. Teixidó-Travesa N, Roig J, Lüders J (2012) The where, when and how of microtubule nucleation – one ring to rule them all. J Cell Sci 125:4445–4456
34. Kollman JM, Merdes A, Mourey L, Agard DA (2011) Microtubule nucleation by gamma-tubulin complexes. Nat Rev Mol Cell Biol 12:709–721. doi:10.1038/nrm3209

35. Mennella V, Keszthelyi B, McDonald KL, Chhun B, Kan F, Rogers GC, Huang B, Agard DA (2012) Subdiffraction-resolution fluorescence microscopy reveals a domain of the centrosome critical for pericentriolar material organization. Nat Cell Biol 14:1159–1168. doi:10.1038/ncb2597
36. Sonnen KF, Schermelleh L, Leonhardt H, Nigg EA (2012) 3D-structured illumination microscopy provides novel insight into architecture of human centrosomes. Biol Open 1:965–976. doi:10.1242/ bio.20122337
37. Lawo S, Hasegan M, Gupta GD, Pelletier L (2012) Subdiffraction imaging of centrosomes reveals higher-order organizational features of pericentriolar material. Nat Cell Biol 14: 1148–1158. doi:10.1038/ncb2591
38. Fu J, Glover DM (2012) Structured illumination of the interface between centriole and pericentriolar material. Open Biol 2:120104. doi:10.1098/rsob.120104
39. Young A, Dictenberg JB, Purohit A, Tuft R, Doxsey SJ (2000) Cytoplasmic dynein-mediated assembly of pericentrin and γ tubulin onto centrosomes. Mol Biol Cell 11:2047–2056
40. Gillingham AK, Munro S (2000) The PACT domain, a conserved centrosomal targeting motif in the coiled-coil proteins AKAP450 and pericentrin. EMBO Rep 1:524–529
41. Levy YY, Lai EY, Remillard SP, Heintzelman MB, Fulton C (1996) Centrin is a conserved protein that forms diverse associations with centrioles and MTOCs in Naegleria and other organisms. Cell Motil Cytoskeleton 33:298–323
42. Salisbury JL (1995) Centrin, centrosomes, and mitotic spindle poles. Curr Opin Cell Biol 7:39–45
43. Lutz W, Lingle WL, McCormick D, Greenwood TM, Salisbury JL (2001) Phosphorylation of centrin during the cell cycle and its role in centriole separation preceding centrosome duplication. J Biol Chem 276:20774–20780
44. Manandhar G, Schatten H, Sutovsky P (2005) Centrosome reduction during gametogenesis and its significance. Biol Reprod 72:2–13
45. Merdes A, Cleveland DA (1998) The role of NuMA in the interphase nucleus. J Cell Sci 111:71–79
46. Sun QY, Schatten H (2006) Multiple roles of NuMA in vertebrate cells: review of an intriguing multi-functional protein. Front Biosci 11:1137–1146
47. Sun Q-Y, Schatten H (2007) Centrosome inheritance after fertilization and nuclear transfer in mammals. Adv Exp Med Biol 591:58–71
48. Saredi A, Howard L, Compton DA (1997) Phosphorylation regulates the assembly of NuMA in a mammalian mitotic extract. J Cell Sci 110:1287–1297
49. Mack GJ, Ou Y, Rattner JB (2000) Integrating centrosome structure with protein composition and function in animal cells. Microsc Res Tech 49:409–419
50. Ou Y, Rattner JB (2004) The centrosome in higher organisms: structure, composition and duplication. Int Rev Cytol 238:119–182
51. Sluder G (2004) Centrosome duplication and its regulation in the higher animal cell. In: Nigg E (ed) Centrosomes in development and disease. Wiley-VCA, Weinheim, pp 167–189
52. Matsumoto Y, Maller JL (2002) Calcium, calmodulin, and CaMKII requirement for initiation of centrosome duplication in Xenopus egg extracts (comment). Science 295:499–502
53. Ohta Y, Ohba T, Miyamoto E (1990) Ca^{2+}/calmodulin-dependent protein kinase II: localization in the interphase nucleus and the mitotic apparatus of mammalian cells. Proc Natl Acad Sci U S A 87:5341–5345
54. Pietromonaco SF, Seluja GA, Elias L (1995) Identification of enzymatically active Ca^{2+}/calmodulin-dependent protein kinase in centrosomes of hematopoietic cells. Blood Cells Mol Dis 21:34–41
55. Fenton B, Glover DM (1993) A conserved mitotic kinase active at late anaphase-telophase in syncytial Drosophila embryos. Nature 363:637–640
56. Donaldson MM, Tavares AAM, Hagan IM, Nigg EA, Glover DM (2001) The mitotic roles of polo-like kinase. J Cell Sci 114:2357–2358
57. Wang Q, Xie S, Chen J, Fukasawa K, Naik U, Traganos F, Darzynkiewicz Z, Jhanwar-Uniyal M, Dai W (2002) Cell cycle arrest and apoptosis by human polo-like kinase 3 is mediated through perturbation of microtubule integrity. Mol Cell Biol 22(10):3450–3459

58. Kais Z, Parvin JD (2012) Centrosome regulation and breast cancer (Chapter 14). In: Schatten H (ed) The centrosome. Springer Science and Business Media, New York
59. Tugendreich S, Tomkiel J, Earnshaw W, Hieter P (1995) CDC27Hs colocalizes with CDC16Hs to the centrosome and mitotic spindle and is essential for the metaphase to anaphase transition. Cell 81:261–268
60. Freed E, Lacey KR, Huie P, Lyapina SA, Deshaies RJ, Stearns T, Jackson PK (1999) Components of an SCF ubiquitin ligase localize to the centrosome and regulate the centrosome duplication cycle. Genes Dev 13:2242–2257
61. Gstaiger M, Marti A, Krek W (1999) Association of human SCF(SKP2) subunit p19(SKP1) with interphase centrosomes and mitotic spindle poles. Exp Cell Res 247:554–562
62. Meraldi P, Nigg EA (2001) Centrosome cohesion is regulated by a balance of kinase and phosphatase activities. J Cell Sci 114:3749–3757
63. Fry AM (2002) The Nek2 protein kinase: a novel regulator of centrosome structure. Oncogene 21:6184–6194
64. Fry AM, Hames RS (2004) The role of the centrosome in cell cycle progression. In: Nigg E (ed) Centrosomes in development and disease. Wiley-VCA, Weinheim, pp 143–166
65. Jackman M, Lindon C, Nigg E, Pines J (2003) Active cyclin B1-Cdk1 first appears on centrosomes in prophase. Nat Cell Biol 5:143–148
66. Golsteyn RM, Mundt KE, Fry AM, Nigg EA (1995) Cell cycle regulation of the activity and subcellular localization of Plk1, a human protein kinase implicated in mitotic spindle function. J Cell Biol 129:1617–1628
67. Merdes A, Ramyar K, Vechio JD, Cleveland DW (1996) A complex of NuMA and cytoplasmic dynein is essential for mitotic spindle assembly. Cell 87:447–458
68. Fry AM, Mayor T, Meraldi P, Stierhof YD, Tanaka K, Nigg EA (1998) C-Nap1, a novel centrosomal coiled-coil protein and candidate substrate of the cell cycle-regulated protein kinase Nek2. J Cell Biol 141:1563–1574
69. Casenghi M, Meraldi P, Weinhart U, Duncan PI, Korner R, Nigg EA (2003) Polo-like kinase 1 regulates Nlp, a centrosome protein involved in microtubule nucleation. Dev Cell 5:113–125. doi:10.1016/S1534-5807(03)00193-X
70. Barr AR, Gergely F (2007) Aurora A: the maker and breaker of spindle poles. J Cell Sci 120:2987–2996
71. Kramer ER, Scheuringer N, Podtelejnikov AV, Mann M, Peters JM (2000) Mitotic regulation of the APC activator proteins CDC20 and CDH1. Mol Biol Cell 11:1555–1569
72. Huang J, Raff JW (1999) The disappearance of cyclin B at the end of mitosis is regulated spatially in Drosophila cells. EMBO J 18:2184–2195
73. Wakefield J, Huang J-Y, Raff JW (2000) Centrosomes have a role in regulating the destruction of cyclin B in early Drosophila embryos. Curr Biol 10:1367–1370
74. Wei Y, Multi S, Yang CR, Ma J, Zhang QH, Wang ZB, Li M, Wei L, Ge ZJ, Zhang CH, Ouyang YC, Hou Y, Schatten H, Sun QY (2011) Spindle assembly checkpoint regulates mitotic cell cycle progression during preimplantation embryo development. PLoS One 6(6):e21557
75. Krämer A, Anderhub S, Maier B (2012) Mechanisms and consequences of centrosome clustering in cancer cells (Chapter 17). In: Schatten H (ed) The centrosome. Springer Science and Business Media, New York
76. Schnackenberg BJ, Palazzo RE (1999) Identification and function of the centrosome centromatrix. Biol Cell 91(6):429–438
77. Schnackenberg BJ, Hull DR, Balczon RD, Palazzo RE (2000) Reconstitution of microtubule nucleation potential in centrosomes isolated from Spisula solidissima oocytes. J Cell Sci 113(Pt 6):943–953
78. Schatten H, Walter M, Mazia D, Biessmann H, Paweletz N, Coffe G, Schatten G (1987) Centrosome detection in sea urchin eggs with a monoclonal antibody against Drosophila intermediate filament proteins: characterization of stages of the division cycle of centrosomes. Proc Natl Acad Sci U S A 84:8488–8492

79. Kwon M, Godinho SA, Chandhok NS, Ganem NJ, Azioune A, Thery M, Pellman D (2008) Mechanisms to suppress multipolar divisions in cancer cells with extra centrosomes. Genes Dev 22:2189–2203
80. Sun Q-Y, Lai L, Wu G, Bonk A, Cabot R, Park K-W, Day B, Prather RS, Schatten H (2002) Regulation of mitogen-activated protein kinase phosphorylation, microtubule organization, chromatin behavior, and cell cycle progression are regulated by protein phosphatases during pig oocyte maturation and fertilization in vitro. Biol Reprod 66(3):580–588
81. Ou X-H, Li S, Xu B-C, Wang Z-B, Quan S, Li M, Zhang Q-H, Ouyang Y-C, Schatten H, Xing F-Q, Sun Q-Y (2010) p38α MAPK is a MTOC-associated protein regulating spindle assembly, spindle length and accurate chromosome segregation during mouse oocyte meiotic maturation. Cell Cycle 9(20):4130–4143
82. Schatten H, Sun Q-Y (2014) Posttranslationally modified tubulins and other cytoskeletal proteins: their role in gametogenesis, oocyte maturation, fertilization and pre-implantation embryo development. In: Sutovsky P (ed) Posttranslational protein modifications in the reproductive system. Springer Science and Business Media, New York
83. Janke C, Bulinski JC (2011) Post-translational regulation of the microtubule cytoskeleton: mechanisms and functions. Nat Rev Mol Cell Biol 12:773–786
84. Rymut SM, Kelley TJ (2015) Broader implications: biological and clinical significance of microtubule acetylation. Cell Health Cytoskelet 7:71–82
85. Schatten G, Simerl y C, Asai DJ, Szöke E, Cooke P, Schatten H (1988) Acetylated α-tubulin in microtubules during mouse fertilization and early development. Dev Biol 130:74–86
86. Asthana J, Kapoor S, Mohan R, Panda D (2013) Inhibition of HDAC6 deacetylase activity increases its binding with microtubules and suppresses microtubule dynamic instability in MCF-7 cells. J Biol Chem 288(31):22516–22526
87. Akella JS, Wloga D, Kim J, Starostina NG, Lyons-Abbott S, Morrissette NS, Dougan ST, Kipreos ET, Gaertig J (2010) MEC-17 is an alpha-tubulin acetyltransferase. Nature 467:218–222. doi:10.1038/nature09324
88. Matsuyama A, Shimazu T, Sumida Y, Saito A, Yoshimatsu Y, Seigneurin-Berny D, Osada H, Komatsu Y, Nishino N, Khochbin S et al (2002) In vivo destabilization of dynamic microtubules by HDAC6-mediated deacetylation. EMBO J 21:6820–6831, PubMed: 12486003
89. Hubbert C, Guardiola A, Shao R, Kawaguchi Y, Ito A, Nixon A, Yoshida M, Wang XF, Yao TP (2002) HDAC6 is a microtubule-associated deacetylase. Nature 417:455–458
90. North BJ, Marshall BL, Borra MT, Denu JM, Verdin E (2003) The human Sir2 ortholog, SIRT2, is an NAD+-dependent tubulin deacetylase. Mol Cell 11:437–444, PubMed: 12620231
91. Zilberman Y, Ballestrem C, Carramusa L, Mazitschek R, Khochbin S, Bershadsky A (2009) Regulation of microtubule dynamics by inhibition of the tubulin deacetylase HDAC6. J Cell Sci 122:3531–3541
92. Haggarty SJ, Koeller KM, Wong JC, Grozinger CM, Schreiber SL (2003) Domain-selective small-molecule inhibitor of histone deacetylase 6 (HDAC6)-mediated tubulin deacetylation. Proc Natl Acad Sci U S A 100:4389–4394
93. Butler KV, Kalin J, Brochier C, Vistoli G, Langley B, Kozikowski AP (2010) Rational design and simple chemistry yield a superior, neuro-protective HDAC6 inhibitor, tubastatin A. J Am Chem Soc 132:10842–10846
94. Verhey KJ, Gaertig J (2007) The tubulin code. Cell Cycle 6:2152–2160, PubMed: 17786050
95. Pan J, Snell W (2007) The primary cilium: keeper of the key to cell division. Cell 129:1255–1257
96. Wheatley DN, Wang AM, Strugnell GE (1996) Expression of primary cilia in mammalian cells. Cell Biol Int 20:73–81
97. D'Angelo A, Franco B (2009) The dynamic cilium in human diseases. Pathogenetics 2(1):3
98. Veland IR, Awan A, Pedersen LB, Yoder BK, Christensen ST (2009) Primary cilia and signaling pathways in mammalian development, health and disease. Nephron Physiol 111:39–53
99. Quarmby LM, Parker JDK (2005) Cilia and the cell cycle? J Cell Biol 169(5):707–710

100. Hildebrandt F, Otto E (2005) Cilia and centrosomes: a unifying pathogenic concept for cystic kidney disease? Nat Rev Genet 6:928–940
101. Davenport JR, Yoder BK (2005) An incredible decade for the primary cilium: a look at a once-forgotten organelle. Am J Physiol Renal Physiol 289:F1159–F1169
102. Michaud EJ, Yoder BK (2006) The primary cilium in cell signaling and cancer. Cancer Res 66:6463–6467
103. Satir P, Christensen ST (2008) Structure and function of mammalian cilia. Histochem Cell Biol 129:687–693
104. Cheng J, Türkel N, Hemati N, Fuller MT, Hunt AJ, Yamashita YM (2008) Centrosome misorientation reduces stem cell division during ageing. Nature 456:599–604
105. Oh J, Lee YD, Amy J, Wagers AJ (2014) Stem cell aging: mechanisms, regulators and therapeutic opportunities. Nat Med 20(8):870–880
106. Schatten H, Chakrabarti A, Hedrick J (1999) Centrosome and microtubule instability in cells during aging. J Cell Biochem 74:229–241
107. Ly DH, Lockhart DJ, Lerner RA, Schultz PG (2000) Mitotic misregulation and human aging. Science 287:2486–2492
108. Miao Y-L, Kikuchi K, Sun Q-Y, Schatten H (2009) Oocyte aging: cellular and molecular changes, developmental potential and reversal possibility. Hum Reprod Update 15(5):573–585
109. Wang Z-B, Schatten H, Sun Q-Y (2011) Why is chromosome segregation error in oocytes increased with maternal aging? Physiology 26(5):314–325
110. Qiao J, Wang ZB, Feng HL, Miao YL, Wang Q, Yu Y, Wei YC, Yan J, Wang WH, Shen W, Sun SC, Schatten H, Sun QY (2014) The root of reduced fertility in aged women and possible therapeutic options: current status and future perspectives. Mol Aspects Med 38:54–85
111. Schatten H, Rawe VY, Sun Q-Y (2012) Cytoskeletal architecture of human oocytes with focus on centrosomes and their significant role in fertilization. In: Nagy ZP, Varghese AC, Agarwal A (eds) Practical manual of in vitro fertilization: advanced methods and novel devices. Humana Press, New York
112. Miao Y-L, Sun Q-Y, Zhang X, Zhao J-G, Zhao M-T, Spate L, Prather RS, Schatten H (2009) Centrosome abnormalities during porcine oocyte aging. Environ Mol Mutagen 50(8):666–671
113. Wang HF, Takenaka K, Nakanishi A et al (2011) BRCA2 and nucleophosmin coregulate centrosome amplification and form a complex with the Rho effector kinase ROCK2. Cancer Res 71:68–77
114. Kim NH, Moon SJ, Prather RS, Day BN (1996) Cytoskeletal alteration in aged porcine oocytes and parthenogenesis. Mol Reprod Dev 43:513–518
115. Alvarez Sedó CA, Schatten H, Combelles C, Rawe VY (2011) The nuclear mitotic apparatus protein NuMA: localization and dynamics in human oocytes, fertilization and early embryos. Mol Hum Reprod 17(6):392–398. doi:10.1093/molehr/gar009
116. Schatten H, Sun Q-Y (2013) Chromosome behavior and spindle formation in mammalian oocytes. In: Trounson A, Gosden R, Eichenlaub-Ritter U (eds) Biology and pathology of the oocyte. Biology & pathology of the Oocyte, 2nd edn
117. Gardner MK, Zanic M, Howard J (2013) Microtubule catastrophe and rescue. Curr Opin Cell Biol 25:14–22
118. Cande Z (1990) Centrosomes: composition and reproduction. Curr Opin Cell Biol 2:301–305
119. Huang B (1990) Genetics and biochemistry of centrosomes and spindle poles. Curr Opin Cell Biol 2:28–32
120. Zhao Y, Fang X, Wang Y, Zhang J, Jiang S, Liu Z, Ma Z, Xu L, Li E, Zhang K (2014) Comprehensive analysis for histone acetylation of human colon cancer cells treated with a novel HDAC inhibitor. Curr Pharm Des 20(11):1866–1873
121. Glozak MA, Seto E (2007) Histone deacetylases and cancer. Oncogene 26:5420–5432
122. Kikuchi K, Naito K, Noguchi J, Shimada A, Kaneko H, Yamashita M, Aoki F, Tojo H, Toyoda Y (2000) Maturation/M-phase promoting factor: a regulator of aging in porcine oocytes. Biol Reprod 63:715–722

123. Kikuchi K, Naito K, Noguchi J, Kaneko H, Tojo H (2002) Maturation/M-phase promoting factor regulates aging of porcine oocytes matured in vitro. Cloning Stem Cells 4:211–222
124. Xu Z, Abbott A, Kopf GS, Schultz RM, Ducibella T (1997) Spontaneous activation of ovulated mouse eggs: time-dependent effects on M-phase exit, cortical granule exocytosis, maternal messenger ribonucleic acid recruitment, and inositol 1,4,5-trisphosphate sensitivity. Biol Reprod 57:743–750
125. Tian XC, Lonergan P, Jeong BS, Evans AC, Yang X (2002) Association of MPF, MAPK, and nuclear progression dynamics during activation of young and aged bovine oocytes. Mol Reprod Dev 62:132–138
126. Fan HY, Sun QY (2004) Involvement of mitogen-activated protein kinase cascade during oocyte maturation and fertilization in mammals. Biol Reprod 70:535–547
127. Tatone C, Carbone MC, Gallo R, Delle Monache S, Di Cola M, Alesse E, Amicarelli F (2006) Age-associated changes in mouse oocytes during postovulatory in vitro culture: possible role for meiotic kinases and survival factor BCL2. Biol Reprod 74:395–402
128. Liang CG, Su YQ, Fan HY, Schatten H, Sun QY (2007) Mechanisms regulating oocyte meiotic resumption: roles of mitogen-activated protein kinase. Mol Endocrinol 21(9):2037–2055
129. Lee JH, Campbell KH (2008) Caffeine treatment prevents age-related changes in ovine oocytes and increases cell numbers in blastocysts produced by somatic cell nuclear transfer. Cloning Stem Cells 10:381–390
130. Boveri T (1914) Zur Frage der Entstehung maligner Tumoren. G. Fisher, Jena
131. Schatten H, Wiedemeier A, Taylor M, Lubahn D, Greenberg NM, Besch-Williford C, Rosenfeld C, Day K, Ripple M (2000) Centrosomes-centriole abnormalities are markers for abnormal cell divisions and cancer in the transgenic adenocarcinoma mouse prostate (TRAMP) model. Biol Cell 92:331–340
132. Lingle WL, Salisbury JL (2000) The role of the centrosome in the development of malignant tumors. Curr Top Dev Biol 49:313–329
133. Lingle WL, Barrett SL, Negron VC, D'Assoro AB, Boeneman K, Liu W, Whitehead CM, Reynolds C, Salisbury JL (2002) Centrosome amplification drives chromosomal instability in breast tumor development. Proc Natl Acad Sci U S A 99:1978–1983
134. Boveri T (2008) Concerning the origin of malignant tumours by Theodor Boveri (Translated and annotated by Henry Harris). J Cell Sci 121(Suppl 1):1–84
135. Duensing S, Lee LY, Duensing A, Basile J, Piboonniyom S, Gonzalez S, Crum CP, Munger K (2000) The human papillomavirus type 16 E6 and E7 oncoproteins cooperate to induce mitotic defects and genomic instability by uncoupling centrosome duplication from the cell division cycle. Proc Natl Acad Sci U S A 97:10002–10007
136. Chan JY (2011) A clinical overview of centrosome amplification in human cancers. Int J Biol Sci 7(8):1122–1144
137. Lingle WL, Lutz WH, Ingle JN, Maihle NJ, Salisbury JL (1998) Centrosome hypertrophy in human breast tumors: implications for genomic stability and cell polarity. Proc Natl Acad Sci U S A 95:2950–2955
138. Liu T, Niu Y, Yu Y, Liu Y, Zhang F (2009) Increased γ-tubulin expression and P16INK4A promoter methylation occur together in preinvasive lesions and carcinomas of the breast. Ann Oncol 20:441–448
139. Niu Y, Liu T, Tse GM, Sun B, Niu R, Li HM, Wang H, Yang Y, Ye X, Wang Y, Yu Q, Zhang F (2009) Increased expression of centrosomal α, γ-tubulin in atypical ductal hyperplasia and carcinoma of the breast. Cancer Sci 100:580–587
140. Kammerer S, Roth RB, Hoyal CR, Reneland R, Marnellos G, Kiechle M, Schwarz-Boeger U, Griffiths LR, Ebner F, Rehbock J, Cantor CR, Nelson MR, Brown A (2005) Association of the NuMA region on chromosome 11q13 with breast cancer susceptibility. Proc Natl Acad Sci U S A 102(6):2004–2009
141. Gehmlich KL, Haren L, Merdes A (2004) Cyclin B degradation leads to NuMA release from dynein/dynactin and from spindle poles. EMBO Rep 5:97–103
142. Lingle WL, Salisbury JL (1999) Altered centrosome structure is associated with abnormal mitoses in human breast tumors. Am J Pathol 155:1941–1951

143. Korzeniewski N, Duensing S (2012) Disruption of centrosome duplication control and induction of mitotic instability by the high-risk human papillomavirus oncoproteins E6 and E7 (Chapter 12). In: Schatten H (ed) The centrosome. Springer Science and Business Media, New York
144. Saladino C, Bourke E, Morrison CG (2012) Centrosomes, DNA damage and aneuploidy (Chapter 13). In: Schatten H (ed) The centrosome. Springer Science and Business Media, New York
145. Yan B, Chng W-J (2012) The role of centrosomes in multiple myeloma (Chapter 15). In: Schatten H (ed) The centrosome. Springer Science and Business Media, New York
146. Ellgaard L, Molinari M, Helenius A (1999) Setting the standards: quality control in the secretory pathway. Science 286:1882–1888
147. Johnston JA, Ward CL, Kopito RR (1998) Aggresomes: a cellular response to misfolded proteins. J Cell Biol 143:1883–1898
148. Wojcik C, DeMartino GN (2003) Intracellular localization of proteasomes. Int J Biochem Cell Biol 35:579–589
149. Kopito RR (2000) Aggresomes, inclusion bodies and protein aggregation. Trends Cell Biol 10:524–530
150. Roth J, Yam GH, Fan J, Hirano K, Gaplovska-Kysela K, Le Fourn V, Guhl B, Santimaria R, Torossi T, Ziak M, Zuber C (2008) Protein quality control: the who's who, the where's and therapeutic escapes. Histochem Cell Biol 129:163–177
151. McNaught KS et al (2002) Impairment of the ubiquitin-proteasome system causes dopaminergic cell death and inclusion body formation in ventral mesencephalic cultures. J Neurochem 81:301–306
152. Ohki R, Nemoto J, Murasawa H, Oda E, Inazawa J, Tanaka N, Taniguchi T (2000) Reprimo, a new candidate mediator of the p53-mediated cell cycle arrest at the G2 phase. J Biol Chem 275:22627–22630
153. Sato N, Maitra A, Fukushima N, van Heek NT, Matsubayashi H, Iacobuzio-Donahue CA, Rosty C, Goggins M (2003) Frequent hypomethylation of multiple genes overexpressed in pancreatic ductal adenocarcinoma. Cancer Res 63:4158–4166
154. Katayama H, Brinkley WR, Sen S (2003) The Aurora kinases: role in cell transformation and tumorigenesis. Cancer Metastasis Rev 22:451–464
155. Goepfert TM, Adigun YE, Zhong L et al (2002) Centrosome amplification and overexpression of aurora A are early events in rat mammary carcinogenesis. Cancer Res 62:4115–4122
156. Wang X, Zhou YX, Qiao W et al (2006) Overexpression of aurora kinase A in mouse mammary epithelium induces genetic instability preceding mammary tumor formation. Oncogene 25:7148–7158
157. Pihan GA, Doxsey SJ (1999) The mitotic machinery is a source of genetic instability in cancer. Semin Cancer Biol 9:289–302
158. D'Assoro AB, Busby R, Suino K et al (2004) Genotoxic stress leads to centrosome amplification in breast cancer cell lines that have an inactive G1/S cell cycle checkpoint. Oncogene 23:4068–4075
159. D'Assoro AB, Busby R, Acu ID et al (2008) Impaired p53 function leads to centrosome amplification, acquired ERalpha phenotypic heterogeneity and distant metastases in breast cancer MCF-7 xenografts. Oncogene 27:3901–3911
160. Shao S, Liu R, Wang Y et al (2010) Centrosomal Nlp is an oncogenic protein that is gene-amplified in human tumors and causes spontaneous tumorigenesis in transgenic mice. J Clin Invest 120:498–507
161. Hayward DG, Clarke RB, Faragher AJ et al (2004) The centrosomal kinase Nek2 displays elevated levels of protein expression in human breast cancer. Cancer Res 64:7370–7376
162. Suizu F, Ryo A, Wulf G et al (2006) Pin1 regulates centrosome duplication, and its overexpression induces centrosome amplification, chromosome instability, and oncogenesis. Mol Cell Biol 26:1463–1479
163. Bergmann S, Royer-Pokora B, Fietze E et al (2005) YB-1 provokes breast cancer through the induction of chromosomal instability that emerges from mitotic failure and centrosome amplification. Cancer Res 65:4078–4087

164. Korzeniewski N, Wheeler S, Chatterjee P et al (2010) A novel role of the aryl hydrocarbon receptor (AhR) in centrosome amplification – implications for chemoprevention. Mol Cancer 9:153
165. Olson JE, Wang X, Pankratz VS, Fredericksen ZS, Vachon CM, Vierkant RA, Cerhan JR, Couch FJ (2011) Centrosome-related genes, genetic variation, and risk of breast cancer. Breast Cancer Res Treat 125(1):221–228. doi:10.1007/s10549-010-0950-8
166. Gergely F, Basto R (2008) Multiple centrosomes: together they stand, divided they fall. Genes Dev 22:2291–2296
167. Schatten H, Walter M, Biessmann H, Schatten G (1988) Microtubules are required for centrosome expansion and positioning while microfilaments are required for centrosome separation in sea urchin eggs during fertilization and mitosis. Cell Motil Cytoskeleton 11(4):248–259
168. Ganem NJ, Godinho SA, Pellman D (2009) A mechanism linking extra centrosomes to chromosomal instability. Nature 460(7252):278–282
169. Inoko A, Matsuyama M, Goto H, Ohmuro-Matsuyama Y, Hayashi Y, Enomoto M, Ibi M, Urano T, Yonemura S, Kiyono T, Izawa I, Inagaki M (2012) Trichoplein and Aurora A block aberrant primary cilia assembly in proliferating cells. J Cell Biol 197(3):391–405
170. Schiff PB, Fant J, Horwitz SB (1979) Promotion of microtubule assembly in vitro by taxol. Nature 277:665–667
171. Schatten G, Schatten H, Bestor T, Balczon R (1982) Taxol inhibits the nuclear movements during fertilization and induces asters in unfertilized sea urchin eggs. J Cell Biol 94:455–465
172. De Brabander M, Geuens G, Nuydens R, Willebrords R, De Mey J (1981) Taxol induces the assembly of free microtubules in living cells and blocks the organizing capacity of the centrosomes and kinetochores. Proc Natl Acad Sci U S A 78:5608–5612
173. Dimitriadis I, Katsaros C, Galatis B (2001) The effect of taxol on centrosome function and microtubule organization in apical cells of *Sphacelaria rigidula* (Phaeophyceae). Phycol Res 49:23–34
174. Shin SJ, Beech JR, Kelly KA (2012) Targeted nanoparticles in imaging: paving the way for personalized medicine in the battle against cancer. Integr Biol 5(1):29–42. doi:10.1039/C2IB20047C
175. Rebacz B, Larsen TO, Clausen MH, Ronnest MH, Loffler H, Ho AD, Krämer A (2007) Identification of griseofulvin as an inhibitor of centrosomal clustering in a phenotype-based screen. Cancer Res 67:6342–6350
176. Schatten H, Schatten G, Petzelt C, Mazia D (1982) Effects of griseofulvin on fertilization and early development of sea urchins. Independence of DNA synthesis, chromosome condensation, and cytokinesis cycles from microtubule-mediated events. Eur J Cell Biol 27:74–87
177. Ho YS, Duh JS, Jeng JH, Wang YJ, Liang YC, Lin CH, Tseng CJ, Yu CF, Chen RJ, Lin JK (2001) Griseofulvin potentiates antitumorigenesis effects of nocodazole through induction of apoptosis and G2/M cell cycle arrest in human colorectal cancer cells. Int J Cancer 91:393–401
178. Panda D, Rathinasamy K, Santra MK, Wilson L (2005) Kinetic suppression of microtubule dynamic instability by griseofulvin: implications for its possible use in the treatment of cancer. Proc Natl Acad Sci U S A 102:9878–9883
179. Uen YH, Liu DZ, Weng MS, Ho YS, Lin SY (2007) NF-kappaB pathway is involved in griseofulvin-induced G2/M arrest and apoptosis in HL-60 cells. J Cell Biochem 101(5):1165–1175
180. Marchetti F, Mailhes JB, Bairnsfather L, Nandy I, London SN (1996) Dose-response study and threshold estimation of griseofulvin induced aneuploidy during female mouse meiosis I and II. Mutagenesis 11:195–200
181. Schatten H (1977) Untersuchungen über die Wirkung von Griseofulvin in Seeigeleiern und in Mammalierzellen. Universität Heidelberg (Effects of griseofulvin on sea urchin eggs and on mammalian cells. University of Heidelberg)
182. Wehland J, Herzog W, Weber K (1977) Interaction of griseofulvin with microtubules, microtubule protein and tubulin. J Mol Biol 111:329–342

183. Grisham LM, Wilson L, Bensch KG (1973) Antimitotic action of griseofulvin does not involve disruption of microtubules. Nature 244:294–296
184. Miao YL, Zhang X, Zhao JG, Spate L, Zhao MT, Murphy CN, Prather RS, Sun QY, Schatten H (2012) Effects of griseofulvin on in vitro porcine oocyte maturation and embryo development. Environ Mol Mutagen 53(7):561–566. doi:10.1002/em.21717
185. Alvarez Sedó CA, Schatten H, Combelles C, Rawe VY (2011) The nuclear mitotic apparatus protein NuMA: localization and dynamics in human oocytes, fertilization and early embryos. Mol Hum Reprod 17(6):392–398
186. Schatten H, Sun Q-Y (2010) The role of centrosomes in fertilization, cell division and establishment of asymmetry during embryo development. Seminars in Cell and Developmental Biology 21:174–184
187. Schatten H, Rawe VY, Sun Q-Y (2012) Cytoskeletal architecture of human oocytes with focus on centrosomes and their significant role in fertilization. In: Practical Manual of In Vitro Fertilization: Advanced Methods and Novel Devices, edited by Zsolt Peter Nagy, Alex C. Varghese, and Ashok Agarwal. Humana Press (Springer Science+Business Media, New York, USA).

Chapter 6
Cytoskeletal Elements and the Reproductive Success in Animals

Alessandra Gallo and Elisabetta Tosti

Abbreviations/Acronyms

AR	Acrosome reaction
CG	Cortical granules
GV	Germinal vesicle
GVBD	Germinal vesicle breakdown
MI	Metaphase I
MII	Metaphase II
ZP	Zona pellucida

Introduction

In animals, sexual reproduction is the biological process by which a new individual is generated through the fusion of the gametes, the spermatozoon and oocyte, that are formed during gametogenesis which in turn is underlined by meiosis, the peculiar process of cell division that provides haploid cells ready for fertilization. A correct maturation and reciprocal activation of gametes are pre-requisites for fertilization and, although their temporal and spatial sequences are not yet fully clarified, they involve numerous cellular structures, molecules, ions and metabolic pathways.

In the cell, the shape and structure are due to the cytoskeleton, a complex set of structures composed of microtubules, microfilaments and intermediate filaments

---

A. Gallo • E. Tosti (✉)
Department of Biology and Evolution of Marine Organisms, Stazione Zoologica Anton Dohrn, Villa Comunale, 80121 Naples, Italy
e-mail: alessandra.gallo@szn.it; tosti@szn.it

H. Schatten (ed.), *The Cytoskeleton in Health and Disease*,
DOI 10.1007/978-1-4939-2904-7_6

that organize cytoplasmic organelles positioning and intracellular compartments, thus generating cell polarity and contractile forces [1–3].

Microfilaments are composed of actin proteins and play a crucial role in structuring the cell surface and plasma membrane during oocyte maturation and fertilization. They also participate in the maintenance of the meiotic spindle near the cortex, the formation of first and second polar body, the pronuclei apposition and cytokinesis.

Microtubules are tubular polymers of tubulin essential for chromosome movements as well as other aspects of motility and cytoplasmic architecture [4]. In most of the oocytes, microtubules are essential for the first and second meiotic division, furthermore they are fundamental for the swimming of the sperm as well as the union of the male and female pronuclei.

Many studies have been devoted to examine the factors that may influence the success of fertilization. This chapter will discuss the modification of gamete ultrastructure during the processes of oocyte and sperm maturation and fertilization, focusing on the crucial role of the cytoskeletal structures in ensuring a successful fertilization and normal embryo formation in some key species of marine invertebrates and mammals.

The Gametes

Oogenesis and spermatogenesis are characterized by meiosis, the unique process of cell division occurring only in gametes, whose goal is the production of haploid cells highly specialized for fertilization.

The Spermatozoon: Structure and Maturation

The primary function of the spermatozoon is to deliver the male genetic material into the oocyte to generate a new diploid individual. The success of fertilization depends on a series of processes based on a correct maturation of the spermatozoon, its transport toward a receptive oocyte and the ability to recognize and fuse with it. In order to perform these functions, the spermatozoon has developed a highly specialized morphology with different structural components each aimed to a specific processes. Basic structure of the spermatozoon is common to almost all the species and includes three major parts [5]: (1) the head, that is the site for recognition and fusion, has a roundish shape and contains a few structures such as the nucleus, scant cytoplasm and the acrosome that is a cap-like structure over the anterior half of the head; (2) the midpiece is located at the base of the head and includes the centrioles, few mitochondria, axoneme base and related anchoring structures; and (3) the tail is a long flagellum composed of an axoneme, a highly organized microtubule-based structure composed of about 250 proteins. A typical structure (9+2) is made up of peripheral doublets of longitudinal microtubules on which are fixed dynein arms and radial spokes and a pair of singlet microtubules located in the central core.

Unlike the oocyte, the spermatozoon undergoes the final step of the maturation process (spermiogenesis) through a dramatic change in shape and morphology passing from a round cell to the characteristic tadpole aspect.

An important feature of spermatogenesis is the change in the cytoskeleton that occurs throughout this pathway. Although primary focus is given to the microtubule cytoskeleton, the importance of actin filaments to the cellular transformation of the male germ cell has also been shown [6].

During spermatogenesis, diverse processes occur such as sequential changes in the nucleus and the acrosome in concert with a prominent bundles of microtubules called the manchette, a high condensation of chromatin in the nucleus and a cytoplasmic remodelling of the sperm body and structures [7, 8].

Actin has been shown to cooperate in various aspects of the spermatogenesis along with myosin, an actin-dependent motor protein. Actin is present in the form of monomer, oligomer and polymer within cells, the latter are called microfilaments and are involved in the shaping and differentiation of spermatids.

Three major cytoskeletal proteins, actin, actin-binding proteins such as spectrin and various tubulins (e.g. a-, b-, g-tubulin), are present in the head of mammalian spermatozoa with a pattern similar for all the species. Changes in localization of cytoskeleton support the image of cytoskeletal proteins as highly dynamic structures participating actively in processes prior to fertilization [9, 10].

During spermatogenesis, the actin cytoskeleton shows active remodeling. Some actin binding or actin regulated proteins have been demonstrated to regulate dynamic changes of the actin-containing structures. Myosin plays also an important role in acrosome biogenesis, vesicle transport, gene transcription and nuclear shaping [11].

The Oocyte: Structure and Maturation

The oocyte is the large cell characterized by a single function of generating a new individual. The general organization of a mature oocyte is similar along the species but sometime shows unique features [12], in fact, the oocyte is surrounded by extracellular membranes that appear to be thin in the sea urchin (the vitelline layer) [13] or tick in ascidians (the chorion) [14] and mammals (the zona pellucida; ZP) [15].

The oocyte plasma membrane marks the borderline between the internal and external compartments, represents a barrier to ions whose passage occurs towards the ion channels, specific proteins located inside the lipid bilayer [16]. Plasma membrane has many extensions, called microvilli, involved in the fusion process [17]. The cortical granules (CG) are round organelles situated in mono and multiple layers under the cell plasma membrane. CG originate from the Golgi complex and contain mainly enzymes and mucopolysaccharides; although they show a high variability in shape and size, they are present in the oocytes of most of the animals, playing a role during oocyte activation when they release their content by exocytosis [18]. The mitochondria are organelles playing key roles in the oxidative cellular energy metabolism. Their number and ultrastructural organization change with cell function and activity, infact oocytes, zygotes, and embryos present particular types of mitochondria [19].

Oocyte maturation is a complex process occurring at the end of oogenesis during which the oocyte completes its growth and undergoes a series of changes that are necessary for ovulation, fertilization and early embryo development. In the stage that precedes maturation, the oocyte is a large round cell with a big nucleus called germinal vesicle (GV) that at very early stage contains decondensed transcriptionally active chromatin [20].

Oocyte maturation involves two different but interlinked processes based on nuclear and cytoplasmic events [21–23].

Nuclear maturation starts when the oocyte under a chemical stimulation resumes meiosis inducing the breakdown of the GV (GVBD), the chromosome condensation and the spindle formation. A second meiotic arrest then occurs at different stages depending on the species such as metaphase I (MI) in ascidians, bivalves and gastropods, whereas mammalian oocytes complete the first round of meiosis with extrusion of the first polar body and, without an interphase, progress to the second meiotic metaphase (MII). Apart from some exceptions, this process is completed upon fertilization in almost all the species studied.

Cytoplasmic maturation is a less clear process that starts at the time of oocyte growth and occurs with the following events: (1) ultrastructural organelle re-organization; (2) molecular modifications of the plasma membrane; (3) differentiation of the calcium signalling machinery and (4) oocyte surface microvilli increase. Ultrastructural reorganization involves the redistribution of microfilaments and microtubules that in turn support the relocation of cytoplasmic organelles such as cortical granules, mitochondria and the Golgi apparatus in several species [18, 20, 24–35]. In particular mitochondria redistribution appears to be functional to oocyte developmental competence and the regulation of normal embryo development [36].

Fertilization

Fertilization is a highly specialized process of cell to cell interaction that marks the creation of a new and unique individual [23, 37]. The main steps of the fertilization process are gametogenesis, gamete reciprocal activation, sperm-oocyte interaction, fusion and syngamy, thereafter these successful events give rise to the beginning of development.

Reciprocal activation of gametes is fundamental for a successfully fertilization. First the oocyte induces sperm activation due mainly to the signals coming from the oocyte investments [38]. After a series of processes, the activated spermatozoon reaches the oocyte exerting its dual function: to transport the male genome into the oocyte and to trigger the quiescent oocyte into activation [39]. The latter is underlined by another series of events, that include electrical, structural and molecular changes, up to the release of oocyte from meiotic arrest [40] giving rise to the zygote, the first cell of the living organism that becomes an embryo after a series of mitotic divisions.

Cytoskeletal Elements Modulating Gamete Activation

The Spermatozoon

Sperm activation triggers in progression: (1) the sperm motility (chemokinesis); (2) the attraction toward the oocyte itself (chemotaxis); (3) the first binding mediated by ligands and receptors on the gamete plasma membrane; (4) the acrosome reaction (AR); (5) the penetration through the extracellular layers; (6) a second binding; and (7) the fusion of the two plasma membranes.

Sperm motility is required for sperm transport toward the oocyte, either in the aquatic environment or the female genital tract. Sperm motility is an essential condition for male fertility and is fully underlined by the tail. A flagellar movement is provided by the sliding of adjacent microtubules thanks to the ATP hydrolysis occurring in the mitochondria located on the close midpiece. In particular, the propagation of a wave is repeated along the flagellum by a mechanochemical cycle of attachment-detachment of dynein arms giving rise to the flagellar sliding and bending [41–43].

Recent proteomic analyses have provided insight into novel cellular and functional aspects of sperm actin isoforms in the axoneme of ascidians [44].

Motility initiation and hyperactivation are also supported by other specific cytoskeletal elements and dynamics such as major sperm protein (MSP) filaments in the nematodes [45, 46] and polymerization of actin in mammals [47, 48].

Chemotaxis is the process by which spermatozoa are attracted by microenvironmental factors mainly released by the oocyte or perioocyte layers [49]. To provide a more efficient motility, chemotaxis generates dramatic movement changes induced by the interaction of external factor with membrane "receptors" and intracellular messengers such as cyclic AMP, ATP, calcium, or pH changes. All the signaling molecules involved in this process are closely arranged in the sperm flagellum controlling dynein-microtubules interaction through a phosphorylation-dephosphorylation process of axonemal proteins [50].

The first contact between the two gametes is the binding of the sperm to the extracellular investments of the oocyte, this is a receptor-ligand interaction with a high degree of species- specificity that allows to prevent fusion of sperm and oocytes of different species. The carbohydrate groups on the oocyte surface function as sperm receptors. The sperm molecules that bind this receptor are not known with certainty, and indeed, there may be several proteins that can serve this function. In mammals the first association of the spermatozoon with the ZP occurs between the zona glycoprotein, ZP3, and sperm receptor, located on the sperm plasma membrane, such as the 95 kDa tyrosine kinase-protein. This interaction induces the AR [51]. The latter is an exocytotic process mediated by calcium occurring at the fusion of the outer acrosomal membrane with the sperm plasma membrane. The breakdown of the fused complex results then in the formation of a highly fusible membrane enabling the spermatozoon to penetrate into the oocyte and fertilize it.

Numerous cytoskeletal elements and proteins appear to be involved in either binding and AR, involving mainly the actin and mediated by numerous second messengers. Data suggest that actin polymerization may represent an important regulatory pathway which is associated with tyrosine phosphorylation in spermatozoa [52].

In echinoderms, the most important event which occurs during the AR is the polymerization of actin, which form the "skeleton" of the acrosomal process, a protuberance formed at the apex of the sperm head supported by a core of actin microfilaments [53, 54]. Apart this peculiar event in echinoderms, many authors have suggested that the acrosomal architecture is supported by a dynamic F-actin skeleton, which probably regulates the differential rate of release of the acrosomal enzymes during AR [55].

Changes and regulation of the sperm actin cytoskeleton in fact, take place during AR; in mammals, polymerization of actin from its globular (G)- monomeric form to filamentous (F)-actin occurs during capacitation, depending on phosphorylation processes. F-actin formation is important for the translocation of phospholipase C from the cytosol to the sperm plasma membrane during capacitation. Before the occurrence of AR, depolymerization of F-actin enables the outer acrosomal membrane to fuse with the plasma membrane [52, 56]. In support of this finding, an important role of actin polymerization has also been shown in human sperm AR since actin is present in the acrosomal area and is lost with the AR [57]. In human and other mammalian spermatozoa, cytoskeletal proteins including spectrin, F-actin and α-tubulin were mostly localized to the apical and the equatorial acrosomal region of the sperm head, and their modification after AR was evidenced suggesting, at least, that they may play more than a role in the development of the AR and priming the spermatozoa for other fertilization events [9, 10].

After AR is completed, the spermatozoon can begin penetration through the extracellular layers. Penetration may involve enzymatic hydrolysis of the extracellular matrix but also requires the forward physical force of sperm motility [58].

More specific structure are present on the sperm head of murids where two very large cytoskeletal structures seem to be involved in binding of the spermatozoon to the outer surface of the ZP and/or in aiding the spermatozoon in ZP penetration at the time of fertilization [59].

The possible role of actin filaments in the penetration of spermatozoa has been evidenced by indirect proofs in mammals since cytochalasin D inhibits sperm penetration and sperm head decondensation [60]. Similar investigations in mammals proved that either actin polymerization [61] and Rho protein(s) regulating actin-based cytoskeletal reorganization are involved in the process leading to sperm incorporation into the oocyte cytoplasm [62].

Once the spermatozoon penetrates the oocyte a fusion of the two plasma membrane occurs.

Sperm penetration occurs vertical to the surface of invertebrate oocytes possessing a jelly coat, whereas in mammals, sperm lies and fuse tangential to the oocyte surface [63]. In the latter, although the exact sperm fusogenic region is not fully established, studies suggest that this is a region overlapping either the equatorial segment or the postacrosomal region depending on the species under study [23]. Despite the importance of this fundamental process, little is known about its molecular basis. Although a number of molecules involved in the binding and fusion have been disclosed [64], indirect evidences supported the Izumo, a sperm-specific member of the immunoglobulin superfamily which relocalization to the equatorial segment after the AR is essential for gamete fusion and the testis-specific serine kinase 6 that plays a role in the changes of Izumo localization through the regulation of actin polymerization [65, 66]. Contrasting data from other authors [61] showed that, although involved in sperm penetration, actin polymerization is not required for plasma membrane gametes fusion in guinea pig.

Very recently, experiments aimed to investigate a role of proteins enriched in the cytoskeletal structures of human spermatozoa demonstrated that signal transducer and activator of transcription 3 (STAT3), present mainly in the flagellar structure, affects sperm functions such as motility parameters, AR and depolarization of mitochondrial membranes [67].

Evidences were presented of an involvement of organellar movements in the ascidian spermatozoon. This lacks an evident acrosome and midpiece but presents a single mitochondrion beside the nucleus in the head that swells at the time of oocyte interaction being translocated to the tail. Such a movement appear to be mediated by an actin-myosin sliding system [68]. To conclude, a very recent computational and experimental approach pointed out that the "actin polymerization" have some important and unique features by linking in a specific way all the intracellular compartments. Thus, it was suggested that actin polymerization could be involved in the signaling coordination of different events and that its functional ablation could compromise spermatozoa ability to complete the capacitation. This study strengthen the idea that the actin cytoskeleton is not only a mechanical support for the sperm cell, but that it exerts a key role in signaling during capacitation [69] (Fig. 6.1).

The Oocyte

In oocyte maturation, GVBD represents the nuclear event strictly related to the first meiotic block resumption and the following meiotic spindle formation. The involvement of cytoskeletal elements to GVBD has been investigated in several species.

In some mammals, actin filaments are distributed in a uniform way just around the oocyte cortex and close to the GV, and undergo a redistribution after the GVBD leading the chromosome to move to a peripheral position [70–74]; however this event does not seem to influence either GVBD or the spindle formation [32]. The presence of a cortical "organizing pole" of microfilaments has been hypothesized in the maturing mouse oocytes especially during centrosome localization,

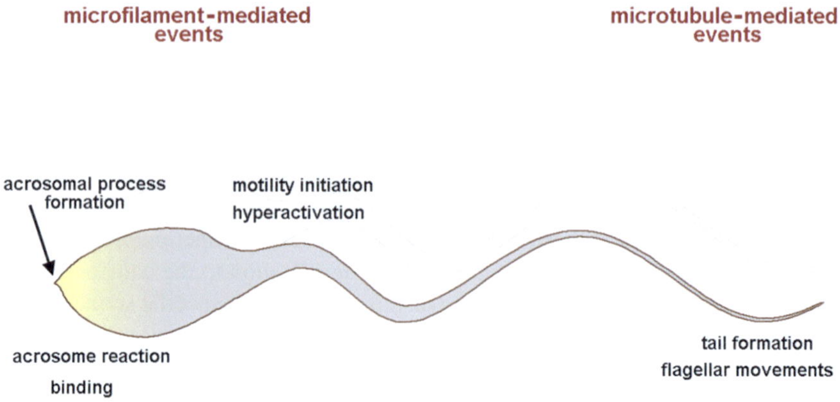

Fig. 6.1 Involvement of cytoskeletal elements during spermatozoon maturation and activation. Microfilaments are involved in the acrosome reaction, formation of the acrosomal process, first binding to the oocyte, motility initiation and hyperactivation. Microtubules are the constituent of sperm flagellum that is formed during the process of spermiogenesis and allow the flagellar movement

spindle (or GV) movement to the oocyte periphery [75]. In other vertebrates, such as amphibians, contrasting data exist on the role of cortical actin microfilaments that appear to be required for anchoring and rotation of the meiotic spindles [76] and the completion of GVBD [77]. In the ascidian *Halocynthia roretzi*, it has been identified the formation of conspicuous actin bundles emanating from the GV during its breakdown [78], and same authors showed that after GVBD a meiotic spindle forms in the center of the oocyte migrating toward the animal pole requiring actin cytoskeleton to support the polarization [79]. Although the latter event occurs normally in the absence of microtubules, cytoskeletal elements interact by each other giving rise to a fruitful interplay; this is the case of actin filament modulation of microtubules functions that drive chromosomes segregation in the mitotic and meiotic spindles, their positioning and orientation, processes that appear to be essential for the asymmetric cell division [80–83]. In fact, the synergy between microfilaments and microtubules has been supported by experiments based on specific chemicals such as cytochalasins [74, 84] in mammals. Similarly in amphibians, data are provided on the involvement of actin filaments in spindle anchorage [76, 85] and of myosin-10 (a phosphoinositide-binding actin-based protein) in association with microtubules in vitro and in vivo, with a specific localization at the point where the meiotic spindle contacts the F-actin-rich cortex [86].

The meiotic spindle consists of bundles of microtubules that emanate from two acentriolar poles and hold chromosomes along the metaphase plate. At meiosis resumption the spindle segregates sister chromatids or homologous chromosomes equally between the pronucleus and the second polar body playing a critical role in the generation of right chromosome segregation [87, 88]. Literature reports the requirement of microtubule associated motor proteins, for the proper distribution of chromosomes or the structural integrity of the mitotic or meiotic spindle [89].

Interestingly, it has been shown that a microtubules perturbation induces negative impact on GVBD and the meiotic resumption [90].

The interaction of the spermatozoon with the oocyte causes a series of physiological changes in the oocyte known as activation. An early event that occurs at fertilization is the change in the oocyte plasma membrane electrical properties [16, 39, 91] and the second main universal event is the massive release of calcium that traverses the oocyte in a wave [92–94], leading also to relocation of the organelles.

Organelles that have been organized in specific sites of the oocyte during growth and maturation undergo a relocation at the oocyte activation. In the ascidian *Styela plicata* we reported a pattern of mitochondria polarization and aggregation in the subcortical cytoplasm during oocyte growth [95]. These data strongly support what occurs in ascidian oocytes at the time of activation when the subcortical mitochondria are transported to the vegetal pole [96, 97]. This process called cytoplasmic segregation, that is necessary for the establishment of cell lines and in turn for determining the embryonic axis, involves myoplasmic actin-filaments network in a first phase whereas in the second phase involves extension of microtubules [14].

On the other hand the requirement of actin in the first phase after sperm contact has been shown [98] since perturbation of fertilization in ascidians with specific channel inhibitors altered either actin filaments and mitochondrial migration after contraction leading to a disturbance in the following cleavage formation. The regulation of mitochondrial translocation by microfilaments and microtubules observed in mammals indicated that either oocyte maturation, fertilization and early embryo development in pigs are associated with changes in active mitochondrial distribution and that this is mediated exclusively by microtubules [99]. However more recent evidence also indicates that the cytoskeleton network is used to shuttle organelles to specific sites within the oocyte cytoplasm [100].

Following the fusion of the spermatozoon with the oocyte plasma membrane, a third event occurs when the oocyte secretes the contents of CG by exocytotic fusions of these vesicles with the oocyte plasma membrane over the entire cell surface, also known as the cortical reaction or CG exocytosis [101]. This peculiar process is followed by an elevation or hardening of the extracellular coat involved in the polispermy prevention in sea urchin and mammals, however it does not occur in ascidians since their oocytes lack CG. Many cell types possess finger-like projections termed microvilli. In the sea urchin an ultrastructural study localized filamentous actin immediately subjacent to the microvilli forming an extensive interconnecting network along the inner surface of the plasma membrane with an organization of this network correlated to the positioning of the underlying CG [102]. That the microfilament assembly is involved in the distribution, movement and exocytosis of CG during maturation and fertilization has been shown by confocal microscopy in the pig oocyte. Here, it was suggested an integral changes in microfilament assembly and CG distribution during oocyte maturation, parthenogenic activation and in vitro fertilization [103]. Similarly in the rat it was supported the role of cytoskeletal cortex as a dynamic network that modulate CG exocytosis by activated actin-associated proteins and/or by activated protein kinase C [104].

As a consequence of cortical reaction the thousands of vesicles fusing with the oocyte surface, add their membranes to the oocyte plasma membrane resulting in an approximate doubling of the amount of membrane on the oocyte surface in a few seconds with the production of a mosaic topography, so the excess surface membrane is therefore accommodated by the elongation of oocyte microvilli [105, 106].

Earlier studies in the 1980 showed a dramatic reorganization occurring in the structure of the oocyte surface due to the mosaic membrane formed after activation and the resulting elongation of numerous short microvilli that covers the surface of the unfertilized oocyte. A localization of actin in the microvilli has been also deeply investigated in sea urchin species showing the formation of bundles of actin filaments in microvilli and in cones [107]. This suggested that this microvillar-associated actin was an organizational state composed of very short filaments arranged in a tight network and that these filament networks were extended beyond the plane of the plasma membrane [108].

In the cortical region of amphibian and rat oocytes it has been shown a significant amount of polymerized actin organized into bundles within the short microvilli covering the oocyte surface [109]. In the sea urchin, morphological studies evidenced two bursts of microvilli elongation concomitantly to sperm entering and incorporation, as a result of a massive polymerization of actin and a new assembly of microfilaments in the oocyte cortex reorganization that was suggested to produce the forces necessary to held firmly the spermatozoon for fusion and subsequently for cytokinesis occurrence [54, 110, 111]. More recent studies in mammals, further support the role of cytoskeletal actin in microvilli formation and their function to capture the sperm cell and bring it into close contact with the oocyte plasma membrane [17]. This data should also support the fact that in some mammalian oocytes the spermatozoon do not normally fuse with the microvillus-free area [112].

That sperm incorporation is a microfilament-dependent process has been shown in Xenopus [113] but this process has been described to also occur through the formation on the oocyte surface of a specific structure named the fertilization cone involving the functioning of actin microfilament organization. This process is related to the overmentioned elongation of microvilli and has been described to occur in the echinoderms [114–117]. In the sheep at the site of sperm head incorporation, the fertilization cone develops above the decondensing male chromatin and is underlined by a submembranous area rich in microfilaments [118].

Once the sperm has entered into the oocyte, the proximal centrosome adjacent to the sperm nucleus may become the center of the sperm aster that brings the male and female pronuclei to the center of the zygote [119]. Aster is a peculiar structure that appears initially after the centriole duplication at the pronuclear stage required for the union of the sperm and oocyte nuclei and is formed by the assembly of the microtubules mainly composed by the g-tubulin which is also needed for the subsequent enlargement and association with the female pronucleus [22, 120, 121]. In the rabbit, earlier studies showed the presence and continuous deposits of tubulin throughout the sperm penetration tunnels and entry point suggesting a role in fertilization, possibly as an enzyme binding or delivery system [122]. More recently it was shown that the microtubules extending from the decondensed sperm head participating in pronuclear migration and organization around the female pronucleus resulted mainly composed by γ-tubulin [123].

Among the events occurring at oocyte activation the change in the ionic permeability of the oocyte due to the generation of a ion current across the plasma membrane and calcium release play a pivotal role. Depending on the species, specific and nonspecific ion currents are involved at the early sperm-oocyte interaction [16, 124]. It is well established that actin filaments are important in ion channel regulation and membrane potential modulation [125, 126], although this does not seem to be the case of ascidian oocytes since actin filaments have no impact on the fertilization current or plasma membrane [98]. Indirect evidences support that spectrin, a major component of the membrane skeleton, is a functional link with membrane channels and transporters [127, 128].

Calcium release is the universal event occurring at fertilization in all species studied [94]. In the ascidians, fertilization at MI initiates a series of dramatic cytoplasmic and cortical reorganizations of the zygote, which occur in two major phases [129]. The first major phase depends on sperm entry which triggers a calcium wave leading in turn to an actomyosin-driven contraction wave. The second major phase of reorganization occurs between meiosis completion and the first cleavage. Sperm aster microtubules and then cortical microfilaments cause the reposition toward the posterior side of the zygote of myoplasm and of domain rich in cortical endoplasmic reticulum and maternal RNAs [130].

The possibility that intracellular calcium signaling could be modulated by the actin cytoskeleton at the time of gamete interaction has been also recently hypothesized in starfish [131, 132] whereas in the sea urchin the calcium-responsive contractility during fertilization is modulated by the myosin II localized to the cortical cytoskeleton. This seems also to influence the fertilization cone absorption and to participate in the remodeling of the cortical actomyosin cytoskeleton during the following first zygotic cell cycle [133]. Finally a coordinated mobilization of intracellular calcium stores and a precise organization of the cytoskeletal network have been shown to be essential for an appropriate activation of the oocyte and chromosome migration during human fertilization [134].

Post-fertilization events include the sperm cell nucleus breakdown and chromatin decondensation that is then surrounded by an envelope forming the male pronucleus. The latter, together with the female pronucleus located just below the extruded polar body, start to move toward the center of the oocyte. These processes are under the influence of factors in the cytoplasm. Emission of the polar body due to meiosis resumption has been shown to be underlined by the formation of a contractile ring of actin in the cleavage furrow of the asymmetric division of the oocyte [70, 135] whereas migration of the pronuclei depends strictly on the microtubules of the sperm aster [63]. Rotation of meiotic spindle is under the control of microfilaments [76, 136] but a peculiar interplay between astral microtubules and cortical actin filaments has been suggested for spindle positioning [137–139] and pronuclear apposition. Although a main role of F-actin in the formation of contractile ring during the first cleavage division has been well documented [140], the cooperation between the two main cytoskeletal elements has also been identified in the Xenopus [141] (Fig. 6.2).

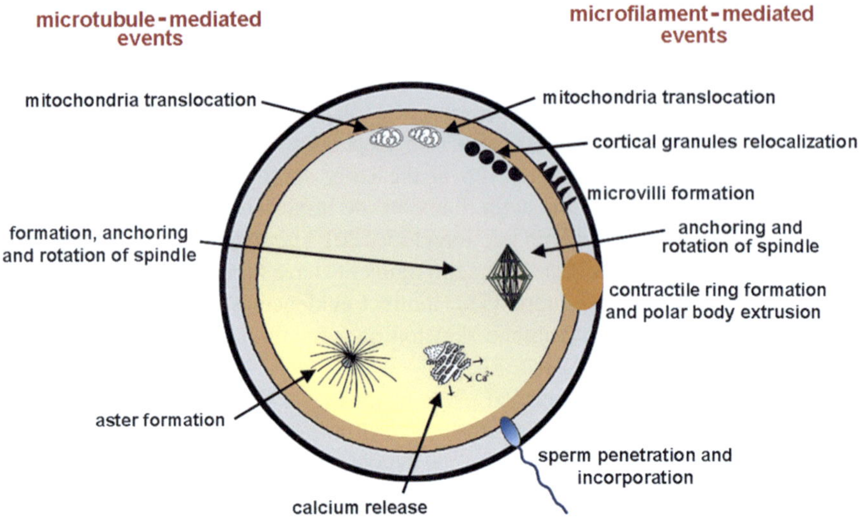

Fig. 6.2 Involvement of cytoskeletal elements during oocyte maturation and activation. Microfilaments are involved in the cortical granules relocalization during growth and the formation of microvilli just after the exocytosis of cortical granules occurring at fertilization. They also participate to formation of contractile ring that in turn give rise to the polar body extrusion. Sperm incorporation and penetration in the oocyte and the following calcium release are also modulated by actin microfilaments. Microtubules are involved in the formation of meiotic spindle and of sperm aster that drives the male pronucleus toward the female one after fertilization. Either microfilaments and microtubules participate to the mitochondria translocation during maturation and to the anchoring and rotation of the meiotic spindle after the sperm-oocyte interaction

The Zygote

Right embryo development relies on the positioning of the cleavage plane which is in turn related to the position of the mitotic spindle. In the mouse zygote it has been demonstrated an accumulation of F-actin surrounding the spindle and that actin network maintains the central spindle position ensuring that the first embryonic mitosis is symmetric [142]. However, in the establishment of the right symmetry in cell divisions during differentiation and subsequent embryo development a central role is also played by the paternal centrosome [143] whose role and involvement of cytoskeletal elements has been previously reported. In fact, in bovine evidence have been provided that γ-tubulin and microtubule dynamics are involved in the migration and centration of the female pronucleus [144]. On the other hand indirect evidences exist that perturbation of tubulin polymerization induces meiotic delay and spindle defects contributing to formation of aneuploid mouse zygotes [145]. By contrast in human zygotes showing abnormal fertilization, no any kind of microtubule alteration with respect to the ploidy level was observed [146]. Also in the invertebrates the cortical actin cytoskeleton undergoes dramatic rearrangements with a level of F-actin decreasing after fertilization and continuing to decrease throughout the first cell cycle of sea urchin [147]. Such a dynamic nature of cortical actin organization during early development demonstrated also that cytokinesis occurs at the point of minimum cortical F-actin content suggesting that these changes do not function in the establishment of the contractile apparatus for cytokinesis, but rather serve other developmental functions [148]. Similarly in ascidians the determinants for unequal cleavage, gastrulation and further developmental events reside in four distinct cortical and cytoplasmic domains localized in the oocyte between fertilization and first divisions [96].

Conclusions

Involvements of cytoskeleton in reproductive processes have received the special attention of many authors. The related large body of literature shows an impressive variation along the species, however, common general characteristics of the process emerge, allowing to depict a general picture of the complex interplay between cytoskeletal elements and the physiology of fertilization. The cytoskeleton has a fundamental role in numerous cellular processes, consequently, it has been shown that abnormalities in the regulation of cytoskeleton dynamics are typical for many pathological states from infection processes up to cancer [149–151]. Fertilization is a multistep process in which all physiologically relevant events are intimately connected with each other and in turn are crucial for the entire process of reproduction; therefore it appears that only the right combination of multifactorial causes brings to a normal embryo and organism development. Here, we have reported that cytoskeletal elements as microfilaments and microtubules are involved in all the steps from the maturation of gametes, their reciprocal activation to the final interaction and the initiation of embryo development. Although sometime results are controversial and come from indirect experimental data, evidence are provided that perturbation of the cytoskeleton, with toxins or heath shock, exerts a wide range of impacts on the entire reproductive process including sperm maturation and motility, oocyte maturation, fertilization and embryo development [152–154]. Studies on human in vitro fertilization evidenced the delicate nature of the oocyte and the instrumental role played in fertilization reinforcing the view that: (1) exposure to mechanical stressors has the potential to compromise oocyte developmental competence; (2) defects in any of the aforementioned reproductive events are lethal to the embryo development and might be causes of infertility; (3) cytoskeletal dynamics perturbation of gametes may be considered a factor of human infertility [124, 155–161]. In this chapter, we wished to bring the general concepts that the major cytoskeletal structures are involved in the reproductive processes. We would like to apologize with the colleagues for not having reported all their valuable studies on animal models such as drosophila, zebrafish, nematodes etc., but given the vastness of the literature on a variety of animal species we have chosen to deal about those species which have always been models for the study of reproduction including either invertebrates and vertebrates.

References

1. Bement WM, Ian Gallicano G, Capco DG (1992) Role of the cytoskeleton during early development. Microsc Res Tech 22:23–48
2. Elinson R, Houliston E (1990) Cytoskeleton in Xenopus oocytes and eggs. Semin Cell Biol 1:349–357
3. Gallicano GI (2001) Composition, regulation, and function of the cytoskeleton in mammalian eggs and embryos. Front Biosci 6:D1089–D1108
4. Longo FJ (1989) Egg cortical architecture. In: Schatten H, Schatten G (eds) The cell biology of fertilization, 1st edn. Academic, San Diego, CA
5. Koch RA, Lambert CC (1990) Ultrastructure of sperm, spermiogenesis, and sperm-egg interactions in selected invertebrates and lower vertebrates which use external fertilization. J Electron Microsc Tech 16:115–154
6. Sperry AO (2012) The dynamic cytoskeleton of the developing male germ cell. Biol Cell 104:297–305
7. McLay DW, Clarke HJ (2003) Remodelling the paternal chromatin at fertilization in mammals. Reproduction 125:625–633
8. Toshimori K, Ito C (2003) Formation and organization of the mammalian sperm head. Arch Histol Cytol 66:383–396
9. Dvořáková K, Moore HD, Šebková N, Paleček J (2005) Cytoskeleton localization in the sperm head prior to fertilization. Reproduction 130:61–69
10. Lie PP, Mruk DD, Lee WM, Cheng CY (2010) Cytoskeletal dynamics and spermatogenesis. Philos Trans R Soc Lond B Biol Sci 365:1581–1592
11. Sun X, Kovacs T, Hu YJ, Yang WX (2011) The role of actin and myosin during spermatogenesis. Mol Biol Rep 38:3993–4001
12. Sathananthan A (1997) Ultrastructure of the human egg. Hum Cell 10:21
13. Oulhen N, Reich A, Wong JL, Ramos I, Wessel GM (2013) Diversity in the fertilization envelopes of echinoderms. Evol Dev 15:28–40
14. Satoh N (1994) Developmental biology of ascidians. Cambridge University Press, Cambridge
15. Gupta SK, Bhandari B, Shrestha A, Biswal BK, Palaniappan C, Malhotra SS, Gupta N (2012) Mammalian zona pellucida glycoproteins: structure and function during fertilization. Cell Tissue Res 349:665–678
16. Tosti E, Boni R (2004) Electrical events during gamete maturation and fertilization in animals and humans. Hum Reprod Update 10:53–65
17. Runge KE, Evans JE, He Z-Y, Gupta S, McDonald KL, Stahlberg H, Primakoff P, Myles DG (2007) Oocyte CD9 is enriched on the microvillar membrane and required for normal microvillar shape and distribution. Dev Biol 304:317–325
18. Liu M (2011) The biology and dynamics of mammalian cortical granules. Reprod Biol Endocrinol 9:149
19. Hyttel P (2011) Electron microscopy of mammalian oocyte development, maturation and fertilization. In: Tosti E, Boni R (eds) Oocyte maturation and fertilization: a long history for a short event. Bentham Science Publisher, Dubai, United Arab Emirates
20. Voronina E, Wessel GM (2003) The regulation of oocyte maturation. Curr Top Dev Biol 58:53–110
21. Eppig JJ (1996) Coordination of nuclear and cytoplasmic oocyte maturation in eutherian mammals. Reprod Fertil Dev 8:485–489
22. Tsafriri A (1979) Mammalian oocyte maturation: model systems and their physiological relevance. Adv Exp Med Biol 112:269–281
23. Yanagimachi R (1994) Mammalian fertilization. Physiol Reprod 1:189–317
24. Albertini DF, Sanfins A, Combelles CM (2003) Origins and manifestations of oocyte maturation competencies. Reprod Biomed Online 6:410–415
25. Connors SA, Kanatsu-Shinohara M, Schultz RM, Kopf GS (1998) Involvement of the cytoskeleton in the movement of cortical granules during oocyte maturation, and cortical granule anchoring in mouse eggs. Dev Biol 200:103–115

26. Ducibella T, Anderson E, Albertini DF, Aalberg J, Rangarajan S (1988) Quantitative studies of changes in cortical granule number and distribution in the mouse oocyte during meiotic maturation. Dev Biol 130:184–197
27. Ferreira E, Vireque A, Adona P, Meirelles F, Ferriani R, Navarro P (2009) Cytoplasmic maturation of bovine oocytes: structural and biochemical modifications and acquisition of developmental competence. Theriogenology 71:836–848
28. Racedo SE, Rawe VY, Niemann H (2012) Dynamic changes of the Golgi apparatus during bovine in vitro oocyte maturation. Reproduction 143:439–447
29. Santella L, De Riso L, Gragnaniello G, Kyozuka K (1999) Cortical granule translocation during maturation of starfish oocytes requires cytoskeletal rearrangement triggered by InsP3-mediated Ca^{2+} release. Exp Cell Res 248:567–574
30. Schorderet-Slatkine S (1972) Action of progesterone and related steroids on oocyte maturation in Xenopus laevis. An in vitro study. Cell Differ 1:179–189
31. Stricker SA (2006) Structural reorganizations of the endoplasmic reticulum during egg maturation and fertilization. Semin Cell Dev Biol 17:303–313
32. Sun QY, Schatten H (2006) Regulation of dynamic events by microfilaments during oocyte maturation and fertilization. Reproduction 131:193–205
33. Suzuki H, Yang X, Foote RH (1994) Surface alterations of the bovine oocyte and its investments during and after maturation and fertilization in vitro. Mol Reprod Dev 38:421–430
34. Tosti E, Boni R, Gallo A, Silvestre F (2013) Ion currents modulating oocyte maturation in animals. Syst Biol Reprod Med 59:61–68
35. Van Blerkom J (1991) Microtubule mediation of cytoplasmic and nuclear maturation during the early stages of resumed meiosis in cultured mouse oocytes. Proc Natl Acad Sci U S A 88:5031–5035
36. Bavister BD, Squirrell JM (2000) Mitochondrial distribution and function in oocytes and early embryos. Hum Reprod 15:189–198
37. Primakoff P, Myles DG (2002) Gamete fusion in mammals. In: Hardy DM (ed) Fertilization, 1st edn. Academic, San Diego, CA
38. Tosti E (1994) Sperm activation in species with external fertilisation. Zygote 2:359–361
39. Dale B (1994) Oocyte activation in invertebrates and humans. Zygote 2:373–377
40. Horner VL, Wolfner MF (2008) Transitioning from egg to embryo: triggers and mechanisms of egg activation. Dev Dyn 237:527–544
41. Inaba K (2003) Molecular architecture of the sperm flagella: molecules for motility and signaling. Zoolog Sci 20:1043–1056
42. Jouannet P, Serres C (1997) The movement of the human spermatozoon. Bull Acad Natl Med 182:1025–1034
43. Shingyoji C (2013) Measuring the regulation of dynein activity during flagellar motility. Methods Enzymol 524:147
44. Nakachi M, Nakajima A, Nomura M, Yonezawa K, Ueno K, Endo T, Inaba K (2011) Proteomic profiling reveals compartment-specific, novel functions of ascidian sperm proteins. Mol Reprod Dev 78:529–549
45. Ma X, Zhao Y, Sun W, Shimabukuro K, Miao L (2012) Transformation: how do nematode sperm become activated and crawl? Protein Cell 3:755–761
46. Roberts TM, Stewart M (2012) Role of major sperm protein (MSP) in the protrusion and retraction of Ascaris sperm. Int Rev Cell Mol Biol 297:265–293
47. Finkelstein M, Megnagi B, Ickowicz D, Breitbart H (2013) Regulation of sperm motility by PIP2(4,5) and actin polymerization. Dev Biol 381:62–72
48. Lin M, Hess R, Aitken R (2002) Induction of sperm maturation in vitro in epididymal cell cultures of the tammar wallaby (*Macropus eugenii*): disruption of motility initiation and sperm morphogenesis by inhibition of actin polymerization. Reproduction 124:107–117
49. Kaupp UB, Kashikar ND, Weyand I (2008) Mechanisms of sperm chemotaxis. Annu Rev Physiol 70:93–117
50. Hozumi A, Padma P, Toda T, Ide H, Inaba K (2008) Molecular characterization of axonemal proteins and signaling molecules responsible for chemoattractant-induced sperm activation in Ciona intestinalis. Cell Motil Cytoskeleton 65:249–267

51. Barros C, Crosby J, Moreno R (1996) Early steps of sperm-egg interactions during mammalian fertilization. Cell Biol Int 20:33–39
52. Brener E, Rubinstein S, Cohen G, Shternall K, Rivlin J, Breitbart H (2003) Remodeling of the actin cytoskeleton during mammalian sperm capacitation and acrosome reaction. Biol Reprod 68:837–845
53. Tilney LG, Inoué S (1982) Acrosomal reaction of Thyone sperm. II. The kinetics and possible mechanism of acrosomal process elongation. J Cell Biol 93:820–827
54. Schatten G, Schatten H (1983) Fertilization and early development of sea urchins. Scan Electron Microsc (Pt 3):1403–1413
55. Zepeda-Bastida A, Chiquete-Felix N, Uribe-Carvajal S, Mujica A (2011) The acrosomal matrix from Guinea pig sperm contains structural proteins, suggesting the presence of an actin skeleton. J Androl 32:411–419
56. Breitbart H, Cohen G, Rubinstein S (2005) Role of actin cytoskeleton in mammalian sperm capacitation and the acrosome reaction. Reproduction 129:263–268
57. Liu D, Martic M, Clarke G, Dunlop M, Baker H (1999) An important role of actin polymerization in the human zona pellucida-induced acrosome reaction. Mol Hum Reprod 5:941–949
58. Miller DJ, Shi X, Burkin H (2002) Molecular basis of mammalian gamete binding. Recent Prog Horm Res 57:37–73
59. Breed WG, Idriss D, Leigh CM, Oko RJ (2009) Temporal deposition and spatial distribution of cytoskeletal proteins in the sperm head of an Australian rodent. Reprod Fertil Dev 21:428–439
60. Rogers B, Bastias C, Coulson RL, Russell LD (1989) Cytochalasin D inhibits penetration of hamster eggs by guinea pig and human spermatozoa. J Androl 10:275–282
61. Sánchez-Gutiérrez M, Contreras RG, Müjica A (2002) Cytochalasin-D retards sperm incorporation deep into the egg cytoplasm but not membrane fusion with the egg plasma membrane. Mol Reprod Dev 63:518–528
62. Kumakiri J, Oda S, Kinoshita K, Miyazaki S (2003) Involvement of Rho family G protein in the cell signaling for sperm incorporation during fertilization of mouse eggs: inhibition by Clostridium difficile toxin B. Dev Biol 260:522–535
63. Schatten G, Schatten H (1987) Cytoskeletal alterations and nuclear architectural changes during mammalian fertilization. Curr Top Dev Biol 23:23–54
64. Talbot P, Shur BD, Myles DG (2003) Cell adhesion and fertilization: steps in oocyte transport, sperm-zona pellucida interactions, and sperm-egg fusion. Biol Reprod 68:1–9
65. Inoue N, Ikawa M, Isotani A, Okabe M (2005) The immunoglobulin superfamily protein Izumo is required for sperm to fuse with eggs. Nature 434:234–238
66. Sosnik J, Miranda PV, Spiridonov NA, Yoon S-Y, Fissore RA, Johnson GR, Visconti PE (2009) Tssk6 is required for Izumo relocalization and gamete fusion in the mouse. J Cell Sci 122:2741–2749
67. Lachance C, Goupil S, Leclerc P (2013) Stattic V, a STAT3 inhibitor, affects human spermatozoa through regulation of mitochondrial activity. J Cell Physiol 228:704–713
68. Lambert CC, Lambert G (1984) The role of actin and myosin in ascidian sperm mitochondrial translocation. Dev Biol 106:307–314
69. Bernabò N, Berardinelli P, Mauro A, Russo V, Lucidi P, Mattioli M, Barboni B (2011) The role of actin in capacitation-related signaling: an in silico and in vitro study. BMC Syst Biol 5:47
70. Azoury J, Verlhac MH, Dumont J (2009) Actin filaments: key players in the control of asymmetric divisions in mouse oocytes. Biol Cell 101:69–76
71. Kim NH, Chung HM, Cha KY, Chung KS (1998) Microtubule and microfilament organization in maturing human oocytes. Hum Reprod 13:2217–2222
72. Longo FJ, Chen D-Y (1985) Development of cortical polarity in mouse eggs: involvement of the meiotic apparatus. Dev Biol 107:382–394
73. Terada Y, Fukaya T, Yajima A (1995) Localization of microfilaments during oocyte maturation of golden hamster. Mol Reprod Dev 41:486–492

74. Wang WH, Abeydeera LR, Prather RS, Day BN (2000) Polymerization of nonfilamentous actin into microfilaments is an important process for porcine oocyte maturation and early embryo development. Biol Reprod 62:1177–1183
75. Calarco PG (2005) The role of microfilaments in early meiotic maturation of mouse oocytes. Microsc Microanal 11:146–153
76. Gard DL, Cha B-J, Roeder AD (1995) F-actin is required for spindle anchoring and rotation in Xenopus oocytes: a re-examination of the effects of cytochalasin B on oocyte maturation. Zygote 3:17–26
77. Okada I, Fujiki S, Iwase S, Abe H (2012) Stabilization of actin filaments prevents germinal vesicle breakdown and affects microtubule organization in Xenopus oocytes. Cytoskeleton 69:312–323
78. Prodon F, Hanawa K, Nishida H (2009) Actin microfilaments guide the polarized transport of nuclear pore complexes and the cytoplasmic dispersal of Vasa mRNA during GVBD in the ascidian *Halocynthia roretzi*. Dev Biol 330:377–388
79. Prodon F, Chenevert J, Sardet C (2006) Establishment of animal-vegetal polarity during maturation in ascidian oocytes. Dev Biol 290:297–311
80. Brunet S, Verlhac MH (2011) Positioning to get out of meiosis: the asymmetry of division. Hum Reprod Update 17:68–75
81. Kwon S, Lim HJ (2011) Small GTPases and formins in mammalian oocyte maturation: cytoskeletal organizers. Clin Exp Reprod Med 38:1–5
82. Li R, Albertini DF (2013) The road to maturation: somatic cell interaction and self-organization of the mammalian oocyte. Nat Rev Mol Cell Biol 14:141–152
83. McNally FJ (2013) Mechanisms of spindle positioning. J Cell Biol 200:131–140
84. Sun QY, Lai L, Park KW, Kühholzer B, Prather RS, Schatten H (2001) Dynamic events are differently mediated by microfilaments, microtubules, and mitogen-activated protein kinase during porcine oocyte maturation and fertilization in vitro. Biol Reprod 64:879–889
85. Ryabova L, Betina M, Vassetzky S (1986) Influence of cytochalasin B on oocyte maturation in Xenopus laevis. Cell Differ 19:89–96
86. Weber KL, Sokac AM, Berg JS, Cheney RE, Bement WM (2004) A microtubule-binding myosin required for nuclear anchoring and spindle assembly. Nature 431:325–329
87. Howe K, FitzHarris G (2013) Recent insights into spindle function in mammalian oocytes and early embryos. Biol Reprod 89:71
88. Yllera-Fernandez MDM, Crozet N, Ahmed-Ali M (1992) Microtubule distribution during fertilization in the rabbit. Mol Reprod Dev 32:271–276
89. Sawin KE, Endow SA (1993) Meiosis, mitosis and microtubule motors. Bioessays 15:399–407
90. Ya R, Downs SM (2012) Perturbing microtubule integrity blocks AMP-activated protein kinase-induced meiotic resumption in cultured mouse oocytes. Zygote 1:1–12
91. Dale B, Wilding M (2011) Ionic events at fertilization. In: Tosti E, Boni R (eds) Oocyte maturation and fertilization: a long history for a short event. Bentham Science Publisher, Dubai, United Arab Emirates
92. Krauchunas AR, Wolfner MF (2012) Molecular changes during egg activation. Curr Top Dev Biol 102:267–292
93. Ramos I, Wessel GM (2013) Calcium pathway machinery at fertilization in echinoderms. Cell Calcium 53:16–23
94. Whitaker M (2006) Calcium at fertilization and in early development. Physiol Rev 86:25–88
95. Bezzaouia A, Gallo A, Silvestre F, Tekaya S, Tosti E (2014) Distribution pattern and activity of mitochondria during oocyte growth and maturation in the ascidian *Styela plicata*. Zygote 22(4):462–469. doi:10.1017/S0967199412000640
96. Roegiers F, Djediat C, Dumollard R, Rouvière C, Sardet C (1999) Phases of cytoplasmic and cortical reorganizations of the ascidian zygote between fertilization and first division. Development 126:3101–3117

97. Sawada T, Osanai K (1985) Distribution of actin filaments in fertilized egg of the ascidian Ciona intestinalis. Dev Biol 111:260–265
98. Tosti E, Romano G, Buttino I, Cuomo A, Ianora A, Miralto A (2003) Bioactive aldehydes from diatoms block the fertilization current in ascidian oocytes. Mol Reprod Dev 66:72–80
99. Sun QY, Wu G, Lai L, Park K, Cabot R, Cheong H, Day B, Prather R, Schatten H (2001) Translocation of active mitochondria during pig oocyte maturation, fertilization and early embryo development in vitro. Reproduction 122:155–163
100. Brevini T, Cillo F, Antonini S, Gandolfi F (2007) Cytoplasmic remodelling and the acquisition of developmental competence in pig oocytes. Anim Reprod Sci 98:23–38
101. Gadella B, Evans J (2011) Membrane fusions during mammalian fertilization. Adv Exp Med Biol 713:65–80
102. Bonder EM, Fishkind DJ, Cotran NM, Begg DA (1989) The cortical actin-membrane cytoskeleton of unfertilized sea urchin eggs: analysis of the spatial organization and relationship of filamentous actin, nonfilamentous actin, and egg spectrin. Dev Biol 134:327–341
103. Kim NH, Day BN, Lee HT, Chung KS (1996) Microfilament assembly and cortical granule distribution during maturation, parthenogenetic activation and fertilisation in the porcine oocyte. Zygote 4:145–149
104. Eliyahu E, Tsaadon A, Shtraizent N, Shalgi R (2005) The involvement of protein kinase C and actin filaments in cortical granule exocytosis in the rat. Reproduction 129:161–170
105. Begg DA, Rebhun LI, Hyatt H (1982) Structural organization of actin in the sea urchin egg cortex: microvillar elongation in the absence of actin filament bundle formation. J Cell Biol 93:24–32
106. Eddy E, Shapiro B (1979) Membrane events of fertilization in the sea urchin. Scan Electron Microsc 3:287–297
107. Tilney LG, Jaffe LA (1980) Actin, microvilli, and the fertilization cone of sea urchin eggs. J Cell Biol 87:771–782
108. Henson JH, Begg DA (1988) Filamentous actin organization in the unfertilized sea urchin egg cortex. Dev Biol 127:338–348
109. Franke W, Rathke P, Seib E, Trendelenburg M, Osborn M, Weber K (1976) Distribution and mode of arrangement of microfilamentous structures and actin in the cortex of the amphibian oocyte. Cytobiologie 14:111–130
110. Burgess DR, Schroeder TE (1977) Polarized bundles of actin filaments within microvilli of fertilized sea urchin eggs. J Cell Biol 74:1032–1037
111. Cline CA, Schatten H, Balczon R, Schatten G (1983) Actin-mediated surface motility during sea urchin fertilization. Cell Motil 3:513–524
112. Talansky BE, Malter HE, Cohen J (1991) A preferential site for sperm-egg fusion in mammals. Mol Reprod Dev 28:183–188
113. Boyle JA, Chen H, Bamburg JR (2001) Sperm incorporation in Xenopus laevis: characterisation of morphological events and the role of microfilaments. Zygote 9:167–181
114. Kyozuka K, Osanai K (1994) Cytochalasin B does not block sperm penetration into denuded starfish oocytes. Zygote 2:103–109
115. Kyozuka K, Osanai K (1988) Fertilization cone formation in starfish oocytes: the role of the egg cortex actin microfilaments in sperm incorporation. Gamete Res 20:275–285
116. Schatten H, Schatten G (1980) Surface activity at the egg plasma membrane during sperm incorporation and its cytochalasin B sensitivity: scanning electron microscopy and time-lapse video microscopy during fertilization of the sea urchin *Lytechinus variegatus*. Dev Biol 78:435–449
117. Yonemura S, Mabuchi I (1987) Wave of cortical actin polymerization in the sea urchin egg. Cell Motil Cytoskeleton 7:46–53
118. Le Guen P, Crozet N, Huneau D, Gall L (1989) Distribution and role of microfilaments during early events of sheep fertilization. Gamete Res 22:411–425
119. Yanagimachi R (2005) Male gamete contributions to the embryo. Ann N Y Acad Sci 1061:203–207
120. Palermo GD, Colombero LT, Rosenwaks Z (1997) The human sperm centrosome is responsible for normal syngamy and early embryonic development. Rev Reprod 2:19–27

121. Schatten G, Simerly C, Schatten H (1985) Microtubule configurations during fertilization, mitosis, and early development in the mouse and the requirement for egg microtubule-mediated motility during mammalian fertilization. Proc Natl Acad Sci U S A 82:4152–4156
122. Stambaugh RL, Nicosia SV (1984) Localization of tubulin and microtubules of in vivo fertilized rabbit oocytes. J Androl 5:259–264
123. Terada Y, Simerly CR, Hewitson L, Schatten G (2000) Sperm aster formation and pronuclear decondensation during rabbit fertilization and development of a functional assay for human sperm. Biol Reprod 62:557–563
124. Swain JE, Pool TB (2008) ART failure: oocyte contributions to unsuccessful fertilization. Hum Reprod Update 14:431–446
125. Cantiello HF (1997) Role of actin filament organization in cell volume and ion channel regulation. J Exp Zool 279:425–435
126. Moccia F (2007) Latrunculin A depolarizes starfish oocytes. Comp Biochem Physiol A Mol Integr Physiol 148:845–852
127. Dubreuil RR (2006) Functional links between membrane transport and the spectrin cytoskeleton. J Membr Biol 211:151–161
128. Machnicka B, Czogalla A, Hryniewicz-Jankowska A, Bogusławska DM, Grochowalska R, Heger E, Sikorski AF (2013) Spectrins: a structural platform for stabilization and activation of membrane channels, receptors and transporters. Biochim Biophys Acta. doi:10.1016/j.bbamem.2013.05.002
129. Brownlee C, Dale B (1990) Temporal and spatial correlation of fertilization current, calcium waves and cytoplasmic contraction in eggs of Ciona intestinalis. Proc R Soc B 239:321–328
130. Sardet C, Paix A, Prodon F, Dru P, Chenevert J (2007) From oocyte to 16-cell stage: cytoplasmic and cortical reorganizations that pattern the ascidian embryo. Dev Dyn 236:1716–1731
131. Chun J, Santella L (2009) Roles of the actin-binding proteins in intracellular Ca^{2+} signalling. Acta Physiol (Oxf) 195:61–70
132. Chun JT, Vasilev F, Santella L (2013) Antibody against the actin-binding protein depactin attenuates Ca^{2+} signaling in starfish eggs. Biochem Biophys Res Commun 441:301–307
133. Stack C, Lucero AJ, Shuster CB (2006) Calcium-responsive contractility during fertilization in sea urchin eggs. Dev Dyn 235:1042–1052
134. Rawe V, Olmedo SB, Nodar F, Doncel G, Acosta A, Vitullo A (2000) Cytoskeletal organization defects and abortive activation in human oocytes after IVF and ICSI failure. Mol Hum Reprod 6:510–516
135. Pickering SJ, Johnson MH, Braude PR, Houliston E (1988) Cytoskeletal organization in fresh, aged and spontaneously activated human oocytes. Hum Reprod 3:978–989
136. Zhu ZY, Chen DY, Li JS, Lian L, Lei L, Han ZM, Sun QY (2003) Rotation of meiotic spindle is controlled by microfilaments in mouse oocytes. Biol Reprod 68:943–946
137. Azoury J, Lee KW, Georget V, Rassinier P, Leader B, Verlhac MH (2008) Spindle positioning in mouse oocytes relies on a dynamic meshwork of actin filaments. Curr Biol 18:1514–1519
138. Maddox AS, Azoury J, Dumont J (2012) Polar body cytokinesis. Cytoskeleton 69:855–868
139. Schaerer-Brodbeck C, Riezman H (2000) Interdependence of filamentous actin and microtubules for asymmetric cell division. Biol Chem 381:815–825
140. Noguchi T, Mabuchi I (2001) Reorganization of actin cytoskeleton at the growing end of the cleavage furrow of Xenopus egg during cytokinesis. J Cell Sci 114:401–412
141. Takayama M, Noguchi T, Yamashiro S, Mabuchi I (2002) Microtuble organization in Xenopus eggs during the first cleavage and its role in cytokinesis. Cell Struct Funct 27:163–171
142. Chew TG, Lorthongpanich C, Ang WX, Knowles BB, Solter D (2012) Symmetric cell division of the mouse zygote requires an actin network. Cytoskeleton 69:1040–1046
143. Schatten H, Sun QY (2010) The role of centrosomes in fertilization, cell division and establishment of asymmetry during embryo development. Semin Cell Dev Biol 21:174–184
144. Morito Y, Terada Y, Nakamura S, Morita J, Yoshimoto T, Murakami T, Yaegashi N, Okamura K (2005) Dynamics of microtubules and positioning of female pronucleus during bovine parthenogenesis. Biol Reprod 73:935–941

145. Mailhes JB, Carabatsos MJ, Young D, London SN, Bell M, Albertini DF (1999) Taxol-induced meiotic maturation delay, spindle defects, and aneuploidy in mouse oocytes and zygotes. Mutat Res 423:79–90
146. Rawe VY, Olmedo SB, Nodar FN, Vitullo AD (2002) Microtubules and parental genome organisation during abnormal fertilisation in humans. Zygote 10:223–228
147. Heil-Chapdelaine RA, Otto JJ (1996) Relative changes in F-actin during the first cell cycle: evidence for two distinct pools of F-actin in the sea urchin egg. Cell Motil Cytoskeleton 34:26–35
148. Wong GK, Allen PG, Begg DA (1997) Dynamics of filamentous actin organization in the sea urchin egg cortex during early cleavage divisions: implications for the mechanism of cytokinesis. Cell Motil Cytoskeleton 36:30–42
149. Döhner K, Sodeik B (2005) The role of the cytoskeleton during viral infection. Curr Top Microbiol Immunol 285:67–108
150. Hall A (2009) The cytoskeleton and cancer. Cancer Metastasis Rev 28:5–14
151. Pardee JD (2009) The actin cytoskeleton in cell motility, cancer, and infection. Morgan & Claypool Publishers, San Rafael, CA
152. Roth Z, Hansen P (2005) Disruption of nuclear maturation and rearrangement of cytoskeletal elements in bovine oocytes exposed to heat shock during maturation. Reproduction 129:235–244
153. Silvestre F, Tosti E (2010) Impact of marine drugs on cytoskeleton-mediated reproductive events. Mar Drugs 8:881–915
154. Tamura AN, Huang TTF, Marikawa Y (2013) Impact of vitrification on the meiotic spindle and components of the microtubule-organizing center in mouse mature oocytes. Biol Reprod 89:112
155. Asch R, Simerly C, Ord T, Ord V, Schatten G (1995) The stages at which human fertilization arrests: microtubule and chromosome configurations in inseminated oocytes which failed to complete fertilization and development in humans. Mol Hum Reprod 1:239–248
156. Hafez E, Goff L, Hafez B (2004) Mammalian fertilization, IVF, ICSI: physiological/molecular parameters, clinical application. Arch Androl 50:69–88
157. Hewitson L, Phil D, Simerly C, Schatten G (2000) Cytoskeletal aspects of assisted fertilization. Semin Reprod Med 18:151–160
158. Salvolini E, Buldreghini E, Lucarini G, Vignini A, Lenzi A, Di Primio R, Balercia G (2012) Involvement of sperm plasma membrane and cytoskeletal proteins in human male infertility. Fertil Steril 99:697–704
159. Simerly C, Wu G-J, Zoran S, Ord T, Rawlins R, Jones J, Navara C, Gerrity M, Rinehart J, Binor Z (1995) The paternal inheritance of the centrosome, the cell's microtubule-organizing center, in humans, and the implications for infertility. Nat Med 1:47–52
160. Terada Y (2007) Functional analyses of the sperm centrosome in human reproduction: implications for assisted reproductive technique. Soc Reprod Fertil Suppl 63:507
161. Zhivkova R, Delimitreva S, Vatev I (2010) Role of oocyte cytoplasmic factors in human IVF failure. Akush Ginekol 49:26

Part III
Focus on Microtubules

Chapter 7
Cytoskeleton and Regulation of Mitochondrial Translocation in Mammalian Eggs

Hiroyuki Suzuki

Introduction

Mammalian oocytes that fully mature in vitro and in vivo are highly desired for animal embryonic engineering, but developmental competence of the oocytes matured in vitro is less efficient than that achieved in vivo [1–8]. The oocyte can only develop to continue proper fetal development if the nuclear and cytoplasmic maturation events are normally coordinated [9]. Nuclear maturational events are characterized by the breakdown of the germinal vesicle (GV), chromosome condensation and segregation, completion of the first meiosis, extrusion of the first polar body (PB1), and arrest at metaphase of the second meiotic division (MII). Cytoplasmic maturation involves accumulation of ribonucleic acids (RNA) and proteins, oocyte growth, organelle positioning and cytoskeletal translocation [4, 8, 10]. These points are of particular importance in the field of the animal biotechnology and in human assisted reproductive technology, where in vitro oocyte maturation, in vitro fertilization, and zygote culture to the blastocyst stage before embryo transfer are routine technologies. This chapter will deal with cytoskeletal changes and the rearrangement of organelles, especially mitochondria, in the oocytes to acquire competence during nuclear and cytoplasmic maturation.

H. Suzuki (✉)
Faculty of Agriculture and Life Science, Hirosaki University,
3 Bunkyo-cho, Hirosaki, Aomori 036-8561, Japan
e-mail: suzuki@hirosaki-u.ac.jp

H. Schatten (ed.), *The Cytoskeleton in Health and Disease*,
DOI 10.1007/978-1-4939-2904-7_7

Cytoskeletal Relocation During Oocyte Maturation

Immature oocytes at the GV stage exhibit a dense subcortical MT array and cortical MFs just beneath the oolemma [11–15], features which are consistent with the interphase cell cycle. At the onset of GV breakdown (GVBD), chromosomes condense and MTs begin to assemble into the spindle at the site of small asters in the mouse, containing microtubule-organizing centers (MTOC; [16]), whereas a microtubular network become poorly-stained in the rest of the ooplasm. Then MTs elongate and the asters migrate towards opposite poles forming a barrel-shaped structure, the meiotic spindle. The spindle fully develops and the highly condensed chromosomes align at the metaphase plate between both poles of the meiotic spindle, referred to as Metaphase I (MI) [17–19]. MFs are responsible for the positioning and orientation of the meiotic spindle, separation of chromosomes, polar body emission, cortical granule exocytosis, sperm incorporation, and pronuclear centration [12, 13, 20–29].

As meiosis proceeds, the chromosomes move toward opposite poles of the meiotic spindle located at the cell cortex (Anaphase I), and the homologous chromosome pairs reach the poles (Telophase I). Then, cytokinesis follows to produce the secondary oocyte and PB1 [30–33]. At the end of MI, a contractile ring is associated with the plasma membrane to create a cleavage furrow that partitions the cell into the oocyte and PB1. The MII oocytes display MFs in the cortex and within the polar body and are characterized by the presence of a microfilament thickening at the cortical region over the meiotic spindle. The contractile ring is a network of actin MFs and myosin II; myosin II is necessary for the contraction of the cytokinetic actomyosin ring [34, 35]. When actin MFs are disrupted, polar body emission is blocked in mice [25], hamsters [36], cattle [37], sheep [38] and pigs [39, 40]. Egg asymmetry relies on constant and dynamic remodeling of MFs [21].

Suzuki et al. [41] have shown that Y-27632, a selective inhibitor of Rho kinase (ROCK), inhibits emission of PB1 in porcine oocytes by disturbing in MF functions. They have also pointed out that ROCK inhibition suppressed GVBD in porcine oocytes, probably not via organization of MFs, suggesting that GVBD may be involved in direct signaling through the Rho-ROCK pathway, which may be different from the pathway including organization of MFs in polar body emission. It has been suggested that RhoA is required for GVBD by the production of intra-oocyte reactive oxygen species (ROS) in mouse oocytes [42]. ROS is also known to mediate the formation of stress fibers [43, 44]. The underlying processes and factors of asymmetry in mammalian oocytes, such as Ran and Rac, have recently been reviewed by Brunet and Verlhac [21].

Cytoskeletal Elements Involved in Oocyte Surface Morphology During Maturation

Scanning electron micrographs of mammalian oocytes show the surface alterations during and after maturation and fertilization in mice, hamsters, cattle, pigs and humans [14, 31, 45–53]. Except for the mouse, the cortex of immature oocytes is characterized by a sparse distribution of microvilli (MV) or a mixed distribution of MV with micro folds, such as tongue-shaped protrusions [53]. After maturation, the vitelline surface is covered with a uniform and dense population of MV. In contrast, mouse oocytes show the polarity in relation to the distribution of MV and the actin cytoskeleton. Longo and Chen [31] have reported that the vitelline surface of mature mouse ova overlying the meiotic apparatus is distinguished by an absence of MV and a thickened layer of actin (MV-free area), whereas the rest of the egg is covered by MV (the MV-rich area). In immature mouse oocytes, the vitelline surface is covered entirely with a dense population of MV that is underlined by a uniform layer of the actin cytoskeleton.

MV are supported by an internal actin bundle to which the phospholipid membrane is anchored. At the core of each microvillus, a bundle of actin filaments, cross-linked by linker proteins such as fimbrin and villin, stabilizes the fingerlike structure [54]. Treatment of cytochalasin B, an inhibitor of actin polymerization, causes partial disappearance of MV from the vitelline surface of mouse oocytes (Fig. 7.1, Suzuki et al., unpublished data). Disassembly of the actin backbone in the MV may account for partial loss of MV from the oocytes.

Furthermore, our previous reports have demonstrated that cumulus morphology, intercellular spaces, and the thick cell projections among cumulus cells change dramatically during oocyte maturation and fertilization [14, 51–53]. MFs are located at the contact surfaces between the cumulus cells, whereas MTs are seen in the cytoplasm of the cumulus cells with cell projections between them [14]. The cumulus-oocyte complexes (COCs) are maintained by delicate cell-to-cell connections among the cumulus cells and with the oocyte [55–57]. The cumulus cells have been shown to be linked to each other, and the innermost layers, the corona radiata cells, often form cytoplasmic processes penetrating through the zona pellucida (ZP), maintaining contact with the vitelline membrane of the oocyte [57–66]. Thus, the cumulus-corona cell mass may play a role in regulating oocyte activity [56, 57, 61, 64]. Likewise, oocytes produce cumulus expansion enabling factor(s) to stimulate expansion of cumulus cells during maturation in rodents [67]. Cumulus cell-oocyte coupling has been shown to be mediated by the cytoskeleton, especially by MFs [14] and occasionally vimentin intermediate filaments (see Chap. 13).

Mitochondria–Cytoskeleton Interactions

Mitochondrial Translocation During Oocyte Maturation

Several studies have examined the organization and distribution of mitochondria during oocyte maturation and early development of embryos [68–73]. In immature oocytes, the mitochondria are primarily aggregated in the cortex and around the GV (Fig. 7.2). The peripheral accumulation of mitochondria has been reported by using of fluorescent probes in immature oocytes in mice [74], hamsters [75], cows [76],

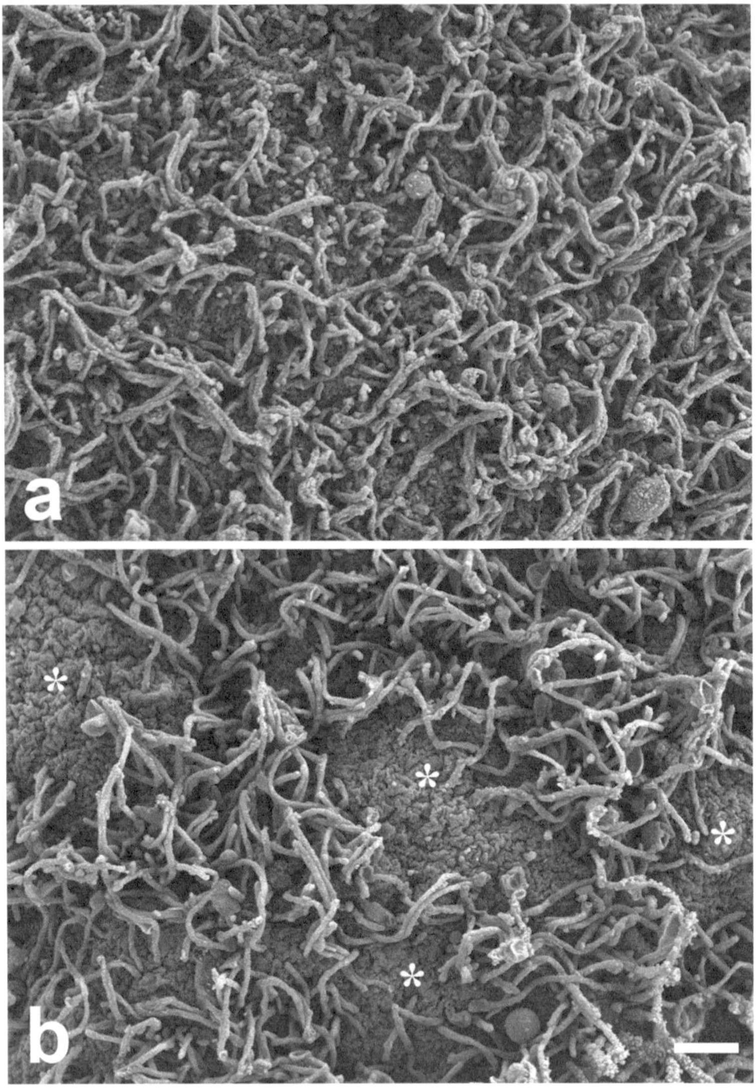

Fig. 7.1 Scanning electron micrographs of mouse oocytes before and after a treatment with cytochalasin B (blocking formation of microfilaments). Bar represents 1 μm. (**a**) A zona-free mouse oocyte at GV stage recovered from ovarian follicles by puncturing with a 25-gauge needle, showing a sparse distribution of microvilli covering the vitelline surface. (**b**) A mouse GV-oocyte treated with 7.5 μg/mL cytochalasin B for 20 min at room temperature (25 °C), showing patchy loss of microvilli (*asterisks*). The vitelline surface with no MV can be seen after mere 20 min-treatment of actin depolymerization

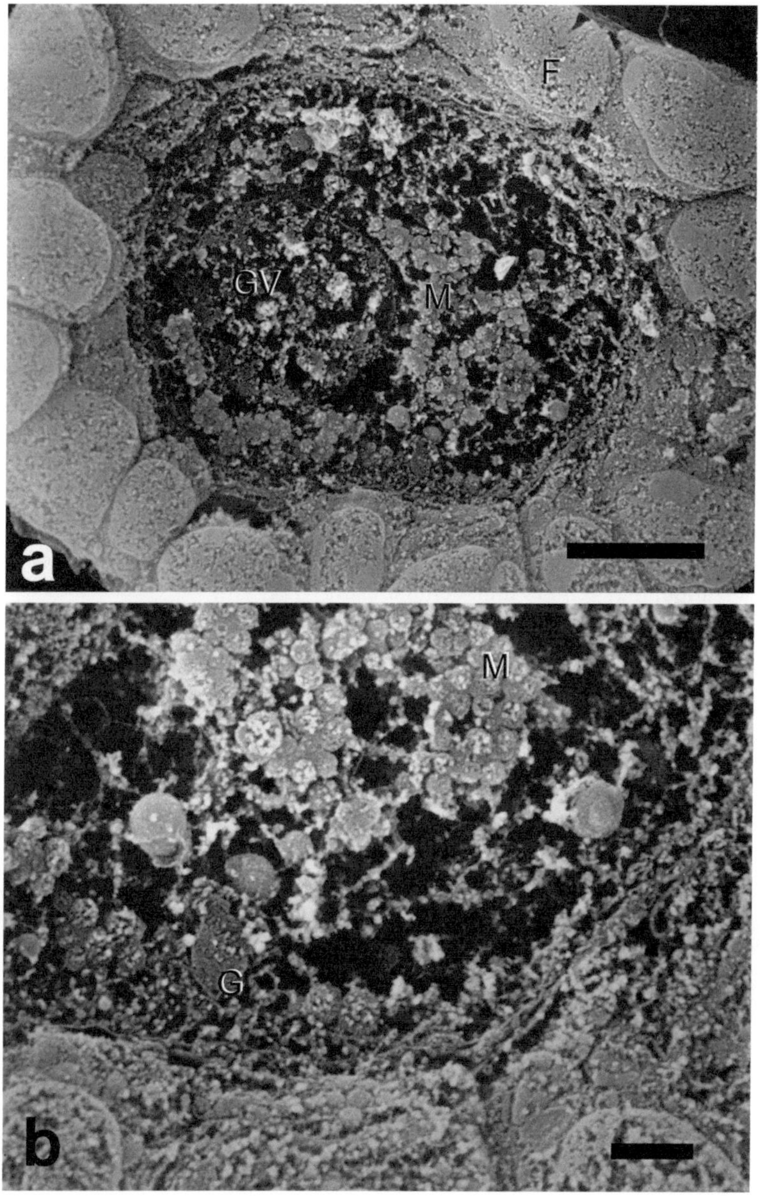

Fig. 7.2 Scanning electron micrograph of fractured surface of a primary follicle after ODO-maceration method for the hamster ovary [65]. Bar represents 5 μm in (**a**) and 1 μm in (**b**). (**a**) Cuboidal and polyhedral follicular epithelial cells (F) closely apposed to an oocyte including the large germinal vesicle (GV). GV is surrounded by mitochondrial cloud (M). (**b**) Higher magnification of (**a**). Note many spherical and lamellae corresponding to mitochondria (M) and Golgi complex (G), respectively, supported by a network of the cytoskeleton. The figure contains images reproduced from the Animal Science Journal (2001, volume 72, pp. 107–116) with the permission of the Japanese Society of Animal Science

pigs [72] and humans [70, 73]. Electron microscopic observations have also shown mitochondrial aggregation at the egg cortex in sheep [77], pigs [78] and cattle [79]. This pattern of distribution may be related to the high energy requirements in the cortex, as oocytes require cumulus cell support at this stage, and intimate association between the oocyte and the cumulus cells is maintained with cumulus cell processes interacting with the vitelline surface [14, 58, 59, 61].

As oocyte maturation proceeds, mitochondria are seen to distribute homogeneously throughout the ooplasm between the time of GVBD and MI. Fluorescent intensity of mitochondria increases more than twice at the MI and MII stages as compared to the GV stage [75], suggesting a high energy requirement for these meiotic processes, which include spindle formation, chromosome condensation and movement, and polar body extrusion [74]. At the MII stage, mitochondria are enriched at the center of the ooplasm and around the meiotic spindle, while they become scarce in the cortex. After fertilization, mitochondria are aggregated around both male and female pronuclei, and the cortical region of the ooplasm has become devoid of active mitochondria [75]. The perinuclear clustering of the mitochondria in zygotes is observed in mice [80], hamsters [68, 75], pigs [72] and humans [71]. The pronuclear migration and apposition to form syngamy, being related to MF reorganization, may require high cellular energy [71, 76].

Cytoskeletal Contributions to Mitochondrial Movement in Oocytes/Embryos

In murine and porcine oocytes, it has been reported that translocation of mitochondria is mediated by MTs, not by MFs [72, 81–83]. On the other hand, studies on hamster 2-cell embryos suggest that MFs play a role in the distribution of mitochondria [68, 84]. However, our previous studies have shown that both MFs and MTs are involved in mitochondrial redistribution in embryos during interphase of the cell cycle [85], and only MFs may function in mitochondrial redistribution during the M-phase of the cell cycle (see section "Cell Cycle-Dependent Dynamics of the Cytoskeleton Involving Mitochondrial Redistribution").

Factors Affecting Cytoskeletal and Mitochondrial Distributions

During in vitro manipulation of oocytes/embryos, micro-environments around them have to be rigorously controlled. However, the immediate cellular environment may influence cytoskeletal distribution and mitochondrial movement. This section will discuss the effects of fluctuations in temperature, culture conditions, and cell cycle progression on the cytoskeletal and mitochondrial distributions.

Ambient Temperature Affects Cytoskeletal Organization in the Oocyte

The cytoskeletal components undergo rearrangements in response to many types of stimuli such as variations in the local extracellular environment, including fluctuations in temperature. MFs and MTs may mutually control the function of the meiotic spindle apparatus to complete chromosome conformation and segregation. Meiotic spindles of oocytes are known to be affected by fluctuations in temperature [86–98].

Effects of Low Temperature on the Oocyte Cytoskeleton

Partial and complete depolymerization of MTs and/or altered morphology of the spindle have been reported in murine [90, 92, 95], bovine [87], porcine [89, 96], ovine [91] and human oocytes [86, 93, 94, 97, 98].

Our previous study has revealed that exposure of porcine MII oocytes to 5 °C produces more detrimental effects on the spindle organization and the cortical MFs than exposure to 18 °C, although the cooling effects may be mild on the assembly of MFs compared to that of MTs [96]. Liu et al. [89] have also reported that disassembly of microtubular spindles in the porcine MII oocytes is quicker at 4 °C than at 24 °C. In mouse oocytes, the recovery of the spindle after rewarming has been reported by Magistrini and Szöllösi [90], Pickering and Johnson [92], and Sun et al. [95]. For example, Magistrini and Szöllösi [90] have observed that most oocyte spindles are restored to normal by rewarming after exposure to 0 °C for 45–60 min. Pickering and Johnson [92] have also found that cooling to 25 °C for 60 min induces complete disassembly of the spindle, but subsequent incubation at 37 °C for 60 min results in recovery of normal spindles. However, spindles of bovine [87], porcine [96], ovine [91] and human [86, 93, 94, 97, 98] oocytes have exhibited only limited recovery after temperature fluctuations of cooling and rewarming. These results suggest that microtubular spindles in mouse oocytes are more stable than those in other animals, and that certain cytoplasmic factor(s) related to the sensitivity of MTs to temperature may be involved in mouse oocytes.

In cryopreservation trials, subzero temperature may cause irreversible damage to the cytoskeleton of GV- and MII-oocytes in cattle [99], pigs [100] and humans [88]. Some kinds of cryoprotectants seem to affect the organization of MFs in the oocytes [101], too. However, Nedambale et al. [99] have found better reorganization of cytoplasmic and spindle MTs in vitrified-thawed bovine oocytes, when post-thaw incubation is extended for 120 min prior to fertilization. To avoid a deleterious effect on the spindle by freezing, immature prophase I oocytes, in which the meiotic spindle is not yet formed, are preferable for cryopreservation of the human oocyte [88, 102].

Effects of Low Temperature on Spindle Morphology

Suzuki et al. [96] have pointed out that cooling affects the MII spindle morphology, changing from normal barrel-shaped to box-shaped spindles, where both spindle poles have broadened in a direction parallel to the equatorial plane. The oocytes with box-shaped spindles show more frequently a shortened interpolar distance after cooling to 18 °C compared to keeping them at 37 °C. Taken together, it is suggested that in the cooling oocytes the polar areas may first become wider, and then the interpolar distance may shorten.

Kinetochore MTs, which are attached to the kinetochores on the centromeres of chromatids, are involved in generating and/or transmitting the force for chromosome movement. Polar MTs are involved in antiparallel interactions within the spindle and are required for keeping the two spindle poles apart [20, 103]. The kinetochore MTs may be more stable and cold-resistant than the polar MTs [104]. We have observed spindles in which MTs have decreased in number and in fluorescence intensity with a short interpolar distance in cooling porcine oocytes [96]. Such 'reduced spindles' may have been the result of disassembly of the polar MTs rather than the kinetochore MTs. In addition, we frequently observed movements of chromosomes toward the cell surface in the oocytes exposed to 5 or 18 °C [96]. Such movement of chromosomes has also been reported for mitosis [103]. Monopolar spindles have also been observed in porcine MII oocytes cooled to 5 °C [96]. In this case, one microtubular spindle is always observed on the side of the outer pole only. The results suggest that the bundles of MTs on the side of the inner pole may disassemble first followed by disassembly of those on the side of the outer pole, eventually resulting in complete disassembly of spindle MTs.

Recently, in a study on network organization of the neuronal cytoskeleton, Spedden et al. [105] have found that the correlation between regions of high tubulin concentration and the high-stiffness areas of the soma decreases dramatically as the temperature is dropped from 37 to 25 °C. Temperature-sensitive MTs or cellular mechanotransduction [106] may also affect mechanical stability of somatic cells.

Effects of Low Temperature on MTOCs

The spindle poles of the oocyte, also known as MTOCs [16, 20, 107, 108]. At low temperature, the polar areas become wider and flattened. The broadened spindle fails to recover to the normal spindle after cooling-rewarming, suggesting that alteration in the MTOC area may be irreversible. It is possible, therefore, that other elements including MTOC-related proteins, such as γ-tubulin and pericentrin [107, 108], have become damaged during cooling-rewarming. Thus, the oocyte spindles failed to recover to the normal status after rewarming, even though some MTs have been repolymerized around the chromosomes and the majority of cortical MFs have been reassembled. The effect of cooling on MTOC-related proteins and the mechanism controlling the organization of MFs and MTs in the mammalian oocytes remain to be elucidated at the cellular and molecular levels.

Effect of Elevated Temperature (Heat Shock) on the Oocyte Cytoskeleton

It is well documented that mammalian females subjected to heat stress have increased embryonic mortality (for review, see [109]). Baumgartner and Chrisman [110] have reported in oocytes derived from heat-stressed mice the increased proportions of abnormal configurations of the chromosomes due to disruption of the spindle. Ju and his colleagues have shown that direct in vitro heat shock at 41 to 42 °C for 2 to 4 h results in aggregation or condensation of metaphase chromosomes in both bovine and porcine oocytes [111, 112]. In some cases, the chromatin of the matured oocytes has separated into several groups after severe heat shock [113]. The reason for the condensation and separation of the chromatin is not clear yet, but it might be due to alteration of the spindle MTs.

It has been shown that exposure of cultured oocytes to elevated temperature (i.e. heat shock) during early stages of maturation interferes with the processes of oocyte maturation. In our previous studies, in vitro matured bovine oocytes could not sustain development when exposed to 43 °C for 45 min in M199 compared to shorter treatments (0, 15, and 30 min of heat shock) [114], with no deleterious effect on the matured oocyte being exposed to lower temperatures (40.5 and 41.5 °C for up to 1 h) [114]. Although exposure at 43 °C for 45 min should be detrimental for bovine oocytes [114], resistance to elevated temperature altered as oocytes mature, become fertilized, and develop [115].

Ju and his colleagues have shown clear depolymerization of the meiotic spindle and the MF density is altered in both bovine and porcine oocytes exposed to heat shock at 41 to 42 °C for 2 to 4 h [112, 113, 116, 117]. Therefore, aggregation or condensation of metaphase chromosomes occurs in these heat-shocked oocytes [111].

Suzuki et al. [118] have demonstrated that, with only a brief heat shock (42 °C for 30 min) to mature bovine oocytes, the surface ultrastructure of the vitelline membranes are changed from a normal MV-dominated pattern to a mixture of MV and lamellipodia-like enlargement of cytoplasmic protrusions. They have suggested that its effect may manifest itself more quickly and to a greater degree in aged oocytes (matured in vitro for 44 h) compared to the young oocytes (matured in vitro for 24 h), too. It is likely that the changes in surface morphology of the oocytes are attributable to the alteration in distribution of the cortical MFs after heat shock. Furthermore, heat shock results in the movement of organelles towards the center of the blastomere [119] and the loss of cortical granules from the vitelline cortex [120]. Taken together, it is suggested that MF and MT structure may be altered in mammalian oocytes subjected to heat shock during maturation. In addition, disruption in resumption of meiosis caused by heat shock may be associated with oocyte apoptosis [121–124].

In Vitro vs. In Vivo Conditions

Due to recent studies, knowledge of the basic cell biology of early embryo development and the in vivo environment has substantially increased. It is now generally accepted that the development of defined embryo culture systems has been one of the key achievements in mammalian embryology. However, changes in culture conditions can alter the organization or structure of mitochondria in the oocytes or embryos of mice [80, 125], hamsters [68, 126–128] and cattle [129, 130]. Bavister and his colleagues have reported that hamster embryos are very sensitive to changes in culture milieu, which may severely affect their viability after embryo transfer (for review, see [131]). For example, exposure of hamster 2-cell embryos to suboptimal culture conditions (addition of glucose and phosphate which disrupt development) causes dispersion of the mitochondria away from the nuclei [84, 126, 127]. In particular, energy substrate requirements for hamster embryo development in vitro are markedly different from those for mouse embryos and details of the mitochondrial metabolism during preimplantation development is still to be resolved.

Suzuki et al. [132] have demonstrated that mitochondrial localization in embryonic cells is changed after the 8-cell stage in the hamster. At the late 8-cell stage, hamster embryos are characterized by a concentration of mitochondria to the cell-to-cell contact region in addition to the perinuclear region in in vivo embryos, but it may not be in in vitro ones. Such heterogeneity in mitochondrial distribution may be associated with energy production/utilization for compaction. In vitro culture may change mitochondrial localization in the cytoplasm of hamster embryos and may be dependent on reduced density of the MF and MT networks. In in vivo hamster embryos, nucleus migrates normally from the central cytoplasm to the apical cytoplasm around the late 8-cell stage [133], but some failure may occur in the mechanism of nuclear outward migration in more than half of the embryos cultured in vitro [132]. Furthermore, in vitro embryonic development is delayed from the late 8-cell to the morula stages, namely around the compaction stage. These data suggest that the process for compaction may require increased energy production in a particular region of each blastomere and that culture conditions which disrupt energy production in the embryo may retard the compaction process.

Cell Cycle-Dependent Dynamics of the Cytoskeleton Involving Mitochondrial Redistribution

The cytoskeleton is remodeled dynamically throughout the cell cycle. During the M phase, MTs are assembled into the spindle and MFs are organized to the contractile rings at cytokinesis [24]. To test the dependence of the cytoskeleton on mitochondrial localization, hamster embryos in interphase or M phase treated with or without assembly inhibitors of the cytoskeleton (20 μM of nocodazole, 5 μM of cytochalasin D or a combined treatment of both inhibitors for 1 h) were centrifuged at 10,000×*g* for 2 min. By combining cytoskeleton assembly inhibitor(s) with high-speed centrifugation, mitochondria dissociated from depolymerized cytoskeletal components should be easily spun down by centrifugation. Centrifugation is effective in eliminating any confusion regarding the notion that mitochondria are gradually spreading in situ even if the cytoskeleton is depolymerized by the specific inhibitors. Our previous results suggest that MTs may strongly associate with mitochondria during interphase but dissociate from them during the M phase [134]. Functional roles of MFs and MTs in the translocation of mitochondria are presented schematically in Fig. 7.3. MTs have primary functions in forming the mitotic spindle to facilitate chromosome segregation during the M phase. MFs are involved in the orientation, positioning and rotation of the spindle [135–137].

It is well known in somatic cells that both cytoskeletal components are included in axonal transport of mitochondria. In the axons, long-distance and fast transport of mitochondria requires MTs, and mitochondria can also move along MFs in short-range movement [138–140]. Mitochondrial transport involves MT-dependent kinesins and dyneins and actin-dependent myosins [140]. All three types of motor proteins are frequently found on the surface of a single mitochondrion and work in concert to correctly maintain mitochondrial localization in the cytoplasm [138]. Furthermore, Lee et al. [141] have reported in HeLa cells that mitochondria in mitotic cells displayed a close association with MFs but not with MTs and intermediate filament keratin. The role of intermediate filaments on organelle positioning will be discussed in Chap. 13. Crosstalk between MFs, MTs and intermediate filaments is an important research theme to understand the relationship between the cytoskeleton and organelle positioning in the cell.

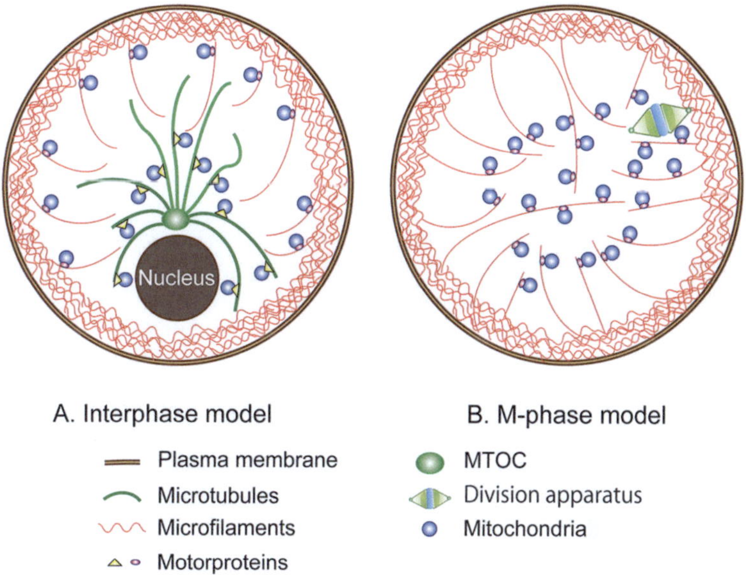

Fig. 7.3 Schematic diagram showing direct interaction of the cytoskeleton with mitochondria according to the cell cycle. The images are reproduced from our previous data [85, 134]. (**a**) Both microtubules and microfilaments are involved in the mitochondrial redistribution of the oocyte/blastomere during interphase. (**b**) The redistribution of mitochondria in the oocyte/blastomere is mainly modulated by microfilaments during the M phase, when the cytoplasmic microtubules are intensively organized into the spindle fibers

Concluding Remarks

Actively anchoring mitochondria at sites where they are needed helps to maintain proper organelle distribution. Organelle anchoring is also a cytoskeleton-based process, where MF- and MT-dependent anchoring occur in the oocyte/embryo depending on the cell cycle progression. Cytoskeletal linker proteins are known but their roles in organelle distribution are not well characterized [142]. Several other proteins may also function as mitochondrial motor adaptors or regulators [143], and a regulatory interaction between cytoskeletal signaling [144] will await further study into their dynamics or relative contributions to oocyte meiosis and embryonic development. Fluctuations in temperature and in vitro culture conditions induce most of the spindle MTs and the cortical MFs to undergo disassembly partially or completely. Such a disassembly is irreversibly in most situations. Therefore, in association with reorganization of MFs and MTs, temperature fluctuations may influence crucial events of the cortical ooplasm, such as normal alignment and segregation of chromosomes, the eccentric anchorage of the spindle, the polar body formation, and the migration of organelles, including mitochondria, and subsequent fertilization and development. These points are of particular importance in the field of animal biotechnology and human assisted reproductive technology, where in vitro oocyte maturation, in vitro fertilization, and zygote culture to the blastocyst stage before embryo transfer are routine technologies.

Acknowledgments This work was supported by JSPS KAKENHI Grant Numbers 17580243, 21580340 and 24580403.

References

1. Greve T, Xu KP, Callesen H, Hyttel P (1987) In vivo development of in vitro fertilized bovine oocytes matured in vivo versus in vitro. J In Vitro Fert Embryo Transf 4:281–285
2. Leibfried-Rutledge ML, Critser ES, Eyestone WH, Northey DL, First NL (1987) Development potential of bovine oocytes matured in vitro or in vivo. Biol Reprod 36:376–383
3. Moor RM, Trounson AO (1977) Hormonal and follicular factors affecting maturation of sheep oocytes in vitro and their subsequent developmental capacity. J Reprod Fertil 49:101–109
4. Sirard MA, Richard F, Blondin P, Robert C (2006) Contribution of the oocyte to embryo quality. Theriogenology 65:126–136
5. Thibault C (1977) Hammond Memorial Lecture. Are follicular maturation and oocyte maturation independent processes? J Reprod Fertil 51:1–15
6. Thibault C, Szöllösi D, Gérard M (1987) Mammalian oocyte maturation. Reprod Nutr Dev 27:865–896
7. van Blerkom J, McGaughey RW (1978) Molecular differentiation of the rabbit ovum. I. During oocyte maturation in vivo and in vitro. Dev Biol 63:139–150
8. Watson AJ (2007) Oocyte cytoplasmic maturation: a key mediator of oocyte and embryo developmental competence. J Anim Sci 85(13 Suppl):E1–E3
9. Eppig JJ (1996) Coordination of nuclear and cytoplasmic oocyte maturation in eutherian mammals. Reprod Fertil Dev 8:485–489

10. Albertini DF, Sanfins A, Combelles CMH (2003) Origins and manifestations of oocyte maturation competencies. Reprod Biomed Online 6:410–415
11. Combelles CMH, Cekleniak NA, Racowsky C, Albetrini DF (2002) Assessment of nuclear and cytoplasmic maturation in in-vitro matured human oocytes. Hum Reprod 17:1006–1016
12. Gumus E, Bulut HE, Kaloglu C (2010) Cytoskeletal changes in oocytes and early embryos during in vitro fertilization process in mice. Anat Histol Embryol 39:51–58. doi:10.1111/j.1439-0264.2009.00982.x
13. Kim NH, Chung HM, Cha KY, Chung KS (1998) Microtubule and microfilament organization in maturing human oocytes. Hum Reprod 13:2217–2222
14. Suzuki H, Jeong B-S, Yang X (2000) Dynamic changes of cumulus-oocyte cell communication during in vitro maturation of porcine oocytes. Biol Reprod 63:723–729
15. Suzuki H, Saito Y (2006) Cumulus cells affect distribution and function of the cytoskeleton and organelle in the porcine oocytes. Reprod Med Biol 5:183–194
16. Maro B, Howlett SK, Webb M (1985) Non-spindle microtubule organizing centers in metaphase II-arrested mouse oocytes. J Cell Biol 101:1665–1672
17. Battaglia DE, Klein NA, Soules MR (1996) Changes in centrosomal domains during meiotic maturation in the human oocyte. Mol Hum Reprod 2:845–851
18. Leader B, Lim H, Carabatsos MJ, Harrington A, Ecsedy J, Pellman D, Maas R, Leder P (2002) Formin-2, polyploidy, hypofertility and positioning of the meiotic spindle in mouse oocytes. Nat Cell Biol 4:921–928
19. Maro B, Verlhac MH (2002) Polar body formation: new rules for asymmetric divisions. Nat Cell Biol 4:E281–E283
20. Brunet S, Maro B (2005) Cytoskeleton and cell cycle control during meiotic maturation of the mouse oocyte: integrating time and space. Reproduction 130:801–811
21. Brunet S, Verlhac MH (2011) Positioning to get out of meiosis: the asymmetry of division. Hum Reprod Update 17:68–75. doi:10.1093/humupd/dmq044
22. Connors SA, Kanatsu-Shinohara M, Schultz RM, Kopf GS (1998) Involvement of the cytoskeleton in the movement of cortical granules during oocyte maturation, and cortical granule anchoring in mouse eggs. Dev Biol 200:103–115
23. DiMaggio AJ Jr, Lonergan TA, Stewart-Savage J (1997) Cortical granule exocytosis in hamster eggs requires microfilaments. Mol Reprod Dev 47:334–340
24. Kunda P, Baum B (2009) The actin cytoskeleton in spindle assembly and positioning. Trends Cell Biol 19:174–179. doi:10.1016/j.tcb.2009.01.006, Epub 2009 Mar 13
25. Maro B, Johnson MH, Pickering SJ, Flach G (1984) Changes in actin distribution during fertilization of the mouse egg. J Embryol Exp Morphol 81:211–237
26. Simerly C, Nowak G, de Lanerolle P, Schatten G (1998) Differential expression and functions of cortical myosin IIA and IIB isotypes during meiotic maturation, fertilization, and mitosis in mouse oocytes and embryos. Mol Biol Cell 9:2509–2525
27. Sun QY, Schatten H (2006) Regulation of dynamic events by microfilaments during oocyte maturation and fertilization. Reproduction 131:193–205
28. Sutovsky P, Oko R, Hewitson L, Schatten G (1997) The removal of the sperm perinuclear theca and its association with the bovine oocyte surface during fertilization. Dev Biol 188:75–84
29. Terada Y, Simerly C, Schatten G (2000) Microfilament stabilization by jasplakinolide arrests oocyte maturation, cortical granule exocytosis, sperm incorporation cone resorption, and cell-cycle progression, but not DNA replication, during fertilization in mice. Mol Reprod Dev 56:89–98
30. Halet G, Carroll J (2007) Rac activity is polarized and regulates meiotic spindle stability and anchoring in mammalian oocytes. Dev Cell 12:309–317
31. Longo FJ, Chen DY (1985) Development of cortical polarity in mouse eggs: involvement of the meiotic apparatus. Dev Biol 107:382–394
32. Maro B, Johnson MH, Webb M, Flach G (1986) Mechanism of polar body formation in the mouse oocyte: an interaction between the chromosomes, the cytoskeleton and the plasma membrane. J Embryol Exp Morphol 92:11–32

33. Wang L, Wang ZB, Zhang X, FitzHarris G, Baltz JM, Sun QY, Liu XJ (2008) Brefeldin A disrupts asymmetric spindle positioning in mouse oocytes. Dev Biol 313:155–166
34. Kneussel M, Wagner W (2013) Myosin motors at neuronal synapses: drivers of membrane transport and actin dynamics. Nat Rev Neurosci 14:233–247. doi:10.1038/nrn3445
35. Glotzer M (2005) The molecular requirements for cytokinesis. Science 307:1735–1739
36. Terada Y, Fukaya T, Yajima A (1995) Localization of microfilaments during oocyte maturation of golden hamster. Mol Reprod Dev 41:486–492
37. Bevacqua RJ, Fernandez-Martin R, Salamone DF (2011) Bovine parthenogenotes produced by inhibition of first or second polar bodies emission. Biocell 35:1–7
38. Le Guen P, Crozet N, Huneau D, Gall L (1989) Distribution and role of microfilaments during early events of sheep fertilization. Gamete Res 22:411–425
39. Suzuki H, Kagawa N, Toyokawa K (2002) Parthenogenetic development and pronuclear migration and cytoskeletal organization of porcine oocytes activated by various stimuli. J Mamm Ova Res 19:96–103
40. Suzuki H, Takashima Y, Toyokawa K (2002) Parthenogenetic development and cytoskeletal distribution of porcine oocytes treated by means of electric pulses and cytochalasin D. J Mamm Ova Res 19:6–11
41. Suzuki H, Koyama K, Kabashima K, Fang J, Matsuzaki M (2011) Temporary inhibition of germinal vesicle breakdown by Rho kinase inhibitor Y-27632 is detrimental to oocyte maturation. J Mamm Ova Res 28:126–130
42. Cheon YP, Kim SW, Kim SJ, Yeom YI, Cheong C, Ha KS (2000) The role of RhoA in the germinal vesicle breakdown of mouse oocytes. Biochem Biophys Res Commun 273:997–1002
43. Huot J, Houle F, Marceau F, Landry J (1997) Oxidative stress-induced actin polymerization mediated by the p38 mitogen-activated protein kinase/heat shock protein 27 pathway in vascular endothelial cells. Circ Res 80:383–392
44. Koo HY, Shin I, Lee ZW, Lee SH, Kim SH, Lee CH, Kang HS, Ha KS (1999) Roles of RhoA and phospholipase A2 in the elevation of intracellular H2O2 by transforming growth factor-beta in Swiss 3T3 fibroblasts. Cell Signal 11:677–683
45. Calafell JM, Nogués C, Ponsà M, Santaló J, Egozcue J (1992) Zona pellucida surface of immature and in vitro matured mouse oocytes: analysis by scanning electron microscopy. J Assist Reprod Genet 9:365–372
46. Familiari G, Nottola SA, Micara G, Aragona C, Motta PM (1988) Is the sperm-binding capability of the zona pellucida linked to its surface structure? A scanning electron microscopic study of human in vitro fertilization. J In Vitro Fert Embryo Transf 5:134–143
47. Familiari G, Nottola SA, Micara G, Aragona C, Motta P (1989) Human in vitro fertilization: the fine three-dimensional architecture of the zona pellucida. In: Motta PM (ed) Developments in ultrastructure of reproduction. Alan R. Liss, New York
48. Jackowski S, Dumont JN (1979) Surface alterations of the mouse zona pellucida and ovum following in vivo fertilization: correlation with the cell cycle. Biol Reprod 20:150–161
49. Motta PM, Nottola SA, Micara G, Familiari G (1988) Ultrastructure of human unfertilized ooytes and polyspermic embryos in an IVF-ET program. In: Jones HW Jr, Schrader C (eds) In vitro fertilization and other assisted reproduction. The New York Academy of Sciences, New York
50. Phillips DM, Shalgi R (1980) Surface architecture of the mouse and hamster zona pellucida and oocyte. J Ultrastruct Res 72:1–12
51. Suzuki H, Fujiwara T, Yang X (1997) Surface ultrastructural characteristics of the hamster oocyte and its investments during in vivo maturation. J Mamm Ova Res 14:191–197
52. Suzuki H, Presicce GA, Yang X (1998) Differential surface ultrastructural characteristics and volumetric dynamics of bovine oocytes during maturation in vivo versus in vitro. J Mamm Ova Res 15:49–62
53. Suzuki H, Yang X, Foote RH (1994) Surface alterations of the bovine oocyte and its investments during and after maturation and fertilization in vitro. Mol Reprod Dev 38:421–430

54. Lodish H, Berk A, Zipursky SL, Matsudaira P, Baltimore D, Darnell J (2000) Intermediate Filaments. In: Lodish H et al (eds) Molecular cell biology, 4th edn. W. H. Freeman, New York
55. Eppig JJ (1982) The relationship between cumulus celloocyte coupling, oocyte meiotic maturation, and cumulus expansion. Dev Biol 89:268–272
56. Eppig JJ, Downs SM (1984) Chemical signals that regulate mammalian oocyte maturation. Biol Reprod 30:1–11
57. Moor RM, Smith MW, Dawson RMC (1980) Measurement of intercellular coupling between oocytes and cumulus cells using intracellular markers. Exp Cell Res 126:15–29
58. Allworth AE, Albertini DF (1993) Meiotic maturation in cultured bovine oocytes is accompanied by remodeling of the cumulus cell cytoskeleton. Dev Biol 158:101–112
59. Anderson E, Albertini DF (1976) Gap junctions between the oocyte and companion follicle cells in the mammalian ovary. J Cell Biol 71:680–686
60. Assey RJ, Hyttel P, Greve T, Purwantara B (1993) Oocyte morphology in dominant and subordinate follicles. Mol Reprod Dev 37:335–344
61. Gilula NB, Epstein ML, Beers WH (1978) Cell-to-cell communication and ovulation. A study of the cumulus-oocyte complex. J Cell Biol 78:58–75
62. Hyttel P (1987) Bovine cumulus-oocyte disconnection in vitro. Anat Embryol (Berl) 176:41–44
63. Laurincík J, Kroslák P, Hyttel P, Pivko J, Sirotkin AV (1992) Bovine cumulus expansion and corona–oocyte disconnection during culture in vitro. Reprod Nutr Dev 32:151–161
64. Laurincík J, Pivko J, Kroslák P (1992) Cumulus oophorus expansion of bovine oocytes cultured in vitro: a SEM and TEM study. Reprod Domest Anim 27:217–228
65. Suzuki H, Fukuda I, Toyokawa K (2001) Follicular epithelial cell-oocyte connection during development of hamster ovarian follicle. Anim Sci J 72:107–116
66. Zamboni L (1974) Fine morphology of the follicle wall and follicle cell-oocyte association. Biol Reprod 10:125–149
67. Vanderhyden BC, Caron PJ, Buccione R, Eppig JJ (1990) Developmental pattern of the secretion of cumulus expansion-enabling factor by mouse oocytes and the role of oocytes in promoting granulosa cell differentiation. Dev Biol 140:307–317
68. Barnett DK, Kimura J, Bavister BD (1996) Translocation of active mitochondria during hamster preimplantation embryo development studied by confocal laser microscopy. Dev Dyn 205:64–72
69. Bavister BD, Squirrell JM (2000) Mitochondrial distribution and function in oocytes and early embryos. Hum Reprod 15(Suppl 2):189–198
70. van Blerkom J, Davis P, Mathwig V, Alexander S (2002) Domains of high-polarized and low-polarized mitochondria may occur in mouse and human oocytes and early embryos. Hum Reprod 17:393–406
71. van Blerkom J, Davis P, Alexander S (2000) Differential mitochondrial distribution in human pronuclear embryos leads to disproportionate inheritance between blastomeres: relationship to microtubular organization, ATP content and competence. Hum Reprod 15:2621–2633
72. Sun QY, Wu GM, Lai L, Park KW, Cabot R, Cheong HT, Day BN, Prather RS, Schatten H (2001) Translocation of active mitochondria during pig oocyte maturation, fertilization and early embryo development in vitro. Reproduction 122:155–163
73. Wilding M, Dale B, Marino M, di Matteo L, Alviggi C, Pisaturo ML, Lombardi L, De Placido G (2001) Mitochondrial aggregation patterns and activity in human oocytes and preimplantation embryos. Hum Reprod 16:909–917
74. van Blerkom J, Runner MN (1984) Mitochondrial reorganization during resumption of arrested meiosis in the mouse oocyte. Am J Anat 171:335–355
75. Suzuki H, Satoh M, Kabashima K (2005) Changes in distribution of active mitochondria during oocyte maturation and fertilization in the hamster. J Mamm Ova Res 22:163–169
76. Stojkovic M, Machado SA, Stojkovic P, Zakhartchenko V, Hutzler P, Gonçalves PB, Wolf E (2001) Mitochondrial distribution and adenosine triphosphate content of bovine oocytes before and after in vitro maturation: correlation with morphological criteria and developmental capacity after in vitro fertilization and culture. Biol Reprod 64:904–909

77. Cran DG, Moor RM, Hay MF (1980) Fine structure of the sheep oocyte during antral follicle development. J Reprod Fertil 59:125–132
78. Cran DG (1985) Qualitative and quantitative structural changes during pig oocyte maturation. J Reprod Fertil 74:237–245
79. Hyttel P, Xu KP, Smith S, Grave T (1986) Ultrastructure of in-vitro oocyte maturation in cattle. J Reprod Fertil 78:615–625
80. Tokura T, Noda Y, Goto Y, Mori T (1993) Sequential observation of mitochondrial distribution in mouse oocytes and embryos. J Assist Reprod Genet 10:417–426
81. Heggeness MH, Simon M, Singer SJ (1978) Association of mitochondria with microtubules in cultured cells. Proc Natl Acad Sci U S A 75:3863–3866
82. van Blerkom J (1991) Microtubule mediation of cytoplasmic and nuclear maturation during the early stages of resumed meiosis in cultured mouse oocytes. Proc Natl Acad Sci U S A 88:5031–5035
83. Nangaku M, Sato-Yoshitake R, Okada Y, Noda Y, Takemura R, Yamazaki H, Hirokawa N (1994) KIF1B, a novel microtubule plus end-directed monomeric motor protein for transport of mitochondria. Cell 79:1209–1220
84. Barnett DK, Clayton MK, Kimura J, Bavister BD (1997) Glucose and phosphate toxicity in hamster preimplantation embryos involves disruption of cellular organization, including distribution of active mitochondria. Mol Reprod Dev 48:227–237
85. Kabashima K, Matsuzaki M, Suzuki H (2007) Both microtubules and microfilaments mutually control the distribution of mitochondria in two-cell embryos of golden hamsters. J Mamm Ova Res 24:120–125
86. Almeida PA, Bolton VN (1995) The effect of temperature fluctuations on the cytoskeletal organization and chromosomal constitution of the human oocyte. Zygote 3:357–365
87. Aman RR, Parks JE (1994) Effects of cooling and rewarming on the meiotic spindle and chromosomes of in vitro matured bovine oocytes. Biol Reprod 50:103–110
88. Baka SG, Toth TL, Veeck LL, Jones HW Jr, Muasher SJ, Lanzendorf SE (1995) Evaluation of the spindle apparatus of in-vitro matured human oocytes following cryopreservation. Hum Reprod 10:1816–1820
89. Liu RH, Sun QY, Li YH, Jiao LH, Wang WH (2003) Effects of cooling on meiotic spindle structure and chromosome alignment within in vitro matured porcine oocytes. Mol Reprod Dev 65:212–218
90. Magistrini M, Szöllösi D (1980) Effects of cold and isopropyl-N-phenylcarbamate on the second meiotic spindle of mouse oocytes. Eur J Cell Biol 22:699–707
91. Moor RM, Crosby IM (1985) Temperature induced abnormalities in sheep oocytes during maturation. J Reprod Fertil 75:467–473
92. Pickering SJ, Johnson MH (1987) The influence of cooling on the organization of the meiotic spindle of the mouse oocytes. Hum Reprod 2:207–216
93. Pickering SJ, Braude PR, Johnson MH, Cant A, Currie J (1990) Transient cooling to room temperature can cause irreversible disruption of the meiotic spindle in the human oocyte. Fertil Steril 54:102–108
94. Sathananthan AH, Trounson A, Freemann L, Brady T (1988) The effects of cooling human oocytes. Hum Reprod 3:968–977
95. Sun XF, Zhang WH, Chen XJ, Xiao GH, Mai WY, Wang WH (2004) Spindle dynamics in living mouse oocytes during meiotic maturation, ageing, cooling and overheating: a study by polarized light microscopy. Zygote 12:241–249
96. Suzuki H, Kumai T, Matsuzaki M (2007) Effect of temperature decline on the cytoskeletal organization of the porcine oocyte. J Mamm Ova Res 24:107–113
97. Wang WH, Meng L, Hackett RT, Odenbourg R, Keefe DL (2001) Limited recovery of meiotic spindles in living human oocytes after cooling–rewarming observed using polarized light microscopy. Hum Reprod 16:2374–2378
98. Zenzes MT, Bielecki R, Casper RF, Leibo SP (2001) Effects of chilling to 0°C on the morphology of meiotic spindles in human metaphase II oocytes. Fertil Steril 75:769–777

99. Nedambale TL, Du F, Xu J, Tian XC, Yang X (2006) Effects of vitrification and post-thawing interval on the cytoskeleton and subsequent fertilization rate of in vitro derived bovine oocytes. South African J Anim Sci 36(Supple 1):42–45
100. Wu C, Rui R, Dai J, Zhang C, Ju S, Xie B, Lu X, Zheng X (2006) Effects of cryopreservation on the developmental competence, ultrastructure and cytoskeletal structure of porcine oocytes. Mol Reprod Dev 73:1454–1462
101. Vincent C, Pickering SJ, Johnson MH, Quick SJ (1990) Dimethylsulphoxide affects the organisation of microfilaments in the mouse oocyte. Mol Reprod Dev 26:227–235
102. Toth TL, Lanzendorf SE, Sandow BA, Veeck LL, Hassen WA, Hansen K, Hodgen GD (1994) Cryopreservation of human prophase I oocytes collected from unstimulated follicles. Fertil Steril 61:1077–1082
103. Inoue S, Salmon ED (1995) Force generation by microtubule assembly/disassembly in mitosis and related movements. Mol Biol Cell 6:1619–1640
104. Rieder CL (1981) The structure of the cold-stable kinetochore fiber in metaphase PtK1 cells. Chromosoma 84:145–158
105. Spedden E, Kaplan DL, Staii C (2013) Temperature response of the neuronal cytoskeleton mapped via atomic force and fluorescence microscopy. Phys Biol 10:056002. doi:10.1088/1478-3975/10/5/056002
106. Ingber DE (2006) Cellular mechanotransduction: putting all the pieces together again. FASEB J 20:811–827
107. Calarco PG (2000) Centrosome precursors in the acentriolar mouse oocyte. Microsc Res Tech 49:428–434
108. Combelles CMH, Albertini DF (2001) Microtubule patterning during meiotic maturation in mouse oocytes is determined by cell cycle-specific sorting and redistribution of γ-tubulin. Dev Biol 239:281–294
109. Bearden HJ, Fuquay JW (1997) Environmental management. In: Bearden HJ, Fuquay JW (eds) Applied animal reproduction, 4th edn. A Simon & Schuster Company, New Jersey
110. Baumgartner AP, Chrisman CL (1981) Cytogenetic analysis of ovulated mouse oocytes following hyperthermic stress during meiotic maturation. Exp Cell Res 132:359–366
111. Tseng JK, Tang PC, Liu CT, Chiou CM, Ju JC (2002) The effect of hyperthermia on the cytoskeleton and development of oocytes. J Agric Assoc China 3:337–354
112. Ju JC, Tseng JK (2004) Nuclear and cytoskeletal alterations of in vitro matured porcine oocytes under hyperthermia. Mol Reprod Dev 68:125–133
113. Ju JC (2005) Cellular responses of oocytes and embryos under thermal stress: hints to molecular signaling. Anim Reprod 2:79–90
114. Ju JC, Parks JE, Yang X (1999) Thermotolerance of IVM-derived bovine oocytes and embryos after short-term heat shock. Mol Reprod Dev 53:336–340
115. Edwards JL, Hansen PJ (1997) Differential responses of bovine oocytes and preimplantation embryos to heat shock. Mol Reprod Dev 46:138–145
116. Tseng JK, Chen CH, Chou PC, Yeh SP, Ju JC (2004) Influences of follicular size on parthenogenetic activation and in vitro heat shock on the cytoskeleton in cattle oocytes. Reprod Domest Anim 39:146–153
117. Ju JC, Jiang S, Tseng JK, Parks JE, Yang X (2005) Heat shock reduces developmental competence and alters spindle configuration of bovine oocytes. Theriogenology 64:1677–1689
118. Suzuki H, Ju JC, Parks JE, Yang X (1998) Surface ultrastructural characteristics of bovine oocytes following heat shock. J Reprod Dev 44:345–351
119. Rivera RM, Kelley KL, Erdos GW, Hansen PJ (2003) Alterations in ultrastructural morphology of two-cell bovine embryos produced in vitro and in vivo following a physiologically relevant heat shock. Biol Reprod 69:2068–2077
120. Andreu-Vázquez C, López-Gatius F, García-Ispierto I, Maya-Soriano MJ, Hunter RHF, López-Béjar M (2010) Does heat stress provoke the loss of a continuous layer of cortical granules beneath the plasma membrane during oocyte maturation? Zygote 18:293–299. doi:10.1017/S0967199410000043

121. Roth Z, Hansen PJ (2004) Involvement of apoptosis in disruption of developmental competence of bovine oocytes by heat shock during maturation. Biol Reprod 71:1898–1906
122. Roth Z, Hansen PJ (2004) Sphingosine 1-phosphate protects bovine oocytes from heat shock during maturation. Biol Reprod 71:2072–2078
123. Roth Z, Hansen PJ (2005) Disruption of nuclear maturation and rearrangement of cytoskeletal elements in bovine oocytes exposed to heat shock during maturation. Reproduction 129:235–244
124. Tseng JK, Tang PC, Ju JC (2006) In vitro thermal stress induces apoptosis and reduces development of porcine parthenotes. Theriogenology 66:1073–1082
125. Muggleton-Harris AL, Brown JJG (1988) Cytoplasmic factors influence mitochondrial reorganization and resumption of cleavage during culture of early mouse embryos. Hum Reprod 3:1020–1028
126. Lane M, Bavister BD (1998) Calcium homeostasis in early hamster preimplantation embryos. Biol Reprod 59:1000–1007
127. Squirrell JM, Lane M, Bavister BD (2001) Altering intercellular pH disrupts development and cellular organization in preimplantation hamster embryos. Biol Reprod 64:1845–1854
128. Ludwig TE, Squirrell JM, Bavister BD (2001) Relationship between development, metabolism, and mitochondrial organization in 2-cell hamster embryos in the presence of low levels of phosphate. Biol Reprod 65:1648–1654
129. Abe H, Otoi T, Tachikawa S, Yamashita S, Satoh T, Hoshi H (1999) Fine structure of bovine morulae and blastocysts in vivo and in vitro. Anat Embryol 199:519–527
130. Krisher RL, Bavister BD (1998) Responses of oocytes and embryos to the culture environment. Theriogenology 49:103–114
131. Bavister BD (2000) Interactions between embryos and the culture milieu. Theriogenology 53:619–626
132. Suzuki H, Satoh M, Kabashima K (2006) Distributions of mitochondria and the cytoskeleton in hamster embryos developed in vivo and in vitro. J Mamm Ova Res 23:128–134
133. Suzuki H, Azuma T, Koyama H, Yang, X (1999) Development of cellular polarity of hamster embryos during compaction. Biol Reprod 61:521–526
134. Kabashima K, Yoshinaga D, Fang J, Matsuzaki M, Suzuki H (2013) Cell cycle-dependent dynamics of cytoskeleton involving mitochondrial redistribution in hamster embryos. Reprod Domest Anim 48:267–271. doi:10.1111/j.1439-0531.2012.02143.x
135. Azoury J, Verlhac MH, Dumont J (2009) Actin filaments: key players in the control of asymmetric divisions in mouse oocytes. Biol Cell 101:69–76. doi:10.1042/BC20080003
136. Webb M, Howlett SK, Maro B (1986) Parthenogenesis and cytoskeletal organization in ageing mouse eggs. J Embryol Exp Morphol 95:131–145
137. Zhu ZY, Chen DY, Li JS, Lian L, Lei L, Han ZM, Sun QY (2003) Rotation of meiotic spindle is controlled by microfilaments in mouse oocytes. Biol Reprod 68:943–946
138. Hollenbeck PJ, Saxon WM (2005) The axonal transport of mitochondria. J Cell Sci 118:5411–5419
139. Morris RL, Hollenbeck PJ (1995) Axonal transport of mitochondria along microtubules and F-actin in living vertebrate neurons. J Cell Biol 131:1315–1326
140. Wu M, Kalyanasundaram A, Zhu J (2013) Structural and biomechanical basis of mitochondrial movement in eukaryotic cells. Int J Nanomedicine 8:4033–4042. doi:10.2147/IJN.S52132
141. Lee S, Kim S, Sun X, Lee J-H, Cho H (2007) Cell cycle-dependent mitochondrial biogenesis and dynamics in mammalian cells. Biochem Biophys Res Commun 357:111–117
142. Ryan SD, Ferrier A, Kothary R (2012) A novel role for the cytoskeletal linker protein dystonin in the maintenance of microtubule stability and the regulation of ER-Golgi transport. Bioarchitecture 2:2–5
143. Sheng ZH, Cai Q (2012) Mitochondrial transport in neurons: impact on synaptic homeostasis and neurodegeneration. Nat Rev Neurosci 13:77–93. doi:10.1038/nrn3156
144. Heng YW, Koh CG (2010) Actin cytoskeleton dynamics and the cell division cycle. Int J Biochem Cell Biol 42:1622–1633. doi:10.1016/j.biocel.2010.04.007

Chapter 8
Tubulin Detyrosination in Epithelial Cells

Sabrina Zink and Ralf Jacob

Introduction

Epithelial cells represent the main contact sites between the internal organism and the outside world. They are characterized by a certain asymmetry of the plasma membrane as well as of the underlying cytoskeletal architecture to fulfil their functions.

The process of epithelial polarization, i.e. the formation of the apical and the basolateral cell pole is established by distinct biosynthetic pathways in which cytoskeletal elements are involved [1].

Epithelial Cells

Epithelial cells cover inner and outer biological surfaces and thus build a natural barrier against the environment. Hence, their functions include protection against mechanical stresses, tightening of body cavities, the resorption and secretion of specific substances and transport processes. A common feature of epithelial cells is the polarity of the plasma membrane with an apical domain facing the lumen and the morphologically and functionally differing basolateral domain facing neighboring cells or the basal lamina [2]. The distinct membrane regions are maintained by tight junctions whereas the adherens junctions and the desmosomes form the actual cell–cell contact in the epithelium. The apical and the basolateral membrane domain are composed of different subsets of proteins and lipids due to their functional specialization.

S. Zink • R. Jacob (✉)
Department of Cell Biology and Cell Pathology, Philipps-Universität Marburg,
Robert-Koch-Str. 6, 35033 Marburg, Germany
e-mail: jacob@staff.uni-marburg.de

H. Schatten (ed.), *The Cytoskeleton in Health and Disease*,
DOI 10.1007/978-1-4939-2904-7_8

This requires directional and highly specific transport processes dependent among others on protein sorting signals [3], motor proteins [4], distinct cytoskeletal tracks [5] and the polarization state of the cell [6, 7].

The polarity of epithelial cells is well maintained by their cytoskeletal organization which is changing significantly during the polarization process.

Microtubule Organization in Epithelial Cells

In non-polar cells microtubule filaments emanate from a centrosome or a microtubule organizing center (MTOC) localized near the nucleus [8]. During the polarization process basic alterations of the microtubular cytoskeleton and the whole intracellular organization occur. After release of the microtubule minus-ends from the centrosome minus-end binding factors anchor the microtubules on cell–cell-contacts or at the apical cell pole [9–11]. In polarized cells the filaments generate a network underneath the apical membrane but they are located at the basal cell pole as well. Furthermore, microtubules can be organized as bundles along the apico-basal axis [12–14]. Remarkably, horizontally stored network microtubules are not consistently oriented whereas minus ends in vertical bundles point to the apical and plus ends to the basolateral membrane [14–18]. In addition to their subcellular orientation, microtubules can be categorized according to their posttranslational modification, building the so called "tubulin code".

The Tyrosination/Detyrosination Cycle

Posttranslational Detyrosination of Tubulin

Posttranslational modifications are a common feature of various proteins including tubulin to generate a required functional diversity. For tubulin acetylation, de-/tyrosination, delta2 modification, polyglycylation and -glutamylation, phosphorylation and palmitoylation are known. Usually the C-Terminus of the alpha tubulin subunit is affected [19–23]. As part of the tyrosination/detyrosination cycle the tyrosine located at the C-terminus of newly synthesized alpha-tubulin is removed by a specific carboxypeptidase resulting in detyrosinated tubulin with a C-terminal glutamate residue [24]. This reaction preferably takes place in polymerized microtubule filaments whereas the opposite re-addition of a tyrosine residue is catalyzed by the tubulin tyrosine-ligase (TTL) on soluble tubulin dimers. Therefore, the tyrosination process is observed only after depolymerisation of microtubules [22, 23, 25–27]. Similar to acetylated tubulin detyrosinated tubulin is a component of stable, persistent microtubules. Thus, the modification is linked to microtubule stability although it is not a consequence of detyrosination. During cell maturation the filaments are exposed increasingly to carboxypeptidases resulting in the removal of C-terminal tyrosine and accordingly to an accumulation of detyrosinated tubulin towards the microtubule minus-ends [28, 29]. In contrast, tyrosinated tubulin is found frequently close to newly formed microtubule plus-ends, where the polymerization usually takes place [30].

Tubulin Tyrosine-Ligase (TTL)

Tubulin tyrosine ligase (TTL) is one of the best-characterized enzymes involved in tubulin modification. It triggers the re-addition of tyrosine to the C-terminus of detyrosinated tubulin [23, 31–33]. The enzyme comprises an N-, a central and a C-terminal domain forming the active site of the enzyme. The ligase recognizes the conserved, positively charged alpha-tubulin surface and a curved conformation of the heterodimer thus forming an elongated structure. Thereby, it caps the tubulin dimer preventing its incorporation into the microtubule filament [33, 34].

Mice deficient for the tubulin tyrosine-ligase (TTL–/–) die within few hours after birth because of neuronal disorganization. This result was confirmed in neuronal cell culture leading to morphogenetic anomalies [35].

The Tubulin Carboxypeptidase (TCP)

In contrast to TTL the tubulin carboxypeptidase (TCP) could not be characterized genetically or biochemically until now. Several researchers were able to show the TCP activity in subcellular fractions [19, 36] and on microtubules in living cells [37]. Furthermore, inhibitors of the enzyme could be identified contributing to a better understanding of the carboxypeptidase mode of action [38, 39].

The Importance of Tubulin Detyrosination for Epithelial Cell Growth and Maintenance

Detyrosinated Tubulin in the Polarisation Process and Protein Trafficking

The two distinct membrane domains of epithelial cells with their specific protein and lipid composition are maintained by directed and highly specific transport and sorting processes.

Vesicular transport along microtubules requires among others motor proteins bound to the filaments and providing the mechanical force for movement. Dependent on the tyrosination/detyrosination state of the microtubule, different motors and accessory proteins are recruited to the filament. Kinesin 1, a plus-end directed motor preferentially binds to detyrosinated tubulin and allows the transport of different molecules on specific microtubules [40–43]. A kinesin1 family member, Kif5C, is known to move slower along, preferentially binds to and decorates detyrosinated microtubules [44]. Jaulin et al. reported in 2007 that Kif5B mediates apical protein transport only in fully polarized cells pointing to a kinesin switch during the polarization process [4]. This could be associated with a switch of posttranslational modifications as was published in the last few years. The results of these articles indicated that the occurrence of posttranslational modifications depends on the polarization state of the cell. Detyrosinated tubulin is increasingly present early in the polarization process in subconfluent cells and shows a decreased incidence in fully polarized cells [6, 7]. Concomitantly, decreasing the levels of detyrosinated tubulin by overexpression of TTL results in a dramatic reduction of apical delivery of gp80. This suggests that certain apical transport pathways in epithelial cells use detyrosinated microtubules as cytoskeletal tracks (Fig. 8.1).

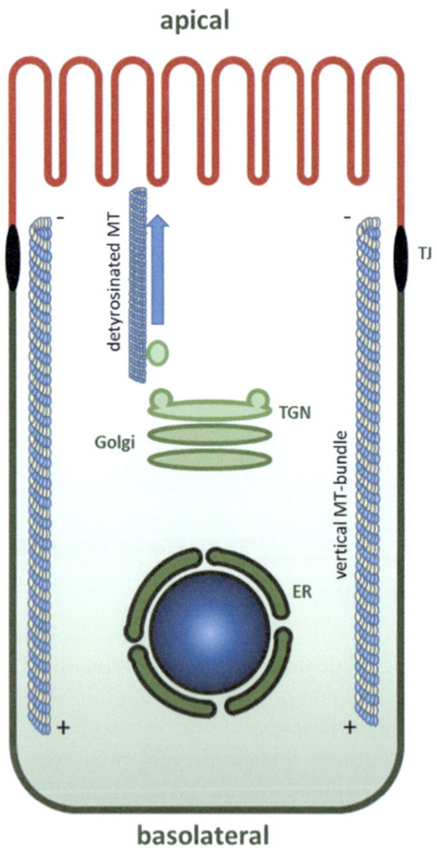

Fig. 8.1 Detyrosinated microtubules serve as tracks for apical trafficking. *Post* Golgi vesicles are transported along detyrosinated microtubules to the apical membrane domain of epithelial cells. Minus (−) and plus (+) ends of vertical bundles are indicated. *ER* endoplasmic reticulum, *MT* microtubules, *TGN trans* Golgi network

Another hint to the importance of tubulin detyrosination in epithelial cell growth was provided by experiments investigating the polarization process in cells with low levels of detyrosinated tubulin. This reduction leads to an early assembly of epithelial MDCK cells into isolated islands which develop a prematurely polarized architecture [7]. This implicates a role of tubulin detyrosination in morphological differentiation from non-polarized cells into an epithelial cell layer.

Detyrosinated Tubulin in Disease and Cancer

Beside its role in growth and maintenance of the integrity of epithelial cells tubulin modifications and associated enzymes are also known to be involved in certain pathomechanisms.

Detyrosinated tubulin, respectively, the level of tubulin tyrosine-ligase is associated with tumor progression, invasiveness and chemoresistance [45–48].

Most human breast cancers are associated with a suppressed TTL activity. The same applies for several human tumors of different tissue origin producing sarcoma growth in mice. In addition, these tumors from TTL^{-} cells essentially contain detyrosinated and delta2 tubulin and were rescued by TTL cDNA. The results suggest a role for TTL as tumor suppressor [46].

In neuroblastoma, the most common solid cancer in childhood, an unfavorable prognosis is mostly associated with deregulation of the tyrosination/detyrosination cycle of tubulin. This is caused by a decreased expression of human TTL and accompanied by an accumulation of delta2 tubulin whereas favorable neuroblastoma cells are positive for tyrosinated, detyrosinated and delta2 tubulin [45].

Compared to normal human prostate epithelial cells, prostate cancer cells exhibit a reduced level of TTL together with an elevated level of detyrosinated tubulin. But in contrast to other cancer cell lines normal cells have significantly higher levels of delta2 tubulin [47].

During the epithelial-to-mesenchymal transition (EMT) epithelial cells lose their polarity and adhesion to acquire a mesenchymal state. This procedure enables cancer cells to become motile and to migrate and invade other tissues. Tumor invasion or metastasis following the EMT is accompanied by TTL downregulation and increased levels of detyrosinated tubulin. Furthermore, the formation of so called microtentacles enriched in detyrosinated tubulin, may enhance the attachment of circulating cancer cells to the endothelium [48]. Interestingly, specific treatment of the cells with parthenolide and costunolide leads to decreased levels of detyrosinated tubulin and with that to a reduced frequency of microtentacles and inhibition of reattachment [49]. In 2007 Fonrose and colleagues proposed that parthenolide inhibits the action of the tubulin carboxypeptidase resulting in reversal of the accumulation of detyrosinated tubulin in tumor cells [39].

Conclusions

In summary, cells manipulate the composition of posttranslationally modified microtubules according to their architecture, morphogenesis and function. Alterations in the ratios of tyrosinated and detyrosinated microtubules will then affect the lifetime, stability and binding capacity to molecular motors. The last point is further strengthened by recent data from the Vale lab, which show that posttranslational tubulin modifications can specifically regulate the activity of distinct dynein and kinesin motors [50]. It is thus attractive to speculate that microtubules enriched in certain posttranslational modification serve as tracks to particular subcellular domains. If this is a general concept of cellular targeting remains an open question and depends on additional studies in the future.

References

1. Rodriguez-Boulan E, Macara IG (2014) Organization and execution of the epithelial polarity programme. Nat Rev Mol Cell Biol 15:225–242
2. Yeaman C, Grindstaff KK, Nelson WJ (1999) New perspectives on mechanisms involved in generating epithelial cell polarity. Physiol Rev 79:73–98
3. Rodriguez-Boulan E, Powell SK (1992) Polarity of epithelial and neuronal cells. Annu Rev Cell Biol 8:395–427
4. Jaulin F, Xue X, Rodriguez-Boulan E, Kreitzer G (2007) Polarization-dependent selective transport to the apical membrane by KIF5B in MDCK cells. Dev Cell 13:511–522
5. Jacob R, Heine M, Alfalah M, Naim HY (2003) Distinct cytoskeletal tracks direct individual vesicle populations to the apical membrane of epithelial cells. Curr Biol 13:607–612
6. Quinones GB, Danowski BA, Devaraj A, Singh V, Ligon LA (2011) The posttranslational modification of tubulin undergoes a switch from detyrosination to acetylation as epithelial cells become polarized 133. Mol Biol Cell 22:1045–1057
7. Zink S, Grosse L, Freikamp A, Banfer S, Muksch F, Jacob R (2012) Tubulin detyrosination promotes monolayer formation and apical trafficking in epithelial cells 1. J Cell Sci 125:5998–6008
8. Vorobjev IA, Nadezhdina ES (1987) The centrosome and its role in the organization of microtubules. Int Rev Cytol 106:227–293
9. Keating TJ, Peloquin JG, Rodionov VI, Momcilovic D, Borisy GG (1997) Microtubule release from the centrosome. Proc Natl Acad Sci U S A 94:5078–5083
10. Ligon LA, Holzbaur EL (2007) Microtubules tethered at epithelial cell junctions by dynein facilitate efficient junction assembly. Traffic 8:808–819
11. Stehbens SJ, Paterson AD, Crampton MS, Shewan AM, Ferguson C, Akhmanova A, Parton RG, Yap AS (2006) Dynamic microtubules regulate the local concentration of E-cadherin at cell-cell contacts 99. J Cell Sci 119:1801–1811
12. Bartolini F, Gundersen GG (2006) Generation of noncentrosomal microtubule arrays. J Cell Sci 119:4155–4163
13. Musch A (2004) Microtubule organization and function in epithelial cells 56. Traffic 5:1–9
14. Van FD, Johnson K, Segbert C, Bossinger O (2004) The C. elegans ezrin-radixin-moesin protein ERM-1 is necessary for apical junction remodelling and tubulogenesis in the intestine. Dev Biol 272:262–276
15. Bacallao R, Antony C, Dotti C, Karsenti E, Stelzer EH, Simons K (1989) The subcellular organization of Madin-Darby canine kidney cells during the formation of a polarized epithelium. J Cell Biol 109:2817–2832

16. Bre MH, Pepperkok R, Hill AM, Levilliers N, Ansorge W, Stelzer EH, Karsenti E (1990) Regulation of microtubule dynamics and nucleation during polarization in MDCK II cells. J Cell Biol 111:3013–3021
17. de Forges H, Bouissou A, Perez F (2012) Interplay between microtubule dynamics and intracellular organization. Int J Biochem Cell Biol 44:266–274
18. Meads T, Schroer TA (1995) Polarity and nucleation of microtubules in polarized epithelial cells. Cell Motil Cytoskeleton 32:273–288
19. Barra HS, Arce CA, Argarana CE (1988) Posttranslational tyrosination/detyrosination of tubulin. Mol Neurobiol 2:133–153
20. Hammond JW, Cai D, Verhey KJ (2008) Tubulin modifications and their cellular functions. Curr Opin Cell Biol 20:71–76
21. MacRae TH (1997) Tubulin post-translational modifications–enzymes and their mechanisms of action. Eur J Biochem 244:265–278
22. Verhey KJ, Gaertig J (2007) The tubulin code. Cell Cycle 6:2152–2160
23. Westermann S, Weber K (2003) Post-translational modifications regulate microtubule function. Nat Rev Mol Cell Biol 4:938–947
24. Argarana CE, Barra HS, Caputto R (1978) Release of [14C]tyrosine from tubulinyl-[14C] tyrosine by brain extract. Separation of a carboxypeptidase from tubulin-tyrosine ligase. Mol Cell Biochem 19:17–21
25. Barra HS, Arce CA, Rodriguez JA, Caputto R (1974) Some common properties of the protein that incorporates tyrosine as a single unit and the microtubule proteins. Biochem Biophys Res Commun 60:1384–1390
26. Webster DR, Gundersen GG, Bulinski JC, Borisy GG (1987) Differential turnover of tyrosinated and detyrosinated microtubules. Proc Natl Acad Sci U S A 84:9040–9044
27. Wehland J, Weber K (1987) Turnover of the carboxy-terminal tyrosine of alpha-tubulin and means of reaching elevated levels of detyrosination in living cells. J Cell Sci 88(Pt 2): 185–203
28. Khawaja S, Gundersen GG, Bulinski JC (1988) Enhanced stability of microtubules enriched in detyrosinated tubulin is not a direct function of detyrosination level. J Cell Biol 106:141–149
29. Schulze E, Asai DJ, Bulinski JC, Kirschner M (1987) Posttranslational modification and microtubule stability. J Cell Biol 105:2167–2177
30. Ahmad FJ, Pienkowski TP, Baas PW (1993) Regional differences in microtubule dynamics in the axon. J Neurosci 13:856–866
31. Ersfeld K, Wehland J, Plessmann U, Dodemont H, Gerke V, Weber K (1993) Characterization of the tubulin-tyrosine ligase. J Cell Biol 120:725–732
32. Murofushi H (1980) Purification and characterization of tubulin-tyrosine ligase from porcine brain. J Biochem 87:979–984
33. Prota AE, Magiera MM, Kuijpers M, Bargsten K, Frey D, Wieser M, Jaussi R, Hoogenraad CC, Kammerer RA, Janke C, Steinmetz MO (2013) Structural basis of tubulin tyrosination by tubulin tyrosine ligase. J Cell Biol 200:259–270
34. Szyk A, Deaconescu AM, Piszczek G, Roll-Mecak A (2011) Tubulin tyrosine ligase structure reveals adaptation of an ancient fold to bind and modify tubulin 122. Nat Struct Mol Biol 18:1250–1258
35. Erck C, Peris L, Andrieux A, Meissirel C, Gruber AD, Vernet M, Schweitzer A, Saoudi Y, Pointu H, Bosc C, Salin PA, Job D, Wehland J (2005) A vital role of tubulin-tyrosine-ligase for neuronal organization. Proc Natl Acad Sci U S A 102:7853–7858
36. Lafanechere L, Job D (2000) The third tubulin pool. Neurochem Res 25:11–18
37. Contin MA, Sironi JJ, Barra HS, Arce CA (1999) Association of tubulin carboxypeptidase with microtubules in living cells. Biochem J 339(Pt 2):463–471
38. Contin MA, Purro SA, Bisig CG, Barra HS, Arce CA (2003) Inhibitors of protein phosphatase 1 and 2A decrease the level of tubulin carboxypeptidase activity associated with microtubules. Eur J Biochem 270:4921–4929
39. Fonrose X, Ausseil F, Soleilhac E, Masson V, David B, Pouny I, Cintrat JC, Rousseau B, Barette C, Massiot G, Lafanechere L (2007) Parthenolide inhibits tubulin carboxypeptidase activity. Cancer Res 67:3371–3378

40. Konishi Y, Setou M (2009) Tubulin tyrosination navigates the kinesin-1 motor domain to axons 1. Nat Neurosci 12:559–567
41. Kreitzer G, Liao G, Gundersen GG (1999) Detyrosination of tubulin regulates the interaction of intermediate filaments with microtubules in vivo via a kinesin-dependent mechanism. Mol Biol Cell 10:1105–1118
42. Liao G, Gundersen GG (1998) Kinesin is a candidate for cross-bridging microtubules and intermediate filaments. Selective binding of kinesin to detyrosinated tubulin and vimentin. J Biol Chem 273:9797–9803
43. Lin SX, Gundersen GG, Maxfield FR (2002) Export from pericentriolar endocytic recycling compartment to cell surface depends on stable, detyrosinated (glu) microtubules and kinesin. Mol Biol Cell 13:96–109
44. Dunn S, Morrison EE, Liverpool TB, Molina-Paris C, Cross RA, Alonso MC, Peckham M (2008) Differential trafficking of Kif5c on tyrosinated and detyrosinated microtubules in live cells. J Cell Sci 121:1085–1095
45. Kato C, Miyazaki K, Nakagawa A, Ohira M, Nakamura Y, Ozaki T, Imai T, Nakagawara A (2004) Low expression of human tubulin tyrosine ligase and suppressed tubulin tyrosination/detyrosination cycle are associated with impaired neuronal differentiation in neuroblastomas with poor prognosis. Int J Cancer 112:365–375
46. Lafanechere L, Courtay-Cahen C, Kawakami T, Jacrot M, Rudiger M, Wehland J, Job D, Margolis RL (1998) Suppression of tubulin tyrosine ligase during tumor growth. J Cell Sci 111(Pt 2):171–181
47. Soucek K, Kamaid A, Phung AD, Kubala L, Bulinski JC, Harper RW, Eiserich JP (2006) Normal and prostate cancer cells display distinct molecular profiles of alpha-tubulin posttranslational modifications. Prostate 66:954–965
48. Whipple RA, Matrone MA, Cho EH, Balzer EM, Vitolo MI, Yoon JR, Ioffe OB, Tuttle KC, Yang J, Martin SS (2010) Epithelial-to-mesenchymal transition promotes tubulin detyrosination and microtentacles that enhance endothelial engagement 1. Cancer Res 70:8127–8137
49. Whipple RA, Vitolo MI, Boggs AE, Charpentier MS, Thompson K, Martin SS (2013) Parthenolide and costunolide reduce microtentacles and tumor cell attachment by selectively targeting detyrosinated tubulin independent from NF-kappaB inhibition. Breast Cancer Res 15:R83
50. Sirajuddin M, Rice LM, Vale RD (2014) Regulation of microtubule motors by tubulin isotypes and post-translational modifications. Nat Cell Biol 16:335–344

Chapter 9
Mutations in *Adenomatous Polyposis Coli*, Their Role in Cytoskeletal Dynamics and Cancer Onset

Alexander E. Davies and Kenneth B. Kaplan

Abbreviations

APC	Adenomatous polyposis coli
CIN	Chromosome instability
GAP	GTPase activating protein
GEF	GTPase exchange factor
LOH	Loss of heterozygosity
Min	Multiple intestinal neoplasia
TSG	Tumor suppressor genes

Introduction

Gross changes in cell shape and migration accompany cancer cell transformation and have reinforced the idea that changes in cytoskeletal dynamics contribute directly to cancer phenotypes. In this view, cytoskeletal changes are downstream of cancer onset and arise directly from deregulation of signaling networks. For example, activation of the c-src proto-oncogene gives rise to changes in the actin

A.E. Davies
Department of Molecular and Cellular Biology and the School of Veterinary Medicine,
University of California, Davis, One Shields Avenue, Davis, CA 95616, USA
e-mail: aedavies@ucdavis.edu

K.B. Kaplan (✉)
Department of Molecular and Cellular Biology, University of California, Davis,
One Shields Avenue, Davis, CA 95616, USA
e-mail: kbkaplan@ucdavis.edu

H. Schatten (ed.), *The Cytoskeleton in Health and Disease*,
DOI 10.1007/978-1-4939-2904-7_9

cytoskeleton. This occurs in part through its activation of the focal-adhesion kinase (FAK), which lies upstream of p21-activated kinases (PAKs) that in turn regulate small GTPases, such as Rho and Rac; as a result, there is a direct increase in actin polymerization [1]. Similarly, activated Ras mutations through the Rac-GTPase exchange protein, Tiam1, also activate the Rac pathway to polymerize actin. A number of functional studies have linked these changes in actin to altered cell–cell contacts and increased cell migration that ultimately contributes to loss of growth control, invasion and metastasis.

Changes in microtubule nucleation or dynamics are also common in cancer cells and associated with the loss of function of a subset of tumor suppressor genes (TSG's) [2]. Deregulation of microtubules can impact signaling pathways that control programmed cell death and contribute to errors in chromosome segregation. This combination of changes may allow tumor cells to acquire and adapt to an aneuploid state (i.e., elevated levels of gene expression and increased genomic instability [3]), a stressed state that results in cell death in non-cancer cells. For example, loss of BRCA1, a TSG found mutated in breast cancers, causes changes in the microtubule cytoskeleton by increasing the number of centrosomes, resulting in multi-polar spindles and chromosome instability [4, 5]. TSGs such as NF1, NF2 as well as LATS1 and LATS2 are also implicated in regulating cytoskeletal organization through microtubule dynamics as well as centrosome number [2, 6–8]. Thus, a variety of TSG mutations affect microtubule organization either by altering their dynamic properties or by changing their organizing centers. Here we will focus on mutations in *adenomatous polyposis coli* (APC) that predispose patients to colorectal cancer and alter microtubule dynamics.

Our work (reviewed below) has shown that mutations in APC result in altered microtubule dynamics *prior* to cancer onset, a role for APC distinct from its regulation of β-catenin and gene transcription. An important implication from this work is that changes in cytoskeletal dynamics that precede cancer onset might have a role in the transformation of normal cells to cancer cells. This idea addresses one of the outstanding questions in cancer; what changes must occur in a normal cell before it takes on the properties we associate with cancer? This is a deceptively difficult question to answer. The identification of cells that represent the first step(s) in cancer onset is technically challenging as these intermediates are presumably short-lived and mature tumors are filled with a myriad of accumulated changes. Insights into the pre-cancer cell state come from observations of so-called "oncogenic stress". This stress is an immediate outcome of the expression of mutants in TSGs or oncogenes. Such mutations often combine with genomic instability and lead to an acute perturbation of overall cellular homeostasis and a state that is antagonistic to cell proliferation [9]. To cope, cells may respond by adjusting pathways that restore homeostasis, providing pre-cancer cells with a heightened ability to adapt and to tolerate mutations that lead to unregulated cell proliferation. Our studies suggest that APC mutants act prior to cancer onset and broadly alter cellular homeostasis by perturbing cytoskeletal organization, causing chromosome aneuploidy and affecting the asymmetric inheritance of information. We propose that these changes may synergize to create a "permissive" cellular environment for cancer transformation and thus may represent a common requirement for all cells destined to become cancer cells.

The Genetics of APC and Colorectal Cancer

In the canonical colorectal cancer model, a single inherited mutation in APC is sufficient to dominantly predispose familial adenomatous polyposis (FAP) patients to colorectal cancer [10]. In FAP patients, the majority of APC alleles result in nonsense mutations that produce a truncated protein. Significantly, similar APC mutations are also found in one of the two alleles in sporadic colorectal tumors [11–13]. The loss of heterozygosity (i.e., LOH, loss of function of the wild type APC allele) observed in all FAP and the majority of sporadic tumors led to the suggestion that loss of APC is the first step in colorectal cancer onset. This observation is consistent with simple tumor suppressor models as described eloquently in Knudson's two-hit hypothesis [14]. Such a model presupposes that LOH at APC favors cell survival and proliferation. Evidence to support the latter prediction comes from both mouse models and from genetic analysis of human patients where full-length APC is already lost in small adenomas [15]. Thus, APC has been described as a "gatekeeper" tumor suppressor gene, which implies that the key recessive phenotype associated with loss of APC is increased cell proliferation [10]. However, there are some complications with this simple model, as detailed below.

The main target of APC in regulating cell proliferation appears to lie in its ability to control the levels of the cell polarity and cell fate determinant, β-catenin. The relevant mechanisms and pathways are well reviewed by others [16, 17] but, briefly, APC forms a multi-protein regulatory complex with Axin and the kinase, GSK3β. Together, this complex modifies β-catenin, a key component of the Wnt-signal transduction pathway [18]. Phosphorylation of β-catenin targets it for interaction with the SCF-β-TrCP ubiquitin ligase complex and proteasomal degradation. Thus, under conditions where Wnt-signaling is turned off, cytosolic β-catenin pools are targeted for phosphorylation and degradation. Under conditions where Wnt signals are "on", the β-catenin degradation pathway is inhibited allowing cytosolic β-catenin to accumulate and to interact with the TCF4 transcription factor. Subsequent nuclear import of β-catenin/TCF4 results in the transcriptional activation of a wide range of direct and indirect targets that include growth regulatory genes such as Myc and cyclin D1 [17, 19–21]. Therefore, an increase in the levels of β-catenin is one of the first molecular hallmarks in cells that have lost APC function and is implicated in driving cellular proliferation.

Paradoxically, loss of APC function and up regulation of β-catenin does not immediately lead to more cell division or promote cell survival, a characteristic shared with other oncogenic and TSG mutations [9]. Mice carrying a homozygous deletion of APC die as embryos with defects in neural tube closure [22, 23]. Moreover, cell death is elevated in naive intestinal epithelium when APC is conditionally knocked-out, a finding that contrasts with the low levels of apoptosis found in colorectal cancers [24–26]. The lack of proliferative advantage in cells without APC in these experiments may be because many of them lack stem cell characteristics. Indeed, targeting APC knockouts specifically to stem cells results in a higher frequency of micro-adenomas compared to knockouts made in both stem cells and transit amplifying cells [27]. However, the very high efficiency of APC gene ablation in these compartments raises additional questions as to the relevance for bona fide in situ cancer onset, especially since loss of heterozygosity at APC is probably a rare event in familial cases and is certainly rare in sporadic cancers. Mutant Cre recombinase that reduces the knockout targeting frequency more faithfully recapitulates human disease. In these animals it is clear that single stem cells deleted for APC require other changes to advance to adenomas, possibly involving crypt fusion to mediate cell expansion [28]. Additionally, the frequency of transformation in cells lacking APC is quite low under these conditions (17 %), suggesting that simple ablation of APC is not sufficient to drive cancer onset [29]. Interestingly, Fischer and colleagues found that engineering these sporadically targeted APC deletions in mice strains that already have a mutant APC encoding for a truncated protein (i.e., APC^{Min} or APC^{1638N}) dramatically increases the rates of cancer transformation in APC null cells. These findings argue strongly that long-term expression of APC mutant proteins predispose cells for cancer transformation following LOH, perhaps by providing a mechanism for cells to withstand the deleterious effects associated with losing APC.

In summary, APC genetics argue for two distinct phenotypes. The recessive phenotype results in the stabilization of β-catenin and in transcriptional re-programming of crypt stem cells. However, there is strong evidence to implicate a mono-allelic mutation in APC in driving cancer onset. Such a phenotype may be related to the truncated APC proteins that frequently arise in tumors or from haplo-insufficiency, as suggested by transgenic mice [30]. We will now review a series of studies on the biochemical and cellular activities of the truncated APC proteins that have shed light on their role in regulating the cytoskeleton and on plausible mechanisms by which they may increase cancer risk.

APC Domain Organization and Cytoskeletal Functions

Despite the initial focus on APC in β-catenin regulation, more comprehensive studies make a strong case that APC regulation of actin and microtubule dynamics is just as important. We will review the evidence for these diverse roles by summarizing the domain organization of APC and their relevant protein interactions. APC encodes for a large polypeptide (~2800 amino acids) with a number of identifiable sub-domains (see Fig. 9.1) [12, 31]. We will first describe the amino (N) terminal domains and their connections to proteins that are involved in regulating actin and β-catenin. We will then discuss the multiple modes of interaction between APC and microtubules, which include both N- and carboxy (C)-terminal domains, and finally make a case that these activities are part of a larger regulatory circuit important during cell migration. Despite our linear approach to reviewing APC domains, it is important to acknowledge these domains are likely to function in the context of important intra-protein interactions that to date remain poorly understood [32]. As diagrammed in Fig. 9.1, the amino terminus of APC (amino acids 1–58), a predicted coiled coil domain, mediates the dimerization of APC [33–35]. A series of heptad

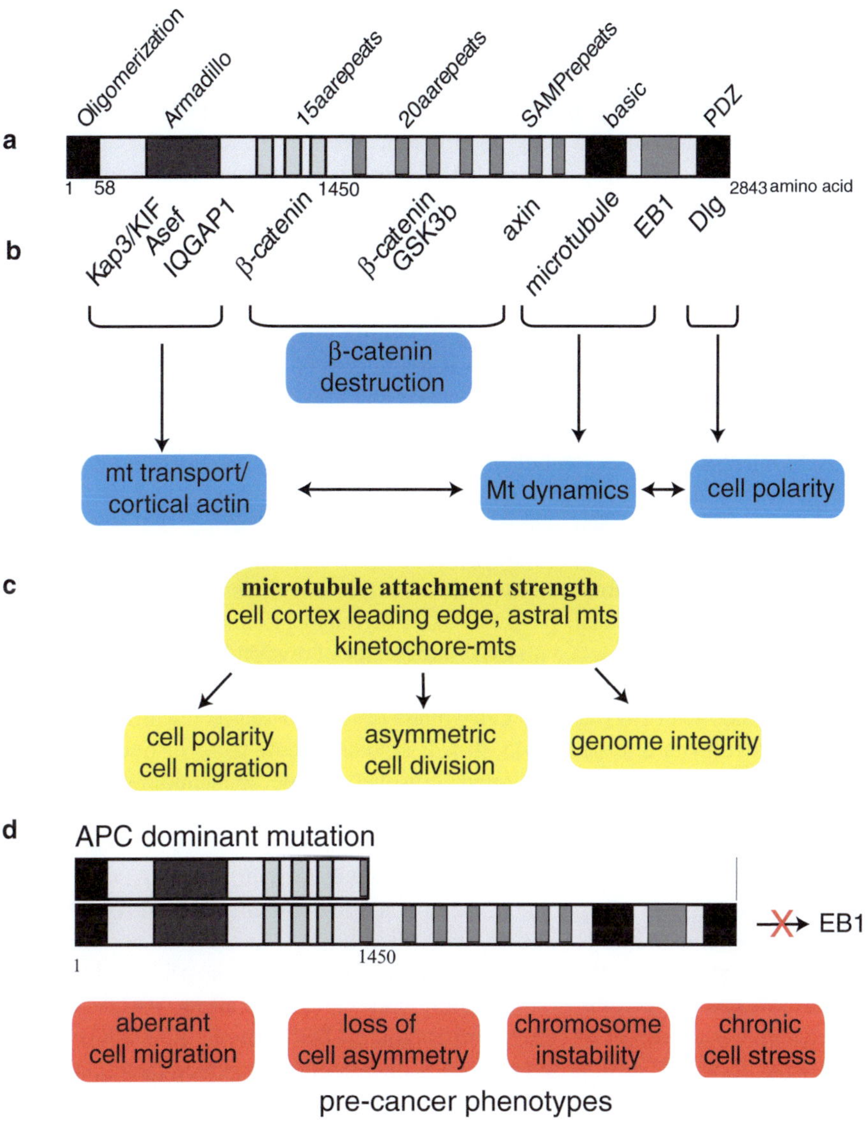

Fig. 9.1 APC domain organization and activities in normal and cancer cells. (**a**) A schematic of protein domains in APC as defined by sequence homology and functional experiments. (**b**) Known protein interactors are aligned below the relevant APC domain that they bind. For the Armadillo domain, each of the multiple interactors are aligned below but this does not imply that they can bind at the same time. The *blue boxes* indicate links between binding partners and the specific indicated cellular functions. (**c**) The text boxes focus on the complex interaction between APC and microtubules. The *yellow boxes* highlight the multiple contexts of microtubule interactions regulated by APC and the downstream cellular pathways that are affected by the indicated microtubule interactions. (**d**) A schematic of the translated product of a typical monoallelic APC mutant and its interaction with full length, wild type APC protein. The *red boxes* highlight the defects observed in cells that express the dominant APC mutants. See text for details and references

repeats (amino acids 250–800) forms a canonical Armadillo domain that interacts with a number of other Armadillo containing proteins, including ASEF, a GEF (*G*TP *e*xchange *f*actor) for Rho, IQGAP1, a GAP (*G*TPase *a*ctivating *p*rotein) for Rac1/Cdc42 and KAP3, a protein linked to microtubule motors [36–38]. A large region of the middle of APC consists of distinct sets of repeats (i.e., 15 amino acid repeats, 20 amino acid repeats and SAMP repeats) that have been implicated in interactions with a variety of protein partners. The 15- and 20-amino acid repeat regions contribute to the interaction between APC and β-catenin (Fig. 9.1b). The 20-amino acid repeats and the SAMP domain link APC to axin and GSK3β [39]. Mutations that eliminate any of the middle domain repeats reduce the affinity of APC for axin and β-catenin, ultimately inhibiting ubiquitination and the degradation of β-catenin. It is not clear how, or if, regulation of β-catenin and actin dynamics are coordinated but it is increasingly clear that the connection to microtubules is integral to all of APC functions.

APC interacts with microtubules via three distinct domains. First, as mentioned, the Armadillo domain in the N-terminus links APC to KAP3 and the kinesin KIF3A/3B. There are two additional microtubule interacting domains in the C-terminus (Fig. 9.1a). First, a stretch of basic amino acids is sufficient to interact with and bundle microtubules in a test-tube but this domain is not sufficient to interact with microtubules in cells [40]. Second, a region of 170 amino acids at the distal C-terminus of APC interacts with EB1, a regulator of microtubule plus-end dynamics [41–44]. APC has been observed to track growing microtubule plus-ends that are also associated with EB1 and the C-terminus of APC increases the affinity of EB1 for microtubules in a test tube [45, 46]. It is unclear how these distinct modes of microtubule interaction are integrated, however unique contributions have been described using APC mutants. For example, the interaction of KAP3 with APC is required for APC to cluster in bundles of peripheral microtubules in MDCK cells [37]. Dominant fragments of KAP3 that interact with kinesin, but not APC, enhance cell migration [37, 47]. Importantly, the interaction between APC and KAP3 is not sufficient for APC to interact directly with microtubules; instead, the accumulation of APC on microtubules requires the C-terminus of APC. It is possible that all three of these modes of microtubule interaction work together to allow APC to regulate the dynamics of a subset of cellular microtubules. For example, KAP3-KIF3A/3B may work to enrich APC at the plus ends of microtubules. Once there, APC can interact with EB1 to promote plus-end growth and the basic region may act to locally bundle microtubules. Although the precise nature of these relationships remains to be defined, it is increasingly clear that these regions work together inside the cell to organize the cytoskeleton at the leading edge of migrating cells.

Indeed, studies of directed cell migration have now shed light on how APC interactions with microtubules are important to both stabilize microtubules and to polymerize actin at the leading edge of the cell. The induction of directed cell migration after monolayer wounding involves reorientation of the microtubule organization center, the rearward movement of the nucleus relative to the direction of cell migration as well as the formation of stable microtubules and actin polymerization at the leading edge of the cell (for review; [48]). Following wounding, lysophosphatidic acid (LPA) in serum induces a signaling pathway that activates the small GTPase, Rho and its effector, mDia, a member of the formin family involved in actin polymerization. This pathway is sufficient to induce stable microtubules (i.e., containing glutamine modified tubulin) at the leading edge of migrating fibroblasts and requires the interaction of APC and EB1 [49]. Interestingly, microtubule stabilization also requires mDia, but it is important to point out that this activity is separable from the role of mDia in inducing actin polymerization [50, 51]. Recent work has suggested that the two activities of mDia oppose each other, meaning that microtubule stabilization may happen only when mDia is released from actin at the leading edge of migrating cells [52]. Thus, microtubules that interact with actin at the leading edge become stabilized because formins are released and interact with microtubule plus end complexes that include APC and EB1. This interplay between APC, actin and microtubule stability has been observed in a variety of settings, including cytokinesis and axon outgrowth and thus may represent a conserved role for APC in these systems [53–55]. Importantly, a similar APC-EB1-formin complex also acts during mitosis to stabilize kinetochore microtubule interactions, suggesting this complex may perform a common function in migrating and dividing cells [56–58]. This is particularly relevant as the region of APC required to form this complex is lost in colorectal cancer mutants that exhibit high rates of chromosome instability and altered microtubule dynamics. We will discuss the implications of disrupting this complex for cancer onset more below.

APC Mutants Act Dominantly to Inhibit Spindle Microtubule Dynamics

Chromosome Instability (CIN) in Colorectal Tumors

Chromosome numbers in a subset of cultured colorectal tumor cell lines show constant instability, an observation that led to the proposal that these tumor cells possess a mutation that affects mitosis [59]. This chromosome instability (CIN) is distinct from other types of cancer cells that have relatively stable aneuploid genomes. Although many reports have suggested that the spindle checkpoint—the feedback pathway that monitors attachment of chromosomes to the mitotic spindle—is altered in some cancers, the report that checkpoint failures are responsible for CIN in colorectal tumor cells has not been born out [60]. Changes in the mitotic checkpoint proteins such as BubR1 seem to be in response to chromosome instability, rather than the cause of it [61, 62]. A major clue to the cause of CIN in colorectal cells comes from the observation that APC associates with kinetochore microtubules that mediate the attachment of chromosomes to the mitotic spindle [63]. Embryonic stem cells from Min (multiple intestinal neoplasia) mice, harboring a single mutant copy of APC that results in a protein truncated after amino acid 850, show an increase in lagging chromosomes compared to wild type litter-mates, but no defect in the spindle checkpoint after challenges with microtubule poisons. Similarly, examination of multiple human colorectal tumor cell lines reveals that only those with APC mutations, and not those with the primary mutations in β-catenin, exhibit chromosome segregation defects. Furthermore, defects in chromosome segregation correlate with aberrant mitotic spindle organization [64]. In fact, colorectal tumor cells exhibit chromosome alignment failures in metaphase, lagging chromosomes in anaphase and mis-positioned spindles. Together, these findings directly implicate APC mutations in the CIN phenotype found in colorectal cancer cells.

Dominant-Negative Action of APC on Microtubule Dynamics

A definitive test of the causal relationship between monoallelic APC mutants and CIN came from reconstituting expression of mutant APC alleles in human cells that only express wild type APC. Inducible expression of an APC mutant that expresses a truncated protein that mimics those found in human colorectal tumors (i.e., APC^{1-1450}) resulted in mitotic spindle defects that recapitulate those observed in tumor cells, including: an increase in lagging chromosomes, a reduction in the number of kinetochore microtubules and defects in spindle positioning [64]. APC^{1-1450} directly binds to microtubule plus-ends in mitosis and is found specifically at kinetochore microtubules. Genetic analysis showed that APC^{1-1450} functions in the same pathway as the microtubule plus-end binding protein, EB1; siRNA of each gene results in identical mitotic defects and combined siRNA of APC and EB1 does not synergize to create more severe mitotic defects [35]. Structure–function studies of APC showed that the oligomerization domain of APC is critical for the dominant activity of APC^{1-1450}. Biochemical analysis confirmed that the truncated forms of APC interact with full length APC via the N-terminal oligomerization domain. The resulting hetero-oligomer (full length APC and truncated APC; see Fig. 9.1d) prevents EB1 from interacting with the full length APC, providing evidence that truncated forms of APC found in cancers act as dominant negatives, interfering with EB1 function and thus microtubule dynamics. Although deletions of the SAMP repeats between amino acids 1020 and 1309 reduce the potency of APC in interfering with microtubule dynamics, alleles that include up to amino acid 638 and remove much of the armadillo domain still have dominant microtubule spindle phenotypes [35]. Measurements of microtubule dynamics in mitotic cells expressing APC^{1-1450} show that microtubule pausing increases by nearly fourfold compared to control cells [35]. These findings provide the clearest evidence that APC directly impacts microtubule dynamics through inhibition of EB1 function. Computational modeling of the fourfold increase in microtubule pausing shows it dramatically increases the time for microtubules to search out and capture kinetochores in mitosis and is sufficient to explain the spindle defects observed in APC mutants [35]. Similar spindle and chromosome segregation defects are observed in cultured cell lines when APC is inhibited by siRNA, consistent with the idea that truncated APC interferes with the normal function of APC [35, 65]. Importantly, β-catenin levels are not affected by dominant APC alleles, suggesting APC regulation of microtubule dynamics is separable from its role in degrading β-catenin.

The mechanism by which APC^{1-1450} acts to dominantly interfere with EB1 function remains unclear. APC^{1-1450} may simply destabilize EB1 interactions with microtubule plus-ends; this is consistent with observed change in EB1 association with microtubule plus-ends in cells expressing APC^{1-1450} [35]. Alternatively, APC^{1-1450} may interfere with the interaction of EB1 and a modifier that is needed for EB1 to promote microtubule growth. Candidate regulators include the formins (e.g., mDia2 or mDia3; see discussion below) as well as the kinase, GSK3-β. Formins are known as molecular stimulators of actin polymerization but, as discussed, in migrating cells they interact with EB1 and APC to form a complex that modulates microtubule stability at the actin rich leading edge [66]. The formin, mDia2, has been shown to slow shrinking microtubules, an activity linked to its role in stabilizing subsets of microtubules at the leading edge of migrating cells [49]. It is possible that formins, EB1 and APC perform a similar function in mitosis. One possibility is that the formin-APC-EB1 complex aids in switching kinetochore-microtubules from growing to pausing during the dynamic process of kinetochore-microtubule attachment. When APC^{1-1450} prevents EB1 from interacting with APC it may be that mDia3 activity is unbalanced, increasing the pause time as observed in measurements of microtubule dynamics in living cells [35, 56]. Consistent with this idea, the formin, mDia3, has been shown to associate with kinetochores and its interaction with EB1 is important for chromosome alignment [56, 58, 67]. It is also possible that GSK3-β is involved separately or in conjunction with formins. In migrating cells, GSK3-β appears to be inhibited by mDia and, GSK3-β inhibitors (e.g. $LiCl_2$) can induce stable microtubules. $LiCl_2$ treatment can also partially rescue the defects in spindle microtubules in cells expressing APC^{1-1450} (Green and Kaplan, unpublished results), suggesting it is also involved in APC regulation of microtubule dynamics. It is possible that APC^{1-1450} directly interferes with the normal regulation of the kinase, although proof of a direct interaction between APC and GSK3-β in mitosis is lacking. In summary, a large amount of evidence from multiple biological contexts argues for an APC-EB1-formin regulatory "hub" that regulates microtubule dynamics. The interesting implication of such a regulatory hub is that APC mutations found in human cancers that compromise this hub may affect both cell migration and chromosomes segregation, potentially contributing to the mis-positioning of stem cells in the crypt as well as to increases in genome instability.

Monoallelic APC Mutants: A Case for an APC-Mediated Cancer Permissive State

A Dominant Role for Monoallelic Mutations in APC in Cancer

The decrease in microtubule dynamics caused by a reduction in APC activity, either through dominant negative mutations found in cancer or by reduction in APC protein levels, raises the important possibility that the microtubule phenotype actively contributes to cancer onset. The fact that we know very little about the first cellular changes that lead to cancer transformation means we cannot say with certainty which changes are important. However, a good deal of evidence argues that chromosome instability and changes in cell polarity that impact asymmetric divisions and cell migration are potent drivers of cancer. In considering the idea that APC mutants actively drive cancer—as opposed to Knudson's passive model where the only import of a single APC mutant is that LOH is more likely—two major questions arise. First, is there evidence that microtubule dynamics are altered in the tissue where cancer develops? Second, how might the resulting changes in microtubules and in information transfer (e.g., chromosome segregation, asymmetric cell divisions) contribute to cancer progression? We will first review the evidence that mitotic errors caused by reduced microtubule dynamics occur at the site of intestinal cancer origin. Second we will discuss both experimental evidence and proposed models for how changes in microtubule dynamics and increased mitotic errors might contribute to cancer initiation and progression.

APC Mutants Perturb the Microtubules and Ploidy in Intestinal Crypt Cells

Mouse models of colorectal cancer provide the most accessible mammalian system to test for the presence of mitotic errors in the tissue where intestinal cancer originates but there are some important caveats to consider. First, APC mutant animals (e.g., Min) develop cancer primarily in their small intestine, as opposed to the large intestine where the disease manifests in humans. Second, engineered knock-outs of APC (i.e. Lox-Cre mediated deletions) give rise to an artificially large numbers of APC null cells, a scenario that does not reflect the normal progression of the disease in humans, where null cells only arise infrequently, as discussed above. Despite these caveats, mouse models faithfully recapitulate at least the early steps of human colorectal cancer, beginning with LOH at APC and elevated levels of β-catenin. Thus, it is relevant that monoallelic APC mutations result in changes in crypt cells that are consistent with inhibition of microtubule dynamics. In APC^{Min} small intestines, spindle orientation was measured as an indication of astral microtubule function in anchoring and orienting the mitotic spindle. Compared to wild type transit amplifying cells that reproducibly orient their spindles parallel to the basement membrane, APC^{Min} animals displayed a nearly random orientation of mitotic spindles [68, 69]. A similar change in spindle orientation was also observed in the stem cell compartment [70]. In vitro, human cells expressing truncated APC exhibit a loss in microtubule-cortical interactions causing spindles to rotate freely; this decrease in cortical attachments also leads to a high rate of spindle collapse, followed by abortive anaphase, mitotic exit and frequent binucleate cells [68]. Consistent with these in vitro mitotic defects, a significant number of tetraploid cells are found in APC^{Min} crypts, whereas none are found in wild type animals. A similar frequency of cells with multi-polar spindles was also observed in APC^{Min} crypts, suggesting that mitotic failure was followed by re-entry into the cell cycle where the two inherited centrosomes were duplicated. Importantly, mitotic defects are observed in cells that retain the full-length copy of APC, meaning they occur prior to the first recognized sign of cancer (i.e., LOH at APC and increased β-catenin). When taken together, the phenotypes observed in the gut are consistent with the ability of APC mutants to inhibit microtubule dynamics and increase the frequency of mitotic errors, prior to cancer onset. The potential for massive changes in ploidy raises the contentious issue of whether and how increased numbers of chromosomes influence cancer onset or progression. It turns out that knowing the fate of single tetraploid cell or single aneuploid cells is not straightforward. In an engineered mouse model for aneuploidy, reduced levels of the mitotic kinesin, CENP-E, result in high rates of tetraploids in cultured MEFs but tetraploids and gains of chromosomes are not readily apparent in mouse tissues. Even more surprisingly, CENP-E deficient animals are resistant to cancer induced by tumor promoters (i.e., chemical or oncogenic), although they eventually succumb to age-related cancers at a high rate [71]. One interpretation of this finding is that aneuploidy is not well tolerated in cells in vivo. Thus, one might conclude that increases in ploidy are not sufficient for cancer initiation. Yet the prevalence of aneuploid and even tetraploid cells in primary tumors suggests that cellular changes are possible to create a permissive state for the survival of aneuploid cells.

The Contribution of Aneuploidy to Genome Damage and Cancer Onset

The idea that aneuploid or tetraploid cells are not by themselves conducive to cancer cell proliferation is also supported by studies of aneuploid cells in culture and by careful analysis of ploidy in cancer cells in situ. Studies of tetraploid cells suggest that these cells experience genomic stress, mostly likely because chromosome replication and segregation become more challenging as ploidy increases [72–74]. Although higher gene dosage has been proposed by some to favor de-regulated cell growth, direct analysis of single cell aneuploids shows that increased gene copy results in "proteomic stress" that is more likely to repress cell division than to induce it [75–77]. The stress experienced by aneuploid cells may help to explain the heterogeneity of ploidies found in most tumors. For example, a high rate of tetraploids are correlated with the most malignant forms of Barrett's esophagus, though tetraploids are by no means dominant in the tumor cell population (10–25 %; [78]). Similarly, in APC^{Min} animals, tetraploids are observed in normal crypt cells but very few are found in early dysplasias (<2 %) [68]; cells in dysplasias from these animals appear to be near diploid, in contrast with human intestinal cancers with APC mutations that exhibit near triploid states and chromosome instability [79, 80]. However, even in the near diploid APC^{Min} tumors, a significant number of aneuploid cells are detected by in situ hybridization, suggesting that mitotic errors persist but that selection favors cells that have reduced their ploidy (Caldwell and Kaplan, unpublished results). Thus, an initial change in chromosome ploidy (e.g., tetraploidy) may represent a transient first step that evolves into alternative genome states more compatible with cell division. A low but constant rate of chromosome instability may persist in a sub-population of cancer cells and contribute to their ability to continue to evolve even though the bulk of the tumor may appear to have a stabilized ploidy. The ubiquity of burdensome aneuploid genomes in cancer cells begs the question of what role if any might these extra chromosomes play in cancer onset or progression.

Most cancer models posit that cancer occurs after a step-by-step acquisition of mutations that gives rise to the mature cancer cell genomic landscape—one filled with 2–8 driver mutations as well as dozens of so-called "passenger" mutations [81, 82]. However, an alternative model has been recently proposed where massive genomic damage occurs all at once. This so-called "all at once" model argues that a catastrophic event is triggered during cell division that results in multiple rearrangements or sequence aberrations limited to small region of the genome (i.e., chromothripsis, chromoanasynthesis and chromsoplexy as reviewed in [83]). Such local rearrangements can result in LOH or expansion of genes and because they are limited to discrete regions of the genome, otherwise lethal outcomes may be avoided [84]. Although the precise DNA dependent events that lead to these chromosomal rearrangements are not yet clear, one possible mechanism to limit the scope of damage is to confine it to chromosomes that are present in micronuclei generated from chromosome mis-segregation events. Micronuclei are similar to a normal nucleus but contain one or several chromosomes that undergo asynchronous DNA replication and accumulate high rates of DNA damage. Importantly, the chromosomes in these micronuclei can be reincorporated into the genome and thus can represent an important mechanism for focusing massive DNA damage on single chromosomes as a result of mitotic errors [85]. This process may explain how artificially induced tetraploid cells that develop into tumors in mice exhibit such massive genomic rearrangements [86]. Tetraploid cells, such as present in APC mutants, may directly lead to chromosome mis-segregation, in part because it is challenging for the mitotic machinery to align and segregate twice as many chromosomes during mitosis [87]. In addition, APC mutants that dominantly inhibit microtubule dynamics also silence the spindle checkpoint in the presence of lagging/mis-attached chromosomes resulting in micronuclei formation [88]. In this model of chromosome damage, single chromosome aneuploidy or tetraploidy caused by mitotic failures represents the first step in a path to restricted chromosome damage. Repeated mitotic failures would in principle increase the chances that a region of the chromosome involved in cell survival and proliferation is targeted, although these ideas remain important predictions to test.

APC Mutants Affect Cortical Microtubule Contacts, Asymmetric Cell Divisions and Cell Homeostasis

Up until now, we have focused on the potential impact of chromosome aneuploidy but there are other significant consequences that arise from the inhibitory effect of APC on microtubule dynamics. As observed in cell culture, APC most dramatically disrupts the highly dynamic interactions between spindle astral microtubules and the cell cortex, which orients the spindle during cell division [35, 68]. Spindle orientation has long been recognized by developmental biologists as critical for determining the specification of cell fate in daughter cells. Spindle orientation and the asymmetric distribution of information during cell division are conserved from budding yeast to complex multi-cellular organisms. This process involves the integration of information from polarity cues at the cell cortex, astral microtubule-mediated forces that position spindles as well as mechanisms that asymmetrically localize cellular fate determinants (for review; [89]). Asymmetric cell divisions are particularly important for specifying stem cell identity and alternative cell fates during stem cell divisions. As discussed, a number of studies demonstrate that APC mutant crypt cells have mis-oriented spindles compared to wild type [68–70]. Although the impact of mis-oriented spindles in crypt cells is not understood, similar disruption of spindle positioning in *Drosophila* neuroblast stem cells results in cancer like phenotypes (i.e., cell proliferation and metastatic properties) that can be indefinitely transplanted to new hosts where they are lethal [90–92]. The majority of cells in these tumors remain neuroblast-like although a constant sub-population of differentiated cells also develops, consistent with the loss of asymmetric inheritance in these cells. The role of asymmetric inheritance varies between stem cell compartments and its role in the gut is not understood. One case where spindle orientation in the gut may matter is in the asymmetric inheritance of the template strand of DNA during replication; labeling studies have shown that the template strand is retained in the crypt stem cells and this asymmetry is eliminated in APC^{Min} crypts [70, 93]. The retention of the template strand in the stem cell has been postulated to protect the integrity of the stem cell genome from mutation that may arise during replication. Another possibility is that epigenetic marks are also asymmetrically inherited [94]. This could mean that loss of spindle positioning might change the distribution of epigenetic marks that influence cell specification after crypt stem cells divide. In fact, we have observed that epigenetic marks are dramatically altered in normal crypts from APC^{Min} animals as well as in normal cells adjacent to dysplasias, although it is unclear if these changes arise from spindle mis-orientation ([111] Fig. 9.2a). Interestingly, we also observe that the stem cell and transit amplifying marker, Musashi [95], is expanded in normal APC^{Min} crypts compared to wild type crypts (Caldwell and Kaplan, unpublished observations, Fig. 9.2b). There may be other explanations for these changes but the high frequency of spindle orientation defects in APC mutant crypts is consistent with the frequency of crypts with atypical levels of Musashi (Caldwell and Kaplan, unpublished observations). Thus, changes in the dynamic properties of astral microtubules in APC mutants may impact cancer onset by changing the asymmetric inheritance of information between crypt stem cells and differentiating progeny. Although these changes do not have to include aneuploidy, aneuploidy may be better tolerated when stem cell-like properties expand in the crypt.

Microtubules have a more general role in organizing sub-cellular structures but the specifics of microtubule interactions with such structures are poorly understood. As discussed earlier, the APC, EB1 and formin complex forms an interface between

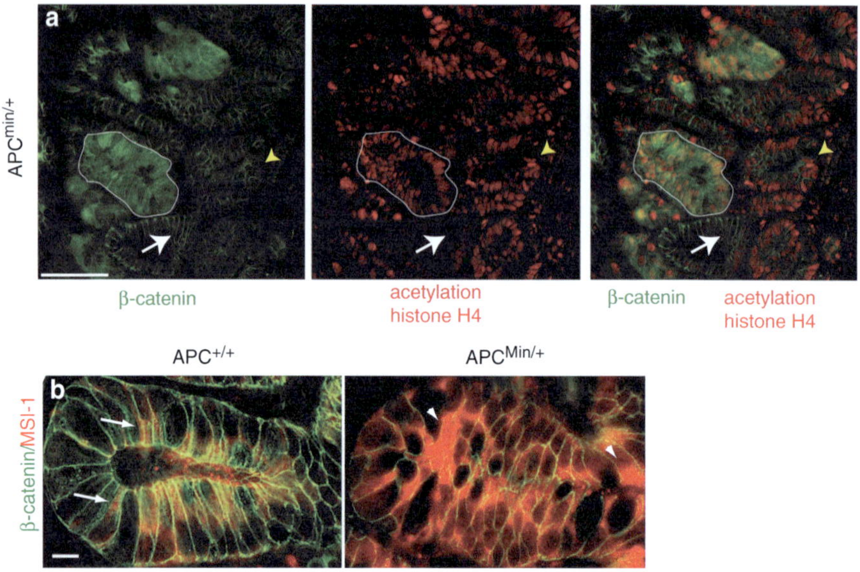

Fig. 9.2 Epigenetic and cell programming altered in $APC^{Min/+}$ cells prior to cancer onset. (**a**) A cross section of an $APC^{Min/+}$ small intestine stained with antibodies to β-catenin (*green*), acetylated histone H4 (*red*; a mark of active gene transcription) and both staining patterns combined, as indicated. The *white outline* delineates an example of a dysplastic region, in this case characterized by elevated levels of both β-catenin and acetylated histone H4 marks. The *white arrows* indicate regions of normal cells with low levels of β-catenin and low levels of acetylated histone H4. *Yellow arrowheads* indicate a region of low β-catenin but elevated levels of acetylated histone H4, similar to levels in the adjacent dysplastic regions. This suggests that epigenetic changes occur prior to up-regulation of β-catenin, perhaps caused by defective inheritance of chromatin. (**b**) Crypts from the small intestine of $APC^{Min/+}$ or $APC^{+/+}$ animals (as indicated) co-stained for β-catenin (*green*) to mark cell boundaries and for the stem-cell and transit amplifying cell marker, Musashi (*red*). *Arrows* in $APC^{+/+}$ crypts show the cells near the base of the crypt that are positive for Musashi. *Arrowheads* show the elevated levels of Musashi in the base of the crypt, extending upward in otherwise normal cells in $APC^{Min/+}$ animals. Scale bar is 100 μm in (**a**) and 10 μm in (**b**)

microtubules and actin that is regulated by the Rho small G protein during cell migration. It is therefore possible that APC not only regulates microtubule stability through the EB1-formin complex but also affects actin polymerization at the cell cortex [96]. This may be particularly relevant during mitosis when microtubule-cortical interactions organize actin at the midzone in preparation for cytokinesis. This dynamic regulation is dramatically demonstrated when centrosomes/spindles are moved by micro-manipulation along the cell cortex, resulting in the stimulation of a dynamic zone of active Rho sufficient to induce furrow ingression [97, 98]. One proposed mechanism for astral microtubule regulation of the cell cortex is suggested by studies in Drosophila S2 cells where EB1 was shown to interact with and transport the Rho GEF to the cell cortex [99]. Changes in cortical actin organization may also have profound effects on cell organization and maintenance of cell polarity.

For example, microtubules and a kinesin motor directly regulate the cell cortex and apical-basolateral polarity through the delivery of the Dlg protein to the cell cortex [100]. Dlg in Drosophila acts as a tumor suppressor and in humans is implicated in cancers through interaction with the oncoviral proteins from Human papillomavirus (HPV) [101]. Thus, changes in the cell cortex can have profound effects on cellular organization as well as cell fate determination. It is therefore possible that APC mutants, by disrupting the EB1-formin complex, change cortical actin and thus impact cell fate.

Is there evidence that APC mutants change cortical actin by changing microtubule dynamics? Major perturbations in actin polymerization can alter epithelial cell polarity. However, characterization of apical-basolateral polarity markers in APC^{Min} crypts suggest that there is no gross loss of polarity, a finding that is perhaps not surprising given the subtle defect in microtubule dynamics [68]. This is also the case in organoid cultures made from APC^{Min} intestinal cells [102]. On the other hand, a closer examination of polarity markers in crypt epithelia shows both apical and basolateral zones are in fact altered in a way that is consistent with subtle perturbations of microtubules and actin. Specifically, ZO-1, an apical marker, no longer displays a contiguous decoration of the apical membrane and β-catenin, a marker of the basolateral membrane, is notably thicker compared to wild type crypt cells (Fig. 9.3a; see arrows). Importantly, a similar thickening of β-catenin at the cortex is observed in cultured cells that express APC mutants or following treatment with low doses of microtubule poisons to inhibit dynamics (Fig. 9.3b and Caldwell and Kaplan, unpublished observations). These changes are accompanied by increased Rho localization at the cortex and enriched actin polymerization, consistent with the idea that APC regulation of microtubule dynamics also affects cortical actin (Fig. 9.3c). Taken together, APC mutants that inhibit microtubule dynamics not only cause defects in the mitotic apparatus but also lead to changes in cortical actin organization that may further affect asymmetric segregation of information, cell polarity, and possibly the stability of cell fate through activation of cell stress pathways.

The complex interplay between the cytoskeleton, organelle function and protein homeostasis raises important new issues when considering cancer-associated mutations that affect cytoskeletal dynamics. Perturbations in both actin and microtubule dynamics have been linked to activation of cell stress pathways, part of an evolutionarily conserved response to maintain cell homeostasis [103–105]. Cancer cells depend on activation of cell stress pathways not only to survive in the face of proteomic stress (both from mutation and gene amplification) but also to adapt in the face of chemotherapeutic induced negative selection [106, 107]. Thus, it is possible that chronic disruption of cytoskeletal dynamics creates a "permissive" state in which pre-cancer cells activate cell stress pathways, allowing adaptation to aneuploidy and other cancer mutations. Consistent with such a model, Hsf1, the main transcriptional regulator of the stress response, has been shown to be required for tumor formation in mice with activated oncogenes [108, 109]. Although the exact role of Hsf1-regulated transcription in cancer is not clear, recent work has linked both cytoskeletal integrity and Hsf1 to stress tolerance and longevity in *C. elegans* [110].

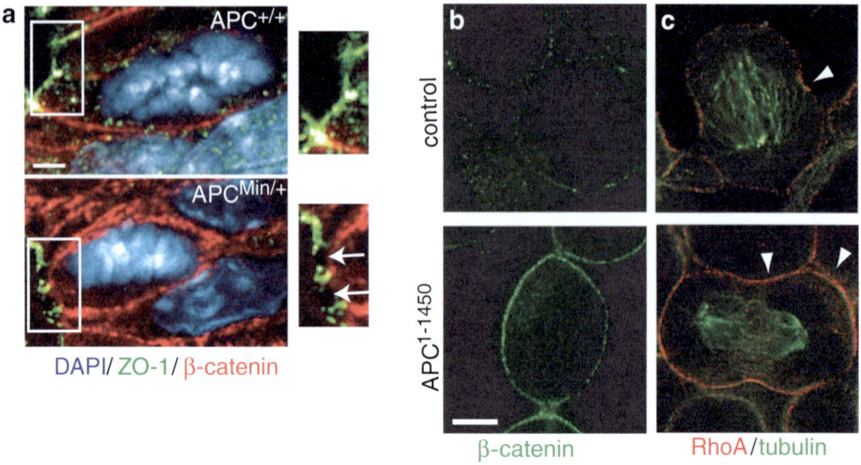

Fig. 9.3 Markers of polarity and actin regulation are mis-localized at the cell cortex in APC mutants. (**a**) A magnified view of crypt cells from $APC^{Min/+}$ and $APC^{+/+}$ animals stained with antibodies to the apical marker, ZO-1 (*green*), the basolateral marker, β-catenin (*red*) and with DAPI staining to reveal chromosomes (*blue*). The *white boxes* indicate regions that are further magnified 1.5-fold in the *insets*. *Arrows* indicate the region of discontiguous ZO-1 staining at the apical region of $APC^{Min/+}$ cells. (**b**) Human kidney epithelial cells (HEK-293) expressing control plasmid or $APC^{1\text{-}1450}$, as indicated and stained with antibodies to β-catenin (*green*). (**c**) Examples of anaphase, HEK-293 cells expressing control or $APC^{1\text{-}1450}$ as indicated and stained with antibodies to RhoA (*red*), a small G protein that regulates actin polymerization, and to tubulin to show the mitotic spindle (*green*). *Arrowheads* indicate the enrichment of RhoA at the cytokinetic furrow of the control anaphase cell and the contiguous enrichment of RhoA along the entire cell cortex of the anaphase cell expressing $APC^{1\text{-}1450}$. All scale bars are 10 μm

It is interesting to speculate that pathways that increase stress tolerance and longevity might also be integral to cancer cell survival. In recent studies in our lab, we have found that APC-induced perturbations in microtubule dynamics lead to activation of these same cell stress pathways [111]. It now becomes important to ask what effect might APC mutations and the resulting chronic perturbation of cytoskeletal dynamics have on cell programming and the permissiveness of cells to oncogenic mutation. Understanding the complex relationships between cytoskeletal regulation, cellular homeostasis and disease progression represents an important challenge for the future of cancer cell biology and APC is an excellent model system to begin to address these issues.

Summary and Future Outlook

The role of the tumor suppressor gene *adenomatous polyposis coli* (APC) in regulating microtubule dynamics has been clearly established and functions independently from its regulation of β-catenin. Inherent in its regulation of microtubule function is the role of APC as a guardian of the genome; down regulation of APC or expression of dominant negative cancer-associated mutations leads to defects in spindle microtubule dynamics, chromosome segregation errors and mitotic failures. The frequency of ploidy changes in normal human intestines expressing an APC mutant is unknown but the large number of cell divisions in the crypt during the lifetime of a human suggests the real possibility that APC mutants allow cells to sample a significant number of aneuploid genomes. Although the latest studies speak against aneuploidy being sufficient for cancer onset, our current model posits that changes in cell programming—possibly caused by chronic cell stress—creates a permissive cell state that can support heightened genome instability and cell proliferation. The challenge remains to understand the molecular and genetic landscape that contributes to a cancer permissive state. Meeting this challenge will require a thorough understanding of how changes in cytoskeletal dynamics affect cell homeostasis and cell programming, how changes in inheritance of information (chromosomes and cell fate determinants) alter cell fate and how the resulting chronic instability combine to allow cells to evolve a cancer state.

References

1. Narumiya S, Tanji M, Ishizaki T (2009) Rho signaling, ROCK and mDia1, in transformation, metastasis and invasion. Cancer Metastasis Rev 28:65–76. doi:10.1007/s10555-008-9170-7
2. Smole Z, Thoma CR, Applegate KT, Duda M, Gutbrodt KL, Danuser G, Krek W (2014) Tumor suppressor NF2/Merlin is a microtubule stabilizer. Cancer Res 74:353–362. doi:10.1158/0008-5472.CAN-13-1334
3. Gordon DJ, Resio B, Pellman D (2012) Causes and consequences of aneuploidy in cancer. Nat Rev Genet 13:189–203. doi:10.1038/nrg3123
4. Xu X, Weaver Z, Linke SP, Li C, Gotay J, Wang XW, Harris CC, Ried T, Deng CX (1999) Centrosome amplification and a defective G2-M cell cycle checkpoint induce genetic instability in BRCA1 exon 11 isoform-deficient cells. Mol Cell 3:389–395
5. Kais Z, Barsky SH, Mathsyaraja H, Zha A, Ransburgh DJR, He G, Pilarski RT, Shapiro CL, Huang K, Parvin JD (2011) KIAA0101 interacts with BRCA1 and regulates centrosome number. Mol Cancer Res 9:1091–1099. doi:10.1158/1541-7786.MCR-10-0503
6. Gregory PE, Gutmann DH, Mitchell A, Park S, Boguski M, Jacks T, Wood DL, Jove R, Collins FS (1993) Neurofibromatosis type 1 gene product (neurofibromin) associates with microtubules. Somat Cell Mol Genet 19:265–274
7. Aylon Y, Michael D, Shmueli A, Yabuta N, Nojima H, Oren M (2006) A positive feedback loop between the p53 and Lats2 tumor suppressors prevents tetraploidization. Genes Dev 20:2687–2700. doi:10.1101/gad.1447006
8. Toji S, Yabuta N, Hosomi T, Nishihara S, Kobayashi T, Suzuki S, Tamai K, Nojima H (2004) The centrosomal protein Lats2 is a phosphorylation target of Aurora-A kinase. Genes Cells 9:383–397. doi:10.1111/j.1356-9597.2004.00732.x
9. Courtois-Cox S, Jones SL, Cichowski K (2008) Many roads lead to oncogene-induced senescence. Oncogene 27:2801–2809. doi:10.1038/sj.onc.1210950
10. Kinzler KW, Vogelstein B (1996) Lessons from hereditary colorectal cancer. Cell 87:159–170. doi:10.1016/S0092-8674(00)81333-1
11. Fearon ER, Vogelstein B (1990) A genetic model for colorectal tumorigenesis. Cell 61:759–767

12. Fearnhead NS, Britton MP, Bodmer WF (2001) The ABC of APC. Hum Mol Genet 10:721–733
13. Caldwell CM, Kaplan KB (2009) The role of APC in mitosis and in chromosome instability. Adv Exp Med Biol 656:51–64
14. Knudson A (1971) Mutation and cancer: statistical study of retinoblastoma. Proc Natl Acad Sci U S A 68:820–823
15. Levy DB, Smith KJ, Beazer-Barclay Y, Hamilton SR, Vogelstein B, Kinzler KW (1994) Inactivation of both APC alleles in human and mouse tumors. Cancer Res 54:5953–5958
16. Polakis P (1999) The oncogenic activation of β-catenin. Curr Opin Genet Dev 9:15–21. doi:10.1016/S0959-437X(99)80003-3
17. Burgess AW, Faux MC, Layton MJ, Ramsay RG (2011) Wnt signaling and colon tumorigenesis—a view from the periphery. Exp Cell Res 317:2748–2758. doi:10.1016/j.yexcr.2011.08.010
18. Stamos JL, Weis WI (2013) The β-catenin destruction complex. Cold Spring Harb Perspect Biol 5:a007898. doi:10.1101/cshperspect.a007898
19. Sansom OJ, Reed KR, Van De Wetering M, Muncan V, Winton DJ, Clevers H, Clarke AR (2005) Cyclin D1 is not an immediate target of beta-catenin following Apc loss in the intestine. J Biol Chem 280:28463–28467. doi:10.1074/jbc.M500191200
20. Tetsu O, McCormick F (1999) Beta-catenin regulates expression of cyclin D1 in colon carcinoma cells. Nature 398:422–426. doi:10.1038/18884
21. He TC, Sparks AB, Rago C, Hermeking H, Zawel L, da Costa LT, Morin PJ, Vogelstein B, Kinzler KW (1998) Identification of c-MYC as a target of the APC pathway. Science 281:1509–1512
22. Moser A, Pitot H, Dove W (1990) A dominant mutation that predisposes to multiple intestinal neoplasia in the mouse. Science 247:322–324
23. Moser AR, Shoemaker AR, Connelly CS, Clipson L, Gould KA, Luongo C, Dove WF, Siggers PH, Gardner RL (1995) Homozygosity for the Min allele of Apc results in disruption of mouse development prior to gastrulation. Dev Dyn 203:422–433. doi:10.1002/aja.1002030405
24. Sansom OJ, Reed KR, Hayes AJ, Ireland H, Brinkmann H, Newton IP, Batlle E, Simon-Assmann P, Clevers H, Näthke IS, Clarke AR, Winton DJ (2004) Loss of Apc in vivo immediately perturbs Wnt signaling, differentiation, and migration. Genes Dev 18:1385–1390. doi:10.1101/gad.287404
25. Bedi A, Pasricha PJ, Akhtar AJ, Barber JP, Bedi GC, Giardiello FM, Zehnbauer BA, Hamilton SR, Jones RJ (1995) Inhibition of apoptosis during development of colorectal cancer. Cancer Res 55:1811–1816
26. Clarke AR (2005) Studying the consequences of immediate loss of gene function in the intestine: APC. Biochem Soc Trans 33:665–666. doi:10.1042/BST0330665
27. Barker N, Ridgway RA, Van Es JH, Van De Wetering M, Begthel H, van den Born M, Danenberg E, Clarke AR, Sansom OJ, Clevers H (2009) Crypt stem cells as the cells-of-origin of intestinal cancer. Nature 457:608–611. doi:10.1038/nature07602
28. Fischer JM, Schepers AG, Clevers H, Shibata D, Liskay RM (2014) Occult progression by Apc-deficient intestinal crypts as a target for chemoprevention. Carcinogenesis 35:237–246. doi:10.1093/carcin/bgt296
29. Fischer JM, Miller AJ, Shibata D, Liskay RM (2012) Different phenotypic consequences of simultaneous versus stepwise Apc loss. Oncogene 31:2028–2038. doi:10.1038/onc.2011.385
30. Cheung AF, Carter AM, Kostova KK, Woodruff JF, Crowley D, Bronson RT, Haigis KM, Jacks T (2010) Complete deletion of Apc results in severe polyposis in mice. Oncogene 29:1857–1864. doi:10.1038/onc.2009.457
31. Näthke IS (2004) The adenomatous polyposis coli protein: the Achilles heel of the gut epithelium. Annu Rev Cell Dev Biol 20:337–366. doi:10.1146/annurev.cellbio.20.012103.094541
32. Li Z, Näthke IS (2005) Tumor-associated NH2-terminal fragments are the most stable part of the adenomatous polyposis coli protein and can be regulated by interactions with COOH-terminal domains. Cancer Res 65:5195–5204. doi:10.1158/0008-5472.CAN-04-4609

33. Day C, Alber T (2000) Crystal structure of the amino-terminal coiled-coil domain of the APC tumor suppressor. J Mol Biol 301:147–156
34. Joslyn G, Richardson D, White R, Alber T (1993) Dimer formation by an N-terminal coiled coil in the APC protein. Proc Natl Acad Sci U S A 90:11109–11113
35. Green RA, Wollman R, Kaplan KB (2005) APC and EB1 function together in mitosis to regulate spindle dynamics and chromosome alignment. Mol Biol Cell 16:4609–4622. doi:10.1091/mbc.E05-03-0259
36. Kawasaki Y, Senda T, Ishidate T, Koyama R, Morishita T, Iwayama Y, Higuchi O, Akiyama T (2000) Asef, a link between the tumor suppressor APC and G-protein signaling. Science 289:1194–1197
37. Jimbo T, Kawasaki Y, Koyama R, Sato R, Takada S, Haraguchi K, Akiyama T (2002) Identification of a link between the tumour suppressor APC and the kinesin superfamily. Nat Cell Biol 4:323–327. doi:10.1038/ncb779
38. Watanabe T, Wang S, Noritake J, Sato K, Fukata M, Takefuji M, Nakagawa M, Izumi N, Akiyama T, Kaibuchi K (2004) Interaction with IQGAP1 links APC to Rac1, Cdc42, and actin filaments during cell polarization and migration. Dev Cell 7:871–883. doi:10.1016/j.devcel.2004.10.017
39. Rubinfeld B, Albert I, Porfiri E, Fiol C, Munemitsu S, Polakis P (1996) Binding of GSK3beta to the APC-beta-catenin complex and regulation of complex assembly. Science 272:1023–1026. doi:10.1126/science.272.5264.1023
40. Deka J, Kuhlmann J, Muller O (1998) A domain within the tumor suppressor protein APC shows very similar biochemical properties as the microtubule-associated protein tau. Eur J Biochem 253:591–597
41. Tirnauer JS, Bierer BE (2000) EB1 proteins regulate microtubule dynamics, cell polarity, and chromosome stability. J Cell Biol 149:761–766
42. Juwana JP, Henderikx P, Mischo A, Wadle A, Fadle N, Gerlach K, Arends JW, Hoogenboom H, Pfreundschuh M, Renner C (1999) EB/RP gene family encodes tubulin binding proteins. Int J Cancer 81:275–284
43. Berrueta L, Kraeft SK, Tirnauer JS, Schuyler SC, Chen LB, Hill DE, Pellman D, Bierer BE (1998) The adenomatous polyposis coli-binding protein EB1 is associated with cytoplasmic and spindle microtubules. Proc Natl Acad Sci U S A 95:10596–10601
44. Su LK, Burrell M, Hill DE, Gyuris J, Brent R, Wiltshire R, Trent J, Vogelstein B, Kinzler KW (1995) APC binds to the novel protein EB1. Cancer Res 55:2972–2977
45. Mimori-Kiyosue Y, Shiina N, Tsukita S (2000) The dynamic behavior of the APC-binding protein EB1 on the distal ends of microtubules. Curr Biol 10:865–868
46. Nakamura M, Zhou XZ, Lu KP (2001) Critical role for the EB1 and APC interaction in the regulation of microtubule polymerization. Curr Biol 11:1062–1067
47. Kroboth K, Newton IP, Kita K, Dikovskaya D, Zumbrunn J, Waterman-Storer CM, Näthke IS (2007) Lack of adenomatous polyposis coli protein correlates with a decrease in cell migration and overall changes in microtubule stability. Mol Biol Cell 18:910–918. doi:10.1091/mbc.E06-03-0179
48. Gundersen GG, Gomes ER, Wen Y (2004) Cortical control of microtubule stability and polarization. Curr Opin Cell Biol 16:106–112. doi:10.1016/j.ceb.2003.11.010
49. Wen Y, Eng CH, Schmoranzer J, Cabrera-Poch N, Morris EJS, Chen M, Wallar BJ, Alberts AS, Gundersen GG (2004) EB1 and APC bind to mDia to stabilize microtubules downstream of Rho and promote cell migration. Nat Cell Biol 6:820–830. doi:10.1038/ncb1160
50. Yang C, Czech L, Gerboth S, Kojima S-I, Scita G, Svitkina T (2007) Novel roles of formin mDia2 in lamellipodia and filopodia formation in motile cells. PLoS Biol 5, e317. doi:10.1371/journal.pbio.0050317
51. Bartolini F, Moseley JB, Schmoranzer J, Cassimeris L, Goode BL, Gundersen GG (2008) The formin mDia2 stabilizes microtubules independently of its actin nucleation activity. J Cell Biol 181:523–536. doi:10.1083/jcb.200709029

52. Bartolini F, Ramalingam N, Gundersen GG (2012) Actin-capping protein promotes microtubule stability by antagonizing the actin activity of mDia1. Mol Biol Cell 23:4032–4040. doi:10.1091/mbc.E12-05-0338
53. Watanabe S, Ando Y, Yasuda S, Hosoya H, Watanabe N, Ishizaki T, Narumiya S (2008) mDia2 induces the actin scaffold for the contractile ring and stabilizes its position during cytokinesis in NIH 3T3 cells. Mol Biol Cell 19:2328–2338. doi:10.1091/mbc.E07-10-1086
54. Kato T, Watanabe N, Morishima Y, Fujita A, Ishizaki T, Narumiya S (2001) Localization of a mammalian homolog of diaphanous, mDia1, to the mitotic spindle in HeLa cells. J Cell Sci 114:775–784
55. Rundle DR, Gorbsky G, Tsiokas L (2004) PKD2 interacts and co-localizes with mDia1 to mitotic spindles of dividing cells: role of mDia1 IN PKD2 localization to mitotic spindles. J Biol Chem 279:29728–29739. doi:10.1074/jbc.M400544200
56. Mao Y (2011) FORMIN a link between kinetochores and microtubule ends. Trends Cell Biol 21:625–629. doi:10.1016/j.tcb.2011.08.005
57. Cheng L, Mao Y (2011) mDia3-EB1-APC: a connection between kinetochores and microtubule plus ends. Commun Integr Biol 4:480–482. doi:10.4161/cib.4.4.15655
58. Cheng L, Zhang J, Ahmad S, Rozier L, Yu H, Deng H, Mao Y (2011) Aurora B regulates formin mDia3 in achieving metaphase chromosome alignment. Dev Cell 20:342–352. doi:10.1016/j.devcel.2011.01.008
59. Lengauer C, Kinzler KW, Vogelstein B (1997) Genetic instability in colorectal cancers. Nature 386:623–627. doi:10.1038/386623a0
60. Cahill DP, Lengauer C, Yu J, Riggins GJ, Willson JK, Markowitz SD, Kinzler KW, Vogelstein B (1998) Mutations of mitotic checkpoint genes in human cancers. Nature 392:300–303. doi:10.1038/32688
61. Baker DJ, Dawlaty MM, Wijshake T, Jeganathan KB, Malureanu L, van Ree JH, Crespo-Diaz R, Reyes S, Seaburg L, Shapiro V, Behfar A, Terzic A, van de Sluis B, van Deursen JM (2012) Increased expression of BubR1 protects against aneuploidy and cancer and extends healthy lifespan. Nat Cell Biol 15:96–102. doi:10.1038/ncb2643
62. Zhao Y, Ando K, Oki E, Ikawa-Yoshida A, Ida S, Kimura Y, Saeki H, Kitao H, Morita M, Maehara Y (2014) Aberrations of BUBR1 and TP53 gene mutually associated with chromosomal instability in human colorectal cancer. Anticancer Res 34:5421–5427
63. Kaplan KB, Burds AA, Swedlow JR, Bekir SS, Sorger PK, Nathke IS (2001) A role for the adenomatous polyposis coli protein in chromosome segregation. Nat Cell Biol 3:429–432. doi:10.1038/35070123
64. Green RA, Kaplan KB (2003) Chromosome instability in colorectal tumor cells is associated with defects in microtubule plus-end attachments caused by a dominant mutation in APC. J Cell Biol 163:949–961. doi:10.1083/jcb.200307070
65. Dikovskaya D, Schiffmann D, Newton IP, Oakley A, Kroboth K, Sansom O, Jamieson TJ, Meniel V, Clarke A, Näthke IS (2007) Loss of APC induces polyploidy as a result of a combination of defects in mitosis and apoptosis. J Cell Biol 176:183–195. doi:10.1083/jcb.200610099
66. Bartolini F, Gundersen GG (2010) Formins and microtubules. Biochim Biophys Acta 1803:164–173. doi:10.1016/j.bbamcr.2009.07.006
67. Yasuda S, Oceguera-Yanez F, Kato T, Okamoto M, Yonemura S, Terada Y, Ishizaki T, Narumiya S (2004) Cdc42 and mDia3 regulate microtubule attachment to kinetochores. Nature 428:767–771. doi:10.1038/nature02452
68. Caldwell CMC, Green RAR, Kaplan KBK (2007) APC mutations lead to cytokinetic failures in vitro and tetraploid genotypes in Min mice. J Cell Biol 178:1109–1120. doi:10.1083/jcb.200703186
69. Fleming ES, Temchin M, Wu Q, Maggio-Price L, Tirnauer JS (2009) Spindle misorientation in tumors from APC(min/+) mice. Mol Carcinog 48:592–598. doi:10.1002/mc.20506
70. Quyn AJ, Appleton PL, Carey FA, Steele RJC, Barker N, Clevers H, Ridgway RA, Sansom OJ, Näthke IS (2010) Spindle orientation bias in gut epithelial stem cell compartments is lost in precancerous tissue. Cell Stem Cell 6:175–181. doi:10.1016/j.stem.2009.12.007

71. Weaver BAA, Silk AD, Montagna C, Verdier-Pinard P, Cleveland DW (2007) Aneuploidy acts both oncogenically and as a tumor suppressor. Cancer Cell 11:25–36. doi:10.1016/j.ccr.2006.12.003
72. Lin H, de Carvalho P, Kho D, Tai CY, Pierre P, Fink GR, Pellman D (2001) Polyploids require Bik1 for kinetochore-microtubule attachment. J Cell Biol 155:1173–1184. doi:10.1083/jcb.200108119
73. Ganem NJ, Storchová Z, Pellman D (2007) Tetraploidy, aneuploidy and cancer. Curr Opin Genet Dev 17:157–162. doi:10.1016/j.gde.2007.02.011
74. Andalis AA, Storchová Z, Styles C, Galitski T, Pellman D, Fink GR (2004) Defects arising from whole-genome duplications in *Saccharomyces cerevisiae*. Genetics 167:1109–1121. doi:10.1534/genetics.104.029256
75. Pfau SJ, Amon A (2012) Chromosomal instability and aneuploidy in cancer: from yeast to man. EMBO Rep 13:515–527. doi:10.1038/embor.2012.65
76. Oromendia AB, Dodgson SE, Amon A (2012) Aneuploidy causes proteotoxic stress in yeast. Genes Dev 26:2696–2708. doi:10.1101/gad.207407.112
77. Tang Y-C, Amon A (2013) Gene copy-number alterations: a cost-benefit analysis. Cell 152:394–405. doi:10.1016/j.cell.2012.11.043
78. Barrett MT, Pritchard D, Palanca-Wessels C, Anderson J, Reid BJ, Rabinovitch PS (2003) Molecular phenotype of spontaneously arising 4N (G2-tetraploid) intermediates of neoplastic progression in Barrett's esophagus. Cancer Res 63:4211–4217
79. Haigis KM, Caya JG, Reichelderfer M, Dove WF (2002) Intestinal adenomas can develop with a stable karyotype and stable microsatellites. Proc Natl Acad Sci U S A 99:8927–8931. doi:10.1073/pnas.132275099
80. Shih IM, Zhou W, Goodman SN, Lengauer C, Kinzler KW, Vogelstein B (2001) Evidence that genetic instability occurs at an early stage of colorectal tumorigenesis. Cancer Res 61:818–822
81. Tomasetti C, Marchionni L, Nowak MA, Parmigiani G, Vogelstein B (2015) Only three driver gene mutations are required for the development of lung and colorectal cancers. Proc Natl Acad Sci U S A 112:118–123. doi:10.1073/pnas.1421839112
82. Vogelstein B, Papadopoulos N, Velculescu VE, Zhou S, Diaz LA, Kinzler KW (2013) Cancer genome landscapes. Science 339:1546–1558. doi:10.1126/science.1235122
83. Zhang C-Z, Leibowitz ML, Pellman D (2013) Chromothripsis and beyond: rapid genome evolution from complex chromosomal rearrangements. Genes Dev 27:2513–2530. doi:10.1101/gad.229559.113
84. Meyerson M, Pellman D (2011) Cancer genomes evolve by pulverizing single chromosomes. Cell 144:9–10. doi:10.1016/j.cell.2010.12.025
85. Crasta K, Ganem NJ, Dagher R, Lantermann AB, Ivanova EV, Pan Y, Nezi L, Protopopov A, Chowdhury D, Pellman D (2012) DNA breaks and chromosome pulverization from errors in mitosis. Nature 482:53–58. doi:10.1038/nature10802
86. Fujiwara T, Bandi M, Nitta M, Ivanova EV, Bronson RT, Pellman D (2005) Cytokinesis failure generating tetraploids promotes tumorigenesis in p53-null cells. Nature 437:1043–1047. doi:10.1038/nature04217
87. Hayashi MT, Karlseder J (2013) DNA damage associated with mitosis and cytokinesis failure. Oncogene 32:4593–4601. doi:10.1038/onc.2012.615
88. Draviam VM, Shapiro I, Aldridge B, Sorger PK (2006) Misorientation and reduced stretching of aligned sister kinetochores promote chromosome missegregation in EB1- or APC-depleted cells. EMBO J 25:2814–2827. doi:10.1038/sj.emboj.7601168
89. Morin X, Bellaiche Y (2011) Mitotic spindle orientation in asymmetric and symmetric cell divisions during animal development. Dev Cell 21:102–119. doi:10.1016/j.devcel.2011.06.012
90. Kelsom C, Lu W (2012) Uncovering the link between malfunctions in Drosophila neuroblast asymmetric cell division and tumorigenesis. Cell Biosci 2:38. doi:10.1186/2045-3701-2-38
91. Beaucher M, Hersperger E, Page-McCaw A, Shearn A (2007) Metastatic ability of Drosophila tumors depends on MMP activity. Dev Biol 303:625–634. doi:10.1016/j.ydbio.2006.12.001

92. Caussinus E, González C (2005) Induction of tumor growth by altered stem-cell asymmetric division in *Drosophila melanogaster*. Nat Genet 37:1125–1129. doi:10.1038/ng1632
93. Potten CS, Owen G, Booth D (2002) Intestinal stem cells protect their genome by selective segregation of template DNA strands. J Cell Sci 115:2381–2388
94. Tran V, Lim C, Xie J, Chen X (2012) Asymmetric division of Drosophila male germline stem cell shows asymmetric histone distribution. Science 338:679–682. doi:10.1126/science.1226028
95. Potten CS, Booth C, Tudor GL, Booth D, Brady G, Hurley P, Ashton G, Clarke R, Sakakibara S-I, Okano H (2003) Identification of a putative intestinal stem cell and early lineage marker; Musashi-1. Differentiation 71:28–41
96. Chesarone MA, DuPage AG, Goode BL (2010) Unleashing formins to remodel the actin and microtubule cytoskeletons. Nat Rev Mol Cell Biol 11:62–74. doi:10.1038/nrm2816
97. Bement WM, Miller AL, von Dassow G (2006) Rho GTPase activity zones and transient contractile arrays. Bioessays 28:983–993. doi:10.1002/bies.20477
98. Miller AL, Bement WM (2009) Regulation of cytokinesis by Rho GTPase flux. Nat Cell Biol 11:71–77. doi:10.1038/ncb1814
99. Rogers SL, Wiedemann U, Häcker U, Turck C, Vale RD (2004) Drosophila RhoGEF2 associates with microtubule plus ends in an EB1-dependent manner. Curr Biol 14:1827–1833. doi:10.1016/j.cub.2004.09.078
100. Siegrist SE, Doe CQ (2005) Microtubule-induced Pins/Galphai cortical polarity in Drosophila neuroblasts. Cell 123:1323–1335. doi:10.1016/j.cell.2005.09.043
101. Humbert PO, Grzeschik NA, Brumby AM, Galea R, Elsum I, Richardson HE (2008) Control of tumourigenesis by the Scribble/Dlg/Lgl polarity module. Oncogene 27:6888–6907. doi:10.1038/onc.2008.341
102. Fatehullah A, Appleton PL, Näthke IS (2013) Cell and tissue polarity in the intestinal tract during tumourigenesis: cells still know the right way up, but tissue organization is lost. Philos Trans R Soc Lond B Biol Sci 368:20130014. doi:10.1098/rstb.2013.0014
103. Launay N, Goudeau B, Kato K, Vicart P, Lilienbaum A (2006) Cell signaling pathways to alphaB-crystallin following stresses of the cytoskeleton. Exp Cell Res 312:3570–3584. doi:10.1016/j.yexcr.2006.07.025
104. Ailenberg M, Silverman M (2003) Cytochalasin D disruption of actin filaments in 3T3 cells produces an anti-apoptotic response by activating gelatinase A extracellularly and initiating intracellular survival signals. Biochim Biophys Acta 1593:249–258
105. Subbaramaiah K, Hart JC, Norton L, Dannenberg AJ (2000) Microtubule-interfering agents stimulate the transcription of cyclooxygenase-2. Evidence for involvement of ERK1/2 AND p38 mitogen-activated protein kinase pathways. J Biol Chem 275:14838–14845
106. Jarosz DF, Lindquist S (2010) Hsp90 and environmental stress transform the adaptive value of natural genetic variation. Science 330:1820–1824. doi:10.1126/science.1195487
107. Kaplan KB, Li R (2012) A prescription for “stress” – the role of Hsp90 in genome stability and cellular adaptation. Trends Cell Biol. doi:10.1016/j.tcb.2012.08.006
108. Whitesell L, Lindquist S (2009) Inhibiting the transcription factor HSF1 as an anticancer strategy. Expert Opin Ther Targets 13:469–478. doi:10.1517/14728220902832697
109. Dai C, Whitesell L, Rogers AB, Lindquist S (2007) Heat shock factor 1 is a powerful multifaceted modifier of carcinogenesis. Cell 130:1005–1018. doi:10.1016/j.cell.2007.07.020
110. Baird NA, Douglas PM, Simic MS, Grant AR, Moresco JJ, Wolff SC, Yates JR, Manning G, Dillin A (2014) HSF-1-mediated cytoskeletal integrity determines thermotolerance and life span. Science 346:360–363. doi:10.1126/science.1253168
111. Davies AE, Kortright K, Kaplan KB. Adenomatous polyposis coli mutants dominantly activate the heat shock pathway through inhibition of microtubule dynamics. Under review at Oncotarget

Chapter 10
Small GTPases Act as Cellular Switches in the Context of Cilia

Yan Li and Jinghua Hu

Introduction

Motile cilia or flagella have long been recognized as organelles involved in cell locomotion, fluid movement, and sexual reproduction [1]. However, another type of cilia featured as non-motile, also known as primary cilia, have not been exposed under the spotlight until the recent two decades. Ironically, as early as in the 1800s, primary cilia have been observed on the surface of mammalian cells and a sensory role has been suggested, but this had been overlooked for the next ~150 years [2]. Recently, the correlation between cilia and human ciliopathies started to attract more and more interest into the field. For example, in 1993, 23 articles with "cilia" or "cilium" as a key word in title were published; whereas in 2014, the incomplete PubMed record already pushed the number to 213 [3]. Currently, primary cilia are found on virtually all eukaryotic cells and considered as signaling center for various extracellular cues. Defects in cilia biogenesis and/or function cause a wide spectrum of complex genetic disorders termed ciliopathies, which affect many vital organs like brain, liver, and kidney [4].

Small GTPases are key cellular switches correlated with various human pathological conditions. The unique feature that small GTPases can tune a particular pathway on and off by simply binding with either GTP or GDP makes them and their

Y. Li
Department of Nephrology and Hypertension, Mayo Clinic, Rochester, MN, USA

Department of Biochemistry and Molecular Biology, Mayo Clinic, Rochester, MN, USA

J. Hu (✉)
Department of Nephrology and Hypertension, Mayo Clinic, Rochester, MN, USA

Mayo Translational PKD Center, Rochester, MN, USA

Department of Biochemistry and Molecular Biology, Mayo Clinic, Rochester, MN, USA
e-mail: hu.jinghua@mayo.edu

H. Schatten (ed.), *The Cytoskeleton in Health and Disease*,
DOI 10.1007/978-1-4939-2904-7_10

binding partners highly favorable therapeutic targets for diverse clinical/preclinical trials. In this chapter, we focus on the role of several groups of small GTPases, including ARFs/ARLs, RABs, and RAN, which are implicated in cilia biogenesis and cilia signaling during the last decade [5]. Understanding of these cellular switches is critical for dissecting how cilia formation and function is regulated in vivo and for providing seminal insights into the role of various small GTPases in the pathogenesis of ciliopathies, and their potential as therapeutic targets.

Cilia and Ciliopathy

The Cilium Is a Microtubule-Based Structure Protruding from the Cell Surface

Although tiny, the cilium is composed of distinct functional domains, each of which holds different protein components and serves distinct functions. Either motile or immotile cilia (primary cilia), are composed of nine outer microtubule doublets, covered tightly by a specialized plasma membrane contiguous with the cell membrane. There is an additional pair of central microtubules in motile cilia (9+2 axonemal configuration), whereas primary cilia exhibit a 9+0 conformation [6]. At the proximal end, the axonemal doublet is attached to the cell body by a modified mother centriole called the basal body, which shares the same structure of a hollow cylinder formed by nine triplet microtubules. When the cell enters into the resting state (G0 or G1 stage of the cell cycle), the mother centriole moves and anchors to the apical membrane and then transforms into the basal body to grow the cilium [7]. Above the basal body, there are transition fibers (TFs, transformed from the distal appendages of the mother centriole) and the transition zone (TZ, the proximal part of the axoneme that contains Y-links). The transition fibers, which look like a 9-bladed propeller-like pinwheel in TEM cross-sections, arise from the B-tubules of the centriole's triplet microtubules and anchor the distal part of the basal body to the plasma membrane [8, 9]. Above TFs, the Y-links of the TZ connect the proximal segments of axonemal microtubules to the ciliary membrane [10, 11] (Fig. 10.1).

The ciliary base also contains other amorphous subcompartments, like the septin ring and nuclear pore-like structures, which contribute to the ciliary trafficking at its base [12, 13]. Along cilia, there are two distinct sub-domains: the Inversin (InV) compartment and the ciliary tip. Recently, several ciliopathy proteins, including NPHP2/Inversin, NPHP3, NPHP9, and ANKS6, have been found to localize specifically to a distinct cilia segment that covers only the proximal segment of the axoneme above the TZ, which is thus termed the InV compartment [14–17]. In the *C. elegans* phasmid cilium, NPHP2 and ARL13B orthologs (NPHP-2 and ARL-13) also localize specifically at the proximal segment of the axoneme and regulate ciliogenesis [15, 18, 19]. Furthermore, the exclusive localization of cyclic nucleotide-gated (CNG) cation channels in the InV domain in *C. elegans* supports the

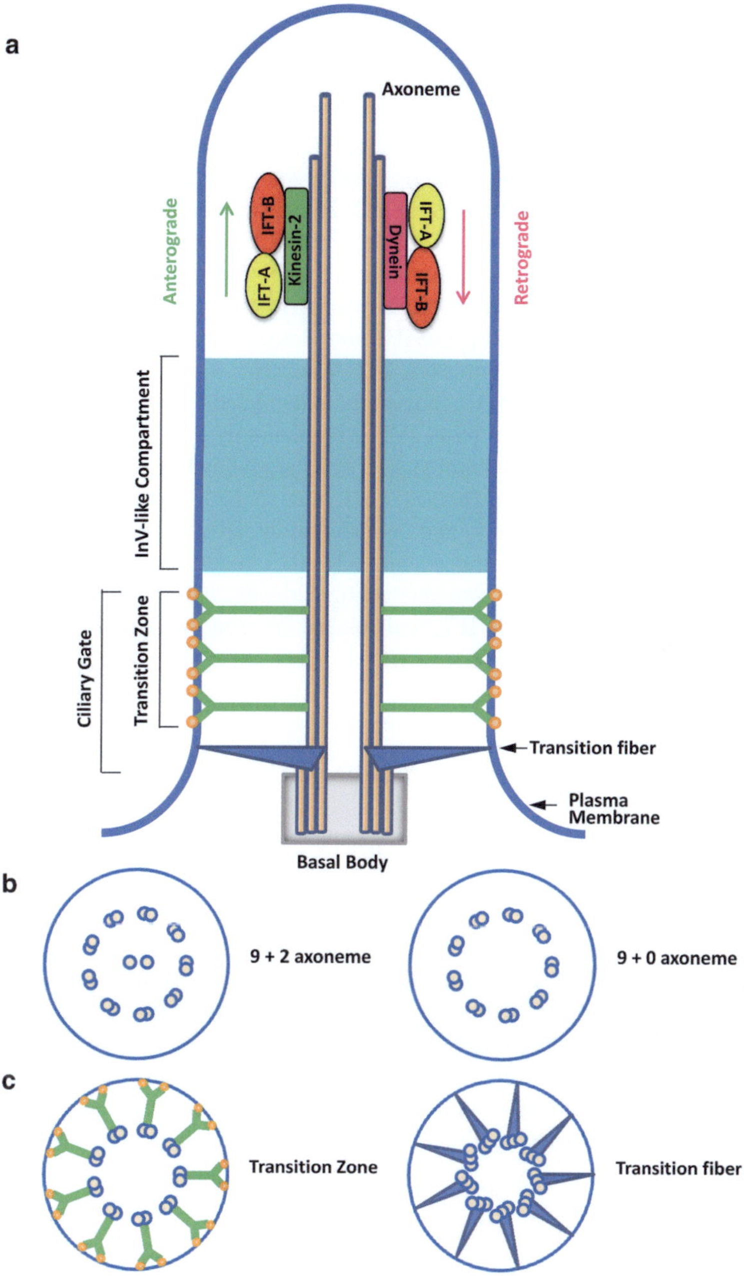

Fig. 10.1 Cilia structure. (**a**) The cilium contains a microtubule-based axoneme that emerges from a centriolar structure called the basal body (depicted as *grey box*). The TFs and TZ depicted schematically here are also proposed as the ciliary gate that is responsible for selecting the ciliary entry for ciliogenic proteins. The TZ is formed by the Y-link structures (shown in *green*) and the ciliary necklace on the ciliary surface (shown as *orange beads*). The *blue box* at the proximal side of TZ is the InV-like compartment with no clear function. The conserved IFT machinery is responsible for both anterograde (from cilia base to its tip) and retrograde (from cilia tip to its base) movement. The motor kinesin-2 plays a role in the anterograde movement, whereas dynein is indispensable for the retrograde movement. (**b**) Transverse section of the axoneme, which is found in two configurations: with a central pair (9+2) in motile cilia and without the central pair (9+0) in primary cilia. (**c**) Transverse section of TFs and TZ, which is shown as 9-bladed propeller-like pinwheel for TFs

assumption that this domain is functionally important [20, 21]. However, the exact role of the InV compartment remains elusive. The ciliary tip, which is the main site for the remodeling of intraflagellar transport (IFT), also encompasses various signaling proteins, including frog olfactory CNG channels and sonic hedgehog signaling proteins such as Gli1/2/3 and Sufu [22–24]. It was found that the BBSome, a group of proteins affected in human Bardet–Biedl syndrome, acts as the key player in regulating IFT assembly and turnaround at the cilia tip [25].

Cilia utilize the highly conserved IFT machinery to bi-directionally move cargo proteins along the axoneme [26]. The IFT complex consists of over 20 proteins that can be divided to two sub-complexes, IFT-A and IFT-B. Specifically, the anterograde (towards the ciliary tip) IFT movement is regulated by kinesin-2, whereas the retrograde (towards the ciliary base) IFT is regulated by dynein [27]. IFT is indispensable for cilia growth and maintenance. IFT cargos (membrane receptors, structural proteins, and signaling molecules) are loaded to the IFT machinery at the cilia base, and are then transported to and unloaded at the cilia tip, where they function in cilia biogenesis or sensory transduction. The IFT machinery can bind signaling molecules, like Glis from the sonic hedgehog pathway, and transport them back to the ciliary base by retrograde IFT, and release them to the cytoplasm for signal transduction [28, 29].

Primary Cilia Are Sensing Antennas of Cells

Mounting evidences demonstrated that the primary cilium acts as cell's antenna, which senses the environmental cues and then transduces signals to trigger cellular responses. Photoreceptor cells and olfactory neurons are well known sensory cells that use cilia to sense light and odor. Primary cilia harbor numerous membrane receptors (GPCRs, tyrosine kinase receptors, ion channels) for many important physiological and developmental signaling pathways, including Sonic hedgehog, PDGF, and canonical and non-canonical Wnt pathways [30–32]. Cilia also function as mechanosensors to sense flow movement, pressure and touch [33–35]. For example, the urine flow in kidney tubules causes cilia to bend, and then lead to Ca^{2+} influx through the ciliary polycystin-1 (PC-1)/polycystin-2 (PC-2) calcium channel complex [33].

Ciliopathies: A Fast Growing Disease Spectrum

Given the fact that primary cilia exist ubiquitously on cell surfaces and regulate various important signaling pathways, it is thus not surprising to see that ciliary dysfunctions cause a wide spectrum of human genetic disorders, now collectively termed ciliopathies. With rapid advancements in the positional cloning of human disease genes, a wide variety of genetic disorders, such as polycystic kidney disease (ADPKD and ARPKD), Bardet–Biedl syndrome (BBS), Joubert syndrome (JBTS), nephronophthisis (NPHP), Oral-Facial-Digital Syndrome (OFD), and Meckel-Gruber syndrome (MKS), have been characterized molecularly as cilia-related diseases, or ciliopathies [36, 37]. Consistent with the presence of cilia on most cell surfaces in the human body, ciliopathies affect many vital organs/systems and share a collection of features that include primarily retinal degeneration, renal cyst and cerebral abnormalities, as well as additional traits, like congenital fibrocystic diseases of the liver and pancreas, deafness, anosmia, cranio-facial defects, situs inversus to infertility, mental retardation, diabetes, obesity and skeletal dysplasia [38, 39]. The total population affected by various ciliopathies is estimated to be at a collective incidence of at least ~1:1000 (~100 ciliopathies × average incidence of 1:100,000) [4]. Despite the physiological and clinical relevance of cilia, the molecular mechanisms that regulate ciliogenesis, cilia function, and the connections between disease gene functions and pathology remain largely elusive. In this regard, the greatest challenges for molecular biologists and clinicians are to understand how cilia form and function; determine the pathogenesis underlying ciliopathies; and design therapies to prevent, delay, or halt disease progression.

Small GTPases Are Vital Switches in Cilia Biogenesis and Function

Small GTPases are a group of GTP hydrolases functioning as key molecular switches in various cellular processes and they are implicated in many human disorders. Small GTPases toggle particular signaling pathways to on/off by cycling through GTP binding and GDP binding. The GTPase is switched on by guanine nucleotide exchange factors (GEFs) that help GDP dissociation, and turned off by GTPase activating proteins (GAPs), which accelerate the hydrolysis of GTP. This unique enzymatic feature presents them together with their regulators/effectors (such as adaptors, motors, kinases, and posttranslational modification enzymes) highly promising therapeutic targets for many human diseases [40, 41]. All small GTPases share a conserved domain including a six β-stranded sheet encircled by five α-helices. GTP and GDP binding sites are highly conserved within most small GTPase family members. Several point mutations can lead to constitutive GTP-bound or GDP-bound forms of small GTPases, which are termed dominant-active (DA) or dominant-negative (DN) mutants [42]. Dominant-active (DA) and dominant-negative (DN) mutants have been widely used as tools to characterize the in vivo functions of small GTPases.

Recently, mounting evidences have pointed out that several small GTPases, including the members in Arf/Arl (Arf-like), Rab and Ran subfamilies, regulate cilia formation and/or cilia signaling. Here, we will review the role of these cilia-related small GTPases and their regulators/effectors in the context of cilia.

Arf/Arls

Arf/Arl GTPases belong to the Ras superfamily, and consist of 29 members, including 6 Arfs (ADP ribosylation factors), Sar1, and over 20 Arf-like proteins (Arls) [43]. Arf proteins are well known for their role in regulating vesicular trafficking and actin remodeling [44, 45], while the role of most Arls remains poorly understood. So far, Arf4 and three Arls (Arl3, Arl6, and Arl13b) have been implicated in cilia physiology. Notably, Arl6 and Arl13b have been identified as ciliopathy genes in humans (Fig. 10.2).

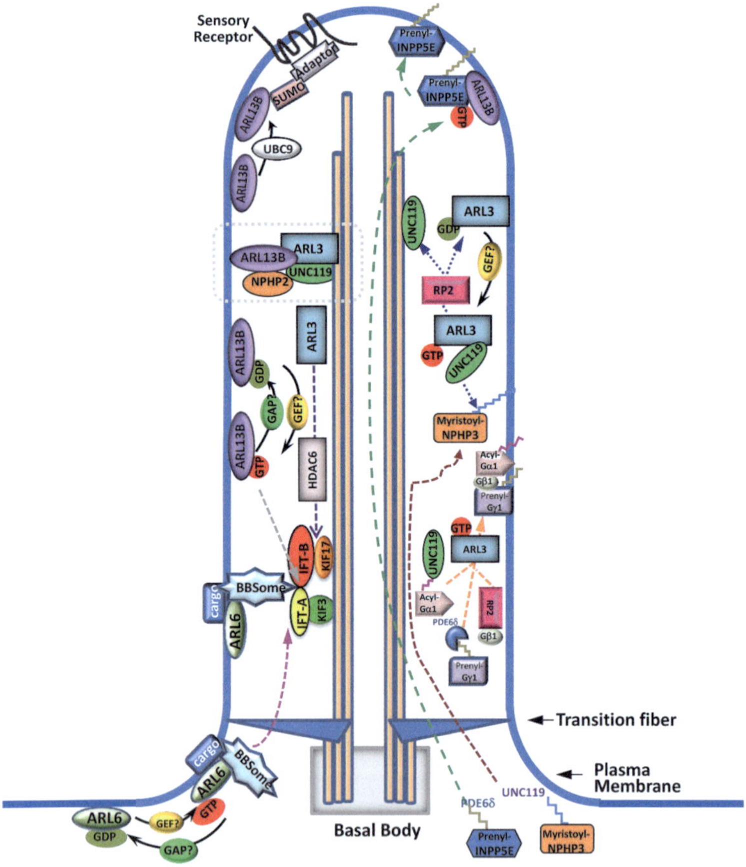

Fig. 10.2 A working model for the role of ciliary Arls. Arl6-GTP recruits the BBSome and cargos (like sensory receptors) into cilia. Arl3 stabilizes the binding between IFT-B and KIF17 depending on HDAC-6; Arl3-GTP can interact with UNC119/myristoylated-NPHP3 and facilitate the release of NPHP3 in cilia, as well as mediate the proper ciliary targeting of transduction to different effectors. Arl-13 strengthens the IFT-A and IFT-B association; Arl13b, Arl3, UNC119 and NPHP2 form a functional module in the InV compartment of cilia; The SUMOylation of Arl13b by UBC9 is indispensable for the ciliary entry of several sensory receptors; Arl13b-GTP binds with and promotes the ciliary targeting of INPP5E. *Dashed lines* indicate that the molecular mechanisms are not clear

Arf4

Arf4 is the only cilia-related Arf small GTPase in the Arf/Arl family [46]. Arf4 mainly resides at the cilia base and facilitates the sorting and targeting of ciliary sensory receptors, but is not required for cilia assembly [47]. Specifically, a complex formed by Arf4, rhodopsin, Arf GAP protein ASAP1, and another small GTPases Rab11 and its effector FIP3, controls the transport of the cargo rhodopsin from the Trans-Golgi-Network to photoreceptor cilia [46, 48]. In a simple working model, Arf4 is probably activated at the TGN budding site by a specific GEF, likely GBF1, and then activated Arf4 recognizes rhodopsin through VxPx motif. After cargo recognition, ASAP1, possibly works together with Rab11 and FIP3, forms a complex with Arf4 and cargo rhodopsin. At this point, ASAP1 likely promotes GTP hydrolysis of Arf4-GTP and releases Arf4 from TGN, and then the remaining complex containing the cargo rhodopsin will be sorted to cilia. The fact that VxPx motif exists in several other cilia sensory proteins, including ADPKD protein polycystin-1 and polycystin-2, the cyclic nucleotide-gated (CNG) channel CNGB1b and other ciliary localized proteins, suggests that VxPx motifs could function as Arf4 binding sites for transport to the cilium [49, 50]. However, a recent study found that Arf4 can also bind the cilia-targeting-sequence of ARPKD protein fibrocystin, which does not contain a VxPx domain, and regulate the efficient transport of newly synthesized fibrocystin from the Golgi to the cilium. Since the clathrin adaptor AP-1, which is identified as one of Arf4 effectors, also plays a role in regulating the proper ciliary localization of non-VxPx domain containing odorant receptor ODR-10 in *C. elegans* [51], it would be interesting to examine whether Arf4 might act through AP-1 and clathrin for the selective ciliary targeting of fibrocystin and ODR-10.

Arl3

Although Arl3 has not been identified as a ciliopathy gene, several evidences suggest a cilia-related role for Arl3. First, a comparative genome study suggested that *Arl3* exclusively exists in the genome of ciliated organisms during evolution [52]; Secondly, Arl3 locates to the connecting cilia in human retina photoreceptor cells [53]; Thirdly, $Arl3^{-/-}$ mice die shortly after birth and display typical ciliopathy phenotypes characterized by cysts in kidney, liver and pancreas and impaired photoreceptor development [54].

Functional analyses suggested that Arl3 acts as a negative regulator for cilia formation. In *Leishmania*, overexpressing the GTP-bound form of Arl-3 induces ciliogenesis defects [55]. In *C. elegans*, depletion of ARL-3 can partially rescue the ciliogenesis defect in *arl-13* mutant worms through a histone deacetylase 6 (HDAC-6)-dependent manner [56]. Similar observations were also reported for mammalian cells in that *Arl3* knockdown shows no effect on ciliogenesis [57], but can rescue the ciliogenesis defect in *Arl-13b*-depleted human RPE cells (unpublished data in the Hu lab).

Arl3 also plays a role in ciliary signaling. *Arl-3* mutant worms are compromised in cilia-mediated male mechanosensory behaviors (response and location of vulva), possibly due to the mislocalization of the mechanosensory receptor polycystin (unpublished data in the Hu lab). Similar observations were reported in a recent paper showing that mammalian Arl3 associates with Rabep1 and GGA1 to regulate the proper sorting of the PC1/PC2 complex to the cilium [58]. Other than polycystins, Arl3 is also involved in regulating several signaling pathways, such as the rhodopsin transduction and the heterotrimeric G-protein transducin traffic in photoreceptor cells, as well as the transportation of Gli3, a key signaling molecule in the Shh pathway [54, 57, 59].

Although several interactors/effectors, such as Golgin-245, CCDC104, C20orf194, and Binder of Arl Two (BART), have been biochemically characterized for ARL3 [60, 61], few have been functionally confirmed to be cilia-related. Notably, HDAC-6 is likely a downstream effector of ARL-3 in regulating the association between IFT complex and IFT motor OSM-3/KIF17 in *C. elegans* [56]. In mammals, activation of HDAC6 promotes cilia disassembly and loss of HDAC6 stabilizes cilia formation in human retinal epithelial cells, suggesting a role for HDAC6 in the Arl3 pathway that is probably highly conserved across ciliated species [62]. Since HDAC6 is a deacetylase, it would be of great interest to identify the ciliary substrate(s) for HDAC6. The X-linked retinitis pigmentosa protein RP2 is another ciliary interactor for Arl3. RP2 acts as GAP for Arl3 and converts ARL3-GTP to ARL3-GDP [59, 60]. In cilia-targeted trafficking, some myristoylated ciliary cargo, such as NPHP3 and transducin α subunit, is first transported to the cilia base by UNC119, and then the association between GTP-bound Arl3 and UNC119 releases the myristoylated cargo to the ciliary membrane. Next, RP2 hydrolyzes Arl3-GTP to Arl3-GDP and this will release UNC119 and promote its removal from cilia [60]. Interestingly, mammalian PDE6δ interacts with GTP-bound ARL-3 to facilitate the dissociation of farnesylated cargo protein from PDE6δ [63, 64].

Arl6

Arl6, also known as BBS3, was the first Arl linked to human ciliopathy [65, 66]. Bardet–Biedl syndrome (BBS) is an autosomal recessive disease, that can be caused by mutations in each of 17 human genes (BBS1, BBS2, BBS3 (ARL6), BBS4, BBS5, BBS6 (MKKS), BBS7, BBS8 (TTC8), BBS9, BBS10, BBS11 (TRIM32), BBS12, MKS1, CEP290, C2ORF86, SDCCAG8, and LZTFL1) [67–69]. BBS is characterized by obesity, polydactyly, mental retardation, retinal degeneration, renal cyst, and learning disabilities [65]. Structural analyses revealed that the mutations identified in BBS3 patients cluster around the ARL6 GTP binding domain and interfere with its GTP binding capacity, indicating the importance of small GTPase activity in its ciliary function [70, 71]. The ciliary role of Arl6 is highly conserved through evolution. Both *Arl6* knockdown zebrafish and *Arl6* knockout mice develop BBS-associated phenotypes [72, 73].

Eight BBS proteins (BBS1, BBS2, BBS4, BBS5, BBS7, BBS8, BBS9 and BBIP10) form a protein complex known as the BBSome, which shares a similar structure with clathrin and other coat protein complexes [74]. GTP-bound Arl6 is suggested to associate with the BBSome and sort membrane proteins, such as melanin concentrating hormone receptor 1 and somatostatin receptor 3, to cilia [74]. Studies in our lab also confirmed that various sensory receptors consistently mislocalize in *arl-6* knockout worms (unpublished data from Hu lab). Loss of Arl6 does not affect BBSome formation. Interestingly, other than the ciliary entry of membrane receptors, Arl6 depletion also regulates the retrograde transport of sensory receptor Smoothened inside cilia [73]. Notably, different from other BBS mutants, both *Arl-6*$^{-/-}$ mice and *Arl-6* mutant *worms* possess normal cilia, suggesting that ARL6 does not always act together with the BBSome in same genetic pathway [75]. Consistent with this assumption, *Arl6* knockout mice develop some unique non-BBSome-related phenotypes, like no overt obesity, severe hydrocephalus, and elevated blood pressure [73]. Taken together, these evidences suggest that the role of Arl6 in the context of cilia is conserved and the variety of symptoms found in BBS3 patients are likely due to the compromised ciliary targeting or removal of various sensory receptors and/or signaling molecules.

Overall, the role of ARL6 in ciliogenesis still remains unclear. The BBSome is thought to regulate IFT integrity and IFT assembly. In *C. elegans*, the BBSome is one integral sub-complex of the IFT machinery. Arl6 is the only Arf/Arl small GTPase that exhibit typical IFT movement inside the cilium, but it actually moves much less frequently when compared with IFT structural components or other BBS proteins, indicating that ARL-6 is probably a cargo of the IFT machinery. Another interesting finding is that overexpression of both the GTP-bound and GDP-bound form of ARL6 in IMCD3 cells influences cilia number and length [71]. The exact determination of the role of ARL6 in ciliogenesis depends on the future characterization of ARL6's GAPs, GEFs, and effectors in cilia.

Arl13b

Arl13b is the other small GTPase identified as a ciliopathy gene. *ARL13b* mutations were identified in families with Joubert Syndrome (JBTS), which exhibit classical ciliopathy phenotypes, like abnormalities in the central nervous system, kidney cyst, retinal impairment, polydactyly, and obesity [76–78]. Through evolution (worm, zebrafish, and mouse), Arl13b protein specifically enrich in cilia (73, 87, 88). So far, solid evidences suggest that Arl13b is equally essential for both cilia formation and signaling. Arl13b mutant zebrafish shows defective cilia in multiple organs [79, 80]. *Arl13b*^{hnn} mice show truncated axoneme and defective B-tubule closing, as well as ciliopathy-related impairments in neural tube patterning, limbs and eyes (88). *arl-13*$^{-/-}$ worms show truncated cilia that are due to the compromised integrity of IFT particles [56, 81]. The GTPase activity of Arl13b is probably required for its function. The R79Q mutation identified in patients partially disrupts the GTPase activity of ARL13B. Moreover, corresponding worm mutant

ARL-13(R83Q) could not fully rescue the ciliary defect of *arl-13*$^{-/-}$ worms [56]. Interestingly, another study suggested that Arl13b likely lacks intrinsic GTP hydrolysis activity and needs to interact with other GTPases to be fully functional [82]. However, this study used a truncated Arl13B from *C. rheinhardtii*, so more evidences are necessary to completely understand the biological function of ARL13B and whether it acts as a GTPase in cilia. To date, no GEF or GAP proteins for Arl13b have been identified in any ciliated species yet.

Arl13b is involved in cilia signaling. *Arl13b*^{hnn} mice display compromised Shh signaling, Bone Morphogenetic Protein (BMP) signaling pathway, as well as abnormally expressed Wnt ligands [83, 84]. Moreover, Shh signaling components are mislocalized in *Arl13b*^{hnn} mice [85]. In *C. elegans*, several chemo- or mechanosensory receptors, such as PKD-2, ODR-10, TAX-2, and OSM-9, are mislocalized in *arl-13*$^{-/-}$ cilia [81]. We found that UBC9, the sole E2 small ubiquitin-like modifier (SUMO)-conjugating enzyme, interacts and SUMOylates Arl13b. Arl13b SUMOylation is required for proper ciliary targeting of polycystin-2 in both worms and mammalian cells [86]. It would be interesting to study whether Arl13b SUMOylation plays a general role in regulating the ciliary entry of other ciliary receptors. Except for sorting sensory receptors into cilia, Arl13b also mediates the ciliary targeting of other proteins, such as inositol polyphosphate-5-phosphatase E (INPP5E), through a function network consisting of Arl13b, INPP5E, PDE6δ, and centrosomal protein 164 (CEP164) in mammalian cells [87].

Arl13b specifically locates in the InV-like compartment in both worm cilia and mammalian cilia (73, 96, 102) [15, 56, 81]. In a recent study, Cevik et al. found that the precise localization of Arl13b to the InV compartment is mediated by cooperation between MKS/NPHP modules and the IFT machinery [15]. However, little is known about the function of the InV compartment. We recently found that Arl3, Arl13b, NPHP2, and UNC119 form a novel protein module and specifically localize in the InV compartment. The association is independent of GTPase activity of either Arl3 or Arl13b. We further found that, in *C. elegans*, ARL-13, UNC-119 and NPHP-2 work synergistically in regulating ciliogenesis, which can be antagonized by ARL-3 (unpublished data from Hu lab). Our discoveries about the functional module in the InV compartment shed lights on the functional crosstalk between two ciliary small GTPases (Arl13b and Arl3) and their interactors (UNC119 and NPHP2), in coordinating ciliogenesis (Table 10.1).

Rabs

Rab GTPases form the largest family of small GTPases in the Ras superfamily, and play general roles in membrane trafficking and membrane fusion [88]. It is thus expected that specific Rabs may be involved in protein transport to primary cilia. Several Rabs have recently been found to regulate cilia biogenesis and/or cilia signaling.

Table 10.1 The ciliary small GTPases and their interactors

Small GTPase	Organism	Function	GEF	GAP	Effector
Arl4	*X. laevis.*	Ciliary protein targeting [46, 48]		ASAP1 [48]	Rab11/FIP3 [46, 48]
	C. elegans	Cilia signaling [51]			AP-1 [51]
Arl3	*C. elegans*	Ciliogenesis [56]			HDAC-6 [56]
	Mouse	Cilia signaling [54, 57]			
	Human	Cilia signaling [58]		RP2 [59]	Rabep1 [58], GGA1 [58], Golgin-25 [61], BART [61], UNC119 [60], CCDC104 [60], C20orf194 [60], PDE6δ [63, 64]
Arl6	*C. elegans*	Cilia signaling [unpublished data]			
	Mouse	Cilia signaling [73]			
	Human	Cilia signaling [74]			The BBSome [74], IFT27 [104]
Arl13b	*C. elegans*	Ciliogenesis [56, 81]; sensory receptor targeting [86]			UBC-9 [86]
	Zebrafish	Ciliogenesis [80]			
	Mouse	Ciliogenesis; cilia signaling [83–85]			
	Human	Ciliogenesis; cilia signaling [77]			UBC-9 [86], INPP5E [87], NPHP2/UNC119 [unpublished data]
Rab8	*C. elegans*	Protein sorting and trafficking [89, 90]			Rabaptin5 [93], Elipas/DYF-11 [93]
	Human		Rabin8 [92]		BBSome [92], Talpid3/Cep290 [96]
Rab11	Human	Cilia signaling [94]			PI3K-C2α [95]
Rab10	Mouse	Cilia signaling [97]			Sec8 [97]
Rab23	Human	Smo turnover in cilia [91, 98]			
IFTA-2	*C. elegans*	Cilia signaling [100]			
IFT27	*Chlamydomonas Reinhardtii*	IFT [101]			
	Trypanosoma brucei	Bidirectional IFT transport [102]			
	Human, mouse	Cilia signaling [103], protein targeting [104]			
Ran	Human, mouse	Ciliary protein import [13, 106–108]			Importin beta-2 [106, 108], nucleoporins [13, 107]

Rabs

Rab8 is the first member found to be cilia-related. Rab8 mediates the trafficking of rhodopsin into cilia in photoreceptor cells through a mechanism involving the exocyst and SNARE proteins [89]. Except for rhodopsin, Rab8 is also connected with the ciliary entry of other cilia membrane proteins, such as the ARPKD causal protein fibrocystin, and Shh receptor Smo [90, 91]. Rab8 tends to function by networking with other proteins. One model proposed that Rab8 can be recruited by the BBSome via its GEF, Rabin8, to regulate the fusion of post-Golgi vesicles at the cilia base [92]. Rab8 also acts together with Rab5 effector Rabaptin5 and IFT component Elipas/DYF-11 to form a bridging structure between the IFT complex/cargo on the cilia membrane [93]. In addition, Rab8 is also involved in the Rab11/Arf4 pathway discussed above to target rhodopsin into cilia, as well as in a pathway in which PI3K-C2α-derived PtdIns3*P* activates Rab11 to regulate cilia formation, Smo ciliary translocation and Shh signaling [94, 95]. In addition to its function in protein targeting, Rab8 may also contribute to cilia assembly. A new study showed that Talpid3, a component of a CP110-containing protein complex, possibly affects ciliogenesis through Rab8 recruitment and/or activation [96]. Except for Rab8, another Rab protein, Rab10, also has cilia-related function. Rab10 directly interacts with the exocyst complex protein Sec8 and colocalizes with the exocyst at the cilia base [97]. In addition, the GAPs of Rab8, Rab17, and Rab23 were found to regulate ciliogenesis [98]. The cellular function of Rab17 is unknown, but Rab23 negatively regulates Shh signaling by negatively controlling Smo turnover in the cilia [91, 99] (Fig. 10.3).

Rab-Like Proteins

Rab-like 5 (IFTA-2) and Rab-like 4 (IFT27), two non-canonical Rab proteins lacking the typical C-terminal prenylation motifs of Rabs, are involved in cilia/flagellar function in *C. elegans* and *C. rheinhardtii* [100, 101]. Unlike other Rabs that remain at the cilia base, both Rabl4/IFT27 and Rabl5/IFTA-2 are part of IFT machinery and can move bidirectionally inside cilia. In *C. elegans*, *ifta-2* null mutants show no defect in ciliogenesis, but affect the activity of ciliary sensory IGF-1-like receptor [100]. In contrast, Rabl4/IFT27 is involved in maintaining the stability of IFT complexes [101]. Intriguingly, a new study found that IFT27, an IFT-B component, is crucial for retrograde transport and regulates the import of both the IFT-A complex and the IFT dynein into the flagellum in a GTPase dependent manner in *Trypanosoma brucei* [102]. Mammalian Rabl4/IFT27 facilitates the ciliary transport of Shh molecules, like patched and Smo, through promoting ARL6 activation and BBSome coat assembly [103, 104]. Consistent with its functional crosstalk with BBS proteins, *Rabl4/IFT27* was recently identified as a ciliopathy gene in BBS patients [105]. This is the first member of the Rab family to be

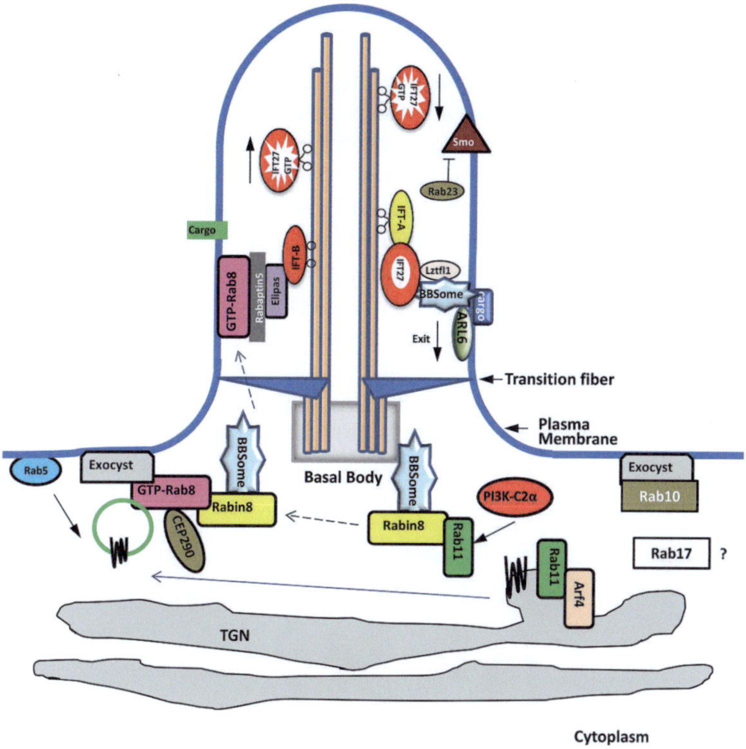

Fig. 10.3 A working model for the role of ciliary Rabs and Rab like proteins. Rab proteins and Rab-like proteins are involved in cilia-targeted trafficking and IFT transport. Rab8 is first activated by Rabin8, and then proposed to help docking and fusion of vesicles bearing transmembrane proteins near the ciliary membrane, including the BBSome. GTP-bound Rab8 in cilia can form a bridge with Rabaptin 5 (Rab5 effector) and Elipas between IFT complex and cargo at the cilia membrane. Rab11 activates the GEF activity of Rabin8 to deactivate Rab8. The activity of Rab11 can be turned on by PI3K-C2α-derived PtdIns3*P*. Rab23 is a negative regulator of Smo signaling and may act on its turnover. Rab-like 4 (IFT27) is indispensable for both anterograde IFT transport and retrograde IFT movement. IFT27 facilitates BBSome/ARL6 complex to exit the cilia in a similar process by which the BBSome mediates cargo entry into cilia

identified as a ciliopathy gene, further highlighting the important role of small GTPases in cilia biophysiology.

Ran

The small GTPase Ran is well known for its role in nuclear transport, mitotic spindle assembly and nuclear envelope assembly. Recently, several observations imply that Ran also facilitates ciliary import. The fact that soluble molecules above a specific size (~40 kDa) are restricted from passively entering the ciliary lumen suggests the existence of a selective ciliary barrier at the cilia base [13]. Recent work demonstrated that the components of the nuclear import machinery, including small GTPase Ran, importin beta-2 and nucleoporins (NUP35, 37, 62, 93, 214, etc.), localize to the ciliary base and may form the ciliary pore complex to regulate selective ciliary entry [13, 106–108]. It was thus proposed that the function of the periciliary base is analogous to the nuclear pore regarding the selective entry of proteins and the exclusion of large soluble proteins [13, 109, 110]. Dishinger et al. showed that Ran and importin-β2 regulate the ciliary entry of kinesin2 motor KIF17 in a similar manner in nuclear transport. Specifically, similar to nuclear transport, a ciliary-cytoplasmic gradient of Ran with high ciliary Ran-GTP level is essential for ciliary import [108]. This phenomenon suggests that the nuclear import could be a useful model for studying the cilia import. Whether and how this pathway regulates the entry of other ciliogenic proteins as well as the molecular composition of the proposed ciliary pore remain to be investigated (Fig. 10.4).

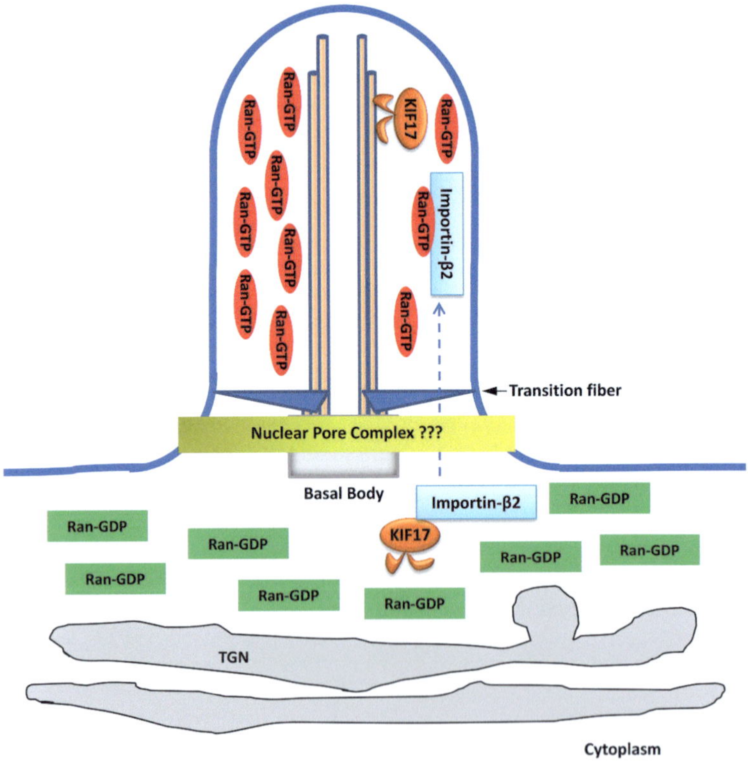

Fig. 10.4 A working model for the ciliary import of KIF17 by Ran. Ran-GDP is enriched in the cytoplasm, whereas Ran-GTP is concentrated in cilia. In the cytoplasm, Kinesin 2 KIF17 forms a complex with importin-β2. Then this complex can shuttle across the ciliary TZ and enter cilia, where high amounts of Ran-GTP cause a dissociation of the KIF-importin-β2 complex

Perspectives

Cilia are specialized organelles. The formation and maintenance are regulated by spatiotemporal integration of cytoskeleton dynamics and polarized membrane trafficking. Because small GTPases usually work as key switches in controlling membrane and cytoskeleton-related cellular processes, it is not surprising that this group of enzymes plays pivotal roles in the context of cilia. However, the molecular mechanisms underlying the correlation between small GTPases and cilia still remains elusive, majorly due to our poor understanding about the GEF, GAP, and effectors of the ciliary small GTPases. The binding partners of small GTPases (such as posttranslational modification enzymes, adaptors, motors, kinases, and phosphatases) can control the localization, activity, and downstream signaling by interacting with and altering different forms of small GTPases [42]. Identifying and characterizing the binding partners of the cilia-related small GTPases would dissect the underlying functional networks. Two good examples would be the functional network among, Rab11, Rab8, and Arf4, which share the same GAP protein Rabin8, as well as the coordination function between ARL-3 and ARL-13 in stabilizing the IFT complex [56, 94].

Characterization of the in vivo function of cilia-related small GTPases and their effectors needs to be pursued in animal models. Unfortunately, due to the essential role of cilia in embryonic development and tissue pattern formation, it is extremely difficult to study cilia biology in mammalian model organisms [111]. To this end, simple ciliated model organisms, like *C. elegans* and *zebrafish*, have emerged as powerful tools for studying cilia biology and human ciliopathies. And there are several reasons. First, cilia structure, the IFT process, and the sensory function are largely conserved through evolution. Secondly, the functions of most human ciliopathy genes are conserved in genetic models. Thirdly, genetic models offer indispensable experimental advantages. For example, powerful genetics toolkits are available for both worm and zebrafish, which include genome-wide mutagenesis screens, transgenesis, and RNAi [112]. In *C. elegans,* a number of simple assays make it possible to quickly test cilia formation and function in live animals. And zebrafish cilia can also be easily observed in both living embryos and adults at high resolution [113, 114]. Overall, combining the genetic tools and the high-throughput methods in genetic models would greatly prompt the identification and characterization of players involved in cilia formation and function.

Our understanding about the correlation between small GTPases and cilia has been advanced significantly in recent decades. Studying the role of ciliary small GTPases will extend our view about how cilia develop and function in normal and pathological states. Most importantly, small GTPases and many of their effectors are highly promising therapeutic candidates. Hopefully, research in this field would bring great benefits regarding new therapeutic invention or phenotype reversion for ciliopathy treatment.

References

1. Haimo LT, Rosenbaum JL (1981) Cilia, flagella, and microtubules. J Cell Biol 91:125s–130s
2. Bloodgood RA (2009) From central to rudimentary to primary: the history of an underappreciated organelle whose time has come. The primary cilium. Methods Cell Biol 94:3–52
3. Wheatley DN (2005) Landmarks in the first hundred years of primary (9+0) cilium research. Cell Biol Int 29:333–339
4. Davis EE, Katsanis N (2012) The ciliopathies: a transitional model into systems biology of human genetic disease. Curr Opin Genet Dev 22:290–303
5. Li Y, Hu J (2011) Small GTPases and cilia. Protein Cell 2:13–25
6. Pedersen LB, Rosenbaum JL (2008) Intraflagellar transport (IFT) role in ciliary assembly, resorption and signalling. Curr Top Dev Biol 85:23–61
7. Kim S, Dynlacht BD (2013) Assembling a primary cilium. Curr Opin Cell Biol 25:506–511
8. Dawe HR, Farr H, Gull K (2007) Centriole/basal body morphogenesis and migration during ciliogenesis in animal cells. J Cell Sci 120:7–15
9. Fisch C, Dupuis-Williams P (2011) Ultrastructure of cilia and flagella – back to the future! Biol Cell 103:249–270
10. Reiter JF, Blacque OE, Leroux MR (2012) The base of the cilium: roles for transition fibres and the transition zone in ciliary formation, maintenance and compartmentalization. EMBO Rep 13:608–618
11. Gilula NB, Satir P (1972) The ciliary necklace. A ciliary membrane specialization. J Cell Biol 53:494–509
12. Hu Q, Milenkovic L, Jin H, Scott MP, Nachury MV, Spiliotis ET, Nelson WJ (2010) A septin diffusion barrier at the base of the primary cilium maintains ciliary membrane protein distribution. Science 329:436–439
13. Kee HL, Dishinger JF, Blasius TL, Liu CJ, Margolis B, Verhey KJ (2012) A size-exclusion permeability barrier and nucleoporins characterize a ciliary pore complex that regulates transport into cilia. Nat Cell Biol 14:431–437
14. Shiba D, Yamaoka Y, Hagiwara H, Takamatsu T, Hamada H, Yokoyama T (2009) Localization of inv in a distinctive intraciliary compartment requires the c-terminal ninein-homolog-containing region. J Cell Sci 122:44–54
15. Cevik S, Sanders AA, Van Wijk E, Boldt K, Clarke L, van Reeuwijk J, Hori Y, Horn N, Hetterschijt L, Wdowicz A, Mullins A, Kida K, Kaplan OI, van Beersum SE, Man Wu K, Letteboer SJ, Mans DA, Katada T, Kontani K, Ueffing M, Roepman R, Kremer H, Blacque OE (2013) Active transport and diffusion barriers restrict Joubert Syndrome-associated ARL13B/ARL-13 to an Inv-like ciliary membrane subdomain. PLoS Genet 9:e1003977
16. Shiba D, Manning DK, Koga H, Beier DR, Yokoyama T (2010) Inv acts as a molecular anchor for Nphp3 and Nek8 in the proximal segment of primary cilia. Cytoskeleton (Hoboken) 67:112–119
17. Hoff S, Halbritter J, Epting D, Frank V, Nguyen TM, van Reeuwijk J, Boehlke C, Schell C, Yasunaga T, Helmstadter M, Mergen M, Filhol E, Boldt K, Horn N, Ueffing M, Otto EA, Eisenberger T, Elting MW, van Wijk JA, Bockenhauer D, Sebire NJ, Rittig S, Vyberg M, Ring T, Pohl M, Pape L, Neuhaus TJ, Elshakhs NA, Koon SJ, Harris PC, Grahammer F, Huber TB, Kuehn EW, Kramer-Zucker A, Bolz HJ, Roepman R, Saunier S, Walz G, Hildebrandt F, Bergmann C, Lienkamp SS (2013) ANKS6 is a central component of a nephronophthisis module linking NEK8 to INVS and NPHP3. Nat Genet 45:951–956
18. Nguyen PA, Liou W, Hall DH, Leroux MR (2014) Ciliopathy proteins establish a bipartite signaling compartment in a *C. elegans* thermosensory neuron. J Cell Sci 127(24): 5317–5330
19. Warburton-Pitt SR, Jauregui AR, Li C, Wang J, Leroux MR, Barr MM (2012) Ciliogenesis in *Caenorhabditis elegans* requires genetic interactions between ciliary middle segment localized NPHP-2 (inversin) and transition zone-associated proteins. J Cell Sci 125:2592–2603

20. Mukhopadhyay S, Lu Y, Shaham S, Sengupta P (2008) Sensory signaling-dependent remodeling of olfactory cilia architecture in *C. elegans*. Dev Cell 14:762–774
21. Wojtyniak M, Brear AG, O'Halloran DM, Sengupta P (2013) Cell- and subunit-specific mechanisms of CNG channel ciliary trafficking and localization in *C. elegans*. J Cell Sci 126:4381–4395
22. Haycraft CJ, Banizs B, Aydin-Son Y, Zhang Q, Michaud EJ, Yoder BK (2005) Gli2 and Gli3 localize to cilia and require the intraflagellar transport protein polaris for processing and function. PLoS Genet 1:e53
23. Flannery RJ, French DA, Kleene SJ (2006) Clustering of cyclic-nucleotide-gated channels in olfactory cilia. Biophys J 91:179–188
24. Matsuzaki O, Bakin RE, Cai X, Menco BP, Ronnett GV (1999) Localization of the olfactory cyclic nucleotide-gated channel subunit 1 in normal, embryonic and regenerating olfactory epithelium. Neuroscience 94:131–140
25. Wei Q, Zhang Y, Li Y, Zhang Q, Ling K, Hu J (2012) The BBSome controls IFT assembly and turnaround in cilia. Nat Cell Biol 14:950–957
26. Orozco JT, Wedaman KP, Signor D, Brown H, Rose L, Scholey JM (1999) Movement of motor and cargo along cilia. Nature 398:674
27. Rosenbaum J (2002) Intraflagellar transport. Curr Biol 12:R125
28. Huangfu D, Anderson KV (2005) Cilia and hedgehog responsiveness in the mouse. Proc Natl Acad Sci U S A 102:11325–11330
29. May SR, Ashique AM, Karlen M, Wang B, Shen Y, Zarbalis K, Reiter J, Ericson J, Peterson AS (2005) Loss of the retrograde motor for IFT disrupts localization of Smo to cilia and prevents the expression of both activator and repressor functions of Gli. Dev Biol 287: 378–389
30. Goetz SC, Anderson KV (2010) The primary cilium: a signalling centre during vertebrate development. Nat Rev Genet 11:331–344
31. Corbit KC, Shyer AE, Dowdle WE, Gaulden J, Singla V, Chen MH, Chuang PT, Reiter JF (2008) Kif3a constrains beta-catenin-dependent Wnt signalling through dual ciliary and non-ciliary mechanisms. Nat Cell Biol 10:70–76
32. Singla V, Reiter JF (2006) The primary cilium as the cell's antenna: signaling at a sensory organelle. Science 313:629–633
33. Praetorius HA, Spring KR (2003) The renal cell primary cilium functions as a flow sensor. Curr Opin Nephrol Hypertens 12:517–520
34. Housset C (2005) Cystic liver diseases. Genetics and cell biology. Gastroenterol Clin Biol 29:861–869
35. Nauli SM, Kawanabe Y, Kaminski JJ, Pearce WJ, Ingber DE, Zhou J (2008) Endothelial cilia are fluid shear sensors that regulate calcium signaling and nitric oxide production through polycystin-1. Circulation 117:1161–1171
36. Badano JL, Mitsuma N, Beales PL, Katsanis N (2006) The ciliopathies: an emerging class of human genetic disorders. Annu Rev Genomics Hum Genet 7:125–148
37. Adams M, Smith UM, Logan CV, Johnson CA (2008) Recent advances in the molecular pathology, cell biology and genetics of ciliopathies. J Med Genet 45:257–267
38. Fry AM, Leaper MJ, Bayliss R (2014) The primary cilium: guardian of organ development and homeostasis. Organogenesis 10:62–68
39. Waters AM, Beales PL (2011) Ciliopathies: an expanding disease spectrum. Pediatr Nephrol 26:1039–1056
40. Benard V, Bokoch GM, Diebold BA (1999) Potential drug targets: small GTPases that regulate leukocyte function. Trends Pharmacol Sci 20:365–370
41. Vigil D, Cherfils J, Rossman KL, Der CJ (2010) Ras superfamily GEFS and GAPS: validated and tractable targets for cancer therapy? Nat Rev Cancer 10:842–857
42. Cherfils J, Zeghouf M (2013) Regulation of small GTPases by GEFs, GAPs, and GDIs. Physiol Rev 93:269–309
43. Gillingham AK, Munro S (2007) The small G proteins of the Arf family and their regulators. Annu Rev Cell Dev Biol 23:579–611

44. Pasqualato S, Renault L, Cherfils J (2002) Arf, Arl, Arp and Sar proteins: A family of GTP-binding proteins with a structural device for 'front-back' communication. EMBO Rep 3:1035–1041
45. Myers KR, Casanova JE (2008) Regulation of actin cytoskeleton dynamics by Arf-family GTPases. Trends Cell Biol 18:184–192
46. Mazelova J, Astuto-Gribble L, Inoue H, Tam BM, Schonteich E, Prekeris R, Moritz OL, Randazzo PA, Deretic D (2009) Ciliary targeting motif VxPx directs assembly of a trafficking module through Arf4. EMBO J 28:183–192
47. Follit JA, San Agustin JT, Jonassen JA, Huang T, Rivera-Perez JA, Tremblay KD, Pazour GJ (2014) Arf4 is required for mammalian development but dispensable for ciliary assembly. PLoS Genet 10:e1004170
48. Deretic D, Williams AH, Ransom N, Morel V, Hargrave PA, Arendt A (2005) Rhodopsin c terminus, the site of mutations causing retinal disease, regulates trafficking by binding to ADP-ribosylation factor 4 (Arf4). Proc Natl Acad Sci U S A 102:3301–3306
49. Geng L, Okuhara D, Yu Z, Tian X, Cai Y, Shibazaki S, Somlo S (2006) Polycystin-2 traffics to cilia independently of polycystin-1 by using an n-terminal RVxP motif. J Cell Sci 119:1383–1395
50. Jenkins PM, Hurd TW, Zhang L, McEwen DP, Brown RL, Margolis B, Verhey KJ, Martens JR (2006) Ciliary targeting of olfactory CNG channels requires the CNGB1b subunit and the kinesin-2 motor protein, KIF17. Curr Biol 16:1211–1216
51. Dwyer ND, Adler CE, Crump JG, L'Etoile ND, Bargmann CI (2001) Polarized dendritic transport and the AP-1 mu1 clathrin adaptor UNC-101 localize odorant receptors to olfactory cilia. Neuron 31:277–287
52. Avidor-Reiss T, Maer AM, Koundakjian E, Polyanovsky A, Keil T, Subramaniam S, Zuker CS (2004) Decoding cilia function: defining specialized genes required for compartmentalized cilia biogenesis. Cell 117:527–539
53. Grayson C, Bartolini F, Chapple JP, Willison KR, Bhamidipati A, Lewis SA, Luthert PJ, Hardcastle AJ, Cowan NJ, Cheetham ME (2002) Localization in the human retina of the X-linked retinitis pigmentosa protein RP2, its homologue cofactor C and the RP2 interacting protein Arl3. Hum Mol Genet 11:3065–3074
54. Schrick JJ, Vogel P, Abuin A, Hampton B, Rice DS (2006) Adp-ribosylation factor-like 3 is involved in kidney and photoreceptor development. Am J Pathol 168:1288–1298
55. Cuvillier A, Redon F, Antoine JC, Chardin P, DeVos T, Merlin G (2000) LdARL-3A, a Leishmania promastigote-specific ADP-ribosylation factor-like protein, is essential for flagellum integrity. J Cell Sci 113(Pt 11):2065–2074
56. Li Y, Wei Q, Zhang Y, Ling K, Hu J (2010) The small GTPases ARL-13 and ARL-3 coordinate intraflagellar transport and ciliogenesis. J Cell Biol 189:1039–1051
57. Lai CK, Gupta N, Wen X, Rangell L, Chih B, Peterson AS, Bazan JF, Li L, Scales SJ (2011) Functional characterization of putative cilia genes by high-content analysis. Mol Biol Cell 22:1104–1119
58. Kim H, Xu H, Yao Q, Li W, Huang Q, Outeda P, Cebotaru V, Chiaravalli M, Boletta A, Piontek K, Germino GG, Weinman EJ, Watnick T, Qian F (2014) Ciliary membrane proteins traffic through the Golgi via a Rabep1/GGA1/Arl3-dependent mechanism. Nat Commun 5:5482
59. Schwarz N, Hardcastle AJ, Cheetham ME (2012) Arl3 and RP2 mediated assembly and traffic of membrane associated cilia proteins. Vision Res 75:2–4
60. Wright KJ, Baye LM, Olivier-Mason A, Mukhopadhyay S, Sang L, Kwong M, Wang W, Pretorius PR, Sheffield VC, Sengupta P, Slusarski DC, Jackson PK (2011) An ARL3-UNC119-RP2 GTPase cycle targets myristoylated NPHP3 to the primary cilium. Genes Dev 25:2347–2360
61. Van Valkenburgh H, Shern JF, Sharer JD, Zhu X, Kahn RA (2001) ADP-ribosylation factors (ARFs) and ARF-like 1 (ARL1) have both specific and shared effectors: characterizing ARL1-binding proteins. J Biol Chem 276:22826–22837

62. Pugacheva EN, Jablonski SA, Hartman TR, Henske EP, Golemis EA (2007) HEF1-dependent Aurora A activation induces disassembly of the primary cilium. Cell 129:1351–1363
63. Zhang H, Li S, Doan T, Rieke F, Detwiler PB, Frederick JM, Baehr W (2007) Deletion of PrBP/delta impedes transport of GRK1 and PDE6 catalytic subunits to photoreceptor outer segments. Proc Natl Acad Sci U S A 104:8857–8862
64. Ismail SA, Chen YX, Rusinova A, Chandra A, Bierbaum M, Gremer L, Triola G, Waldmann H, Bastiaens PI, Wittinghofer A (2011) Arl2-GTP and Arl3-GTP regulate a GDI-like transport system for farnesylated cargo. Nat Chem Biol 7:942–949
65. Chiang AP, Nishimura D, Searby C, Elbedour K, Carmi R, Ferguson AL, Secrist J, Braun T, Casavant T, Stone EM, Sheffield VC (2004) Comparative genomic analysis identifies an ADP-ribosylation factor-like gene as the cause of Bardet-Biedl syndrome (BBS3). Am J Hum Genet 75:475–484
66. Fan Y, Esmail MA, Ansley SJ, Blacque OE, Boroevich K, Ross AJ, Moore SJ, Badano JL, May-Simera H, Compton DS, Green JS, Lewis RA, van Haelst MM, Parfrey PS, Baillie DL, Beales PL, Katsanis N, Davidson WS, Leroux MR (2004) Mutations in a member of the Ras superfamily of small GTP-binding proteins causes Bardet-Biedl syndrome. Nat Genet 36:989–993
67. Zaghloul NA, Katsanis N (2009) Mechanistic insights into Bardet-Biedl syndrome, a model ciliopathy. J Clin Invest 119:428–437
68. Marion V, Stutzmann F, Gerard M, De Melo C, Schaefer E, Claussmann A, Helle S, Delague V, Souied E, Barrey C, Verloes A, Stoetzel C, Dollfus H (2012) Exome sequencing identifies mutations in LZTFL1, a BBSome and smoothened trafficking regulator, in a family with Bardet–Biedl syndrome with situs inversus and insertional polydactyly. J Med Genet 49:317–321
69. Otto EA, Hurd TW, Airik R, Chaki M, Zhou W, Stoetzel C, Patil SB, Levy S, Ghosh AK, Murga-Zamalloa CA, van Reeuwijk J, Letteboer SJ, Sang L, Giles RH, Liu Q, Coene KL, Estrada-Cuzcano A, Collin RW, McLaughlin HM, Held S, Kasanuki JM, Ramaswami G, Conte J, Lopez I, Washburn J, Macdonald J, Hu J, Yamashita Y, Maher ER, Guay-Woodford LM, Neumann HP, Obermuller N, Koenekoop RK, Bergmann C, Bei X, Lewis RA, Katsanis N, Lopes V, Williams DS, Lyons RH, Dang CV, Brito DA, Dias MB, Zhang X, Cavalcoli JD, Nurnberg G, Nurnberg P, Pierce EA, Jackson PK, Antignac C, Saunier S, Roepman R, Dollfus H, Khanna H, Hildebrandt F (2010) Candidate exome capture identifies mutation of SDCCAG8 as the cause of a retinal-renal ciliopathy. Nat Genet 42:840–850
70. Kobayashi T, Hori Y, Ueda N, Kajiho H, Muraoka S, Shima F, Kataoka T, Kontani K, Katada T (2009) Biochemical characterization of missense mutations in the Arf/Arl-family small GTPase Arl6 causing Bardet-Biedl syndrome. Biochem Biophys Res Commun 381:439–442
71. Wiens CJ, Tong Y, Esmail MA, Oh E, Gerdes JM, Wang J, Tempel W, Rattner JB, Katsanis N, Park HW, Leroux MR (2010) Bardet-Biedl syndrome-associated small GTPase ARL6 (BBS3) functions at or near the ciliary gate and modulates Wnt signaling. J Biol Chem 285:16218–16230
72. Yen HJ, Tayeh MK, Mullins RF, Stone EM, Sheffield VC, Slusarski DC (2006) Bardet-Biedl syndrome genes are important in retrograde intracellular trafficking and Kupffer's vesicle cilia function. Hum Mol Genet 15:667–677
73. Zhang Q, Nishimura D, Seo S, Vogel T, Morgan DA, Searby C, Bugge K, Stone EM, Rahmouni K, Sheffield VC (2011) Bardet-Biedl syndrome 3 (Bbs3) knockout mouse model reveals common BBS-associated phenotypes and Bbs3 unique phenotypes. Proc Natl Acad Sci U S A 108:20678–20683
74. Jin H, White SR, Shida T, Schulz S, Aguiar M, Gygi SP, Bazan JF, Nachury MV (2010) The conserved Bardet-Biedl syndrome proteins assemble a coat that traffics membrane proteins to cilia. Cell 141:1208–1219
75. Blacque OE, Reardon MJ, Li C, McCarthy J, Mahjoub MR, Ansley SJ, Badano JL, Mah AK, Beales PL, Davidson WS, Johnsen RC, Audeh M, Plasterk RH, Baillie DL, Katsanis N, Quarmby LM, Wicks SR, Leroux MR (2004) Loss of *C. elegans* bbs-7 and bbs-8 protein

function results in cilia defects and compromised intraflagellar transport. Genes Dev 18:1630–1642
76. Cantagrel V, Silhavy JL, Bielas SL, Swistun D, Marsh SE, Bertrand JY, Audollent S, Attie-Bitach T, Holden KR, Dobyns WB, Traver D, Al-Gazali L, Ali BR, Lindner TH, Caspary T, Otto EA, Hildebrandt F, Glass IA, Logan CV, Johnson CA, Bennett C, Brancati F, Valente EM, Woods CG, Gleeson JG (2008) Mutations in the cilia gene ARL13B lead to the classical form of Joubert syndrome. Am J Hum Genet 83:170–179
77. Parisi MA (2009) Clinical and molecular features of Joubert syndrome and related disorders. Am J Med Genet C Semin Med Genet 151C:326–340
78. Thomas S, Cantagrel V, Mariani L, Serre V, Lee JE, Elkhartoufi N, de Lonlay P, Desguerre I, Munnich A, Boddaert N, Lyonnet S, Vekemans M, Lisgo SN, Caspary T, Gleeson J, Attie-Bitach T (2015) Identification of a novel ARL13B variant in a Joubert syndrome-affected patient with retinal impairment and obesity. Eur J Hum Genet 23(5):621–627
79. Sun Z, Amsterdam A, Pazour GJ, Cole DG, Miller MS, Hopkins N (2004) A genetic screen in zebrafish identifies cilia genes as a principal cause of cystic kidney. Development 131:4085–4093
80. Duldulao NA, Lee S, Sun Z (2009) Cilia localization is essential for in vivo functions of the Joubert syndrome protein ARL13B/scorpion. Development 136:4033–4042
81. Cevik S, Hori Y, Kaplan OI, Kida K, Toivenon T, Foley-Fisher C, Cottell D, Katada T, Kontani K, Blacque OE (2010) Joubert syndrome ARL13B functions at ciliary membranes and stabilizes protein transport in *Caenorhabditis elegans*. J Cell Biol 188:953–969
82. Miertzschke M, Koerner C, Spoerner M, Wittinghofer A (2014) Structural insights into the small g-protein ARL13B and implications for Joubert syndrome. Biochem J 457:301–311
83. Caspary T, Larkins CE, Anderson KV (2007) The graded response to sonic hedgehog depends on cilia architecture. Dev Cell 12:767–778
84. Horner VL, Caspary T (2011) Disrupted dorsal neural tube BMP signaling in the cilia mutant Arl13b hnn stems from abnormal Shh signaling. Dev Biol 355:43–54
85. Larkins CE, Aviles GD, East MP, Kahn RA, Caspary T (2011) Arl13b regulates ciliogenesis and the dynamic localization of Shh signaling proteins. Mol Biol Cell 22:4694–4703
86. Li Y, Zhang Q, Wei Q, Zhang Y, Ling K, Hu J (2012) Sumoylation of the small GTPase Arl-13 promotes ciliary targeting of sensory receptors. J Cell Biol 199:589–598
87. Humbert MC, Weihbrecht K, Searby CC, Li Y, Pope RM, Sheffield VC, Seo S (2012) ARL13B, PDE6D, and CEP164 form a functional network for INPP5E ciliary targeting. Proc Natl Acad Sci U S A 109:19691–19696
88. Stenmark H, Olkkonen VM (2001) The Rab GTPase family. Genome Biol 2:REVIEWS3007
89. Moritz OL, Tam BM, Hurd LL, Peranen J, Deretic D, Papermaster DS (2001) Mutant Rab8 impairs docking and fusion of Rhodopsin-bearing post-Golgi membranes and causes cell death of transgenic Xenopus rods. Mol Biol Cell 12:2341–2351
90. Follit JA, Li L, Vucica Y, Pazour GJ (2010) The cytoplasmic tail of fibrocystin contains a ciliary targeting sequence. J Cell Biol 188:21–28
91. Boehlke C, Bashkurov M, Buescher A, Krick T, John AK, Nitschke R, Walz G, Kuehn EW (2010) Differential role of Rab proteins in ciliary trafficking: Rab23 regulates smoothened levels. J Cell Sci 123:1460–1467
92. Nachury MV, Loktev AV, Zhang Q, Westlake CJ, Peranen J, Merdes A, Slusarski DC, Scheller RH, Bazan JF, Sheffield VC, Jackson PK (2007) A core complex of bbs proteins cooperates with the GTPase Rab8 to promote ciliary membrane biogenesis. Cell 129:1201–1213
93. Omori Y, Zhao C, Saras A, Mukhopadhyay S, Kim W, Furukawa T, Sengupta P, Veraksa A, Malicki J (2008) Elipsa is an early determinant of ciliogenesis that links the IFT particle to membrane-associated small GTPase Rab8. Nat Cell Biol 10:437–444
94. Knodler A, Feng S, Zhang J, Zhang X, Das A, Peranen J, Guo W (2010) Coordination of Rab8 and Rab11 in primary ciliogenesis. Proc Natl Acad Sci U S A 107:6346–6351
95. Franco I, Gulluni F, Campa CC, Costa C, Margaria JP, Ciraolo E, Martini M, Monteyne D, De Luca E, Germena G, Posor Y, Maffucci T, Marengo S, Haucke V, Falasca M, Perez-Morga

D, Boletta A, Merlo GR, Hirsch E (2014) PI3K class II α controls spatially restricted endosomal PtdIns3P and Rab11 activation to promote primary cilium function. Dev Cell 28:647–658

96. Kobayashi T, Kim S, Lin YC, Inoue T, Dynlacht BD (2014) The CP110-interacting proteins Talpid3 and Cep290 play overlapping and distinct roles in cilia assembly. J Cell Biol 204:215–229
97. Babbey CM, Bacallao RL, Dunn KW (2010) Rab10 associates with primary cilia and the exocyst complex in renal epithelial cells. Am J Physiol Renal Physiol 299:F495–F506
98. Yoshimura S, Egerer J, Fuchs E, Haas AK, Barr FA (2007) Functional dissection of Rab GTPases involved in primary cilium formation. J Cell Biol 178:363–369
99. Eggenschwiler JT, Bulgakov OV, Qin J, Li T, Anderson KV (2006) Mouse Rab23 regulates hedgehog signaling from smoothened to Gli proteins. Dev Biol 290:1–12
100. Schafer JC, Winkelbauer ME, Williams CL, Haycraft CJ, Desmond RA, Yoder BK (2006) IFTA-2 is a conserved cilia protein involved in pathways regulating longevity and Dauer formation in *Caenorhabditis elegans*. J Cell Sci 119:4088–4100
101. Qin H, Wang Z, Diener D, Rosenbaum J (2007) Intraflagellar transport protein 27 is a small g protein involved in cell-cycle control. Curr Biol 17:193–202
102. Huet D, Blisnick T, Perrot S, Bastin P (2014) The GTPase IFT27 is involved in both anterograde and retrograde intraflagellar transport. Elife 3:e02419
103. Eguether T, San Agustin JT, Keady BT, Jonassen JA, Liang Y, Francis R, Tobita K, Johnson CA, Abdelhamed ZA, Lo CW, Pazour GJ (2014) IFT27 links the BBSome to IFT for maintenance of the ciliary signaling compartment. Dev Cell 31:279–290
104. Liew GM, Ye F, Nager AR, Murphy JP, Lee JS, Aguiar M, Breslow DK, Gygi SP, Nachury MV (2014) The intraflagellar transport protein IFT27 promotes BBSome exit from cilia through the GTPase ARL6/BBS3. Dev Cell 31:265–278
105. Aldahmesh MA, Li Y, Alhashem A, Anazi S, Alkuraya H, Hashem M, Awaji AA, Sogaty S, Alkharashi A, Alzahrani S, Al Hazzaa SA, Xiong Y, Kong S, Sun Z, Alkuraya FS (2014) IFT27, encoding a small GTPase component of IFT particles, is mutated in a consanguineous family with Bardet-Biedl syndrome. Hum Mol Genet 23:3307–3315
106. Hurd TW, Fan S, Margolis BL (2011) Localization of retinitis pigmentosa 2 to cilia is regulated by importin beta2. J Cell Sci 124:718–726
107. Fan S, Whiteman EL, Hurd TW, McIntyre JC, Dishinger JF, Liu CJ, Martens JR, Verhey KJ, Sajjan U, Margolis B (2011) Induction of Ran Gtp drives ciliogenesis. Mol Biol Cell 22:4539–4548
108. Dishinger JF, Kee HL, Jenkins PM, Fan S, Hurd TW, Hammond JW, Truong YN, Margolis B, Martens JR, Verhey KJ (2010) Ciliary entry of the kinesin-2 motor KIF17 is regulated by importin-beta2 and RanGTP. Nat Cell Biol 12:703–710
109. Satir P, Christensen ST (2007) Overview of structure and function of mammalian cilia. Annu Rev Physiol 69:377–400
110. Obado SO, Rout MP (2012) Ciliary and nuclear transport: different places, similar routes? Dev Cell 22:693–694
111. Essner JJ, Vogan KJ, Wagner MK, Tabin CJ, Yost HJ, Brueckner M (2002) Conserved function for embryonic nodal cilia. Nature 418:37–38
112. Barr MM (2005) Caenorhabditis elegans as a model to study renal development and disease: sexy cilia. J Am Soc Nephrol 16:305–312
113. Inglis PN, Ou G, Leroux MR, Scholey JM (2006) The sensory cilia of Caenorhabditis elegans. In: WormBook (ed) The C. elegans research community. WormBook. doi:10.1895/wormbook.1.126.1. http://www.wormbook.org
114. Drummond I (2009) Studying cilia in zebrafish. Methods Cell Biol 93:197–217

Part IV
Focus on Intermediate Filaments

Chapter 11
Desmin Plays Dual Structural and Regulatory Functions Through Its Interaction with Partners in Muscle

Zhenlin Li, Ara Parlakian, Jacqueline Gao-Li, Mathias Mericskay, and Onnik Agbulut

Abbreviations

AAV	Adeno-associated virus
AGE	Advanced glycation end products
AKAP	A-kinase anchoring protein
ART	ADP-ribosyl transferase
CDK1	Cyclin-dependent kinase 1
DRM	Desmin-related myopathy
EBS-MD	Epidermolysis bullosa simplex-muscular dystrophy
EDL	Extensor digitorum longus
ER	Endoplasmic reticulum
GGA	Geranylgeranylacetone
GSK3	Glycogen synthase kinase 3
HSP	Heat shock protein
I/R	Ischemia–reperfusion
IF	Intermediate filament
JNK	c-Jun N-terminal kinases
KO	Knock-out
LCR	Locus control region
MAPK	Mitogen-activated protein kinase
MFM	Myofibrillar myopathy
MHC	Myosin heavy chain
OGA	O-GlcNAcase
O-GlcNAc	O-linked β-*N*-acetylglucosamine

Z. Li (✉) • A. Parlakian • J. Gao-Li • M. Mericskay • O. Agbulut
Sorbonne Universités, UPMC Univ Paris 06, Institut de Biologie Paris-Seine (IBPS), UMR CNRS 8256/INSERM ERL U1164, Biological Adaptation and Ageing, 75005 Paris, France
e-mail: zhenlin.li@upmc.fr

H. Schatten (ed.), *The Cytoskeleton in Health and Disease*,
DOI 10.1007/978-1-4939-2904-7_11

OGT	O-GlcNAc transferase
OPN	Osteopontin
PAK	p21-activated kinase
PKA	Protein kinase A
PKC	Protein kinase C
PKG	Protein kinase G
PTM	Post-translational modification
SMC	Smooth muscle cell
SPRY	SPIa/ryanodine receptor
SRF	Serum response factor
TUNEL	Terminal deoxynucleotidyl transferase dUTP nick end labeling
UPS	Ubiquitin-proteasome system
XLCNM	X-linked centronuclear myopathy

Desmin was first termed as such by Lazarides and Hubbard in 1976 [1] for its linking role in muscle (from the Greek delta epsilon sigma mu os = link, bond). It is a major component of intermediate filaments (IF) in cardiac, skeletal and smooth muscles. Other IF subunits such as vimentin, synemin, nestin, syncolin, cytokeratins and lamins are also present either during muscle development and/or in the maintenance of mature muscle tissues. Desmin is also present in other types of cells, such as liver stellate cells [2], vascular pericytes [3], cardiac purkinje fibers [4], sertoli cells [5]. Desmin is used as a marker for identification of tumor origin [6, 7], for proteinuria in early stages of membranous nephropathy in elderly patients [8] and podocyte injury [9]. Desmin is a potential circulating biomarker of human heart failure [10]. Here we review the results concerning the regulation and function of desmin gene and its involvement in human diseases.

Regulation of the Desmin Gene

Desmin is encoded by a single gene in mammals whereas two copies of genes are present in some non-mammalian vertebrates such as Xenopus and zebrafish. Desmin, one of the earliest muscle-specific proteins to appear during myogenesis, is present in the replicating myoblasts [11]. The desmin gene is expressed as early as embryonic day 7.5 (E7.5) in the precardiac area and on day 9 in the myotomes and the smooth muscle cells of the mouse embryo. It has been suggested that the level of expression of the desmin gene and the accumulation of desmin filaments during muscle differentiation is regulated at the transcriptional and/or posttranscriptional level but not at the translational level [12]. The regulation of human, mouse, and hamster desmin genes have been investigated [13–18]. Some in vitro experiments have shown that the promoter region upstream of the transcription initiation site of the desmin gene is sufficient to confer low-level, muscle-specific gene expression. A negative region is located between −693 and −228 bp in the human gene. Active expression of the human and mouse desmin gene depends on a 280 bp muscle-specific enhancer located at −1 kb. This skeletal enhancer contains two regions, a myoblast-specific region containing the Sp1, Krox, and Mb sites that functions only in myoblasts and a myotube-specific region containing MyoD1, MEF2, and a Mt site responsible for gene expression in myotubes [13, 19]. The results from transgenic mice have shown that the 1.2-kb upstream region of the human desmin gene containing muscle-specific enhancer directs the expression of the reporter gene LacZ in the myotome of the mouse embryo at E9.0 and in the skeletal muscle during embryonic and postnatal development, but not in adult muscle [20]. The MEF2 binding site of the mouse skeletal enhancer is also active in the heart, although only in the right ventricle in transgenic mice [21]. The 4-kb upstream region of the mouse desmin promoter directs expression of a LacZ gene throughout the heart from E7.5, and in skeletal muscle and vascular smooth muscle cells from E9.5. The distal fragment (−4005/−2495) harboring a CArG/octamer overlapping element that can bind both the serum response factor (SRF) and an Oct-like factor is active in arterial smooth muscle cells, but not in venous smooth muscle cells or in the heart in vivo. LacZ expression is abolished when mutations are introduced into this CArG element that prevents the binding of SRF and/or the Oct-like factor. These data suggest that SRF and the Oct-like factor may cooperate to drive its specific expression in arterial smooth muscle cells [22]. A remarkable progress in the study of desmin gene regulation came from the identification of a muscle-specific locus control region (LCR) of desmin gene [23]. LCRs are tissue-specific transcription regulators that can generate dominantly a transcriptionally active chromatin structure and confer highly reproducible, physiological transgene expression. The human desmin LCR consists of five regions of muscle-specific DNase I hypersensitivity (HS) localized between −9 and −18 kb 5′ of desmin that are highly conserved between humans and other mammals. All transgenic mice containing 220-kb 5′-flanking and 10-kb 3′-flanking sequences of the human desmin gene direct the complete pattern of tissue-specific gene expression in all muscle cell types and reproduce the human development expression profile [24].

The first epigenetic investigation of the muscle-specific LCR of desmin gene has been performed on 500 kb of human chromosome 2q35 [25]. They have shown that the desmin gene cluster lies within a hyperacetylated transcriptionally competent muscle specific gene domain and that H3K4me2 and H3K4me3 mark the LCR. In addition, H3K27me3 marks the CpG island of desmin only in non-muscle cells. As H3K27me3 correlates with inactive genes, this methylation could contribute to the silencing of desmin gene in non-muscle tissues. It has been shown that the histone acetylation in myoblasts has been established at an earlier time point before myoblast differentiation. Similar histone acetylation has been observed in myotubes, suggesting that the increase of desmin expression in myotubes could result from an enhancement in transcription factor binding activity rather than a change in overall chromatin architecture. The only significant difference between myoblast and myotube modification patterns is that myotubes have 20–30-fold higher H3K4me2 and H3K4me3 histone methylation at desmin 3′ of the first exon than the myoblasts.

In conclusion, as shown in Fig. 11.1, the desmin gene is controlled by a combination of different transcription control regions in muscle tissues. In order to obtain a physiological tissue-specific concentration of desmin, a muscle-specific LCR is required which cooperates with different regions and associates with epigenetic regulation.

Consequences of the Absence of Desmin in Mice

Two independent desmin knock-out (KO) mice were obtained in 1996 by using exon 1 interruption strategy [26, 27]. The desmin KO mice are viable and fertile. The muscle development in the embryo is not affected by the absence of desmin. No anatomical or behavioral defects are notable at birth. However, after birth, mice lacking desmin show defects in all types of muscle (Fig. 11.2). They suffer from a cardiomyopathy, skeletal myopathy, and smooth muscle dysfunction. These disorders reduce the life span of the desmin KO mice. We summarize here the defects in the different muscle tissues.

Loss of Desmin Leads to a Dilated Cardiomyopathy

Most severe defects are observed in the heart of the desmin KO mice, correlating with the highest expression of desmin in the cardiomyocytes. Loss of desmin leads to cardiomyocyte hypertrophy and a dilated cardiomyopathy characterized by extensive cardiomyocyte death, calcification, multiple ultrastructural alterations, cardiac contraction and conduction disturbances [26–31]. Expression of a wild-type desmin specifically within cardiomyocyte of desmin KO mice indicates that defects in the desmin KO heart are due to an intrinsic cardiomyocytes defect rather than compromised coronary circulation [32].

Degenerating cardiomyocytes with Z-disk streaming, hypercontraction, and disorganized myofibrils were observed as early as 5 days post-partum in the hearts of desmin KO mice. Electron microscopy analysis indicated that the sarcotubular system is often dilated and intercalated disks, the contact regions between neighboring

Fig. 11.1 (continued) expression in the mouse embryo from E7.5 (**b**), E8.5 (**c**) to E9.5 (**d**). LacZ expression is evident at E7.5 in the cardiogenic progenitors and at E9.5 in the myotomes and the vessels of the embryo. (**e**) Expression of human desmin 1.2 kb-nlacZ transgene in E12.5 mouse embryo is observed only in skeletal muscle. (**f**) Skeletal and cardiac expression of LacZ in the E12.5 embryo in which Lacz is inserted into exon 1 of desmin gene. (**g**) Faint staining was used to see external vessels of E15.5 embryo of mouse desmin 4 kb-LacZ transgenic mice. Expression of LacZ in vessels is indicated by *arrowheads*. *Abbreviations*: *ao* aorta, *bc* bulbus cordis, *cv* common ventricle, *da* dorsal aortae, *dm* dermamyotomal bud, *hl* hindlimbs, *h* heart, *pm* precardiogenic mesoderm, *so* somite, *ua* umbilical artery, *va* vitelline artery, *v* vessels

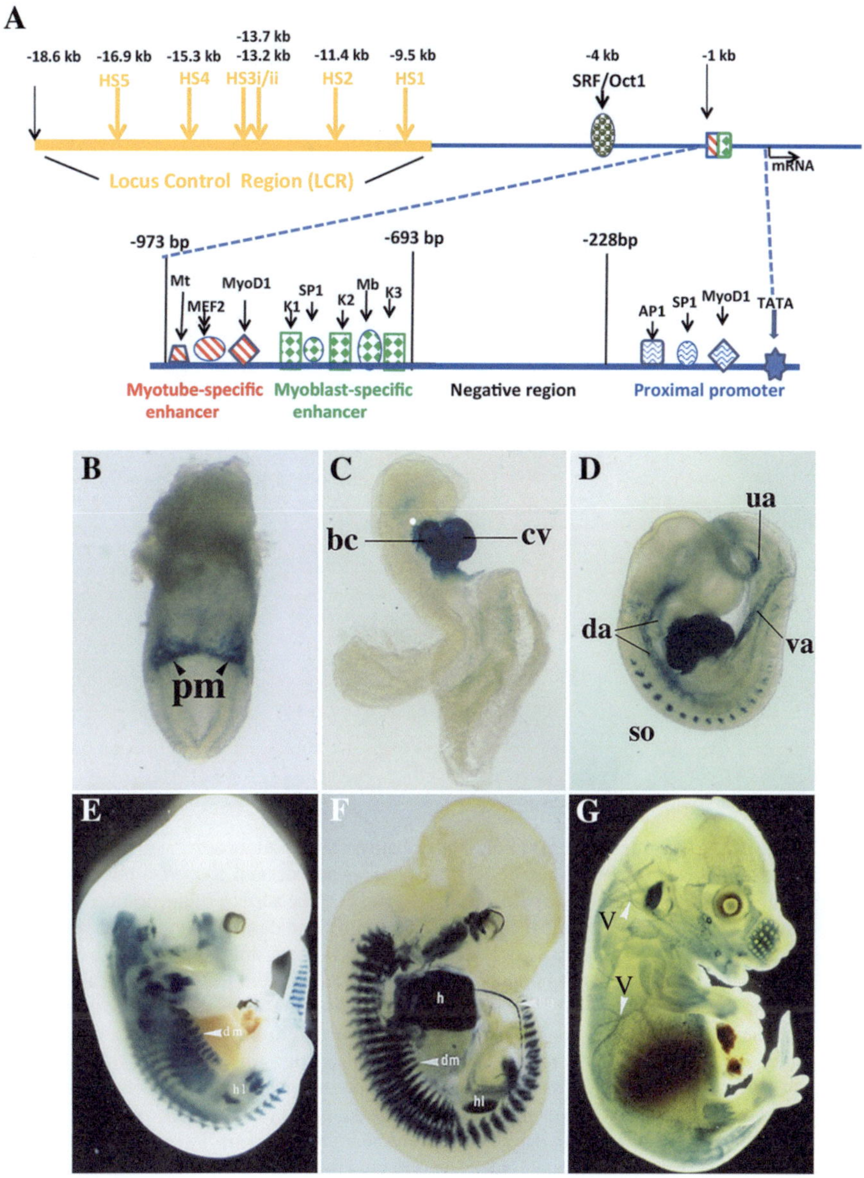

Fig. 11.1 Regulation of the desmin gene. (**a**) Schematic representation of the desmin promoter. The proximal promoter which contains one MyoD1, one SP1 and one AP1 binding sites and gives a low-level muscle-specific expression, is followed by a negative region. High-level expression depends on a muscle-specific enhancer which contains a myotube-specific enhancer harboring one MyoD1, one MEF2 and one Mt sites and a myoblast-specific enhancer having three Krox (K1–K3), one Mb and one SP1 binding sites. A SRF/Oct1-like site around −4 kb in mouse desmin promoter is required for the expression of desmin in arteries. The locus control region (LCR) is represented by *yellow line*. The positions of five DNase I hypersensitivity (HS) sites (HS1–5) are indicated by *vertical arrows*. (**b–d**) Developmental patterns of mouse desmin 4 kb-LacZ

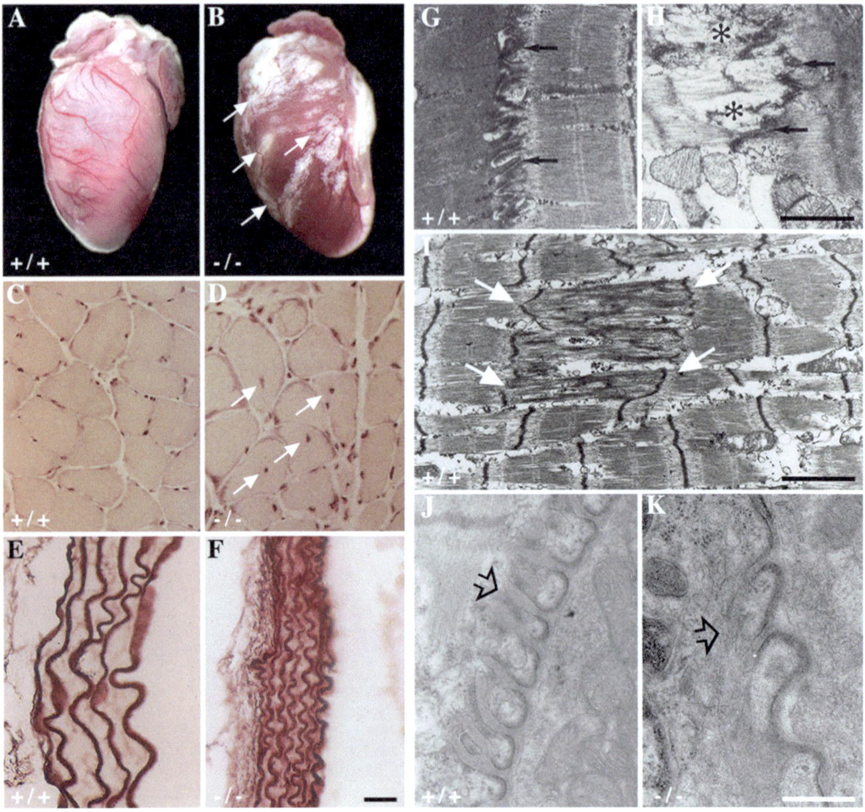

Fig. 11.2 Mice lacking desmin lead to cardiomyopathy, skeletal myopathy, and smooth muscle defects. (**a** and **b**) Hearts from 2-month old control (+/+) and desmin KO (–/–) mice. Note the presence of calcification region (*white regions*, see *arrows*) at the surface of the desmin KO heart. (**c** and **d**) Skeletal muscle sections from 2-month old control and desmin KO mice are stained by Hematoxylin and Eosin. Note the presence of central nuclei in the desmin KO muscle (see *arrows*). (**e** and **f**) Sections of aorta from 3-month old control (+/+) and desmin KO (–/–) mice are stained with orcein and hematoxylin. Note the distances between the layers are reduced for the desmin KO mice compared to the control. (**g**) Intercalated disk (*arrows*) with a typical zig-zag pattern in a heart of control (+/+) mice. (**h**) Disorganization of intercalated disk (*arrows*) and myofibrils (*) in the free right wall of desmin KO mice. (**i**) Disorganization of Z-disk and myofibrils in the soleus muscle of adult desmin KO mice (*arrows*). (**j**) The neuromuscular junctions (*arrows*) with regular folded form the control (+/+) mice. (**k**) The neuromuscular junctions (*arrows*) are markedly disorganized with abnormal postjunctional folds were observed in desmin KO mice. Bars: 2 mm (**a** and **b**); 25 μm (**c**–**f**); 2 μm (**g**–**i**); 0.5 μm (**j** and **k**)

cardiomyocytes, is altered. The configuration of many intercalated disks in the desmin KO hearts is abnormal, with slender, elongated contact regions accompanied with a severely altered sarcolemmal membrane including few intermyofibrillar regions, fasciae adherens, and few desmosomes. Calcified lesions appeared on the outer surface of the heart 2–3 weeks after birth. Extensive fibrosis was observed near the calcified region accompanied with up-regulation of osteopontin (OPN), decorin, transforming growth factor-β1, collagen and angiotensin-converting enzyme [33].

The mitochondrial defects can be detected very early in development of desmin KO mice. The position, distribution morphology, proliferation, and function of mitochondria are all abnormal in mice lacking desmin [27, 31, 34]. It has been observed that the mitochondria are swollen and accumulate abnormally in some area, and there is extensive proliferation of mitochondria in a significant fraction of the myocytes, particularly after work overload. The rate of mitochondrial respiration in situ, measured using saponin-skinned muscle fibers, is significantly altered in desmin KO mice [34, 35]. Conventional microtubule (MT)-dependent molecular motor kinesin which is associated with the outer mitochondrial membrane remains bound to the purified organelle in vitro. However, desmin KO hearts lose this association both in vitro and in vivo [36]. The fact that kinesin is associated with mitochondria in the normal mouse heart raises the possibility that the distribution of mitochondria in muscle cells may be more dynamic than previously thought. The positioning of mitochondria in restricted subcellular domains in the heart is a dynamic event, involving the IF desmin, molecular motor kinesin, and very probably the MT network. The creatine kinase activity of the mitochondria in desmin KO hearts is three times greater than that of the wild-type mice heart. It has been suggested that the disruption of Z-lines and myofibrils that occurs in the desmin KO heart may result in the lack of a kinesin- and MT-dependent association of mitochondria with the myofibrils, and an insufficient supply of the ATP required for the contraction of myofibrils. The increased mitochondrial creatine kinase activity in the desmin KO heart may also be a way of compensating for the lack of mitochondria close to the contractile process. But desmin KO hearts have significantly less cytochrome c and the Bcl-2 is relocated. Thus, the relocation of mitochondria away from sites where large amounts of ATP are used could lead to the generation of preapoptotic signals (release of cytochrome c and translocation of Bcl-2) [36], driving cardiomyocyte degeneration and calcinosis of the heart in desmin KO mice. Alternatively, this cardiomyocyte degeneration could be due to cell membrane damage leading to the leakage of cell components from the cell and an influx of Ca^{2+} into the cell. The increased osmolarity inside the cells might, as a result of swelling, give rise to the rounded appearance of the mitochondria. The dense bodies seen in some of the mitochondria in young animals are probably due to increased calcium loading of the mitochondria, and would thus be the first sign of calcification. Analysis of heart mitochondrial proteome by 2-D electrophoresis associated with mass spectrometry has identified a series of proteins that are different between desmin KO and wild-type mice heart. These proteins are involved in different pathways linked or not to mitochondrial function. The most significant changes were observed in ketone body and acetate metabolism, NADH shuttle proteins, amino-acid metabolism proteins and respiratory enzymes. Several of these changes are consistent with the known phenotype of desmin deficiency [37]. All these results indicate that the desmin network influences the position, movement, and respiratory activity of mitochondria in cardiac muscle.

Mice Lacking Desmin Develop a Progressive Skeletal Myopathy

Mice lacking desmin develop a skeletal myopathy [26, 27, 38]. Alterations begin to appear after birth, particularly in weight-bearing muscles such as the soleus and continually used muscles such as the diaphragm. The organization of myofibers is irregular, with misaligned myofibrils, Z-disk streaming, focal degeneration and disorganization of the mitochondria [26, 27, 38]. Most of the costameres are lost in the tibialis anterior and extensor digitorum longus (EDL), but not quadriceps and gastrocnemius muscles of desmin KO mice. This may explain partially why some muscles are more susceptible to damage than others. The reason for this difference is not clear, but it is possible to link to the differences in the costamere's components and organization. Surprisingly, some cytokeratins contribute to the structure of the costameres in skeletal muscle [39]. The absence of desmin delays and alters muscle regeneration, and causes adipocytes accumulation [40]. This myofibrillogenesis during regeneration is often abortive and shows signs of disorganization [38, 40, 41]. The cycles of degeneration and regeneration result in increased relative amounts of slow myosin heavy chain (MHC) and a relative decrease in fast MHC in the desmin KO mice. The neuromuscular junctions are markedly disorganized with abnormal postjunctional folded forms [40]. Lack of desmin resulted in changes in the subcellular distribution of synemin, but not of plectin and nestin [42, 43].

The lack of desmin adversely affects the ability of mice to engage in both chronic and acute bouts of endurance running exercise [44]. Desmin knockout muscles generate lower stress and are less vulnerable to injury compared with wild-type muscles [45]. Skinned muscle fibers from desmin KO mice develop also less active force, as do intact soleus muscles [46]. During high-frequency fatigue, tension of isolated desmin KO EDL muscles declined faster than wild-type muscle, and this deficiency was offset by adding caffeine that stimulates the release of Ca^{2+} from the sarcoplasmic reticulum [47].

It has been proposed that desmin may play a role in regulating the optimal arrangement of sarcomeres and sarcomere number regulation within the muscle [48]. Wider filament spacing and a slight increase in fiber volume in muscles of desmin KO mice are observed [49, 50]. With a higher sarcomeric length (more than 2.90 μm), skeletal muscles from desmin KO mice display an increased myofibrillar mobility [51]. The existence of a limit on the extension of desmin suggests a mechanism for the recruitment of desmin into a network of force transmission. Connectivity in wild-type fibers was significantly greater when compared to desmin-null fibers, demonstrating a requirement of desmin in functional connection between myofibrils [50].

In contrast to the finding that young and adult desmin KO fibers are more compliant or equally compliant as wild-type [52], young and adult desmin KO fiber bundles have higher stress and are stiffer compared with wild-type fiber bundles [53]. Additionally, desmin KO fiber bundles exhibited increasing modulus with age, a trend that was absent in wild-type bundles [53]. Stretched skeletal muscles of desmin KO mice develop greater passive stiffness than control muscles [54–56].

This increase in passive stiffness is not due to changes of titin modification, but to the absence of desmin and a possible change in extraconnective tissue such as increased collagen.

During chronic stimulation of denervated muscles, desmin plays a role in muscle nuclei alignment within muscle [57]. Linkage of desmin intermediate filaments to muscle nuclei was strongly suggested based on extensive loss of nuclei positioning in the absence of desmin during passive fiber loading [48] and the results from the double desmin and nesprin 1 KO mice [58]. Nesprin 1 is another protein thought to link the nucleus to the cytoskeletal network.

Smooth Muscle Defects in Desmin KO Mice

Vimentin and desmin are major constituents of the IF networks in smooth muscle cells (SMCs). Another IF component, synemin, is also present in SMCs [59]. SMCs are able to adjust their contraction/relaxation status by reorganizing the actin cytoskeleton and the IF network in response to external stimulation (for reviews see [60]). SMCs from the respiratory, digestive, and urogenital systems contain both desmin and vimentin. Higher concentration of vimentin is present in elastic arteries such as aorta whereas a gradient in desmin expression in the arterial tree is observed; the desmin content increased from the elastic artery aorta, via the muscular mesenteric artery to the resistance-sized mesenteric microarteries [61].

The results from the study of desmin KO mice demonstrated that desmin is essential to maintain proper viscoelastic properties, structure and mechanical strength of the vascular wall [62]. In fact, the distensibility is lower and the viscosity of the arterial wall is increased in the desmin KO mice. The vascular wall of desmin KO mice had less mechanical strength as evidenced by lower intraluminal pressure required to rupture the carotid artery wall in vitro in desmin KO mice. The perimeter of finger-like cell projections was smaller in desmin KO mice, indicating that the cells lost some of their connections to the extracellular matrix. In the bladder hypertrophy model induced by a partial obstruction of the urethra, it has been shown that desmin muscle is not needed for growth of SMCs, but has a role in the maintenance of wall structure [63].

There was no difference in the passive or active circumference/stress relationships in the aorta and mesenteric arteries between desmin KO and wild-type mice, which contain more vimentin and relatively little desmin. However, the microarteries of the desmin KO mice, containing high-level of desmin, developed lower passive and active stress [61]. Similar results were obtained with visceral smooth muscle like the vas deferens and urinary bladder, which paralleled a relative increase in the quantity of the basic, essential myosin light chain [64]. During bladder hypertrophy, desmin KO bladders had slightly lower passive stress and significantly lower active stress compared to wild-type mice [63].

The large compliance arteries use their elastic properties to absorb the energy produced by the ejection of blood by the heart at each systole, while resistance arteries use their muscular tone and endothelial relaxing capacity to regulate the local blood flow in organs. Both the dilation and contraction of the vascular smooth muscle is decreased in resistance arteries but not in large, compliance arteries [65]. Agonist-induced muscular tone is reduced in the resistance mesenteric arteries. Flow (shear stress), acetylcholine-induced endothelium-dependent dilation, and endothelium-independent dilation are all lower in resistance mesenteric arteries. In addition, the changes in blood flow cause an exaggerated structural adaptation in the resistance arteries of mice lacking desmin. The overadaptation might be due to a defect in the capacity of the cytoskeleton to rearrange after stimulation [66]. The fact that the concentration of desmin is higher in small arteries may explain why a lack of desmin affects mainly the contractility of resistance arteries.

Desmin-Related Myopathy

Desmin-related myopathy (DRM) (also called: Desminopathies, desmin storage myopathy, desmin myopathy) is a clinically heterogeneous group of disorders encompassing myopathies, cardiomyopathies, conduction disease, respiratory deficiency, smooth muscle disorders and combinations of these disorders. DRM is one type of myofibrillar myopathy (MFM) which is characterized by the loss of desmin filament network and abnormal aggregates of desmin form in muscle fibers (for reviews see [67–72]). Publication of desmin KO mice in 1996 promoted the screening for desmin mutations in myopathies. The first description of desmin mutation causing myopathy was in 1998 by Goldfarb [73]. Up to today, more than 60 hereditary or sporadic nucleotide mutations or deletions/insertions of desmin gene linked to DRM are reported (Fig. 11.3) [73–115]. Eight mutations are found in the head domain of desmin gene, seven in the coil 1A, seven in the coil 1B, thirty in the coil 2 region, seventeen in the tail domain. In these desmin myopathies, the age of clinical symptoms onset is very variable and span from 3-year-old to late adulthood, with high frequency between 20- and 40-year-old. Some mutations result in only cardiomyopathy, and other skeletal muscle defect, but the majority of mutations lead to both skeletal and cardiac phenotypes. The smooth muscle defects such as dysfunction of intestine or bladder are observed in some DRM patients. The majority of hereditary desmin mutations are autosomal dominant, few mutations follow autosomal recessive mode, and few others are present as sporadic forms. The most frequent mutations are missense mutations leading to the change of amino acids. For example, 13 residues in the coil 2 (274, 335, 337, 345, 350, 355, 357, 360, 370, 377, 385, 389, 392) are replaced by proline which is a potent helix breaker and creates a kink in the rod domain. Several splice site mutations leading to the loss of exon 3 (Asp214_Glu245 del) [83, 92, 116] or leading to the insertion of 16 amino acids (AQPGLGQAECAMDPVT) within the tail domain of desmin, at the amino acid residue 428 [78], small in-frame deletion of 1, 3 or 7 amino acids, and frame shift mutation producing truncated desmin have been found in several families.

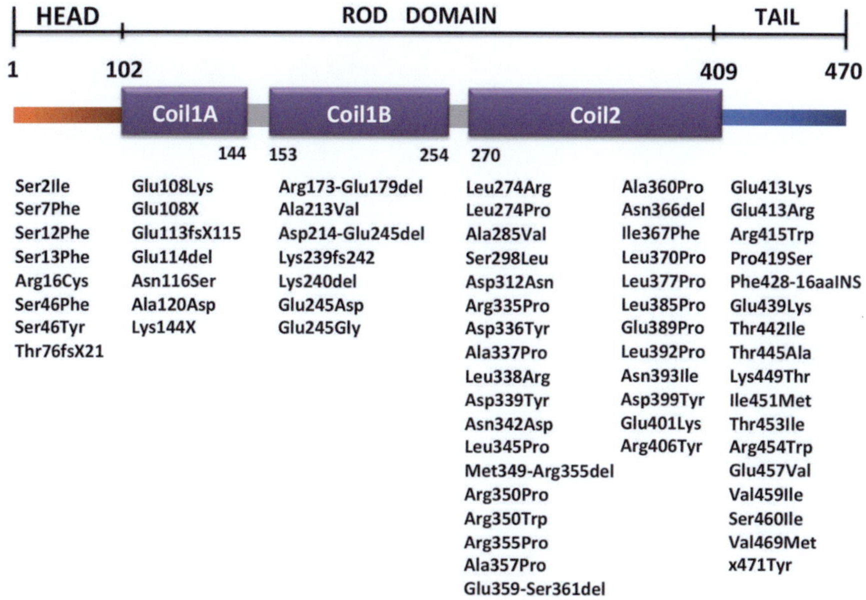

Fig. 11.3 Desmin mutations in desminopathies. Desmin is composed of head domain, rod domain and tail domain. The structure of desmin protein is schematized by colored bars for head (*orange*) and tail (*blue*) domains and boxes for the rod domain. The number indicates the position of amino acid at the domain borders. Desmin mutations are indicated under the corresponding regions of desmin (head and tail domains, coil1A, coil1B and coil2 regions). More desmin mutations are found in Coil2 and tail domain

The pathogenic effects of mutated desmin have been investigated using cell culture, in vitro filament assembly experiments and transgenic mouse models. In vitro assembly studies with wild-type and mutant desmin proteins revealed that the in vitro assembly process of mutated desmin is already disturbed at the unit length filament level and that further association reactions generate huge, tightly packed protein aggregates [117–119]. Transfection of mutated desmin cDNA into cells with or without endogenous IF filaments revealed that the majority of desmin mutants cannot form correct filaments, segregates and/or disrupts the endogenous vimentin-containing or desmin-containing network, and forms the aggregates (Fig. 11.4). Loss of IF network and/or presence of aggregates in cells have been considered as a major reason of the pathological development. These mutations can interfere with the position and movement of cellular organelles such as mitochondria, lead to the dysfunction of protein quality control such as the ubiquitin-proteasome system (UPS) and autophagy, and influence the cytoskeleton organization. It should be noted that some desmin mutations do not disrupt the IF filament formation (Fig. 11.4). It is assumed that these mutations disrupt also the interaction of desmin with its protein partners. Some amino acid change such as

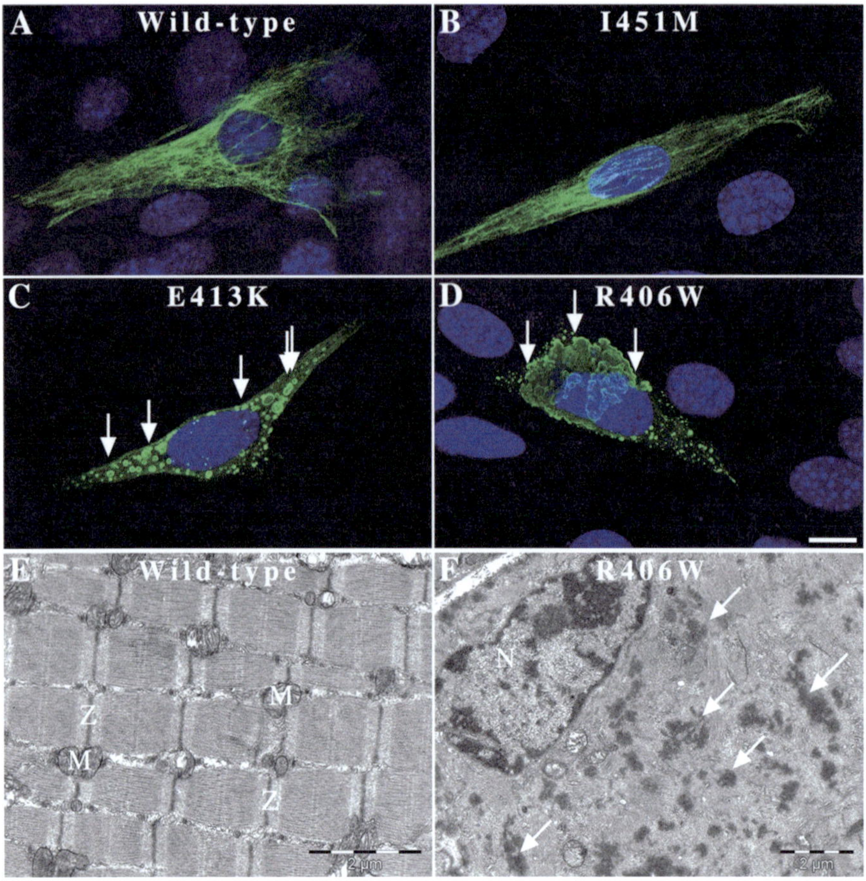

Fig. 11.4 Functional and ultrastructural analyses of desmin mutants. (**a–d**) Functional analysis of different desmin mutants were realized after expression of different desmin mutants in C2C12 muscle cells. Immunostaining of desmin (*green fluorescence*) 48 h after transfection of wild-type desmin (**a**); I451M (**b**); E413K (**c**) and R406W (**d**) desmin mutants. Note filamentous network organization in C2C12 cells when transfected with I451M desmin mutant. In contrast, E413K and R406W desmin mutants form desmin-containing aggregates which are distributed throughout the cytoplasm. Nuclei are stained in *blue* by DAPI. (**e** and **f**). Ultrastructural perturbations induced by desmin mutants in tibialis anterior muscles of wild-type (**e**) and R406W desmin (**f**) expressing mice. Wild-type and mutant desmin were expressed in muscle of wild-type mice using AAV vectors. All analyses were realized 1-month after AAV vectors injection. Note that expression of wild-type desmin does not seem to modify muscle ultrastructure (**a**) whereas expression of R406W mutant of desmin induces the appearance of granulofilamentous electron-dense material (see *arrows*) located at the perinuclear regions. *N* nucleus, *M* mitochondria, *Z* Z-disk. Bar: 5 μm (**a–d**); 2 μm

A213V has been considered as polymorphism or rare variant by some investigators due to the variable penetrance. Rare variants such as K241E and H326R have been identified. Analysis of mutated nucleotides should be taken with precaution. New generation sequencing provides further accuracy to analyze mutated nucleotides.

Desmin Protein Partners and Their Involvement in Desmin-Related Myopathy

Desmin functions are carried out by its interaction with partners. More and more partners that bind to desmin directly have been identified during the last 30 years. We describe here desmin partners belonging to the IF family and the other partners related to the membrane-associated proteins, sarcomeric and cytoskeletal-associated proteins, chaperons, calcium-binding proteins and signaling proteins. The other partners concerning apoptosis-related proteins, posttranslational modifying proteins and nucleic acids have been described in previous reviews [67, 120, 121].

IF Partners

It has been reported that desmin can interact with the type V nuclear B-type lamin in cardiomyocytes and myoblasts [122, 123]. In the cytoplasm, desmin can interact or copolymerize with vimentin, nestin, synemin, syncoilin. Vimentin is highly expressed in the precursor or myoblast of striated muscles and progressively replaced by desmin during myogenesis so that very little vimentin is present in the mature striated muscle [124, 125]. In the smooth muscle vimentin and desmin are co-expressed [60, 126, 127]. Nestin is expressed in some muscle during early development and its expression is decreased in the mature muscle, but remains high at myotendinous junctions and neuromuscular junctions [43, 128]. Syncoilin is present in mature muscle but is enriched at myotendinous junctions and neuromuscular junctions [129, 130]. Synemin, initially co-purified with desmin and vimentin in chicken muscle and identified as IF-associated protein is present in striated muscle and smooth muscle [131]. Several isoforms of synemin or syncoilin have been identified [59, 132, 133]. Nestin, syncoilin and synemin have a unique property in that they cannot self-assemble to form filaments, they necessarily assemble into heteropolymers. However, some studies demonstrated that synemin and syncoilin do not seem to participate in the formation of mixed filaments, but bind to pre-formed desmin or vimentin IFs [134–136]. The localization of these proteins is altered in the desmin KO mice [42, 43, 59]. Synemin interacts with α-actinin, plectin 1, zyxin, and 3 components of the dystrophin-associated protein complex such as dystrophin, utrophin and α-dystrobrevin [134, 137–140]. Alpha-synemin, but not β-synemin, interacts with vinculin, metavinculin, and talin, suggesting that these isoforms may have different roles [140–142]. Synemin participates in focal adhesion dynamics and cell migration [140, 143, 144]. Synemin has an additional role as an A-kinase anchoring protein (AKAP) [145]. Recently, synemin has been shown to play a role in skeletal muscle hypertrophy via participation in the regulation of PKA activity [146]. Syncoilin also binds to α-dystrobrevin, a component of the dystrophin-associated protein complex. Synemin and syncoilin seem to play a linker role between desmin IF and the dystrophin-associated protein complex. The latter is involved in mechanotransduction, so synemin and syncoilin should play a role in mechanical stress signaling transmission from the membrane to the nucleus through the desmin IF network. The synemin and syncoilin KO mouse models have shown that they are required for generating maximum isometric stress in skeletal muscle [146, 147]. No mutation has been found in nestin, synemin and syncoilin genes in myopathy patients up to date.

Heat Shock Proteins (HSP) and Chaperones

Alpha B-Crystallin

The αB-crystallin (CRYAB), a member of the small HSP, has been firstly shown to bind to desmin in cardiac muscle [148]. We have identified an R120G missense mutation in CRYAB in a DRM family [149]. This R120G CRYAB has an irregular quaternary structure with an absence of a clear central cavity, reduced chaperone-like function, decreased apparent dissociation constant and increased binding capacity, resulting in desmin filament aggregation [150, 151]. Transgenic mice overexpressing high-level R120G CRYAB exhibit 100 % mortality in early adulthood while the wild-type CRYAB expression was relatively benign, with no increases in mortality. Transgenic mice with modest R120G CRYAB result in a phenotype strikingly similar to that observed for the desmin-related cardiomyopathies with CRYAB and desmin accumulations within cardiac muscle, mitochondrial deficiencies, activation of apoptosis, increased endoplasmic reticulum (ER) stress and heart failure [152, 153]. Recently, it has been demonstrated that 28-week-old R120G CRYAB transgenic mice exhibit cardiac arrhythmias accompanied with decreased expression of SERCA2, phospholamban, ryanodine receptor 2 and calsequestrin 2 [154].

The results from Jeffrey Robbins laboratory indicated that CRYAB protects the desmin filaments via preventing abnormal desmin protein from aggregating adversely. The double transgenic mouse hearts containing desmin mutant (D7-des) and R120G-CRYAB transgenes have significantly higher levels of aberrant desmin aggregates than the D7-des transgenic hearts [155]. They further show that expression of CRYAB (R120G) leads to amyloid oligomer formation, which are characteristic of neurodegenerative diseases. These oligomeric amyloid intermediates are present also in cardiomyocytes derived from many human dilated and hypertrophic cardiomyopathies [156].

AlphaB-crystallin has three serine residues (positions 19, 45, and 59) that can be phosphorylated under various conditions. CRYAB is a direct target of the p38 mitogen-activated protein kinase (MAPK) cascade. p38 MAPK plays a pro-survival role and confers protection of cardiomyocytes during myocardial infarction [157]. Inhibition of p38 MAPK with SB203580 in hypoxic adult cardiomyocytes results in the formation of desmin aggregates. In a mouse model of dilated cardiomyopathy triggered by cardiomyocyte-specific serum response factor (SRF), muscle creatine kinase is the primary down-regulated protein followed by α-actin and α-tropomyosin down-regulation leading to a decrease of polymerized F-actin. We found that the early response to these defects was an increase in the amount of desmin intermediate filaments and phosphorylation of the αB-crystallin chaperon. Desmin and αB-crystallin progressively lose their striated pattern and accumulate at the intercalated disk and the sarcolemma, respectively [158].

HSP25

HSP25, a 25-kDa heat shock protein, binds directly to desmin [159]. The interaction between HSP25 and desmin is strengthened by the phosphorylation of HSP25 at serine residue 15. It has been shown that H_2O_2 pretreatment induces activation of p38 MAPK and subsequent HSP25/27 phosphorylation and translocation attenuates desmin degradation by calpain-1 activation in ischemia-reperfused hearts [160]. These results indicate that HSP25 contributes to the desmin cytoskeletal organization.

Sarcomeric and Cytoskeletal-Associated Proteins

Plectin

Plectin (>500 kDa) is a multifunctional and widely expressed cytoskeletal linker protein which binds actin, MT and IF. It has been suggested that plectin plays a role in myofibrillogenesis, and its association with Z-discs is an early event in the lateral alignment of myofibrils that precedes the formation of the intermyofibrillar desmin cytoskeleton [161]. Plectin interacts directly with desmin via its carboxyl-terminal IF-binding domain and acts as a universal mediator of desmin IF anchorage at the sarcolemma and Z-disks by linking adjacent Z-disks, tethering desmin filaments into subsarcolemmal dense plaques containing dystrophin and vinculin, preventing individual myofibrils from disruption during contraction and ensuring effective force generation [162–166]. Plectin plays a linker function between desmin IFs and the mitochondrial surface and could be important for the positioning and shape formation, in particular branching, of mitochondrial organelles in striated muscle tissues [164]. Alternative splicing of plectin transcripts gives rise to more than eight protein isoforms differing only in small N-terminal sequences (5–180 residues), which fulfill distinct functions in different cell types and tissues. Four isoforms (plectins 1, 1b, 1d, and 1f) are found at substantial levels in muscle tissue. Plectins 1 and 1f are sarcolemma-associated isoforms, whereas plectin 1d localizes exclusively to Z-disks [165, 167]. Using conditional gene targeting in mice, the group of Gerhard Wiche has shown that plectin 1d and 1f integrate fibers by specifically targeting and linking desmin IFs to Z-disks and costameres, whereas plectin 1b establishes a linkage to mitochondria [168].

Mutations of the human plectin gene as well as the targeted inactivation of its murine analog cause a generalized blistering skin disorder and muscular dystrophy [167–169]. The most common disease caused by plectin deficiency is epidermolysis bullosa simplex-muscular dystrophy (EBS-MD), a rare autosomal-recessive skin blistering disorder with late-onset muscular dystrophy. EBS-MD patients and plectin-deficient mice display pathologic desmin-positive protein aggregates, degenerated myofibrils, and severe mitochondrial abnormalities, the hallmarks of MFM.

Nebulin

Nebulin is a giant F-actin binding protein that co-extends along the length of actin thin filaments and is expressed specifically in skeletal muscle tissues [170]. The central coiled-coil region (150–263) of desmin interacts with nebulin's C-terminal repeats M163-M170 [171]. The targeted mutation (K190A) in the desmin coil 1B region, or desminopathy-causing desmin mutations in the head, rod, or tail domains (S46F, E245D, and T453I) exhibit significantly delayed filament assembly kinetics when bound to nebulin and destabilizes actin thin filaments [79, 172, 173]. A mechanism has been proposed a mechanism by which mutant desmin slows desmin remodeling in myocytes by retaining nebulin near the Z-disks. The molecular etiology of desminopathy for some filament-forming desmin mutants could be resulting from subtle deficiencies in their association with nebulin.

Calponin

Desmin interacts with smooth muscle-specific protein calponin, a major actin-, tropomyosin-, and calmodulin-binding protein that has been identified. The N-terminal 22 kDa fragment of calponin interacts with the central a-helical rod domain of the desmin in a concentration-dependent manner in vitro. The addition of calmodulin or S100 to the mixture of calponin and desmin causes removal of calponin from the desmin filaments and inhibits bundle formation in the presence of Ca^{2+}. Calponin might bridge IFs with actin in the vicinity of dense bodies [174–176].

Membrane-Associated Proteins

Caveolin-3

It has been shown that desmin is associated with caveolin-3, a major component of caveolar microdomains in myogenic cells [177]. In patients suffering myopathy with a heterozygous A337P mutation of desmin, aggregates of vesicular and tubular structures contain caveolae. As caveolae occur in the Golgi complex and are transported to the cell surface, the presence of caveolin-3 in aggregates suggests that the accumulation of multiple proteins in desminopathies could be partially due to inhibited intracellular trafficking of caveolae to the sarcolemma [178].

Polycystin-1

IF proteins such as vimentin, cytokeratin 8, cytokeratin 18 and desmin were found to interact with polycystin-1 [179]. These interactions are mediated by coiled-coil motifs in polycystin-1 and IF proteins. Polycystin-1, which is a large membrane-associated protein, plays a role in mechanosensation by regulating calcium signaling and nitric oxide release in response to fluid shear stress in endothelial cells [180]. In vascular SMCs, the polycystins regulate the activity of the stretch-activated cation channels [181]. Polycystin-1 is required for maintaining the structural integrity of the vasculature as well as epithelium [182]. Polycystin-1 mutation is found in a majority of patients with autosomal dominant polycystic kidney disease [183].

Desmoplakin

The association of the desmosome component desmoplakin with desmin C-terminal, rod portion and tail domain depends on sequences within the linker region and the B and C subdomains of C-terminal extremity of desmoplakin [184, 185]. Desmosomes are intercellular adhesive complexes that anchor the IF cytoskeleton to the cell membrane in cardiac muscle cells. Desmoplakin–desmin interactions are important for maintenance of the cytoarchitecture in cardiomyocytes. Mutations impairing this interaction in either the C-terminus of desmoplakin or the desmin tail lead to cardiomyopathy [184].

Signaling Proteins

Myotubularin

Myotubularin MTM1, a phosphoinositide 3-phosphatase, is mutated in X-linked centronuclear myopathy (XLCNM; myotubular myopathy). MTM1 interacts with desmin in vitro and in vivo [186]. XLCNM-causing mutations in MTM1 result in loss of MTM1-desmin interaction, formation of desmin-positive aggregates and abnormal mitochondrial positioning, shape, dynamics, and function in both mouse and human skeletal muscles [186]. Conditional deletion of the MTM1 gene via Cre-LoxP strategy in adult muscle fibers has demonstrated that myotubularin is required during adulthood [187]. Viral gene transfer of phosphatase-dead myotubularin mutants (MTM1-C375S) significantly improves most histological signs of XLCNM displayed by a MTM1-null mouse, improves muscle performance and restores the localization of nuclei, triad alignment, and the desmin intermediate filament network, while it does not normalize phosphoinositide 3-phosphatase level, supporting a phosphatase-independent role of MTM1 in maintaining normal muscle performance and organelle positioning in skeletal muscle [188].

Myospryn

The amino-terminal domain of desmin interacts directly with the 24-amino acid-long carboxyl-terminal end of the SPRY (SPIa/ryanodine receptor) domain of the tripartite motif-like protein myospryn (also called cardiomyopathy-associated 5 or TRIM76) [189]. Myospryn, originally identified as an associated partner to the biogenesis of lysosome-related organelle complex 1 protein dysbindin, colocalizes with desmin at the periphery of the nucleus of mouse neonatal cardiomyocytes and predominantly at intercalated disks and costameres of adult cardiomyocytes. Myospryn colocalizes also with the makers of endoplasmic reticulum (KDEL receptor) and Golgi sorting machinery (TGN38) and lysosomal markers such as cathepsin D. In the absence of desmin, proper perinuclear localization of myospryn is lost and shows a diffuse cytoplasmic localization. In addition, extensive mislocalization of the lysosomes is found, suggesting a potential role of desmin IF in vesicle trafficking and in lysosomes and lysosome-related organelle biogenesis and/or positioning. Myospryn is also an AKAP protein and contains three PKA RII subunit anchoring motifs in relation to the TRIM region [190], raising the possibility that together with desmin and other cytoskeletal and signaling proteins, it can participate in the subcellular targeting of PKA activity in striated muscle. Disturbance in these highly coordinated signaling pathways is thought to compromise efficient maintenance of structure–function integrity of muscle, leading to different cardiac and skeletal myopathies [191].

Calcium-Binding Proteins

S100A1/S100B

S100A1 and S100B are members of the S100 family of calcium-binding proteins. S100A1, which inhibits MT protein assembly and promotes MT disassembly, interacts with desmin in the presence of a few μM concentrations of Ca^{2+} [192]. The C-terminal extension of S100A1 or S100B represents a critical part of the sites that recognize the N-terminal domain of desmin (a portion of a coiled-coil helix 2A). S100A1 and S100B inhibit the assembly of desmin into IFs and stimulate the disassembly of preformed desmin IFs. The activity of S100A1 and S100B is modulated by annexin VI, a member of a family of Ca^{2+}-dependent phospholipid and membrane-binding proteins. Annexin VI blocks the ability of S100A1 and S100B to inhibit the assembly of desmin into IF in a Ca^{2+}- and dose-dependent manner. S100A1 and S100B may be considered as Ca^{2+}-dependent regulators of the assembly of two important elements of the cytoskeleton, IFs and MTs, and, potentially, of MT- and IF-based activities [192, 193].

Post-translational Modifications of Desmin

Post-translational modifications (PTMs) of desmin are described in several recent reviews [120, 121]. The PTMs of desmin can be carried out both by enzymatic and non-enzymatic reactions. Phosphorylation of desmin is mostly studied. The known enzymes that can phosphorylate desmin include protein kinase A and C (PKA, PKC) [194–196], cyclin-dependent kinase 1 (CDK1) [197], p21-activated kinase (PAK) [198], Rho-associated kinase (Rho-kinase) [199], Aurora kinase B (Aurora-B) [200] and glycogen synthase kinase 3 (GSK3) [201]. All these kinases function as serine/threonine kinase. Phosphorylation is one of the regulatory mechanisms involved in the assembly–disassembly of desmin. Phosphorylation of desmin induces the disassembly of desmin filaments and prevents the polymerization of soluble protofilaments. The biopsy of DRM patients is frequently associated with the presence of desmin-positive aggregates in which desmin is hyperphosphorylated [85, 202]. A number of desmin missense mutations affect serine and threonine, some of them are the potential phosphorylatable sites, for example, S7F and S13F mutations [85, 203]. Whether these mutations affect the phosphorylated state or structural changes needs further investigation. Recently, we found desmin phosphorylation at the serine 28 in the mouse model of dilated cardiomyopathy induced by conditional inactivation of transcription factor SRF in cardiomyocytes in which desmin is not mutated but increasingly disorganized during the time course of the disease [158]. Recently, this serine 28 has been shown to be the target site in vivo of GSK3. Desmin is phosphorylated at serine 28 and 32, this modified desmin contains a consensus sequence for GSK3 (SXXXS(p), where S is Ser, X is any other amino acid and 'p' is a phosphate group) [201]. They confirmed for the first time that GSK3 phosphorylates desmin both in in vitro model and in human heart failure samples. They observed increased levels of selectively phosphorylated and cleaved desmin in a canine pacing model of dyssynchronous heart failure when compared with either controls or animals treated with cardiac resynchronization therapy. Desmin-positive oligomers were also increased in dyssynchronous heart failure hearts compared with controls. There is a relationship between phosphorylation and cleavage of desmin and cardiac toxicity due to desmin amyloid-like oligomers formation. Decrease of amyloid properties by treatment with cardiac resynchronization therapy or an anti-amyloid small molecule opens a novel potential therapeutic strategy for cardiomyopathy [201].

Another type of PTM of proteins is monosaccharide O-linked β-N-acetylglucosamine (O-GlcNAc). This attachment of O-GlcNAc to Ser/Thr residues of proteins is frequently considered to be analogous to protein phosphorylation in that it is a highly dynamic, reversible, and tightly regulated enzyme-catalyzed process. The level of O-GlcNAcylation is regulated by O-GlcNAc transferase (OGT; uridinediphospho-N-acetylglucosamine: polypeptide β-*N*-acetylglucosaminyltransferase), which catalyzes O-GlcNAc synthesis and attachment, and O-GlcNAcase (OGA; β-*N*-acetyl-D-glucosaminidase), which catalyzes its removal. The O-GlcNAc modification was demonstrated to be involved in skeletal muscle atrophy [204]. O-GlcNAc desmin has been identified [205] and is involved in cardiac ischemia–reperfusion (I/R) injury [206]. After I/R, there was a marked loss of both cytosolic and nuclear O-GlcNAcylation and disruption of Z-line structures. OGA inhibition largely preserved structural integrity and attenuated the loss of O-GlcNAcylation. Desmin immunofluorescence following ischemia–reperfusion injury is decreased and its localization is altered. Increased O-GlcNAc level which attenuates the loss of desmin, prevents its structural disorganization and could be an important contributing factor to the improved functional recovery and decreased injury.

ADP-ribosylation of desmin has been identified. Arginine-specific mono (ADP-ribosyl transferase, ART1) is responsible for ADP-ribosylation of desmin [207]. ADP-ribosylation results in potent inhibition of desmin's ability to assemble into filaments [208] and phosphorylation of desmin by PKA. Arginine 48 and 68 of desmin's head domain have been shown to be sites of modification, with arginine 48 being the major ADP-ribosylation site [209].

Ubiquitylation of desmin by ubiquitin ligase TRIM32 is involved in the loss of thin filaments during atrophy [210]. Desmin was shown to be ubiquitinated by TRIM32 and then degraded. Other substrates of TRIM32 include actin, tropomyosin and α-actinin. Mutations in TRIM32 are associated with myopathies, such as limb-girdle muscular dystrophy [211–214] and sarcotubular myopathy [215]. A TRIM32 KO mouse displays myopathic changes, replicating the phenotype of limb-girdle muscular dystrophy [216]. An interesting interplay between desmin ubiquitylation and phosphorylation has also been reported in the same study. During atrophy, TRIM32 ubiquitinates preferentially phosphorylated desmin, promoting its solubilization and degradation. It is unclear which kinases could be responsible for the increased desmin phosphorylation during atrophy.

PTMs of desmin include also modifications that do not require enzymes, such as glycation (or non-enzymatic glycosylation), oxidation and nitration. Glycation is a reaction between the carbonyl group of reducing sugars and the primary amino groups (typically lysine and arginine residues) of proteins. Such glycated proteins can react further with intermediate compounds produced through oxidative or non-oxidative pathways to form stable end products or advanced glycation end products (AGEs). Desmin is a preferential target of AGE in the context of dilated cardiomyopathy leading to heart failure, both in mice and in humans [158]. Increased oxidative stress in cardiomyopathy and heart failure due to the dysfunction of mitochondria can increase formation of highly reactive dicarbonyls, such as glyoxal and methylglyoxal, generated from triose sugar oxidation or lipid peroxidation and subsequent AGE adducts formation on proteins.

Desmin-Related Myopathy Mouse Models and Therapeutic Potentials

Desmin-Related Myopathy Mouse Models

Three transgenic mice expressing mutated desmin in cardiomyocytes have been published. The first mouse model expresses a 7-amino acid deletion (R173 through E179) desmin (D7-des) mutation linked to DRM, published in 2001 [217]. Expression of the dominant negative D7-des mutant protein using α-MHC promoter leads to the appearance of aberrant intrasarcoplasmic and electron-dense granular filamentous aggregates in cardiac tissue that are characteristic of human desmin-related cardiomyopathy. Further analysis indicated that this desmin mutant mouse model shows compromised ability of the heart to respond to β-agonist stimulation, impaired UPS proteolytic function at the level of entry of ubiquitinated proteins into the 20S proteasome [218], increased autophage flux associated with up-regulation of p62, a mediator between the aberrant aggregates and autophagosomes [219]. The latter has been proposed as an adaptive response to overexpression of misfolded proteins. The second mouse model expressing a desmin missense mutation L345P in skeletal and cardiac muscle using desmin promoter was published in 2008 [220]. In this model, mutated desmin gene is expressed at low-level and no protein aggregates were detected. However, L345P desmin transgenic mice show mitochondrial swelling vacuolization, increased mitochondrial Ca^{2+} level in skeletal and cardiac myocytes, decreased motor function, hypertrophy of the left ventricular wall and decreased left ventricular chamber dimension. Soleus muscle has impaired contractile function and recovery from fatigue. The third model is a mouse expressing a desmin missense mutation I451M linked to human dilated cardiomyopathy in cardiomyocytes using α-MHC promoter [221]. The mutant desmin loses its Z-disk localization but it can still associate with the intercalated disks, which, however, have an altered architecture, resembling other examples of dilated cardiomyopathy.

Recently, R349P (corresponding human desmin missense mutation R350P) desmin knock-in mice have been obtained [222]. Compared to the classical transgenic mouse model, this R349P knock-in mouse model has the advantage that the expression of mutant desmin gene is under the control of the endogenous gene regulatory elements and presents better at the protein level the situation of the heterozygous human mutation carriers. These mice develop age-dependent desmin-positive protein aggregation pathology, skeletal muscle weakness, dilated cardiomyopathy, as well as cardiac arrhythmias and conduction defects. The results from this model underline the importance of extrasarcomeric IF network disruption, but not the presence of protein aggregates, in the development of desminopathies. Disruption of the extrasarcomeric IF network increases mechanical vulnerability of muscle fibers and finally impacts the entire organ and subsequently causes myopathy and cardiomyopathy [222].

Alternatively, desmin-related myopathy mouse models can be created by using adeno-associated virus (AAV) vectors carrying mutated desmin cDNA (R406W or E413K) [223]. The main advantage of this strategy is that it is easy to quickly obtain a large number of viral vectors allowing the expression of various desmin mutants and to study the effects of their expression in one genetic background. Expression of desmin mutants in mouse muscles induces morphological changes of muscle fibers (irregular shape and size), disruption of Z-disks and the appearance of desmin accumulations around the nuclei (for R406W) or in subsarcolemmal regions of fibers (for E413K) (Fig. 11.4). Both desmin mutants studied here induce a decrease in muscle force generation capacity.

In addition, a desmin-related myopathy mouse model has been created by expressing R120G-αB-crystallin in cardiomyocytes [153]. High-level expression of CRYAB R120G is deleterious and results in 100 % mortality in early adulthood. Modest expression levels resulted in a phenotype that was strikingly similar to that observed for the desmin-related cardiomyopathies with aberrant desmin and CRYAB aggregation, myofibril misalignment and cardiac hypertrophy.

Therapeutic Potentials

There is no efficient treatment for desmin-related myopathies as of today; some strategies have been tested in vitro or in vivo. These strategies aim to reduce the oxidative and ER stress, and apoptotic signals, to increase chaperones activity, to improve UPS and autophagy activity, and to correct mitochondrial defects.

Antioxidant Treatments

Pre-treatment with *N*-acetyl-L-cysteine, an antioxidant, of myoblasts expressing desmin D399Y prevented desmin aggregation in vitro in response to redox-associated stress (H_2O_2 and cadmium chloride), suggesting its potential therapeutic function for desminopathies [224]. The treatment of R120G CRYAB transgenic mice with oxypurinol, an inhibitor of superoxide-generating enzyme xanthine oxidase that is upregulated in this model of DRM, restored mitochondrial function and morphology by preventing excessive production of reactive oxygen species (ROS), but unimproved cardiac contractility and compliance remained [225].

Heat Shock Proteins (HSP) and Chaperones Activity

The CRYAB R120G DRM model has been used to test the effects of the HSP inducer geranylgeranylacetone (GGA) [226]. Oral administration of GGA results in reduced amyloid oligomer levels and aggregates, decreased heart size and less interstitial fibrosis, as well as improved cardiac function and survival when compared to untreated CRYAB R120G mice. These effects could be linked to a reduction in amyloid oligomer and aggregate formation due to an overexpression of HSPB8. Treatment with GGA as well as overexpression of HSPB8 also inhibit cytochrome c release from mitochondria, activation of caspase-3 and TUNEL-positive cardiomyocyte death in the CRYAB R120G mice [226]. Gerhard Wiche and his collaborators demonstrated that the treatment with 4-phenylbutyrate, a chemical chaperone, in plectin-deficient myotubes as well as in plectin-deficient mice results in remarkable amelioration of the pathological phenotypes [227]. Hsp27 has been shown to suppress the formation of inclusion bodies induced by expression of CRYAB R120G in cell culture. When CRYAB R120G and Hsp27 are transiently co-expressed in HeLa cells, the amount of CRYAB R120G in the soluble fraction was greater when compared to expression of CRYAB R120G alone [228].

Anti-apoptotic Agents

The earliest ultrastructural defects are observed in mitochondria and the capacity of mitochondria to resist exposure to calcium is diminished in desmin KO mice. Overexpression of bcl-2, an anti-apoptotic gene, in the desmin null heart results in correction of mitochondrial defects, reduced occurrence of fibrotic lesions in the myocardium, prevention of cardiac hypertrophy, restoration of cardiomyocyte ultra-structure, partial rescue of the capacity to calcium exposure and significant improvement of cardiac function [229]. Overexpression of bcl-2 in R120G CRYAB hearts prolongs transgenic mouse survival by 20 % associated with decreased mitochondrial abnormalities, restoration of cardiac function, decreased protein aggregation and apoptosis. However, inhibition of apoptotic signaling resulted in the upregulation of autophagy and alternative death pathways, the net result being increased necrosis [230]. More than twofold increase in cardiomyocyte autophagic activity was observed in R120G CRYAB hearts. This increase of autophagy has been shown to be an adaptive response to toxic aggregates [231].

Exercise and Autophagy

It seems that voluntary exercise restores normal level of metallomembrane endopeptidase neprilysin that is decreased in control CRYAB R120G mice, and slows the progression to heart failure in the CRYAB R120G DRM model [232]. Both overexpression of autophagic gene autophagy-related 7 (Atg7) and voluntary exercise that upregulate autophagy ameliorate desmin-related cardiomyopathy by the decrease of interstitial fibrosis and cardiac hypertrophy, reduction of intracellular aggregates, amelioration of ventricular dysfunction and prolongation of life span [233].

Decrease of Fibrosis

Osteopontin (OPN) promotes cardiac dysfunction in the desmin KO model. Double KO mice (Des$^{-/-}$:OPN$^{-/-}$) improve left ventricular function, paralleled to a reduction in fibrosis. The diminished fibrotic response in the absence of OPN could be in part mediated by a dramatic reduction of galectin-3 secretion by OPN-deficient infiltrating macrophages [234].

Channel Opener Nicorandil Treatment

The cardioprotective effect of nicorandil, a mitochondrial ATP-sensitive potassium channel opener, has been shown to reduce mitochondrial impairment and apoptotic cell death and prolonged survival of CRYAB R120G transgenic mice [235]. Nicorandil treatment inhibits the increase in BAX, the decrease in bcl2, the activation of caspase-3 and apoptotic cell death in mutant CRYAB R120G transgenic mice. This treatment also prevents ventricular tachyarrhythmia with increased expression of total and phosphorylated connexin-43 in CRYAB R120G transgenic mouse hearts [236].

Modulation of Proteasome Activity

Sildenafil treatment significantly increases myocardial protein kinase G (PKG) activity and reduces myocardial accumulation of CRYAB R120G, ubiquitin conjugates, and aberrant protein aggregates in CRYAB R120G transgenic mouse hearts. This could be linked to PKG activation by sildenafil that positively regulates proteasome activities and proteasome-mediated degradation of misfolded proteins, likely through posttranslational modifications of proteasome subunits. This may be a new mechanism underlying the benefit of PKG stimulation in treating cardiac diseases. Stimulation of PKG by drugs such as sildenafil administration is potentially a new therapeutic strategy to treat cardiac proteinopathies [237]. Overexpression of UBC9, a SUMO-conjugating enzyme, enhances UPS function in cardiomyocytes while knockdown of UBC9 by siRNA causes significant accumulations of aggregated protein. It was shown that UBC9 reduces preamyloid oligomer content in cardiomyocyte and its activity can be exploited to reduce toxic levels of misfolded or aggregated proteins in cardiomyopathy [238].

Conclusions and Perspectives

Since desmin was identified in 1976, significant progress has been made regarding the regulation of desmin gene expression and the role of desmin in muscle homeostasis and the development of DRM. It has been shown that the cooperation between muscle-specific LCR and different transcription control regions such as muscle-specific enhancer is required to obtain a physiological concentration of desmin in different muscle tissues. Although desmin is not necessary for myogenic commitment, differentiation, or the fusion of skeletal muscle, it is essential for the structural integrity and function of muscles. Mice lacking desmin develop cardiomyopathy, skeletal myopathy, and smooth muscle defects. It is clear that the desmin network influences the position, movement, and respiratory activity in situ of mitochondria. Desmin is important for sarcomere alignment and maintaining cell compliance, optimal excitation–contraction coupling, neuromuscular junction integrity, cell–matrix interaction, vascular myogenic tone control, and generation of passive and active tension in muscles. More than 60 desmin mutations have been identified in patients. These mutations can interfere with the position and movement of cellular organelles such as mitochondria, lead to the dysfunction of protein quality control such as UPS and autophagy, and influence the cytoskeleton organization. For the mutations which do not disrupt the IF filament formation, the pathology could come from defects of interactions between desmin and its partners. More and more partners that bind to desmin have been identified during the last 30 years, and these partners are involved in diverse biological processes such as intracellular trafficking, organelle biogenesis and/or positioning, chaperone activity, cytoskeleton organization, calcium homeostasis, mechanosensing, and signaling pathways. From these data, it is reasonable to suggest that in addition of its structural role, desmin may act as a signaling platform for the integration of mechanotransduction signals from the outside to the inside of organelles such as mitochondria or the nucleus. Recently, it has been suggested that desmin filaments form a stress-transmitting and stress-signaling network and are required for JNK-mediated stress sensing in muscle [239]. This stress-transmitting and stress-signaling network should include its partners such as synemin and syncoilin. Synemin has been shown recently to play a role in skeletal muscle hypertrophy via participation in the regulation of PKA activity [146]. Synemin and syncoilin play a linker role between desmin IF and the dystrophin-associated protein mechanotransduction complex. It has been reported that desmin can act as a novel regulator of microRNA in airway smooth muscle hypertrophy through ErK1/2Egr-1//miR-26a/GSK-3β pathway [240] and could be the direct or indirect target of microRNA [241, 242]. Further studies are needed to better understand the function of the desmin-containing stress-sensing network, the influence of epigenetic factors involved in chromatin remodeling, microRNA and long non-coding RNA on the expression and function of desmin, PTMs of proteins in health and disease situations, and the effects of different desmin mutation variants on energy metabolism for which few investigations have been pursued as of now.

The recent R349P desmin knock-in mouse model provides an important advancement in understanding the molecular pathogenesis of desminopathies. These studies emphasized that the loss of the IF network rather than the presence of aggregates in cells is a major reason for the pathological development. However, more desmin mutation knock-in mouse models including the mutations that do not interfere with the filament formation in vitro are needed to evaluate the real effect of protein aggregates on the development of diseases and decipher sequential molecular events leading from mutant desmin to progressive muscle damage at the different stages of lifespan related to the patients. There is no efficient treatment for desmin-related myopathies up to date; the strategies aiming to reduce the oxidative and ER stress, and apoptotic signals, to increase chaperones activity, and to improve UPS and autophagy activity have been *tested* in vitro and in vivo. Future studies on the desmin mutant knock-in mouse models will provide useful information for the design of potential therapies.

References

1. Lazarides E, Hubbard BD (1976) Immunological characterization of the subunit of the 100 A filaments from muscle cells. Proc Natl Acad Sci U S A 73:4344–4348
2. Burt AD, Robertson JL, Heir J, MacSween RN (1986) Desmin-containing stellate cells in rat liver; distribution in normal animals and response to experimental acute liver injury. J Pathol 150:29–35. doi:10.1002/path.1711500106
3. Hellström M, Kalén M, Lindahl P, Abramsson A, Betsholtz C (1999) Role of PDGF-B and PDGFR-beta in recruitment of vascular smooth muscle cells and pericytes during embryonic blood vessel formation in the mouse. Development 126:3047–3055
4. Kjörell U, Thornell LE (1982) Identification of a complex between alpha-actinin and the intermediate filament subunit skeletin in bovine heart Purkinje fibres. Eur J Cell Biol 28:139–144
5. Rogatsch H, Jezek D, Hittmair A, Mikuz G, Feichtinger H (1996) Expression of vimentin, cytokeratin, and desmin in Sertoli cells of human fetal, cryptorchid, and tumour-adjacent testicular tissue. Virchows Arch 427:497–502
6. Altmannsberger M, Weber K, Droste R, Osborn M (1985) Desmin is a specific marker for rhabdomyosarcomas of human and rat origin. Am J Pathol 118:85–95
7. Ma Y, Peng J, Liu W, Zhang P, Huang L, Gao B, Shen T, Zhou Y, Chen H, Chu Z, Zhang M, Qin H (2009) Proteomics identification of desmin as a potential oncofetal diagnostic and prognostic biomarker in colorectal cancer. Mol Cell Proteomics 8:1878–1890. doi:10.1074/mcp.M800541-MCP200
8. Maruyama M, Sugiyama H, Sada K, Kobayashi M, Maeshima Y, Yamasaki Y, Makino H (2007) Desmin as a marker of proteinuria in early stages of membranous nephropathy in elderly patients. Clin Nephrol 68:73–80
9. Herrmann A, Tozzo E, Funk J (2012) Semi-automated quantitative image analysis of podocyte desmin immunoreactivity as a sensitive marker for acute glomerular damage in the rat puromycin aminonucleoside nephrosis (PAN) model. Exp Toxicol Pathol 64:45–49. doi:10.1016/j.etp.2010.06.004
10. Chugh S, Ouzounian M, Lu Z, Mohamed S, Li W, Bousette N, Liu PP, Gramolini AO (2013) Pilot study identifying myosin heavy chain 7, desmin, insulin-like growth factor 7, and annexin A2 as circulating biomarkers of human heart failure. Proteomics 13:2324–2334. doi:10.1002/pmic.201200455
11. Kaufman SJ, Foster RF (1988) Replicating myoblasts express a muscle-specific phenotype. Proc Natl Acad Sci U S A 85:9606–9610
12. Capetanaki YG, Ngai J, Lazarides E (1984) Characterization and regulation in the expression of a gene coding for the intermediate filament protein desmin. Proc Natl Acad Sci U S A 81:6909–6913
13. Gao J, Li Z, Paulin D (1998) A novel site, Mt, in the human desmin enhancer is necessary for maximal expression in skeletal muscle. J Biol Chem 273:6402–6409

14. Van de Klundert FA, Jansen HJ, Bloemendal H (1994) A proximal promoter element in the hamster desmin upstream regulatory region is responsible for activation by myogenic determination factors. J Biol Chem 269:220–225
15. Li H, Capetanaki Y (1993) Regulation of the mouse desmin gene: transactivated by MyoD, myogenin, MRF4 and Myf5. Nucleic Acids Res 21:335–343
16. Li H, Capetanaki Y (1994) An E box in the desmin promoter cooperates with the E box and MEF-2 sites of a distal enhancer to direct muscle-specific transcription. EMBO J 13: 3580–3589
17. Li Z, Colucci E, Babinet C, Paulin D (1993) The human desmin gene: a specific regulatory programme in skeletal muscle both in vitro and in transgenic mice. Neuromuscul Disord 3:423–427
18. Li ZL, Paulin D (1991) High level desmin expression depends on a muscle-specific enhancer. J Biol Chem 266:6562–6570
19. Li Z, Paulin D (1993) Different factors interact with myoblast-specific and myotube-specific enhancer regions of the human desmin gene. J Biol Chem 268:10403–10415
20. Li Z, Marchand P, Humbert J, Babinet C, Paulin D (1993) Desmin sequence elements regulating skeletal muscle-specific expression in transgenic mice. Development 117:947–959
21. Kuisk IR, Li H, Tran D, Capetanaki Y (1996) A single MEF2 site governs desmin transcription in both heart and skeletal muscle during mouse embryogenesis. Dev Biol 174:1–13. doi:10.1006/dbio.1996.0046
22. Mericskay M, Parlakian A, Porteu A, Dandré F, Bonnet J, Paulin D, Li Z (2000) An overlapping CArG/octamer element is required for regulation of desmin gene transcription in arterial smooth muscle cells. Dev Biol 226:192–208. doi:10.1006/dbio.2000.9865
23. Tam JLY, Triantaphyllopoulos K, Todd H, Raguz S, de Wit T, Morgan JE, Partridge TA, Makrinou E, Grosveld F, Antoniou M (2006) The human desmin locus: gene organization and LCR-mediated transcriptional control. Genomics 87:733–746. doi:10.1016/j.ygeno.2006.01.009
24. Raguz S, Hobbs C, Yagüe E, Ioannou PA, Walsh FS, Antoniou M (1998) Muscle-specific locus control region activity associated with the human desmin gene. Dev Biol 201:26–42. doi:10.1006/dbio.1998.8964
25. Lindahl Allen M, Koch CM, Clelland GK, Dunham I, Antoniou M (2009) DNA methylation-histone modification relationships across the desmin locus in human primary cells. BMC Mol Biol 10:51. doi:10.1186/1471-2199-10-51
26. Li Z, Colucci-Guyon E, Pinçon-Raymond M, Mericskay M, Pournin S, Paulin D, Babinet C (1996) Cardiovascular lesions and skeletal myopathy in mice lacking desmin. Dev Biol 175:362–366. doi:10.1006/dbio.1996.0122
27. Milner DJ, Weitzer G, Tran D, Bradley A, Capetanaki Y (1996) Disruption of muscle architecture and myocardial degeneration in mice lacking desmin. J Cell Biol 134:1255–1270
28. Balogh J, Merisckay M, Li Z, Paulin D, Arner A (2002) Hearts from mice lacking desmin have a myopathy with impaired active force generation and unaltered wall compliance. Cardiovasc Res 53:439–450
29. Milner DJ, Taffet GE, Wang X, Pham T, Tamura T, Hartley C, Gerdes AM, Capetanaki Y (1999) The absence of desmin leads to cardiomyocyte hypertrophy and cardiac dilation with compromised systolic function. J Mol Cell Cardiol 31:2063–2076
30. Schrickel JW, Stöckigt F, Krzyzak W, Paulin D, Li Z, Lübkemeier I, Fleischmann B, Sasse P, Linhart M, Lewalter T, Nickenig G, Lickfett L, Schröder R, Clemen CS (2010) Cardiac conduction disturbances and differential effects on atrial and ventricular electrophysiological properties in desmin deficient mice. J Interv Card Electrophysiol 28:71–80. doi:10.1007/s10840-010-9482-8
31. Thornell L, Carlsson L, Li Z, Mericskay M, Paulin D (1997) Null mutation in the desmin gene gives rise to a cardiomyopathy. J Mol Cell Cardiol 29:2107–2124
32. Weisleder N, Soumaka E, Abbasi S, Taegtmeyer H, Capetanaki Y (2004) Cardiomyocyte-specific desmin rescue of desmin null cardiomyopathy excludes vascular involvement. J Mol Cell Cardiol 36:121–128

33. Mavroidis M, Capetanaki Y (2002) Extensive induction of important mediators of fibrosis and dystrophic calcification in desmin-deficient cardiomyopathy. Am J Pathol 160:943–952. doi:10.1016/S0002-9440(10)64916-4
34. Milner DJ, Mavroidis M, Weisleder N, Capetanaki Y (2000) Desmin cytoskeleton linked to muscle mitochondrial distribution and respiratory function. J Cell Biol 150:1283–1298
35. Kay L, Li Z, Mericskay M, Olivares J, Tranqui L, Fontaine E, Tiivel T, Sikk P, Kaambre T, Samuel JL, Rappaport L, Usson Y, Leverve X, Paulin D, Saks VA (1997) Study of regulation of mitochondrial respiration in vivo. An analysis of influence of ADP diffusion and possible role of cytoskeleton. Biochim Biophys Acta 1322:41–59
36. Lindén M, Li Z, Paulin D, Gotow T, Leterrier JF (2001) Effects of desmin gene knockout on mice heart mitochondria. J Bioenerg Biomembr 33:333–341
37. Fountoulakis M, Soumaka E, Rapti K, Mavroidis M, Tsangaris G, Maris A, Weisleder N, Capetanaki Y (2005) Alterations in the heart mitochondrial proteome in a desmin null heart failure model. J Mol Cell Cardiol 38:461–474. doi:10.1016/j.yjmcc.2004.12.008
38. Li Z, Mericskay M, Agbulut O, Butler-Browne G, Carlsson L, Thornell LE, Babinet C, Paulin D (1997) Desmin is essential for the tensile strength and integrity of myofibrils but not for myogenic commitment, differentiation, and fusion of skeletal muscle. J Cell Biol 139:129–144
39. O'Neill A, Williams MW, Resneck WG, Milner DJ, Capetanaki Y, Bloch RJ (2002) Sarcolemmal organization in skeletal muscle lacking desmin: evidence for cytokeratins associated with the membrane skeleton at costameres. Mol Biol Cell 13:2347–2359. doi:10.1091/mbc.01-12-0576
40. Agbulut O, Li Z, Périé S, Ludosky MA, Paulin D, Cartaud J, Butler-Browne G (2001) Lack of desmin results in abortive muscle regeneration and modifications in synaptic structure. Cell Motil Cytoskeleton 49:51–66. doi:10.1002/cm.1020
41. Agbulut O, Li Z, Mouly V, Butler-Browne GS (1996) Analysis of skeletal and cardiac muscle from desmin knock-out and normal mice by high resolution separation of myosin heavy-chain isoforms. Biol Cell 88:131–135
42. Carlsson L, Li ZL, Paulin D, Price MG, Breckler J, Robson RM, Wiche G, Thornell LE (2000) Differences in the distribution of synemin, paranemin, and plectin in skeletal muscles of wild-type and desmin knock-out mice. Histochem Cell Biol 114:39–47
43. Carlsson L, Li Z, Paulin D, Thornell LE (1999) Nestin is expressed during development and in myotendinous and neuromuscular junctions in wild type and desmin knock-out mice. Exp Cell Res 251:213–223. doi:10.1006/excr.1999.4569
44. Haubold KW, Allen DL, Capetanaki Y, Leinwand LA (2003) Loss of desmin leads to impaired voluntary wheel running and treadmill exercise performance. J Appl Physiol (1985) 95:1617–1622. doi:10.1152/japplphysiol.00408.2003
45. Sam M, Shah S, Fridén J, Milner DJ, Capetanaki Y, Lieber RL (2000) Desmin knockout muscles generate lower stress and are less vulnerable to injury compared with wild-type muscles. Am J Physiol Cell Physiol 279:C1116–C1122
46. Balogh J, Li Z, Paulin D, Arner A (2003) Lower active force generation and improved fatigue resistance in skeletal muscle from desmin deficient mice. J Muscle Res Cell Motil 24:453–459
47. Wieneke S, Stehle R, Li Z, Jockusch H (2000) Generation of tension by skinned fibers and intact skeletal muscles from desmin-deficient mice. Biochem Biophys Res Commun 278:419–425. doi:10.1006/bbrc.2000.3810
48. Shah SB, Peters D, Jordan KA, Milner DJ, Fridén J, Capetanaki Y, Lieber RL (2001) Sarcomere number regulation maintained after immobilization in desmin-null mouse skeletal muscle. J Exp Biol 204:1703–1710
49. Balogh J, Li Z, Paulin D, Arner A (2005) Desmin filaments influence myofilament spacing and lateral compliance of slow skeletal muscle fibers. Biophys J 88:1156–1165. doi:10.1529/biophysj.104.042630
50. Shah SB, Davis J, Weisleder N, Kostavassili I, McCulloch AD, Ralston E, Capetanaki Y, Lieber RL (2004) Structural and functional roles of desmin in mouse skeletal muscle during passive deformation. Biophys J 86:2993–3008. doi:10.1016/S0006-3495(04)74349-0

51. Shah SB, Su F-C, Jordan K, Milner DJ, Fridén J, Capetanaki Y, Lieber RL (2002) Evidence for increased myofibrillar mobility in desmin-null mouse skeletal muscle. J Exp Biol 205:321–325
52. Shah SB, Love JM, O'Neill A, Lovering RM, Bloch RJ (2012) Influences of desmin and keratin 19 on passive biomechanical properties of mouse skeletal muscle. J Biomed Biotechnol 2012:704061. doi:10.1155/2012/704061
53. Meyer GA, Lieber RL (2012) Skeletal muscle fibrosis develops in response to desmin deletion. Am J Physiol Cell Physiol. doi:10.1152/ajpcell.00441.2011
54. Anderson J, Joumaa V, Stevens L, Neagoe C, Li Z, Mounier Y, Linke WA, Goubel F (2002) Passive stiffness changes in soleus muscles from desmin knockout mice are not due to titin modifications. Pflugers Arch 444:771–776. doi:10.1007/s00424-002-0875-0
55. Anderson J, Li Z, Goubel F (2001) Passive stiffness is increased in soleus muscle of desmin knockout mouse. Muscle Nerve 24:1090–1092
56. Anderson J, Li Z, Goubel F (2002) Models of skeletal muscle to explain the increase in passive stiffness in desmin knockout muscle. J Biomech 35:1315–1324
57. Ralston E, Lu Z, Biscocho N, Soumaka E, Mavroidis M, Prats C, Lømo T, Capetanaki Y, Ploug T (2006) Blood vessels and desmin control the positioning of nuclei in skeletal muscle fibers. J Cell Physiol 209:874–882. doi:10.1002/jcp.20780
58. Chapman MA, Zhang J, Banerjee I, Guo LT, Zhang Z, Shelton GD, Ouyang K, Lieber RL, Chen J (2014) Disruption of both nesprin 1 and desmin results in nuclear anchorage defects and fibrosis in skeletal muscle. Hum Mol Genet 23:5879–5892. doi:10.1093/hmg/ddu310
59. Xue ZG, Cheraud Y, Brocheriou V, Izmiryan A, Titeux M, Paulin D, Li Z (2004) The mouse synemin gene encodes three intermediate filament proteins generated by alternative exon usage and different open reading frames. Exp Cell Res 298:431–444. doi:10.1016/j.yexcr.2004.04.023
60. Tang DD (2008) Intermediate filaments in smooth muscle. Am J Physiol Cell Physiol 294:C869–C878. doi:10.1152/ajpcell.00154.2007
61. Wede OK, Löfgren M, Li Z, Paulin D, Arner A (2002) Mechanical function of intermediate filaments in arteries of different size examined using desmin deficient mice. J Physiol 540:941–949
62. Lacolley P, Challande P, Boumaza S, Cohuet G, Laurent S, Boutouyrie P, Grimaud JA, Paulin D, Lamazière JM, Li Z (2001) Mechanical properties and structure of carotid arteries in mice lacking desmin. Cardiovasc Res 51:178–187
63. Scott RS, Li Z, Paulin D, Uvelius B, Small JV, Arner A (2008) Role of desmin in active force transmission and maintenance of structure during growth of urinary bladder. Am J Physiol Cell Physiol 295:C324–C331. doi:10.1152/ajpcell.90622.2007
64. Sjuve R, Arner A, Li Z, Mies B, Paulin D, Schmittner M, Small JV (1998) Mechanical alterations in smooth muscle from mice lacking desmin. J Muscle Res Cell Motil 19:415–429
65. Loufrani L, Matrougui K, Li Z, Levy BI, Lacolley P, Paulin D, Henrion D (2002) Selective microvascular dysfunction in mice lacking the gene encoding for desmin. FASEB J 16:117–119. doi:10.1096/fj.01-0505fje
66. Loufrani L, Li Z, Lévy BI, Paulin D, Henrion D (2002) Excessive microvascular adaptation to changes in blood flow in mice lacking gene encoding for desmin. Arterioscler Thromb Vasc Biol 22:1579–1584
67. Clemen CS, Herrmann H, Strelkov SV, Schröder R (2013) Desminopathies: pathology and mechanisms. Acta Neuropathol 125:47–75. doi:10.1007/s00401-012-1057-6
68. Goldfarb LG, Olivé M, Vicart P, Goebel HH (2008) Intermediate filament diseases: desminopathy. Adv Exp Med Biol 642:131–164
69. Hnia K, Ramspacher C, Vermot J, Laporte J (2014) Desmin in muscle and associated diseases: beyond the structural function. Cell Tissue Res. doi:10.1007/s00441-014-2016-4
70. Paulin D, Huet A, Khanamyrian L, Xue Z (2004) Desminopathies in muscle disease. J Pathol 204:418–427. doi:10.1002/path.1639
71. Selcen D (2011) Myofibrillar myopathies. Neuromuscul Disord 21:161–171. doi:10.1016/j.nmd.2010.12.007

72. Van Spaendonck-Zwarts K, van Hessem L, Jongbloed JDH, de Walle HEK, Capetanaki Y, van der Kooi AJ, van Langen IM, van den Berg MP, van Tintelen JP (2010) Desmin-related myopathy: a review and meta-analysis. Clin Genet. doi:10.1111/j.1399-0004.2010.01512.x
73. Goldfarb LG, Park KY, Cervenáková L, Gorokhova S, Lee HS, Vasconcelos O, Nagle JW, Semino-Mora C, Sivakumar K, Dalakas MC (1998) Missense mutations in desmin associated with familial cardiac and skeletal myopathy. Nat Genet 19:402–403. doi:10.1038/1300
74. Arias M, Pardo J, Blanco-Arias P, Sobrido M-J, Arias S, Dapena D, Carracedo A, Goldfarb LG, Navarro C (2006) Distinct phenotypic features and gender-specific disease manifestations in a Spanish family with desmin L370P mutation. Neuromuscul Disord 16:498–503. doi:10.1016/j.nmd.2006.05.011
75. Bär H, Goudeau B, Wälde S, Casteras-Simon M, Mücke N, Shatunov A, Goldberg YP, Clarke C, Holton JL, Eymard B, Katus HA, Fardeau M, Goldfarb L, Vicart P, Herrmann H (2007) Conspicuous involvement of desmin tail mutations in diverse cardiac and skeletal myopathies. Hum Mutat 28:374–386. doi:10.1002/humu.20459
76. Bergman JEH, Veenstra-Knol HE, van Essen AJ, van Ravenswaaij CMA, den Dunnen WFA, van den Wijngaard A, van Tintelen JP (2007) Two related Dutch families with a clinically variable presentation of cardioskeletal myopathy caused by a novel S13F mutation in the desmin gene. Eur J Med Genet 50:355–366. doi:10.1016/j.ejmg.2007.06.003
77. Cao L, Hong D, Zhu M, Li X, Wan H, Hong K (2013) A novel heterozygous deletion–insertion mutation in the desmin gene causes complete atrioventricular block and mild myopathy. Clin Neuropathol 32:9–15. doi:10.5414/NP300514
78. Cetin N, Balci-Hayta B, Gundesli H, Korkusuz P, Purali N, Talim B, Tan E, Selcen D, Erdem-Ozdamar S, Dincer P (2013) A novel desmin mutation leading to autosomal recessive limb-girdle muscular dystrophy: distinct histopathological outcomes compared with desminopathies. J Med Genet 50:437–443. doi:10.1136/jmedgenet-2012-101487
79. Conover GM, Henderson SN, Gregorio CC (2009) A myopathy-linked desmin mutation perturbs striated muscle actin filament architecture. Mol Biol Cell 20:834–845. doi:10.1091/mbc.E08-07-0753
80. Dagvadorj A, Goudeau B, Hilton-Jones D, Blancato JK, Shatunov A, Simon-Casteras M, Squier W, Nagle JW, Goldfarb LG, Vicart P (2003) Respiratory insufficiency in desminopathy patients caused by introduction of proline residues in desmin c-terminal alpha-helical segment. Muscle Nerve 27:669–675. doi:10.1002/mus.10370
81. Dagvadorj A, Olivé M, Urtizberea J-A, Halle M, Shatunov A, Bönnemann C, Park K-Y, Goebel HH, Ferrer I, Vicart P, Dalakas MC, Goldfarb LG (2004) A series of West European patients with severe cardiac and skeletal myopathy associated with a de novo R406W mutation in desmin. J Neurol 251:143–149. doi:10.1007/s00415-004-0289-3
82. Dalakas MC, Park KY, Semino-Mora C, Lee HS, Sivakumar K, Goldfarb LG (2000) Desmin myopathy, a skeletal myopathy with cardiomyopathy caused by mutations in the desmin gene. N Engl J Med 342:770–780. doi:10.1056/NEJM200003163421104
83. Dunand M, Lobrinus JA, Jeannet P-Y, Behin A, Claeys KG, Selcen D, Kuntzer T (2009) Confirmation that abnormal desmin accumulation and migration are due to a desmin gene mutation in a familial cardiomyopathy and distal myopathy. Neuromuscul Disord 19:802. doi:10.1016/j.nmd.2009.07.013
84. Fidziańska A, Kotowicz J, Sadowska M, Goudeau B, Walczak E, Vicart P, Hausmanowa-Petrusewicz I (2005) A novel desmin R355P mutation causes cardiac and skeletal myopathy. Neuromuscul Disord 15:525–531. doi:10.1016/j.nmd.2005.05.006
85. Goldfarb LG, Dalakas MC (2009) Tragedy in a heartbeat: malfunctioning desmin causes skeletal and cardiac muscle disease. J Clin Invest 119:1806–1813. doi:10.1172/JCI38027
86. Goudeau B, Dagvadorj A, Rodrigues-Lima F, Nédellec P, Casteras-Simon M, Perret E, Langlois S, Goldfarb L, Vicart P (2001) Structural and functional analysis of a new desmin variant causing desmin-related myopathy. Hum Mutat 18:388–396. doi:10.1002/humu.1210
87. Goudeau B, Rodrigues-Lima F, Fischer D, Casteras-Simon M, Sambuughin N, de Visser M, Laforet P, Ferrer X, Chapon F, Sjöberg G, Kostareva A, Sejersen T, Dalakas MC, Goldfarb LG, Vicart P (2006) Variable pathogenic potentials of mutations located in the desmin alpha-helical domain. Hum Mutat 27:906–913. doi:10.1002/humu.20351

88. Hedberg C, Melberg A, Kuhl A, Jenne D, Oldfors A (2012) Autosomal dominant myofibrillar myopathy with arrhythmogenic right ventricular cardiomyopathy 7 is caused by a DES mutation. Eur J Hum Genet 20:984–985. doi:10.1038/ejhg.2012.39
89. Henderson M, De Waele L, Hudson J, Eagle M, Sewry C, Marsh J, Charlton R, He L, Blakely EL, Horrocks I, Stewart W, Taylor RW, Longman C, Bushby K, Barresi R (2013) Recessive desmin-null muscular dystrophy with central nuclei and mitochondrial abnormalities. Acta Neuropathol 125:917–919. doi:10.1007/s00401-013-1113-x
90. Hong D, Wang Z, Zhang W, Xi J, Lu J, Luan X, Yuan Y (2011) A series of Chinese patients with desminopathy associated with six novel and one reported mutations in the desmin gene. Neuropathol Appl Neurobiol 37:257–270. doi:10.1111/j.1365-2990.2010.01112.x
91. Kaminska A, Strelkov SV, Goudeau B, Olivé M, Dagvadorj A, Fidzianska A, Simon-Casteras M, Shatunov A, Dalakas MC, Ferrer I, Kwiecinski H, Vicart P, Goldfarb LG (2004) Small deletions disturb desmin architecture leading to breakdown of muscle cells and development of skeletal or cardioskeletal myopathy. Hum Genet 114:306–313. doi:10.1007/s00439-003-1057-7
92. Kostareva A, Gudkova A, Sjoberg G, Kiselev I, Moiseeva O, Karelkina E, Goldfarb L, Schlyakhto E, Sejersen T (2006) Desmin mutations in a St. Petersburg cohort of cardiomyopathies. Acta Myol 25:109–115
93. Li D, Tapscoft T, Gonzalez O, Burch PE, Quiñones MA, Zoghbi WA, Hill R, Bachinski LL, Mann DL, Roberts R (1999) Desmin mutation responsible for idiopathic dilated cardiomyopathy. Circulation 100:461–464
94. Lorenzon A, Beffagna G, Bauce B, De Bortoli M, Li Mura IEA, Calore M, Dazzo E, Basso C, Nava A, Thiene G, Rampazzo A (2013) Desmin mutations and arrhythmogenic right ventricular cardiomyopathy. Am J Cardiol 111:400–405. doi:10.1016/j.amjcard.2012.10.017
95. Maddison P, Damian MS, Sewry C, McGorrian C, Winer JB, Odgerel Z, Shatunov A, Lee HS, Goldfarb LG (2012) Clinical and myopathological characteristics of desminopathy caused by a mutation in desmin tail domain. Eur Neurol 68:279–286. doi:10.1159/000341617
96. McDonald KK, Stajich J, Blach C, Ashley-Koch AE, Hauser MA (2012) Exome analysis of two limb-girdle muscular dystrophy families: mutations identified and challenges encountered. PLoS One 7, e48864. doi:10.1371/journal.pone.0048864
97. Miyamoto Y, Akita H, Shiga N, Takai E, Iwai C, Mizutani K, Kawai H, Takarada A, Yokoyama M (2001) Frequency and clinical characteristics of dilated cardiomyopathy caused by desmin gene mutation in a Japanese population. Eur Heart J 22:2284–2289. doi:10.1053/euhj.2001.2836
98. Muñoz-Mármol AM, Strasser G, Isamat M, Coulombe PA, Yang Y, Roca X, Vela E, Mate JL, Coll J, Fernández-Figueras MT, Navas-Palacios JJ, Ariza A, Fuchs E (1998) A dysfunctional desmin mutation in a patient with severe generalized myopathy. Proc Natl Acad Sci U S A 95:11312–11317
99. Nalini A, Gayathri N, Richard P, Cobo A-M, Urtizberea JA (2013) New mutation of the desmin gene identified in an extended Indian pedigree presenting with distal myopathy and cardiac disease. Neurol India 61:622–626. doi:10.4103/0028-3886.125269
100. Olivé M, Armstrong J, Miralles F, Pou A, Fardeau M, Gonzalez L, Martínez F, Fischer D, Martínez Matos JA, Shatunov A, Goldfarb L, Ferrer I (2007) Phenotypic patterns of desminopathy associated with three novel mutations in the desmin gene. Neuromuscul Disord 17:443–450. doi:10.1016/j.nmd.2007.02.009
101. Piñol-Ripoll G, Shatunov A, Cabello A, Larrodé P, de la Puerta I, Pelegrín J, Ramos FJ, Olivé M, Goldfarb LG (2009) Severe infantile-onset cardiomyopathy associated with a homozygous deletion in desmin. Neuromuscul Disord 19:418–422. doi:10.1016/j.nmd.2009.04.004
102. Pruszczyk P, Kostera-Pruszczyk A, Shatunov A, Goudeau B, Dramiñska A, Takeda K, Sambuughin N, Vicart P, Strelkov SV, Goldfarb LG, Kamiñska A (2007) Restrictive cardiomyopathy with atrioventricular conduction block resulting from a desmin mutation. Int J Cardiol 117:244–253. doi:10.1016/j.ijcard.2006.05.019
103. Schröder R, Goudeau B, Simon MC, Fischer D, Eggermann T, Clemen CS, Li Z, Reimann J, Xue Z, Rudnik-Schöneborn S, Zerres K, van der Ven PFM, Fürst DO, Kunz WS, Vicart P (2003) On noxious desmin: functional effects of a novel heterozygous desmin insertion

mutation on the extrasarcomeric desmin cytoskeleton and mitochondria. Hum Mol Genet 12:657–669

104. Sharma S, Mücke N, Katus HA, Herrmann H, Bär H (2009) Disease mutations in the "head" domain of the extra-sarcomeric protein desmin distinctly alter its assembly and network-forming properties. J Mol Med. doi:10.1007/s00109-009-0521-9
105. Sjöberg G, Saavedra-Matiz CA, Rosen DR, Wijsman EM, Borg K, Horowitz SH, Sejersen T (1999) A missense mutation in the desmin rod domain is associated with autosomal dominant distal myopathy, and exerts a dominant negative effect on filament formation. Hum Mol Genet 8:2191–2198
106. Strach K, Sommer T, Grohé C, Meyer C, Fischer D, Walter MC, Vorgerd M, Reilich P, Bär H, Reimann J, Reuner U, Germing A, Goebel HH, Lochmüller H, Wintersperger B, Schröder R (2008) Clinical, genetic, and cardiac magnetic resonance imaging findings in primary desminopathies. Neuromuscul Disord 18:475–482. doi:10.1016/j.nmd.2008.03.012
107. Sugawara M, Kato K, Komatsu M, Wada C, Kawamura K, Shindo PS, Yoshioka PN, Tanaka K, Watanabe S, Toyoshima I (2000) A novel de novo mutation in the desmin gene causes desmin myopathy with toxic aggregates. Neurology 55:986–990
108. Taylor MRG, Slavov D, Ku L, Di Lenarda A, Sinagra G, Carniel E, Haubold K, Boucek MM, Ferguson D, Graw SL, Zhu X, Cavanaugh J, Sucharov CC, Long CS, Bristow MR, Lavori P, Mestroni L (2007) Prevalence of desmin mutations in dilated cardiomyopathy. Circulation 115:1244–1251. doi:10.1161/CIRCULATIONAHA.106.646778
109. Van Tintelen JP, Van Gelder IC, Asimaki A, Suurmeijer AJH, Wiesfeld ACP, Jongbloed JDH, van den Wijngaard A, Kuks JBM, van Spaendonck-Zwarts KY, Notermans N, Boven L, van den Heuvel F, Veenstra-Knol HE, Saffitz JE, Hofstra RMW, van den Berg MP (2009) Severe cardiac phenotype with right ventricular predominance in a large cohort of patients with a single missense mutation in the DES gene. Heart Rhythm 6:1574–1583. doi:10.1016/j.hrthm.2009.07.041
110. Tse H-F, Ho JCY, Choi S-W, Lee Y-K, Butler AW, Ng K-M, Siu C-W, Simpson MA, Lai W-H, Chan Y-C, Au K-W, Zhang J, Lay KWJ, Esteban MA, Nicholls JM, Colman A, Sham PC (2013) Patient-specific induced-pluripotent stem cells-derived cardiomyocytes recapitulate the pathogenic phenotypes of dilated cardiomyopathy due to a novel DES mutation identified by whole exome sequencing. Hum Mol Genet 22:1395–1403. doi:10.1093/hmg/dds556
111. Vernengo L, Chourbagi O, Panuncio A, Lilienbaum A, Batonnet-Pichon S, Bruston F, Rodrigues-Lima F, Mesa R, Pizzarossa C, Demay L, Richard P, Vicart P, Rodriguez M-M (2010) Desmin myopathy with severe cardiomyopathy in a Uruguayan family due to a codon deletion in a new location within the desmin 1A rod domain. Neuromuscul Disord 20:178–187. doi:10.1016/j.nmd.2010.01.001
112. Vrabie A, Goldfarb LG, Shatunov A, Nägele A, Fritz P, Kaczmarek I, Goebel HH (2005) The enlarging spectrum of desminopathies: new morphological findings, eastward geographic spread, novel exon 3 desmin mutation. Acta Neuropathol 109:411–417. doi:10.1007/s00401-005-0980-1
113. Wahbi K, Béhin A, Charron P, Dunand M, Richard P, Meune C, Vicart P, Laforêt P, Stojkovic T, Bécane HM, Kuntzer T, Duboc D (2012) High cardiovascular morbidity and mortality in myofibrillar myopathies due to DES gene mutations: a 10-year longitudinal study. Neuromuscul Disord 22:211–218. doi:10.1016/j.nmd.2011.10.019
114. Walter MC, Reilich P, Huebner A, Fischer D, Schröder R, Vorgerd M, Kress W, Born C, Schoser BG, Krause KH, Klutzny U, Bulst S, Frey JR, Lochmüller H (2007) Scapuloperoneal syndrome type Kaeser and a wide phenotypic spectrum of adult-onset, dominant myopathies are associated with the desmin mutation R350P. Brain 130:1485–1496. doi:10.1093/brain/awm039
115. Yuri T, Miki K, Tsukamoto R, Shinde A, Kusaka H, Tsubura A (2007) Autopsy case of desminopathy involving skeletal and cardiac muscle. Pathol Int 57:32–36. doi:10.1111/j.1440-1827.2007.02053.x
116. Park KY, Dalakas MC, Goebel HH, Ferrans VJ, Semino-Mora C, Litvak S, Takeda K, Goldfarb LG (2000) Desmin splice variants causing cardiac and skeletal myopathy. J Med Genet 37:851–857

117. Bär H, Mücke N, Kostareva A, Sjöberg G, Aebi U, Herrmann H (2005) Severe muscle disease-causing desmin mutations interfere with in vitro filament assembly at distinct stages. Proc Natl Acad Sci U S A 102:15099–15104. doi:10.1073/pnas.0504568102
118. Bär H, Mücke N, Ringler P, Müller SA, Kreplak L, Katus HA, Aebi U, Herrmann H (2006) Impact of disease mutations on the desmin filament assembly process. J Mol Biol 360:1031–1042. doi:10.1016/j.jmb.2006.05.068
119. Bär H, Schopferer M, Sharma S, Hochstein B, Mücke N, Herrmann H, Willenbacher N (2010) Mutations in desmin's carboxy-terminal "tail" domain severely modify filament and network mechanics. J Mol Biol 397:1188–1198. doi:10.1016/j.jmb.2010.02.024
120. Snider NT, Omary MB (2014) Post-translational modifications of intermediate filament proteins: mechanisms and functions. Nat Rev Mol Cell Biol 15:163–177. doi:10.1038/nrm3753
121. Winter DL, Paulin D, Mericskay M, Li Z (2014) Posttranslational modifications of desmin and their implication in biological processes and pathologies. Histochem Cell Biol 141:1–16. doi:10.1007/s00418-013-1148-z
122. Lockard VG, Bloom S (1993) Trans-cellular desmin-lamin B intermediate filament network in cardiac myocytes. J Mol Cell Cardiol 25:303–309. doi:10.1006/jmcc.1993.1036
123. Tolstonog GV, Sabasch M, Traub P (2002) Cytoplasmic intermediate filaments are stably associated with nuclear matrices and potentially modulate their DNA-binding function. DNA Cell Biol 21:213–239. doi:10.1089/10445490252925459
124. Fürst DO, Osborn M, Weber K (1989) Myogenesis in the mouse embryo: differential onset of expression of myogenic proteins and the involvement of titin in myofibril assembly. J Cell Biol 109:517–527
125. Gard DL, Lazarides E (1980) The synthesis and distribution of desmin and vimentin during myogenesis in vitro. Cell 19:263–275
126. Schmid E, Osborn M, Rungger-Brändle E, Gabbiani G, Weber K, Franke WW (1982) Distribution of vimentin and desmin filaments in smooth muscle tissue of mammalian and avian aorta. Exp Cell Res 137:329–340
127. Small JV, Gimona M (1998) The cytoskeleton of the vertebrate smooth muscle cell. Acta Physiol Scand 164:341–348
128. Sejersen T, Lendahl U (1993) Transient expression of the intermediate filament nestin during skeletal muscle development. J Cell Sci 106(Pt 4):1291–1300
129. Newey SE, Howman EV, Ponting CP, Benson MA, Nawrotzki R, Loh NY, Davies KE, Blake DJ (2001) Syncoilin, a novel member of the intermediate filament superfamily that interacts with alpha-dystrobrevin in skeletal muscle. J Biol Chem 276:6645–6655. doi:10.1074/jbc.M008305200
130. Poon E, Howman EV, Newey SE, Davies KE (2002) Association of syncoilin and desmin: linking intermediate filament proteins to the dystrophin-associated protein complex. J Biol Chem 277:3433–3439. doi:10.1074/jbc.M105273200
131. Granger BL, Lazarides E (1980) Synemin: a new high molecular weight protein associated with desmin and vimentin filaments in muscle. Cell 22:727–738
132. Kemp MW, Edwards B, Burgess M, Clarke WT, Nicholson G, Parry DAD, Davies KE (2009) Syncoilin isoform organization and differential expression in murine striated muscle. J Struct Biol 165:196–203. doi:10.1016/j.jsb.2008.11.002
133. Titeux M, Brocheriou V, Xue Z, Gao J, Pellissier JF, Guicheney P, Paulin D, Li Z (2001) Human synemin gene generates splice variants encoding two distinct intermediate filament proteins. Eur J Biochem 268:6435–6449
134. Bellin RM, Sernett SW, Becker B, Ip W, Huiatt TW, Robson RM (1999) Molecular characteristics and interactions of the intermediate filament protein synemin. Interactions with alpha-actinin may anchor synemin-containing heterofilaments. J Biol Chem 274:29493–29499
135. Chourbagi O, Bruston F, Carinci M, Xue Z, Vicart P, Paulin D, Agbulut O (2011) Desmin mutations in the terminal consensus motif prevent synemin-desmin heteropolymer filament assembly. Exp Cell Res 317:886–897. doi:10.1016/j.yexcr.2011.01.013
136. Hirako Y, Yamakawa H, Tsujimura Y, Nishizawa Y, Okumura M, Usukura J, Matsumoto H, Jackson KW, Owaribe K, Ohara O (2003) Characterization of mammalian synemin, an

intermediate filament protein present in all four classes of muscle cells and some neuroglial cells: co-localization and interaction with type III intermediate filament proteins and keratins. Cell Tissue Res 313:195–207. doi:10.1007/s00441-003-0732-2
137. Bhosle RC, Michele DE, Campbell KP, Li Z, Robson RM (2006) Interactions of intermediate filament protein synemin with dystrophin and utrophin. Biochem Biophys Res Commun 346:768–777. doi:10.1016/j.bbrc.2006.05.192
138. Hijikata T, Nakamura A, Isokawa K, Imamura M, Yuasa K, Ishikawa R, Kohama K, Takeda S, Yorifuji H (2008) Plectin 1 links intermediate filaments to costameric sarcolemma through beta-synemin, alpha-dystrobrevin and actin. J Cell Sci 121:2062–2074. doi:10.1242/jcs.021634
139. Mizuno Y, Thompson TG, Guyon JR, Lidov HG, Brosius M, Imamura M, Ozawa E, Watkins SC, Kunkel LM (2001) Desmuslin, an intermediate filament protein that interacts with alpha-dystrobrevin and desmin. Proc Natl Acad Sci U S A 98:6156–6161. doi:10.1073/pnas.111153298
140. Sun N, Huiatt TW, Paulin D, Li Z, Robson RM (2010) Synemin interacts with the LIM domain protein zyxin and is essential for cell adhesion and migration. Exp Cell Res 316:491–505. doi:10.1016/j.yexcr.2009.10.015
141. Sun N, Critchley DR, Paulin D, Li Z, Robson RM (2008) Identification of a repeated domain within mammalian alpha-synemin that interacts directly with talin. Exp Cell Res 314:1839–1849. doi:10.1016/j.yexcr.2008.01.034
142. Sun N, Critchley DR, Paulin D, Li Z, Robson RM (2008) Human alpha-synemin interacts directly with vinculin and metavinculin. Biochem J 409:657–667. doi:10.1042/BJ20071188
143. Jing R, Pizzolato G, Robson RM, Gabbiani G, Skalli O (2005) Intermediate filament protein synemin is present in human reactive and malignant astrocytes and associates with ruffled membranes in astrocytoma cells. Glia 50:107–120. doi:10.1002/glia.20158
144. Jing R, Wilhelmsson U, Goodwill W, Li L, Pan Y, Pekny M, Skalli O (2007) Synemin is expressed in reactive astrocytes in neurotrauma and interacts differentially with vimentin and GFAP intermediate filament networks. J Cell Sci 120:1267–1277. doi:10.1242/jcs.03423
145. Russell MA, Lund LM, Haber R, McKeegan K, Cianciola N, Bond M (2006) The intermediate filament protein, synemin, is an AKAP in the heart. Arch Biochem Biophys 456:204–215. doi:10.1016/j.abb.2006.06.010
146. Li Z, Parlakian A, Coletti D, Alonso-Martin S, Hourdé C, Joanne P, Gao-Li J, Blanc J, Ferry A, Paulin D, Xue Z, Agbulut O (2014) Synemin acts as a regulator of signalling molecules during skeletal muscle hypertrophy. J Cell Sci 127:4589–4601. doi:10.1242/jcs.143164
147. Zhang J, Bang M-L, Gokhin DS, Lu Y, Cui L, Li X, Gu Y, Dalton ND, Scimia MC, Peterson KL, Lieber RL, Chen J (2008) Syncoilin is required for generating maximum isometric stress in skeletal muscle but dispensable for muscle cytoarchitecture. Am J Physiol Cell Physiol 294:C1175–C1182. doi:10.1152/ajpcell.00049.2008
148. Bennardini F, Wrzosek A, Chiesi M (1992) Alpha B-crystallin in cardiac tissue. Association with actin and desmin filaments. Circ Res 71:288–294
149. Vicart P, Caron A, Guicheney P, Li Z, Prévost MC, Faure A, Chateau D, Chapon F, Tomé F, Dupret JM, Paulin D, Fardeau M (1998) A missense mutation in the alphaB-crystallin chaperone gene causes a desmin-related myopathy. Nat Genet 20:92–95. doi:10.1038/1765
150. Bova MP, Yaron O, Huang Q, Ding L, Haley DA, Stewart PL, Horwitz J (1999) Mutation R120G in alphaB-crystallin, which is linked to a desmin-related myopathy, results in an irregular structure and defective chaperone-like function. Proc Natl Acad Sci U S A 96:6137–6142
151. Perng MD, Wen SF, van den Ijssel P, Prescott AR, Quinlan RA (2004) Desmin aggregate formation by R120G alphaB-crystallin is caused by altered filament interactions and is dependent upon network status in cells. Mol Biol Cell 15:2335–2346. doi:10.1091/mbc.E03-12-0893
152. Maloyan A, Sanbe A, Osinska H, Westfall M, Robinson D, Imahashi K, Murphy E, Robbins J (2005) Mitochondrial dysfunction and apoptosis underlie the pathogenic process in alpha-B-crystallin desmin-related cardiomyopathy. Circulation 112:3451–3461. doi:10.1161/CIRCULATIONAHA.105.572552

153. Wang X, Osinska H, Klevitsky R, Gerdes AM, Nieman M, Lorenz J, Hewett T, Robbins J (2001) Expression of R120G–αB-crystallin causes aberrant desmin and αB-crystallin aggregation and cardiomyopathy in mice. Circ Res 89:84–91. doi:10.1161/hh1301.092688
154. Jiao Q, Sanbe A, Zhang X, Liu J-P, Minamisawa S (2014) αB-Crystallin R120G variant causes cardiac arrhythmias and alterations in the expression of Ca(2+)-handling proteins and endoplasmic reticulum stress in mice. Clin Exp Pharmacol Physiol 41:589–599. doi:10.1111/1440-1681.12253
155. Wang X, Klevitsky R, Huang W, Glasford J, Li F, Robbins J (2003) AlphaB-crystallin modulates protein aggregation of abnormal desmin. Circ Res 93:998–1005. doi:10.1161/01.RES.0000102401.77712.ED
156. Sanbe A, Osinska H, Saffitz JE, Glabe CG, Kayed R, Maloyan A, Robbins J (2004) Desmin-related cardiomyopathy in transgenic mice: a cardiac amyloidosis. Proc Natl Acad Sci U S A 101:10132–10136. doi:10.1073/pnas.0401900101
157. Mitra A, Ray A, Datta R, Sengupta S, Sarkar S (2014) Cardioprotective role of P38 MAPK during myocardial infarction via parallel activation of α-crystallin B and Nrf2. J Cell Physiol 229:1272–1282. doi:10.1002/jcp.24565
158. Diguet N, Mallat Y, Ladouce R, Clodic G, Prola A, Tritsch E, Blanc J, Larcher J-C, Delcayre C, Samuel J-L, Friguet B, Bolbach G, Li Z, Mericskay M (2011) Muscle creatine kinase deficiency triggers both actin depolymerization and desmin disorganization by advanced glycation end products in dilated cardiomyopathy. J Biol Chem 286:35007–35019. doi:10.1074/jbc.M111.252395
159. Kawano F, Fujita R, Nakai N, Terada M, Ohira T, Ohira Y (2012) HSP25 can modulate myofibrillar desmin cytoskeleton following the phosphorylation at Ser15 in rat soleus muscle. J Appl Physiol (1985) 112:176–186. doi:10.1152/japplphysiol.00783.2011
160. Blunt BC, Creek AT, Henderson DC, Hofmann PA (2007) H2O2 activation of HSP25/27 protects desmin from calpain proteolysis in rat ventricular myocytes. Am J Physiol Heart Circ Physiol 293:H1518–H1525. doi:10.1152/ajpheart.00269.2006
161. Schröder R, Fürst DO, Klasen C, Reimann J, Herrmann H, van der Ven PF (2000) Association of plectin with Z-discs is a prerequisite for the formation of the intermyofibrillar desmin cytoskeleton. Lab Invest 80:455–464
162. Hijikata T, Murakami T, Imamura M, Fujimaki N, Ishikawa H (1999) Plectin is a linker of intermediate filaments to Z-discs in skeletal muscle fibers. J Cell Sci 112(Pt 6):867–876
163. Hijikata T, Murakami T, Ishikawa H, Yorifuji H (2003) Plectin tethers desmin intermediate filaments onto subsarcolemmal dense plaques containing dystrophin and vinculin. Histochem Cell Biol 119:109–123. doi:10.1007/s00418-003-0496-5
164. Reipert S, Steinböck F, Fischer I, Bittner RE, Zeöld A, Wiche G (1999) Association of mitochondria with plectin and desmin intermediate filaments in striated muscle. Exp Cell Res 252:479–491. doi:10.1006/excr.1999.4626
165. Rezniczek GA, Konieczny P, Nikolic B, Reipert S, Schneller D, Abrahamsberg C, Davies KE, Winder SJ, Wiche G (2007) Plectin 1f scaffolding at the sarcolemma of dystrophic (mdx) muscle fibers through multiple interactions with beta-dystroglycan. J Cell Biol 176:965–977. doi:10.1083/jcb.200604179
166. Schröder R, Warlo I, Herrmann H, van der Ven PF, Klasen C, Blümcke I, Mundegar RR, Fürst DO, Goebel HH, Magin TM (1999) Immunogold EM reveals a close association of plectin and the desmin cytoskeleton in human skeletal muscle. Eur J Cell Biol 78:288–295. doi:10.1016/S0171-9335(99)80062-4
167. Winter L, Wiche G (2013) The many faces of plectin and plectinopathies: pathology and mechanisms. Acta Neuropathol 125:77–93. doi:10.1007/s00401-012-1026-0
168. Konieczny P, Fuchs P, Reipert S, Kunz WS, Zeöld A, Fischer I, Paulin D, Schröder R, Wiche G (2008) Myofiber integrity depends on desmin network targeting to Z-disks and costameres via distinct plectin isoforms. J Cell Biol 181:667–681. doi:10.1083/jcb.200711058
169. Schröder R, Kunz WS, Rouan F, Pfendner E, Tolksdorf K, Kappes-Horn K, Altenschmidt-Mehring M, Knoblich R, van der Ven PFM, Reimann J, Fürst DO, Blümcke I, Vielhaber S, Zillikens D, Eming S, Klockgether T, Uitto J, Wiche G, Rolfs A (2002) Disorganization of the

desmin cytoskeleton and mitochondrial dysfunction in plectin-related epidermolysis bullosa simplex with muscular dystrophy. J Neuropathol Exp Neurol 61:520–530
170. Labeit S, Kolmerer B (1995) The complete primary structure of human nebulin and its correlation to muscle structure. J Mol Biol 248:308–315
171. Bang M-L, Gregorio C, Labeit S (2002) Molecular dissection of the interaction of desmin with the C-terminal region of nebulin. J Struct Biol 137:119–127. doi:10.1006/jsbi.2002.4457
172. Baker LK, Gillis DC, Sharma S, Ambrus A, Herrmann H, Conover GM (2013) Nebulin binding impedes mutant desmin filament assembly. Mol Biol Cell 24:1918–1932. doi:10.1091/mbc.E12-11-0840
173. Conover GM, Gregorio CC (2011) The desmin coil 1B mutation K190A impairs nebulin Z-disc assembly and destabilizes actin thin filaments. J Cell Sci 124:3464–3476. doi:10.1242/jcs.087080
174. Fujii T, Takagi H, Arimoto M, Ootani H, Ueeda T (2000) Bundle formation of smooth muscle desmin intermediate filaments by calponin and its binding site on the desmin molecule. J Biochem 127:457–465
175. Mabuchi K, Li B, Ip W, Tao T (1997) Association of calponin with desmin intermediate filaments. J Biol Chem 272:22662–22666
176. Wang P, Gusev NB (1996) Interaction of smooth muscle calponin and desmin. FEBS Lett 392:255–258
177. Mermelstein CS, Martins ER, Portilho DM, Costa ML (2007) Association between the muscle-specific proteins desmin and caveolin-3 in muscle cells. Cell Tissue Res 327:343–351. doi:10.1007/s00441-006-0296-z
178. Shinde A, Nakano S, Sugawara M, Toyoshima I, Ito H, Tanaka K, Kusaka H (2008) Expression of caveolar components in primary desminopathy. Neuromuscul Disord 18:215–219. doi:10.1016/j.nmd.2007.12.006
179. Xu GM, Sikaneta T, Sullivan BM, Zhang Q, Andreucci M, Stehle T, Drummond I, Arnaout MA (2001) Polycystin-1 interacts with intermediate filaments. J Biol Chem 276:46544–46552. doi:10.1074/jbc.M107828200
180. AbouAlaiwi WA, Takahashi M, Mell BR, Jones TJ, Ratnam S, Kolb RJ, Nauli SM (2009) Ciliary polycystin-2 is a mechanosensitive calcium channel involved in nitric oxide signaling cascades. Circ Res 104:860–869. doi:10.1161/CIRCRESAHA.108.192765
181. Sharif-Naeini R, Folgering JHA, Bichet D, Duprat F, Lauritzen I, Arhatte M, Jodar M, Dedman A, Chatelain FC, Schulte U, Retailleau K, Loufrani L, Patel A, Sachs F, Delmas P, Peters DJM, Honoré E (2009) Polycystin-1 and -2 dosage regulates pressure sensing. Cell 139:587–596. doi:10.1016/j.cell.2009.08.045
182. Kim K, Drummond I, Ibraghimov-Beskrovnaya O, Klinger K, Arnaout MA (2000) Polycystin 1 is required for the structural integrity of blood vessels. Proc Natl Acad Sci U S A 97:1731–1736
183. Cornec-Le Gall E, Audrézet M-P, Meur YL, Chen J-M, Férec C (2014) Genetics and pathogenesis of autosomal dominant polycystic kidney disease: 20 years on. Hum Mutat 35:1393–1406. doi:10.1002/humu.22708
184. Lapouge K, Fontao L, Champliaud M-F, Jaunin F, Frias MA, Favre B, Paulin D, Green KJ, Borradori L (2006) New insights into the molecular basis of desmoplakin- and desmin-related cardiomyopathies. J Cell Sci 119:4974–4985. doi:10.1242/jcs.03255
185. Meng JJ, Bornslaeger EA, Green KJ, Steinert PM, Ip W (1997) Two-hybrid analysis reveals fundamental differences in direct interactions between desmoplakin and cell type-specific intermediate filaments. J Biol Chem 272:21495–21503
186. Hnia K, Tronchère H, Tomczak KK, Amoasii L, Schultz P, Beggs AH, Payrastre B, Mandel JL, Laporte J (2011) Myotubularin controls desmin intermediate filament architecture and mitochondrial dynamics in human and mouse skeletal muscle. J Clin Invest 121:70–85. doi:10.1172/JCI44021
187. Joubert R, Vignaud A, Le M, Moal C, Messaddeq N, Buj-Bello A (2013) Site-specific Mtm1 mutagenesis by an AAV-Cre vector reveals that myotubularin is essential in adult muscle. Hum Mol Genet 22:1856–1866. doi:10.1093/hmg/ddt038

188. Amoasii L, Bertazzi DL, Tronchère H, Hnia K, Chicanne G, Rinaldi B, Cowling BS, Ferry A, Klaholz B, Payrastre B, Laporte J, Friant S (2012) Phosphatase-dead myotubularin ameliorates X-linked centronuclear myopathy phenotypes in mice. PLoS Genet 8, e1002965. doi:10.1371/journal.pgen.1002965
189. Kouloumenta A, Mavroidis M, Capetanaki Y (2007) Proper perinuclear localization of the TRIM-like protein myospryn requires its binding partner desmin. J Biol Chem 282:35211–35221. doi:10.1074/jbc.M704733200
190. Reynolds JG, McCalmon SA, Tomczyk T, Naya FJ (2007) Identification and mapping of protein kinase A binding sites in the costameric protein myospryn. Biochim Biophys Acta 1773:891–902. doi:10.1016/j.bbamcr.2007.04.004
191. Tsoupri E, Capetanaki Y (2013) Myospryn: a multifunctional desmin-associated protein. Histochem Cell Biol 140:55–63. doi:10.1007/s00418-013-1103-z
192. Garbuglia M, Verzini M, Sorci G, Bianchi R, Giambanco I, Agneletti AL, Donato R (1999) The calcium-modulated proteins, S100A1 and S100B, as potential regulators of the dynamics of type III intermediate filaments. Braz J Med Biol Res 32:1177–1185
193. Garbuglia M, Verzini M, Donato R (1998) Annexin VI binds S100A1 and S100B and blocks the ability of S100A1 and S100B to inhibit desmin and GFAP assemblies into intermediate filaments. Cell Calcium 24:177–191
194. O'Connor CM, Gard DL, Lazarides E (1981) Phosphorylation of intermediate filament proteins by cAMP-dependent protein kinases. Cell 23:135–143
195. Rasmussen H, Takuwa Y, Park S (1987) Protein kinase C in the regulation of smooth muscle contraction. FASEB J 1:177–185
196. Wu SC, Solaro RJ (2007) Protein kinase C zeta. A novel regulator of both phosphorylation and de-phosphorylation of cardiac sarcomeric proteins. J Biol Chem 282:30691–30698. doi:10.1074/jbc.M703670200
197. Kusubata M, Tokui T, Matsuoka Y, Okumura E, Tachibana K, Hisanaga S, Kishimoto T, Yasuda H, Kamijo M, Ohba Y (1992) p13suc1 suppresses the catalytic function of p34cdc2 kinase for intermediate filament proteins, in vitro. J Biol Chem 267:20937–20942
198. Ohtakara K, Inada H, Goto H, Taki W, Manser E, Lim L, Izawa I, Inagaki M (2000) p21-activated kinase PAK phosphorylates desmin at sites different from those for Rho-associated kinase. Biochem Biophys Res Commun 272:712–716. doi:10.1006/bbrc.2000.2854
199. Van Eyk JE, Arrell DK, Foster DB, Strauss JD, Heinonen TY, Furmaniak-Kazmierczak E, Côté GP, Mak AS (1998) Different molecular mechanisms for Rho family GTPase-dependent, Ca2+-independent contraction of smooth muscle. J Biol Chem 273:23433–23439
200. Kawajiri A, Yasui Y, Goto H, Tatsuka M, Takahashi M, Nagata K-I, Inagaki M (2003) Functional significance of the specific sites phosphorylated in desmin at cleavage furrow: Aurora-B may phosphorylate and regulate type III intermediate filaments during cytokinesis coordinatedly with Rho-kinase. Mol Biol Cell 14:1489–1500. doi:10.1091/mbc.E02-09-0612
201. Agnetti G, Halperin VL, Kirk JA, Chakir K, Guo Y, Lund L, Nicolini F, Gherli T, Guarnieri C, Caldarera CM, Tomaselli GF, Kass DA, Van Eyk JE (2014) Desmin modifications associate with amyloid-like oligomers deposition in heart failure. Cardiovasc Res 102:24–34. doi:10.1093/cvr/cvu003
202. Rappaport L, Contard F, Samuel JL, Delcayre C, Marotte F, Tomè F, Fardeau M (1988) Storage of phosphorylated desmin in a familial myopathy. FEBS Lett 231:421–425
203. Pica EC, Kathirvel P, Pramono ZAD, Lai P-S, Yee W-C (2008) Characterization of a novel S13F desmin mutation associated with desmin myopathy and heart block in a Chinese family. Neuromuscul Disord 18:178–182. doi:10.1016/j.nmd.2007.09.011
204. Cieniewski-Bernard C, Mounier Y, Michalski J-C, Bastide B (2006) O-GlcNAc level variations are associated with the development of skeletal muscle atrophy. J Appl Physiol (1985) 100:1499–1505. doi:10.1152/japplphysiol.00865.2005
205. Cieniewski-Bernard C, Bastide B, Lefebvre T, Lemoine J, Mounier Y, Michalski J-C (2004) Identification of O-linked N-acetylglucosamine proteins in rat skeletal muscle using

two-dimensional gel electrophoresis and mass spectrometry. Mol Cell Proteomics 3: 577–585. doi:10.1074/mcp.M400024-MCP200
206. Laczy B, Marsh SA, Brocks CA, Wittmann I, Chatham JC (2010) Inhibition of O-GlcNAcase in perfused rat hearts by NAG-thiazolines at the time of reperfusion is cardioprotective in an O-GlcNAc-dependent manner. Am J Physiol Heart Circ Physiol 299:H1715–H1727. doi:10.1152/ajpheart.00337.2010
207. Huang HY, Graves DJ, Robson RM, Huiatt TW (1993) ADP-ribosylation of the intermediate filament protein desmin and inhibition of desmin assembly in vitro by muscle ADP-ribosyltransferase. Biochem Biophys Res Commun 197:570–577. doi:10.1006/bbrc.1993.2517
208. Yuan J, Huiatt TW, Liao CX, Robson RM, Graves DJ (1999) The effects of mono-ADP-ribosylation on desmin assembly–disassembly. Arch Biochem Biophys 363:314–322
209. Zhou H, Huiatt TW, Robson RM, Sernett SW, Graves DJ (1996) Characterization of ADP-ribosylation sites on desmin and restoration of desmin intermediate filament assembly by de-ADP-ribosylation. Arch Biochem Biophys 334:214–222. doi:10.1006/abbi.1996.0449
210. Cohen S, Zhai B, Gygi SP, Goldberg AL (2012) Ubiquitylation by Trim32 causes coupled loss of desmin, Z-bands, and thin filaments in muscle atrophy. J Cell Biol 198:575–589. doi:10.1083/jcb.201110067
211. Frosk P, Weiler T, Nylen E, Sudha T, Greenberg CR, Morgan K, Fujiwara TM, Wrogemann K (2002) Limb-girdle muscular dystrophy type 2H associated with mutation in TRIM32, a putative E3-ubiquitin-ligase gene. Am J Hum Genet 70:663–672. doi:10.1086/339083
212. Kudryashova E, Struyk A, Mokhonova E, Cannon SC, Spencer MJ (2011) The common missense mutation D489N in TRIM32 causing limb girdle muscular dystrophy 2H leads to loss of the mutated protein in knock-in mice resulting in a Trim32-null phenotype. Hum Mol Genet 20:3925–3932. doi:10.1093/hmg/ddr311
213. Saccone V, Palmieri M, Passamano L, Piluso G, Meroni G, Politano L, Nigro V (2008) Mutations that impair interaction properties of TRIM32 associated with limb-girdle muscular dystrophy 2H. Hum Mutat 29:240–247. doi:10.1002/humu.20633
214. Shieh PB, Kudryashova E, Spencer MJ (2011) Limb-girdle muscular dystrophy 2H and the role of TRIM32. Handb Clin Neurol 101:125–133. doi:10.1016/B978-0-08-045031-5.00009-8
215. Schoser BGH, Frosk P, Engel AG, Klutzny U, Lochmüller H, Wrogemann K (2005) Commonality of TRIM32 mutation in causing sarcotubular myopathy and LGMD2H. Ann Neurol 57:591–595. doi:10.1002/ana.20441
216. Kudryashova E, Wu J, Havton LA, Spencer MJ (2009) Deficiency of the E3 ubiquitin ligase TRIM32 in mice leads to a myopathy with a neurogenic component. Hum Mol Genet 18:1353–1367. doi:10.1093/hmg/ddp036
217. Wang X, Osinska H, Dorn GW, Nieman M, Lorenz JN, Gerdes AM, Witt S, Kimball T, Gulick J, Robbins J (2001) Mouse model of desmin-related cardiomyopathy. Circulation 103:2402–2407
218. Liu J, Tang M, Mestril R, Wang X (2006) Aberrant protein aggregation is essential for a mutant desmin to impair the proteolytic function of the ubiquitin-proteasome system in cardiomyocytes. J Mol Cell Cardiol 40:451–454. doi:10.1016/j.yjmcc.2005.12.011
219. Zheng Q, Su H, Ranek MJ, Wang X (2011) Autophagy and p62 in cardiac proteinopathy. Circ Res 109:296–308. doi:10.1161/CIRCRESAHA.111.244707
220. Kostareva A, Sjöberg G, Bruton J, Zhang S-J, Balogh J, Gudkova A, Hedberg B, Edström L, Westerblad H, Sejersen T (2008) Mice expressing L345P mutant desmin exhibit morphological and functional changes of skeletal and cardiac mitochondria. J Muscle Res Cell Motil 29:25–36. doi:10.1007/s10974-008-9139-8
221. Mavroidis M, Panagopoulou P, Kostavasili I, Weisleder N, Capetanaki Y (2008) A missense mutation in desmin tail domain linked to human dilated cardiomyopathy promotes cleavage of the head domain and abolishes its Z-disc localization. FASEB J 22:3318–3327. doi:10.1096/fj.07-088724
222. Clemen CS, Stöckigt F, Strucksberg K-H, Chevessier F, Winter L, Schütz J, Bauer R, Thorweihe J-M, Wenzel D, Schlötzer-Schrehardt U, Rasche V, Krsmanovic P, Katus HA,

Rottbauer W, Just S, Müller OJ, Friedrich O, Meyer R, Herrmann H, Schrickel JW, Schröder R (2014) The toxic effect of R350P mutant desmin in striated muscle of man and mouse. Acta Neuropathol. doi:10.1007/s00401-014-1363-2

223. Joanne P, Chourbagi O, Hourdé C, Ferry A, Butler-Browne G, Vicart P, Dumonceaux J, Agbulut O (2013) Viral-mediated expression of desmin mutants to create mouse models of myofibrillar myopathy. Skelet Muscle 3:4. doi:10.1186/2044-5040-3-4
224. Segard B-D, Delort F, Bailleux V, Simon S, Leccia E, Gausseres B, Briki F, Vicart P, Batonnet-Pichon S (2013) N-acetyl-L-cysteine prevents stress-induced desmin aggregation in cellular models of desminopathy. PLoS One 8, e76361. doi:10.1371/journal.pone.0076361
225. Maloyan A, Osinska H, Lammerding J, Lee RT, Cingolani OH, Kass DA, Lorenz JN, Robbins J (2009) Biochemical and mechanical dysfunction in a mouse model of desmin-related myopathy. Circ Res 104:1021–1028. doi:10.1161/CIRCRESAHA.108.193516
226. Sanbe A, Daicho T, Mizutani R, Endo T, Miyauchi N, Yamauchi J, Tanonaka K, Glabe C, Tanoue A (2009) Protective effect of geranylgeranylacetone via enhancement of HSPB8 induction in desmin-related cardiomyopathy. PLoS One 4, e5351. doi:10.1371/journal.pone.0005351
227. Winter L, Staszewska I, Mihailovska E, Fischer I, Goldmann WH, Schröder R, Wiche G (2014) Chemical chaperone ameliorates pathological protein aggregation in plectin-deficient muscle. J Clin Invest 124:1144–1157. doi:10.1172/JCI71919
228. Ito H, Kamei K, Iwamoto I, Inaguma Y, Tsuzuki M, Kishikawa M, Shimada A, Hosokawa M, Kato K (2003) Hsp27 suppresses the formation of inclusion bodies induced by expression of R120G alpha B-crystallin, a cause of desmin-related myopathy. Cell Mol Life Sci 60:1217–1223. doi:10.1007/s00018-003-3024-9
229. Weisleder N, Taffet GE, Capetanaki Y (2004) Bcl-2 overexpression corrects mitochondrial defects and ameliorates inherited desmin null cardiomyopathy. Proc Natl Acad Sci U S A 101:769–774. doi:10.1073/pnas.0303202101
230. Maloyan A, Sayegh J, Osinska H, Chua BHL, Robbins J (2010) Manipulation of death pathways in desmin-related cardiomyopathy. Circ Res 106:1524–1532. doi:10.1161/CIRCRESAHA.109.212639
231. Tannous P, Zhu H, Johnstone JL, Shelton JM, Rajasekaran NS, Benjamin IJ, Nguyen L, Gerard RD, Levine B, Rothermel BA, Hill JA (2008) Autophagy is an adaptive response in desmin-related cardiomyopathy. Proc Natl Acad Sci U S A 105:9745–9750. doi:10.1073/pnas.0706802105
232. Maloyan A, Gulick J, Glabe CG, Kayed R, Robbins J (2007) Exercise reverses preamyloid oligomer and prolongs survival in alphaB-crystallin-based desmin-related cardiomyopathy. Proc Natl Acad Sci U S A 104:5995–6000. doi:10.1073/pnas.0609202104
233. Bhuiyan MS, Pattison JS, Osinska H, James J, Gulick J, McLendon PM, Hill JA, Sadoshima J, Robbins J (2013) Enhanced autophagy ameliorates cardiac proteinopathy. J Clin Invest 123:5284–5297. doi:10.1172/JCI70877
234. Psarras S, Mavroidis M, Sanoudou D, Davos CH, Xanthou G, Varela AE, Panoutsakopoulou V, Capetanaki Y (2012) Regulation of adverse remodelling by osteopontin in a genetic heart failure model. Eur Heart J 33:1954–1963. doi:10.1093/eurheartj/ehr119
235. Sanbe A, Marunouchi T, Yamauchi J, Tanonaka K, Nishigori H, Tanoue A (2011) Cardioprotective effect of nicorandil, a mitochondrial ATP-sensitive potassium channel opener, prolongs survival in HSPB5 R120G transgenic mice. PLoS One 6, e18922. doi:10.1371/journal.pone.0018922
236. Matsushita N, Hirose M, Sanbe A, Kondo Y, Irie Y, Taira E (2014) Nicorandil improves electrical remodelling, leading to the prevention of electrically induced ventricular tachyarrhythmia in a mouse model of desmin-related cardiomyopathy. Clin Exp Pharmacol Physiol 41:89–97. doi:10.1111/1440-1681.12185
237. Ranek MJ, Terpstra EJM, Li J, Kass DA, Wang X (2013) Protein kinase g positively regulates proteasome-mediated degradation of misfolded proteins. Circulation 128:365–376. doi:10.1161/CIRCULATIONAHA.113.001971

238. Gupta MK, Gulick J, Liu R, Wang X, Molkentin JD, Robbins J (2014) Sumo E2 enzyme UBC9 is required for efficient protein quality control in cardiomyocytes. Circ Res 115: 721–729. doi:10.1161/CIRCRESAHA.115.304760
239. Palmisano MG, Bremner SN, Hornberger TA, Meyer GA, Domenighetti AA, Shah SB, Kiss B, Kellermayer M, Ryan AF, Lieber RL (2014) Muscle intermediate filaments form a stress-transmitting and stress-signaling network in muscle. J Cell Sci. doi:10.1242/jcs.142463
240. Mohamed JS, Hajira A, Li Z, Paulin D, Boriek AM (2011) Desmin regulates airway smooth muscle hypertrophy through early growth-responsive protein-1 and microRNA-26a. J Biol Chem 286:43394–43404. doi:10.1074/jbc.M111.235127
241. Sadegh MK, Ekman M, Rippe C, Uvelius B, Swärd K, Albinsson S (2012) Deletion of Dicer in smooth muscle affects voiding pattern and reduces detrusor contractility and neuroeffector transmission. PLoS One 7, e35882. doi:10.1371/journal.pone.0035882
242. Ye X, Zhang HM, Qiu Y, Hanson PJ, Hemida MG, Wei W, Hoodless PA, Chu F, Yang D (2014) Coxsackievirus-induced miR-21 disrupts cardiomyocyte interactions via the down-regulation of intercalated disk components. PLoS Pathog 10, e1004070. doi:10.1371/journal.ppat.1004070

Chapter 12
Desmin Filaments and Desmin-Related Myopathy

Xuejun Wang

Introduction

At the histological level, muscle tissue is one of the four basic tissue types in the body, with the other three being epithelial, connective, and nervous tissues. Muscle tissue consists primarily of myocytes (muscle cells). Based on the morphology and functionality, muscle is further classified into three types: skeletal muscle, cardiac muscle, and smooth muscle. Skeletal and cardiac muscles together are also known as striated muscle as their myocytes or muscle fibers display microscopic striations. Among all cell types, myocytes especially striated myocytes have the highest content of cytoskeleton. This is because the primary function of myocytes is to generate mechanical force through contraction. The contractile apparatus (myofibril) in myocytes is primarily formed by the thick filament and the thin filament which are considered cytoskeleton. In addition to myofibrillar cytoskeleton, myocytes contain extra-myofibrillar cytoskeleton, including microtubules, microfilaments, and intermediate filaments (IFs), similar to non-muscle cells. In general IFs are formed by tissue-specific IF proteins. The most prominent IF protein expressed in muscle tissue is desmin.

Desmin was initially purified in 1976 from smooth muscle (chicken gizzard) [1]. As a cytoskeletal protein, desmin is a highly insoluble. Most buffers that solubilize myosin and the majority of actin are unable to solubilize desmin in muscle tissue; but desmin becomes soluble in presence of urea. Even in presence of 8M urea under a variety of conditions, desmin co-migrates with actin during purification. The employment of sodium dodecyl sulfate-polyacrylamide gel electrophoresis (SDS-PAGE) helped separate desmin from actin, leading to biochemical purification of desmin proteins. The first antibodies against desmin were then produced using

X. Wang (✉)
Division of Basic Biomedical Sciences, Sanford School of Medicine
of the University of South Dakota, 414 East Clark Street, Lee Medical
Sciences Building, Vermillion, SD 57069, USA
e-mail: xuejun.wang@usd.edu

H. Schatten (ed.), *The Cytoskeleton in Health and Disease*,
DOI 10.1007/978-1-4939-2904-7_12

the purified desmin protein, which allowed the spatial arrangement and structure of desmin filaments in striated and smooth muscles to be initially characterized via immunofluorescence staining. In striated muscle, desmin filaments were localized in between myofibrils at the z line level where they appear to link adjacent myofibrils to each other and to the sarcolemma. In cardiac muscle, desmin protein was also found to be enriched in the desmosome-like structure at the intercalated disc which couples adjacent cardiomyocytes head-to-head [1]. Lazarides and Hubbard termed the protein desmin (from the Greek, δεσμós = link, bond) to indicate the linking function that this molecule might have in muscle cells based on its immunofluorescence localization [1]. They named this protein as desmin for also denoting that this protein, besides being the subunit of the 10 nm filaments, is also a component of muscle desmosome-like structure in muscle cells [1].

In the past four decades, significant progresses are made in understanding of the structure and function of desmin protein and desmin IFs. Mouse genetic modification studies reveal desmin is not required for embryonic myogenesis but is essential to postnatal maintenance and function of striated muscles. The recent renaissance of research into desmin IFs was triggered by the first identification of mutations in the desmin gene (*DES*) and the genes encoding the partners of desmin IFs, such as αB-crystallin (*CRYAB*), in human familial desmin-related myopathy (DRM) in 1998 [2–4]. DRM is heterogeneous group of myopathies characterized pathologically by the presence of desmin-containing aberrant protein aggregates in myocytes. Since aberrant protein aggregation is a common process in all proteinopathies, insight gained from studying DRM pathogenesis serves also to improve our understanding of protein quality control and proteinopathy in general.

The Desmin Gene

Approximately 5 years after Capetanaki et al. had first identified the *DES* in chicken [5], Paulin and colleague cloned and characterized the human *DES* in 1989 [6]. In the human genome, there is a single copy of *DES*, which is 8.4 kbp in length, consists of nine exons and eight introns, and is localized to chromosome 2q35 [6, 7]. Northern blot analysis revealed that a single desmin mRNA of 2.2 kb is expressed in human striated and smooth muscles, which translates a protein of 470 amino acids (~53 kDa) [1, 6, 8]. From zebrafish to humans, DES is highly conserved.

A 280-bp muscle-specific enhancer located between −693 and −973 bp upstream of the transcription initiation site of human *DES* has been shown to confer high level expression of *DES* in both myoblasts and myotubes [9]. Further analysis revealed that this enhancer contains two different regions, one active in myoblasts and the other in myotubes. The myotube-specific region contains two E-box elements, providing the binding sites for MyoD1 and muscle-specific enhancer factor-2 (MEF-2), respectively; both are required for full enhancer activity. The myoblast-specific region is downstream of the myotube enhancer region, containing a region with homology to the M-CAT motif (at −587) and multiple regions harboring a GC-rich sequence sharing homology with the Krox binding site [10]. This combination of promoter regulation may help explain why DES, but not contractile proteins, is expressed in myoblasts. A transgenic mouse model harboring a LacZ gene under the control of a promoter containing these regulatory regions shows transgene expression exclusively in skeletal muscle but not cardiac and smooth muscles [11], indicating that these regulatory elements are only sufficient to drive *DES* expression in skeletal muscle and that other regulatory regions must exist for cardiac and smooth muscle expression.

In mice, the first 85-bp upstream of the transcription initiation site of mouse *DES* contains an E box (E1), sufficient to drive low level of muscle-specific expression. An enhancer located between nucleotides −798 and −976 harbors another E box (E2) and a MEF-2 binding site, which drives high levels of *DES* expression. Both MyoD and myogenin are capable of binding to the E1 and E2; and MEF-2C, a myocyte-restricted member of the MEF-2 family can bind to the MEF-2 site in the enhancer [12]. Subsequent studies using mouse transgenesis demonstrated that a single MEF-2C binding site within 1kbp 5'-flanking sequence of mouse *DES* is sufficient to direct appropriate temporal expression of desmin in both cardiac and skeletal muscles during mouse embryogenesis [13].

In murine embryogenesis, desmin is detected as early as 8.25 days post coitum (d.p.c.) in the ectoderm, and then in the heart rudiment at 8.5 d.p.c., with elevated expression in cardiogenic cells thereafter [14]. Desmin protein begins to accumulate in somites at 9 d.p.c. and progresses in a rostro-caudal gradient with somatic maturation [14]. Desmin-positive myofibers are found in limb buds by 14 d.p.c. The levels of desmin expression in cardiac, skeletal, and smooth muscles stay high throughout subsequent embryonic development and into postnatal life [11]. Desmin is one of the earliest muscle-specific genes expressed during development; desmin protein has substantially accumulated in myogenic cells before myosin, actin, and titin begin to form the contractile apparatus [15–17].

Desmin Protein Structure and Filament Formation

Desmin belongs to the type III IF protein family. Other members of this family include vimentin, glial fibrillary acidic protein (GFAP), and peripherin. As illustrated in Fig. 12.1, a desmin protein molecule is comprised of three domains: an α-helical rod domain containing 304 amino acid residues, flanked by globular amino- and carboxyl-terminal structures which are known as the head and tail domains, respectively. The α-helical rod domain is interrupted by three short non-helical linker regions, resulting in four consecutive α-helical segments (1A, 1B, 2A, and 2B). The helical segments of the rod are highly conserved, sharing a sequence characteristic of a 7-residue (heptad) repeat pattern with a typical sequence of hydrophobic and hydrophilic amino acids. Each heptad forms a full helical turn and the heptad repeats of the α-helix in a desmin monomer guide two monomers into formation of a coiled-coil dimer, the basic unit of the filament. The equivalence of the eighth heptad of the 2B segment is interrupted by an insertion of four extra residues referred to as a 'stutter' (Fig. 12.1) [18]. The 'stutter' appears to be an obligatory

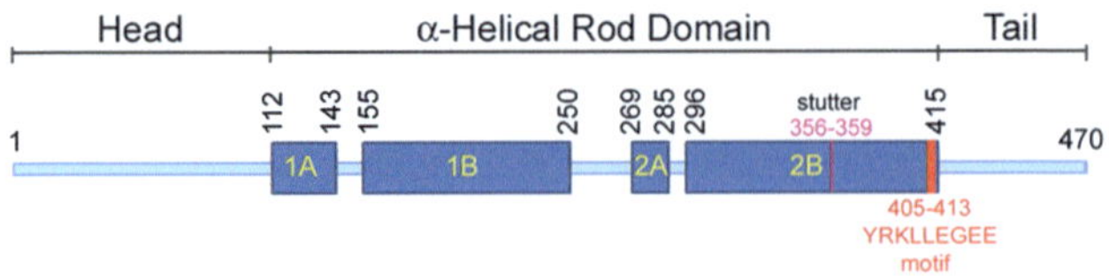

Fig. 12.1 A schematic illustration of desmin protein structure. Desmin protein consists of three domains: an α-helical rod domain containing 304 amino acid residues, flanked by globular amino- and carboxyl-terminal structures which are known as the head and tail domains, respectively. The α-helical rod domain is separated by three short non-helical linker regions, resulting in four consecutive α-helical segments (1A, 1B, 2A, and 2B). The heptad repeat pattern of the 2B helical segment is interrupted by an insertion of 4 amino acid residues at position 356–359, known as a stutter, and an YRKLLEGEE motif, marked by the *purple* and *orange boxes*, respectively in the illustration

feature of all IF proteins and its position is strictly conserved among different IF proteins. "Stutterless" desmin monomers created experimentally by insertion of the three "missing" amino acid to restore a continuous heptad repeat, lose the ability to anneal into longer filaments during filament elongation [19]. The presence of the stutter results in a slight unwinding of the coiled coil in the stutter vicinity. The local unwinding is apparently essential to the proper assembly of the filament. The YRKLLEGEE motif at the C-terminal end of the 2B segment is another well-studied structure. The coiled-coil interaction starts to loosen in this motif so that the two α-helices gradually separate and eventually bend away from each other at the EGEE level [20]. In vitro filament assembly studies reveal that the YRKLLEGEE motif directs the proper formation of tetramers and dictate the number of subunits per filament cross section. Approximately 30 % of the 'tail' domain is constituted by β-sheets, with the remainder of the domain bearing predominantly random structure and lacking the heptad repeat pattern. The tail domain participates in the longitudinal head-to-tail assembly of tetramers [21] and is involved in the control of lateral packing, as well as stabilization and elongation of the higher order filament structures [22, 23]. Another major function of the tail is to interact with other cytoskeletal proteins, which helps establish a cytoplasmic IF network [24].

The Desmin IF Network

In mature striated muscle, desmin IFs link myofibrils to each other at the z-disc level, to the sarcolemma, and to the nuclear envelope (Fig. 12.2) [25]. Desmin filaments surround each myofibril at the z-disc level where they interact with α-actinin

Fig. 12.2 (continued) the nuclear pore. The longitudinal component of desmin filament networks lay in the inter-myofibrillar space where they interact with and, help the positioning of, mitochondria (**b**) A confocal micrograph of a mouse ventricular myocardial section immunofluorescence-stained for desmin (*green*). Desmin staining displays a striated pattern and is enriched at the intercalated discs

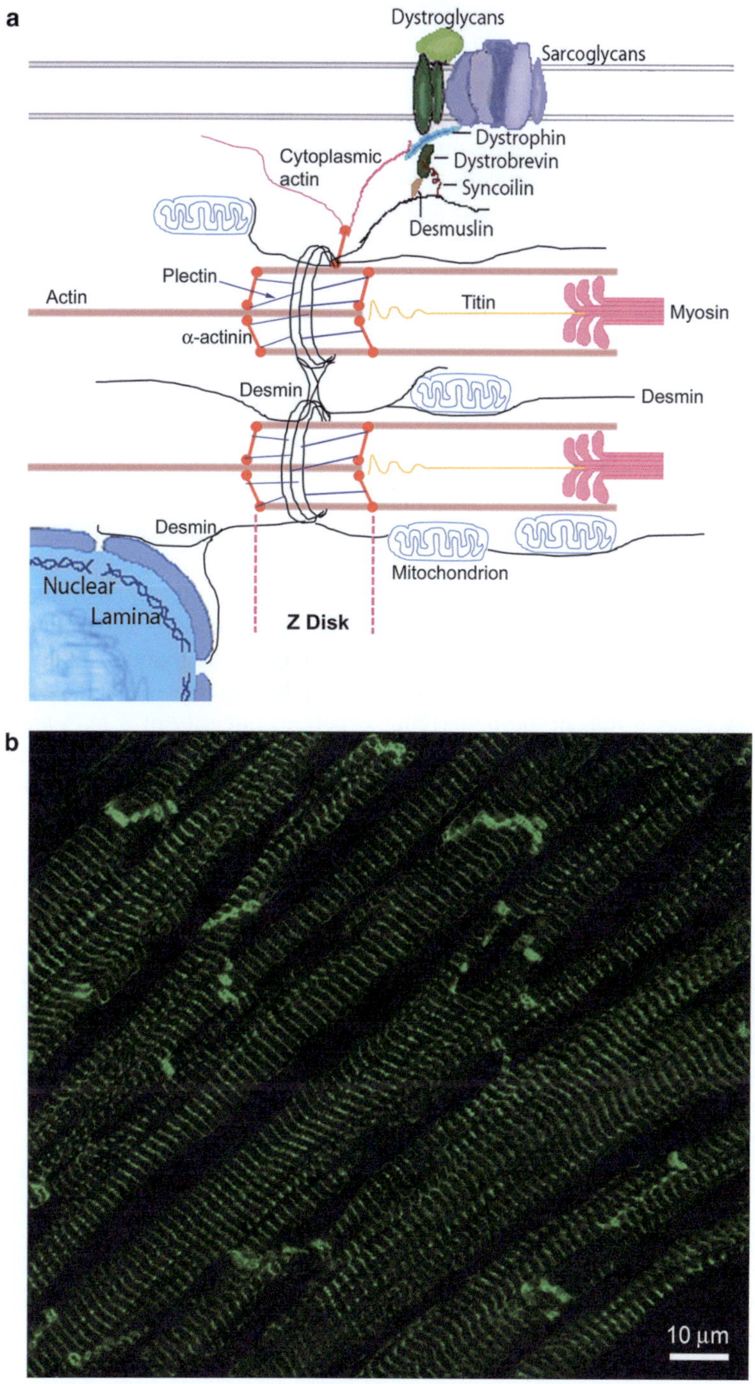

Fig. 12.2 Desmin filament distribution in striated muscle. (**a**) A schematic illustration of desmin filaments in relation to the z-disc of myofibrils, dystrophin complex, the nuclear envelop, and mitochondria. Desmin filaments surround each myofibril at the z-disc level and interact with α-actinin through plectin; desmin filaments interact with dytrobrevin in the dystrophin complex via syncoilin and desmuslin; and desmin filaments insert into the nuclear envelop at the proximity of

through plectin (isoforms 1d and 1f) [26]; desmin IFs interact with dytrobrevin in the dystrophin complex via syncoilin and desmuslin [27], thereby connecting to dystroglycans and sarcoglycans on the cell membrane; desmin IFs also associate with integrin via interaction with subsarcolemmal cytoskeleton and costameres; and lastly desmin filaments insert into the nuclear envelop at the proximity of the nuclear pore. Plectin (isoform 1b) also mediates the interaction of desmin filaments with mitochondria [26]. The longitudinal components of the desmin network insert into the desmosomes of the intercalated disc via desmoplakin. Desmin is highly enriched in desmosomes. Desmin showed a normal intracellular distribution but failed to localize at the intercalated discs in the myocardium from a patient with homozygous C-terminal truncation of desmoplakin [28]. Some data suggest that desmin filaments interact with the nuclear lamina through the nuclear pore [25, 29]. The postulated physical interaction is difficult to prove but a functional relationship is well implicated. For example, transgenic ablation of the gene encoding lamin A/C, the major proteins forming the nuclear lamina, produced a typical dilated cardiomyopathy (DCM) with markedly altered distribution of the desmin IF network and its relationship with the nuclear pores [30], suggesting an interaction between desmin IFs and lamin A/C and potential significance of this interaction in mechanotransduction. The unique and ubiquitous distribution of the desmin IF network has led to the hypothesis that the desmin IF network plays a role in the underlying structural integrity of a muscle cell, as well as participating in the signaling processes necessary for integration of cellular responses to external and internal stimuli [29]. However, this hypothesis has not been fully tested.

Like IFs in epithelia, desmin filaments in myocytes are mainly organized by plectin, a 500 kDa cytolinker protein. A study of conditional knockout of the plectin gene in striated muscles in mice demonstrates that plectin deficiency prevents desmin filaments from attaching to Z-disks, costameres, mitochondria, and the nuclear envelop, causing the formation of desmin aggregates of distinct morphology and in distinct cytoplasmic compartments, depending on which plectin isoforms are missing [26]. Striated muscle expresses two major plectin isoforms, plectin 1d and 1f, which specifically target and link desmin filaments to Z-disks and costameres, while plectin 1b forges a linkage to mitochondria [26]. On the other hand, desmin filaments also act as a scaffold for its interacting partner proteins; hence loss of desmin or altered distribution of desmin can cause changes in the expression and distribution of its partner proteins in the cell. For example, in desmin null mice, syncoilin was markedly decreased in skeletal muscle, disappeared from sarcomeric z-lines and neuromuscular junctions, and relocated from the sub-sarcolemmal cytoskeleton to the cytoplasm [31]; however, immunofluorescence microscopy revealed that knockout of syncoilin did not appear to discernibly change the distribution of desmin in striated muscle cells [32].

Desmin Loss of Function

As one of the earliest expressed genes during embryonic myogenesis, *DES* was shown by earlier studies using cell culture systems to be required for myocyte differentiation and myogenesis [33, 34]; however, this proposition is not proven by subsequent in vivo desmin loss of function studies. Two mouse models of germline knockout of *DES* were generated by two independent groups and both showed that desmin-null mice are viable and fertile [35, 36]. Studies of these desmin-null mice have yielded significant insight into the function of desmin IFs. Desmin is dispensable for the formation of skeletal, cardiac and smooth muscles during embryonic development but desmin null mice do display postnatal progressive structural disintegration and functional impairment in all three types of muscle [35, 36]. Adult desmin-null mice often show a slightly smaller body size, tend to be less strong and quicker to become fatigued, and show a significantly shortened lifespan (~12-months vs. 2 years in wild type), compared with their wild type littermate controls [35–39]. These general abnormalities observed in desmin-null mice are underlined by specific pathologies occurred in cardiac, skeletal, and smooth muscles, including life-threatening cardiomyopathy.

Consequence of Desmin Loss of Function in Cardiac Muscle

Desmin-null mice develop cardiomyopathy, the primary cause of their premature death. A prominent myocardial lesion in desmin-null hearts is focal cardiomyocyte degeneration, necrosis, macrophage infiltration, calcification, and fibrosis. Yellowish-white lesions ranging from single or multiple spots to confluent areas corresponding to calcification at the surface of the heart are macroscopically visible as early as 2 weeks postnatal. These lesions seem to affect more frequently the free wall of the right ventricle and the right ventricle side of the interventricular septum. At the microscopic level, cardiomyocyte degeneration can be detected as early as postnatal day 5 and throughout the heart [39]. Electron microscopy discerns alterations in the intercalated disc and sarcolemma of cardiomyocytes. Myofibril disorganization and abnormal nuclear and mitochondrial morphology and positioning are also detected via ultrastructural examination of the desmin-null hearts [40]. Cardiac injury caused by desmin deficiency triggers reactivation of the fetal gene program and the loss of cardiomyocytes renders the remaining cardiomyocytes to undergo hypertrophy and impairs cardiac mechanical function; ultimately, desmin-null mice develop DCM and congestive heart failure [38]. Cardiomyocyte-restricted overexpression of desmin completely rescued the cardiac pathology in desmin null mice [41], demonstrating that cardiac lesions observed in desmin-null mice are cardiomyocyte autonomous, not secondary to vascular abnormalities. Nevertheless, expression of desmin is high in microarteries, where desmin plays a role in generating both passive and active tension [42].

According to some reports, the earliest ultrastructural defects of desmin-null cardiomyocytes are observed in mitochondria. In addition to mitochondrial clumping, extensive mitochondrial proliferation, swelling, and matrix degeneration are observed in a fraction of cardiomyocytes deficient of desmin, particularly after exercise. Functionally, no difference was discerned in the in vitro maximal rates of respiration in isolated cardiac mitochondria from desmin-null and wild-type mice; however, ADP-stimulated mitochondrial respiration in situ using saponin skinned muscle fibers was significantly reduced in cardiac and soles muscles from desmin-null mice, compared with wild type control [43]. Bcl-2 is an integral outer membrane protein of mitochondria that protects against apoptosis without affecting mitochondrial function. By cross-breeding desmin-null mice with Bcl2 overexpression transgenic mice, Weisleder et al. found that mitochondrial abnormalities in desmin-null hearts were remarkably ameliorated by Bcl-2 overexpression and, more intriguingly, that the correction of mitochondrial defects was associated with reduction in fibrotic lesions in the myocardium, prevention of cardiac hypertrophy, restoration of cardiomyocyte ultrastructure, and significant improvement of cardiac function. Furthermore, loss of desmin was found to diminish the capacity of mitochondria to resist exposure to calcium, a defect that was also partially restored by Bcl-2 overexpression [44]. These findings demonstrate that desmin filaments are essential for the positioning and functioning of mitochondria and mitochondrial dysfunction is a major cause of cardiomyopathy in desmin-null mice.

Consequence of Desmin Loss of Function in Skeletal Muscle

Skeletal muscle formation starts with the progenitor cell commitment to the myogenic lineage, myoblast proliferation, differentiation, and fusion to form first primary, and then secondary myotubes. As the primary myotube forms, new generations of myoblasts cluster around the primary myotube and use it as a scaffold to form secondary myotubes. Detailed analyses of myogenesis in mouse embryos have revealed that somites and myotomes form normally and mononucleate muscle precursor cells migrate normally in the absence of desmin. The assembly of sarcomeres and myofibrils does not seem to differ between desmin-null and wild type littermate embryos [37]. However, like cardiac muscle, alterations in skeletal muscle are discernible in desmin-null mice soon after birth. These pathological changes including focal areas of muscle degeneration, regeneration involving satellite cell activation and formation of new fibers, and fibrosis are most prominent in the highly solicited muscles, such as soleus (a weight-bearing muscle) or diaphragm and tongue (both very active muscles). Faulty myofibrillogenesis was frequently observed in regenerating myotubes and fibers from postnatal day 11 up to 12 weeks, suggesting that the desmin filaments are not required in embryonic muscle formation but are needed for a proper myofibrillar assembly during postnatal regeneration. This is perhaps because additional mechanical stress is applied to these muscles after birth as they regenerate, which would not occur in utero. At 2 months of age, the force generated by the soleus of desmin-null mice is significantly less than in the control mice. At 5 months, the soleus of desmin-null mice is no longer able to respond to the stimulation and generate very little force. Defective mitochondrial activity may be an underlying cause; both NADH staining and ultrastructural examinations show large accumulations of mitochondria in muscle fibers deficient of desmin. Hence, it is generally concluded that desmin is not required for the proliferation and commitment of the early myoblasts to the myogenic lineage or for their migration, fusion, and subsequent organization of the muscle fiber before birth; however, after birth, it is not only essential to maintaining structural integrity in mature muscles but also important for muscle regeneration.

Consequence of Desmin Loss of Function in Smooth Muscle

Desmin protein was first isolated from smooth muscle [1]. It was determined later on that in smooth muscle cells, desmin is associated with the dense body, the smooth muscle analogue of the z-disk of striated muscle. Wede et al. analyzed the mechanical property of the aorta, the mesenteric artery, and resistance arteries in desmin-null mice [42]. For aorta and mesenteric artery, passive or active circumference- stress relations were not different between desmin-null and wild type mice. Both passive and active stresses in the microarteries were lower in the desmin-null group. Thus, desmin filaments do not seem to play a major role in the mechanical properties of the large elastic and muscular arteries where desmin expression is relatively low. In the microarteries, which contain a greater amount of desmin protein, desmin IFs show discernible contribution to both passive and active tension.

Asthma is one of the common chronic obstructive pulmonary diseases with airway inflammation and abnormal airway smooth muscle contraction, which is due to an intrinsic abnormality of the airway smooth muscle cells (ASMCs). ASMC hyperplasia and hypertrophy are key determinants of airway remodeling and hyperresponsiveness, characteristic of severe asthma and other chronic obstructive pulmonary diseases. Analysis of bronchial biopsies of asthmatic patients reveals a negative correlation between desmin expression in ASMCs and airway hyperresponsiveness [45], suggesting an important role for desmin in ASMC homeostasis. Indeed, experiments comparing desmin-null mice with wild type mice demonstrate that desmin is a load-bearing protein that increases the stiffness of the airways and the lung and modulates airway contractile response [46]. A subsequent study shows that desmin deficiency induces hypertrophy of ASMCs via up-regulation of microRNA-26a (miR-26a) which targets glycogen synthase kinase-3β, demonstrating a novel role for desmin as an anti-hypertrophic protein necessary for ASMC homeostasis and identifying desmin as a novel regulator of microRNA [47].

The role of desmin IFs in smooth muscle hypertrophy has also been examined in a urinary bladder growth model created by partial obstruction of the urethra.

The results indicate that desmin is not required for urinary bladder smooth muscle growth but plays a role in active force transmission and maintenance of wall structural integrity during growth of urinary bladder [48].

Taken together, the desmin loss of function studies have provided strong support for the hypothesis proposed originally by Lazarides [49] that desmin distributed in the intracellular space functions to link the Z disks together and to the membrane, and is important to maintain structural integrity of the muscle.

Desmin-Related Myopathy (DRM)

Desmin-related myopathy (DRM), also known as desmin myopathy, is a heterogeneous group of myopathies that have a shared pathological characteristic: the presence of desmin-positive aberrant protein aggregates in muscle cells. Biopsies of the affected skeletal and cardiac muscles display intrasarcoplasmic areas containing amorphous eosinophilic deposits which are immunostaining positive for desmin. Electron microscopic examination reveals that these abnormal structures at the ultrastructural level display as electron-dense granular or granulofilamentous material in the intermyofibrillar space; physical contact is sometimes observed between the granulofilamentous material and the z line of the sarcomere. Myofibril organization is apparently altered as evidenced by wavy z lines and z disc widening/streaming (Fig. 12.3) [50].

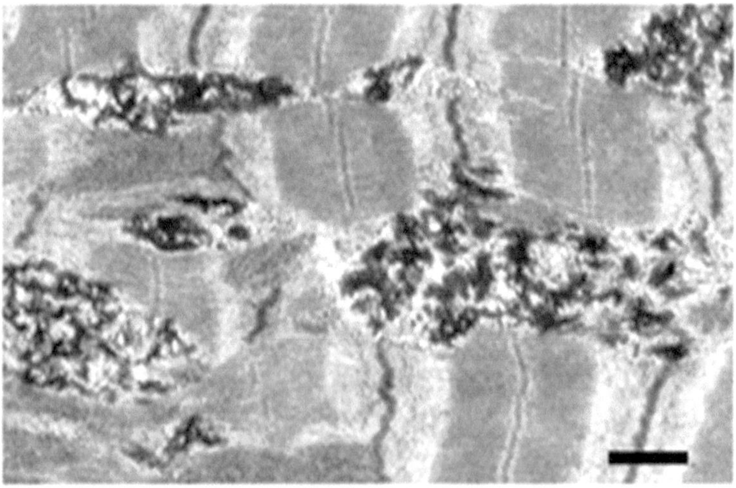

Fig. 12.3 An electron microphotograph of skeletal muscle biopsy from a human DRM patient heterozygous for a missense mutation (Q348P) in the desmin gene. Note that the myofibril organization shows dramatic changes and that Z line widening/streaming as well as accumulation of granular-filamentous materials are prominent. Scale bar = 1 μm. (Adopted from Fichna JP et al. PLoS One 2014; 9: e115470) [50]

DRM belongs to myofibrillar myopathy which is an even more heterogeneous group of muscle disorders featured by the presence of myofibrillar proteins positive inclusions and myofibrillar disintegration and disorganization. These inclusions are desmin-positive as well. Typically, DRM presents with muscle weakness initially in distal muscles, which slowly spreads to affect truncal, neck-flexor, facial, bulbar and respiratory muscles [51]. DRM can present as isolated skeletal myopathy or isolated cardiomyopathy but more often it presents in the form of combined skeletal and cardiac myopathy, with smooth muscle sometime being involved as well [52]. Cardiomyopathy in DRM is known as desmin-related cardiomyopathy (DRC). DRC can present phenotypes of hypertrophic, dilated, or restrictive cardiomyopathies. Conduction blocks and arrhythmias resulting in sudden death are observed as a major clinical manifestation of DRC. Age of disease onset varies from early childhood to mid-aged adult, seemingly depending on the type of inheritance, location of the causative mutation, and the gene mutated. Perhaps for the same reasons, the rate of disease progression is not quite uniformed.

Human Genetics of DRM

The pattern of inheritance pattern in familial DRM includes autosomal dominant or autosomal recessive. However, many DRM cases have no family history, at least some of which de novo *DES* mutations are identified. The first batch of *DES* mutations were reported in 1998 by Goldfarb et al. [2]; through genetic linkage analysis, they associated mutations in the highly conserved carboxyl-terminal end of the rod domain (2B segment) with two families with desmin-related cardiac and skeletal myopathy. They identified a heterozygous A337P mutation in a family with an adult-onset DRM and compound heterozygosity for two other mutations, A360P and N393I, in a second family with childhood-onset aggressive course of DRM. Approximately 1 month after Goldfarb's report, a putative 7-amino acid (R173-E179) deletion in the 1B segment of desmin rod domain was identified in a patient with generalized myopathy by MunÄoz-MaÂrmol et al. [3]. A wild type allele of *DES* was not found in this patient's genome; hence, this patient might be either hemizygous or homozygous for this mutation [3]. As summarized in Fig. 12.4, a large number of mutations have been identified in *DES*, including point substitutions, insertion, small in-frame deletions and a larger exon-skipping deletion. Most mutations are located in the highly conserved alpha-helical rod domain of desmin although mutations in the head and tail domains as well as the linker regions are also common.

Many of the missense mutations result in replacement of the original amino acid into proline, a known helix breaker. Since the helical structure in the rod domain is essential to the orderly polymerization of desmin protein molecules to form desmin filaments, mutations that destroy the α-helixes are believed to impair desmin filament formation. Indeed, studies of mutant desmin transfected cell cultures demonstrate that mutant desmin proteins are often incapable of assembling normal IF but

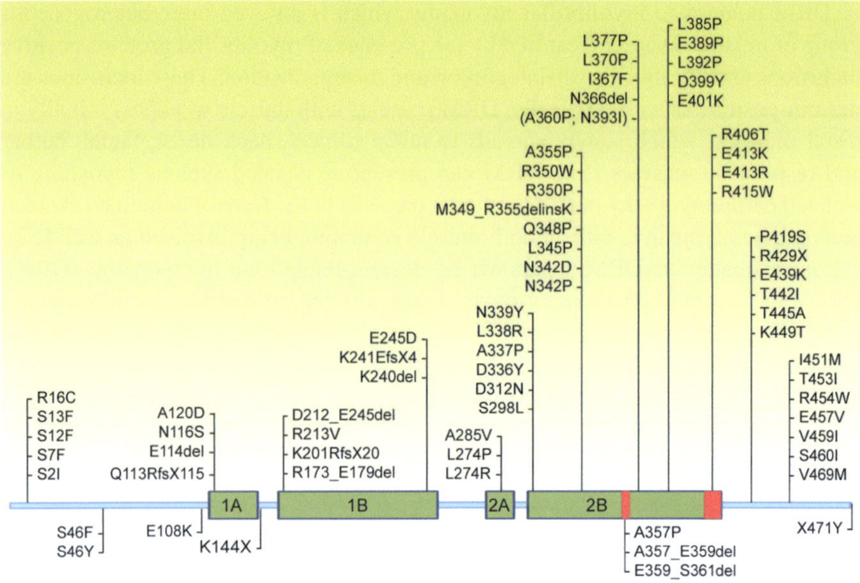

Fig. 12.4 A summary of DES mutations in relation to DES protein domain structure. A total of 67 mutations are identified so far with 8 in the head, 14 in the tail, and 45 in the rod domains

are usually able to disrupt a pre-existing filamentous network in a dominant-negative manner (see section "Disruption of the Desmin Filament Network in DRM").

Point substitutions can also result in premature stop codon and thereby truncation of the protein, in addition to missense mutant proteins. Insertions cause frameshift and truncation. All these situations are observed in DRM-associated *DES* mutations. In frame deletion can result from not only a small deletion in a coding exon but also splice mutations in introns of *DES* (Fig. 12.4). For example, splice site mutations in intron 2 or 3 that flank exon 3 result in deletion of 32 amino acids encoded by exon 3 [53]. Disease caused by mutations in *DES* is known as desminopathy to differentiate from the DRM caused by mutations in other genes.

In a large French pedigree with DRM, no mutation in *DES* was detected but Vicart et al. have identified an R120G missense mutation in the αB-crystallin gene (*CRYAB*) in this pedigree [4]. CRYAB was initially found in the lens of eyes over a century ago but studies published in 1980s have unraveled that CRYAB is also constitutively expressed in many non-lenticular tissues, especially in cardiac and skeletal muscles [54]. CRYAB is a highly conserved protein. Human CRYAB contains 175 amino acid residues with a molecular weight of 20 kDa. It turns out CRYAB is a member of the small heat shock protein family, is one of the most expressed cytosolic non-myofibrillar proteins in cardiomyocytes [54]. CRYAB had been shown to interact with desmin and actin in myocytes, especially under stress conditions. Muscle fibers from patients harboring CRYAB^{R120G} were shown to contain aberrant protein aggregates that are immunopositive for both DES and CRYAB.

Table 12.1 Human disease-linked *CRYAB* mutations

Nucleotide change	Amino acid change	Type of inheritance	Muscle affected	References
c.358A > G	R120G	AD	HCM, SkM, PPC	[4]
c.325G > C	D109H	AD	HCM, SkM, PPC	[55]
c.460G > A	G154S	AD	DCM	[56]
c.470G > A	R157H	AD	DCM	[57]
c.451C > T	Q151X	AD	SkM	[58]
c.464delCT	L155fs_163X	AD	SkM	[58]
c.60delC	S21Afs_24X	AR	SkM	[59]
c.343delT	S115Pfs_129X	AR	SkM	[60]
c.450delA	K150Nfs_184X	AD	PPC	[61]
c.58C > T	P20S	AD	PPC	[62]
c.59C> G	P20R	AD	PPC	[63]
c.557G > A	A171T	AD	cataract	[64]

AD autosomal dominant, *AR* autosomal recessive, *SkM* skeletal myopathy, *HCM* hypertrophic cardiomyopathy, *DCM* dilated cardiomyopathy, *PPC* posterior polar cataract

Hence, CRYAB^{R120G} represents the first mutation in molecular chaperones linked to muscle disease. To date, several additional mutations in *CRYAB* have been linked to DRM. Since the identification of the R120G missense mutation, at least nine additional *CRYAB* mutations have been reported to associate with human disease (Table 12.1), including additional missense mutations as well as truncation and extension due to frame shift resulting from nucleotide deletion. Most of the mutations identified so far are autosomal dominant but autosomal recessive inheritance is also seen in two truncation mutations that linked to families with primarily skeletal myopathy. Patients with *CRYAB* mutations can develop isolated cardiomyopathy (primarily DCM), isolated skeletal myopathy (myofibrillar myopathy), isolated posterior polar cataract type 2 (PPC2), or all three in combination. Overall, all diseases caused by mutations in *CRYAB* is also known as αB-crystallinopathy.

In addition to *DES* and *CRYAB* mutations, mutations in a number of other genes that encode partner proteins of DES, such as myotilin, Z-band alternatively spliced PDZ-containing protein (ZASP), filamin C (FLNC), Bcl-2-associated athanogene-3 (BAG3) [65], are linked to myofibrillar myopathy to which DRM belongs.

Transgenic Mouse Models of DRM

The mechanisms by which DRM-linked genetic mutations cause pathology in DRM patients have been investigated primarily using mouse transgenics and cell cultures. The author of this Chapter and his colleague created the first mouse model of DRC by cardiomyocyte-restricted overexpression of a murine 7-amino-acid (R172-E178) deletion mutant desmin (known as D7-des) [66], which is the mouse homologue of human R173-E179 deletion mutant DES linked to DRM [3]. The D7-des transgenic

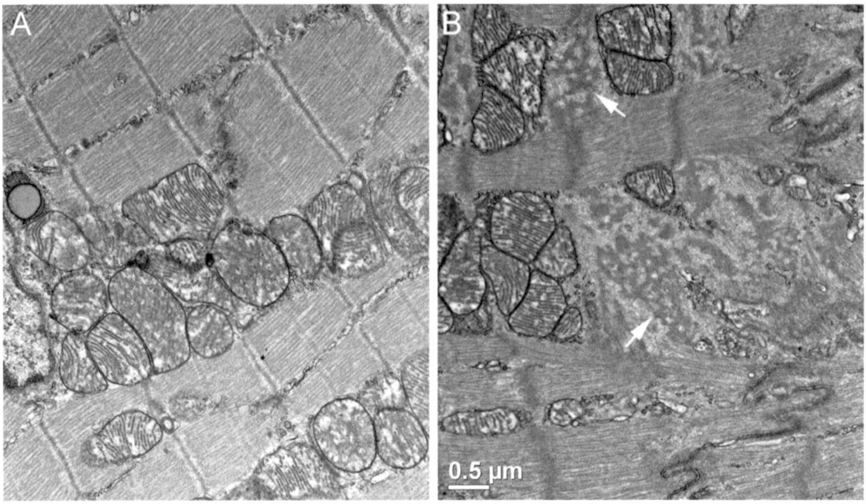

Fig. 12.5 Ultrastructural analysis of DRC mouse myocardium. Longitudinal sections of myocardial specimen from an adult non-transgenic mouse (**a**) and a littermate D7-des transgenic mouse (**b**) were analyzed. *Arrows point* to aberrant granular-filamentous desmin aggregates in the intermyofibril space

mice displays the characteristic intrasarcoplasmic electron-dense granulofilamentous materials in the intermyofibrillar space in the heart (Fig. 12.5) and marked cardiac hypertrophy and cardiac dysfunction, recapitulating main aspects of human DRC. Subsequently, several additional transgenic mouse models expressing other *DES* mutations have also been reported [67, 68]. Notably, an R349P desmin knock-in mouse model, which harbors the ortholog of the most frequently occurring human *DES* missense mutation R350P, has recently been generated [69], providing an animal model arguably most closely mimicking human DRM for investigating the pathogenesis of this mutant desmin in vivo.

Wang et al. also developed the first transgenic mouse model of $\text{CRYAB}^{\text{R120G}}$-based DRC via the mouse *myh6* promoter-driven cardiac-specific overexpression of murine cDNA encoding $\text{CRYAB}^{\text{R120G}}$ [70]. In multiple stable lines of the $\text{CRYAB}^{\text{R120G}}$ transgenic mice, aberrant protein aggregates immune-positive for DES and CRYAB are detected in cardiomyocytes throughout the heart. The rate of disease progression in this model depends on transgene expression level which is transgene copy number dependent in the *myh6* promoter-driven transgenics. In a stable line harboring three copies of the transgene, DRC progression can be divided into three distinct stages. At 1 month of age, these $\text{CRYAB}^{\text{R120G}}$ mice show no apparent cardiac morphological and functional changes except for the presence of aberrant protein aggregates characteristic of DRM in cardiomyocytes; at 3 months, concentric cardiac hypertrophy is clearly discernible with compensated systolic function but impaired diastolic function; by 6 months, typical DCM and congestive heart failure are developed and the mice die prematurely around this age with an average lifespan of

~6 to 7 months. The disease progression in a stable transgenic line harboring one copy of the transgene is much slower [70]. Six years after the report of this first CRYAB^{R120G} mouse model, Rajasekaran et al. described another similarly developed CRYAB^{R120G} transgenic mouse model in which the cDNA of human *CRYAB* was used [71]. In fact, the amino acid sequence of CRYAB is highly conserved between mice and humans with both being 175 amino acid residues long with only 4 amino acids being different but all quite distal to the R120G mutation site. Not surprisingly, this transgenic mouse expressing human CRYAB^{R120G} displays exactly the same pathology as the strain expressing murine CRYAB^{R120G}. More recently, a mouse strain with the R120G mutation knocked in the exact locus of the mouse *cryab* gene, which should mimic most closely the human genetic alteration, has been reported. Initial characterization of this knock-in mouse shows that CRYAB^{R120G} is capable of causing skeletal myopathy and cataract in a dominant manner [72].

Disruption of the Desmin IF Network in DRM

As described in earlier sections, desmin-null mice develop generalized myopathy affecting skeletal, cardiac and smooth muscle structure and functions. Some of the desmin loss-of-function phenotypes resemble the manifestations of DRM [69]. This suggests that loss or disruption of the normal desmin IF network is likely an important pathogenic mechanism of DRM-linked *DES* mutations.

Based on a "time-lapse" electron microscopy of the in vitro assembly of desmin filaments from denatured purified desmin proteins, the assembly process can be divided into four stages: tetramer formation, unit-length filament (ULF) formation, longitudinal annealing and radial compaction, and IF network formation [73]. A seminal study by Bar et al. has analyzed 14 DES missense mutations and showed that two of them (A213V, E245D), both residing in the 1B section of the rod domain, can assemble into morphologically normal IFs that are indistinguishable from IFs formed by wild type DES, whereas four mutations residing in the 2B section (A360P, Q389P, N393I, and D399Y) can also form seemingly normal IFs but these IFs display subtle and yet discernible alterations in morphology and physical property [73]. Moreover, the remaining eight mutants interfere with the assembly process at distinct stages. The L385P and R406W mutants can yield apparently normal ULF but show defective longitudinal annealing and radial compaction; filament assemblies formed by A337P, N342D, or A357P -DES show enhanced stickiness and eventually lead to large aggregates; and for L345P, R350P, or L370P-DES, the assembly can be initiated and progresses to the ULF state but after stalling at ULF briefly, the ULF-like structure rapidly breaks down into small aggregates [73]. When transfected to cultured non-myocyte cells which express no endogenous IF proteins, the mutants with in vitro assembly defects produce dot-like aggregates whereas the mutants that can form IFs in vitro yield a seemingly normal IF network in the cellular context. This corroborates well the in vitro assembly findings. Since all these tested mutants are DRM causing mutants which lead to formation of aberrant desmin aggregates in myocytes of DRM patients, these in vitro findings suggest that aberrant protein aggregation and disruption of desmin filaments are an intrinsic property to only some of the DRM-linked DES mutants but not to others. The latter may give rise to the disease phenotype only in the natural physiological context of cytoskeletal organization and function in myocytes. For example, pathogenic posttranslational modifications (PTMs) triggered by the mutation might not occur in vitro or in non-natural hosting cells.

Indeed, desmin is subject to a number of known PTMs, including phosphorylation, ADP-ribosylation, and ubiquitination as well as non-enzymatic modifications such as glycation, oxidation and nitration [74]. These PTMs are likely crucial to its conformation and function. Many of the *DES* mutations directly remove or add amino acid residues that can be modified by, for example, phosphorylation (Ser, Thr, Tyr) or ubiquitination (Lys) (Fig. 12.4), which could potentially alter the PTMs of the mutant DES and thereby alter desmin filament assembly/disassembly and/or desmin protein stability under physiological or stress conditions. It was recently reported that phosphorylation of desmin triggers Trim32-mediated ubiquitination and degradation of desmin during muscle atrophy [75]. Increased desmin phosphorylation, which likely promotes desmin filament disassembly, has been reported in DRM [76]. Oxidized and nitrated desmin proteins were also found in affected muscle of DRM patients [77]. In addition, a missense mutation (I451M) at the C-terminus of DES, which is the first *DES* mutation identified in human idiopathic DCM [78], is found to promote proteolytic cleavage at its head domain and abolish its association with the z-disc in mouse hearts [68]. Therefore, DRM *DES* mutations may cause desmin filament disruption in multiple ways.

Similar to what observed in desmin-null mice, myofibril misalignment and mitochondrial dislocation and dysfunction are all seen in desminopathy human muscles and mouse models of desminopathy [67]. Mitochondrial dysfunction and cell death through a mitochondrial pathway are also observed in $CRYAB^{R120G}$-based DRC mouse hearts [79, 80]. This supports the notion that disruption of the desmin IF network contributes to pathogenesis in DRM. Nevertheless, it is not ruled out that gained toxicity from the disease-linked mutant proteins might have actually played a greater role in causing mitochondrial dysfunction and cell death than loss of desmin function.

Overburdened Protein Quality Control in DRM

Like other cells, myocytes especially cardiomyocytes possess multi-layered protein quality control (PQC) mechanisms serving to minimize the level and toxicity of misfolded proteins. First, with the help from chaperones, the cell attempts to unfold and refold a misfolded polypeptide, and if the repairing effort fails, the misfolded protein is then referred to as terminally misfolded protein and targeted for degradation by primarily the ubiquitin-proteasome system (UPS). The UPS is responsible for targeted degradation of most cellular proteins that are either abnormal or normal but no longer needed. The UPS does so by two main steps: first, covalent attachment of a chain of ubiquitin (Ub) to the substrate protein molecule via a process known as ubiquitination which is catalyzed sequentially by the Ub activating enzyme (E1), Ub conjugating enzyme (E2), and Ub ligase (E3); and second, the ubiquitinated protein will be shuttled to and recognized and degraded by the 26S proteasome, an ATP-dependent multi-units protease complex with its peptidase activity sequestered in the interior chamber of the 20S core subcomplex [81]. When escaped from or overwhelmed the surveillance of chaperones and the UPS, misfolded proteins undergo aberrant aggregation via hydrophobic interaction to give rise to highly active and toxic soluble oligomers and insoluble aggregates. The protein aggregates, which are enriched in ubiquitinated proteins but inaccessible by the proteasome, are generally believed to be degraded by macroautophagy. Macroautophagy (commonly known as autophagy) is a cellular process that sequesters a portion of cytoplasm by forming a double-membraned vacuole, known as an autophagosome; the latter fuses with the lysosome to form the autolysosome where the delivered cytoplasmic content is degraded by lysosomal enzymes [82].

Both *DES* and *CRYAB* are highly expressed genes in skeletal and cardiac myocytes, with their expression being further induced under stress conditions; therefore, one or both alleles of mutated *DES* or *CRYAB* in the genome will result in a considerably high level of mutant proteins, which would conceivably increase PQC burden more dramatically than other low-expressing genes. Indeed, a recent study showed that the total desmin protein level in muscle biopsies from desminopathy humans could be reach a level higher than 3.5-folds of that of the control individuals and the soluble desmin proteins could be increased by ~15-folds in human desminopathic skeletal muscles [50].

The fact that CRYAB^{R120G} causes DRM represents a remarkable real-world illustration for an indispensable role of chaperones in muscle health. Overexpression of wild type CRYAB suppresses aberrant aggregation of DRM-linked mutant desmin in cultured cardiomyocytes and, conversely, expressing CRYAB^{R120G} exacerbates D7-des aggregation and pathogenesis in mouse hearts [83]. Biochemical studies have revealed that the R120G mutation causes CRYAB misfolding, compromises the chaperone function of CRYAB, and alters its interaction with IFs [84, 85]. In mouse hearts overexpressing CRYAB^{R120G}, two types of protein aggregates are observed with electron microscopy. Type I has a homogenous appearance with clear boundaries and is, as revealed by immunogold staining, positive for CRYAB but not desmin; however, type II shows the amorphous appearance of the electron-dense granular-filamentous structures that are similar to the desmin aggregates observed in DRM patients, immunopositive for both desmin and CRYAB [70]. This observation suggests that the formation of desmin aggregates is due to lacking protection of functional CRYAB rather than attraction of misfolded CRYAB^{R120G}. Hence, in the case of CRYAB mutations, both loss of a critical chaperone of desmin and the increased burden to remove misfolded CRYAB^{R120G} are likely in play.

Both myocardial total ubiquitinated proteins and in vitro proteasome peptidase activities are significantly increased in both D7-des and CRYAB^{R120G} transgenic mice, indicative of UPS dysfunction in the heart of the DRC mouse models [86, 87].

To better assess UPS performance in vivo, an inverse UPS function reporter mouse model was created by ubiquitous and constitutive expression of a transgenic modified green fluorescence protein (GFP) with carboxyl fusion of a known ubiquitination signal sequence degron CL1, referred to as GFPdgn. In GFPdgn transgenic mice, an increase in GFPdgn proteins in absence of changes in protein synthesis would indicate a decreased UPS performance and vice versa [88]. Taking advantage of this reporter mouse, myocardial UPS functional insufficiency is revealed in both DRC transgenic mice [86, 87]. It remains to be investigated whether desmin loss-of-function directly impairs UPS function but loss-of-function of CRYAB is unlikely the cause of UPS impairment seen in CRYAB^{R120G} transgenic hearts because knockout of *CRYAB* in GFPdgn transgenic mice via cross-breeding failed to increase myocardial GFPdgn protein levels [86]. Aberrant protein aggregation impairs proteasome function in cultured cells [89]. This provides a reason for UPS function impairment in both D7-des and CRYAB^{R120G} expressing hearts; indeed, inhibition of aberrant aggregation of D7-des or CRYAB^{R120G} by either genetic or pharmacological means markedly attenuated these DRM-linked mutant proteins from impairing UPS performance in cultured cardiomyocytes [86, 87, 90]. These studies provide the first demonstration of UPS impairment by aberrant protein aggregation in intact animals. A major pathogenic role for proteasome functional insufficiency (PFI) in DRC or proteinopathies in general has further been demonstrated by subsequent studies showing that enhancing proteasome function by either genetic or pharmacological means can significantly reduce the prevalence of protein aggregates in cardiomyocytes, slow down disease progression, and delay the premature death of CRYAB^{R120G} mice [91, 92].

Ultrastructural examination of human DRM muscle biopsies shows increased abundance of autophagic vacuoles in the myocytes affected, suggesting that the alternative proteolytic pathway for PQC in the cell, autophagy might be activated in DRM muscles [93]. Indeed, activation of autophagy in cardiomyocytes has been observed in several mouse models of DRC [94, 95]. This activation is compensatory for increased proteolytic stress because genetic suppression of autophagy was shown to exacerbate pathology and disease progression in a CRYAB^{R120G}-based DRC mouse model [94]. Despite of increased autophagic flux, further activation of autophagy by overexpression of Atg7 was shown to protect against the toxicity of CRYAB^{R120G} in both cultured cardiomyocytes and DRC transgenic mice [96, 97]. Taken together, these studies suggest that the autophagosomal-lysosomal pathway is activated but is inadequate and this inadequacy, just like UPS inadequacy, contributes to DRM pathogenesis.

Overburdening PQC is certainly regarded as gained toxicity from DRM-linked mutant proteins. Another example of gained toxicity resulting from CRYAB^{R120G} is redox disturbance, on which conflicting reports exist though. Work from Robbins' group shows that increased oxidative stress derived from mitochondrial malfunction contributes to DRC pathogenesis in CRYAB^{R120G} transgenic mice [80]; however, intriguing data primarily from Benjamin's group support a major pathogenic role for increased reductive stress in the heart of a similar CRYAB^{R120G} mouse model [71, 98, 99]. A unified explanation for such controversy is currently lacking but is

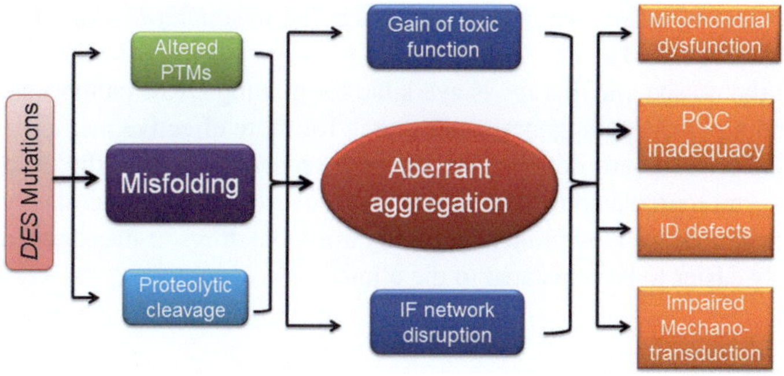

Fig. 12.6 A scheme of pathogenic processes of desminopathy. *PTMs* posttranslational modifications, *IF* intermediate filament, *PQC* protein quality control, *ID* intercalate disc

desperately needed because the completely opposite therapeutic strategies would be otherwise implicated.

In summary, a number of mutually non-exclusive mechanisms are potentially taken by DRM-linked mutations to cause DRM. It is likely that genetic mutations per se or its secondary effects (e.g., proteolytic processing, altered PTMs, etc.) increase the level of toxic desmin proteins in affected myocytes and disrupt the desmin IF network, which in turn either individually or in combination causes PQC inadequacy, mitochondrial dysfunction, intercalated disc defects, and perhaps impaired mechano-transduction. So far, most mechanistic studies support a central role for aberrant protein aggregation (not necessarily the final insoluble aggregates) in DRM pathogenesis (Fig. 12.6).

Experimental Therapeutic Exploration for DRM

The CRYAB^{R120G}-based DRC mouse model not only has been extensively utilized for pathogenic studies but has been facilitating therapeutic exploration as well. Sanbe et al. have reported that preamyloid oligomers (PAO) are increased cardiomyocytes of CRYAB^{R120G}-based DRC mice and voluntary exercise can significantly reduce myocardial PAO and remarkably slow down disease progression and delay premature death of mice in this mouse model [100, 101]. Zheng et al. show that oral administration of high dose of doxycycline leads to reduction of aberrant protein aggregation, ameliorates cardiac pathology, and delays premature death in CRYAB^{R120G}-based DRC mice [102]. Ranek et al. demonstrate that phosphodiesterase five specific inhibitor sildenafil administrated via osmotic minipumps can also reduce myocardial aberrant CRYAB aggregation and slow down disease progression in the CRYAB^{R120G}-based DRC mice through likely PKG activation and thereby priming the proteasome [92, 103]. More recently, McLendon et al. show that histone deacetylase (HDAC) inhibition with SAHA further increases myocardial levels of acetylated tubulin and cardiac autophagic flux, reduces protein aggregates in cardiomyocytes, and attenuates cardiac dysfunction in the CRYAB^{R120G}-based DRC mice [104].

Currently, no specific therapy is available for treating DRM patients and further understanding DRM pathogenesis and search for more effective measures to intervene DRM or DRC are doubtlessly urgently needed. However, the experimental treatments summarized above not only target known pathogenic factors revealed by basic research but also use clinically readily available drugs or measures, rendering it relatively easier to be translated to the clinic.

Concluding Remarks

It is clear that desmin filaments play an indispensable role in maintaining the structural integrity and mechanical function of muscle tissues after birth but it is also equally certain that function of the desmin IF network remains to be fully understood. Moreover, the pathogenic mechanisms of DRM are far from fully understood and experimental research to target these known mechanisms has just begun to emerge, further effort is warranted. For instance, improving the degradation of the DRM-causing mutant proteins currently is stalled at grossly enhancing the proteasome or autophagy, a thorough understanding of the specific factors that suppress specifically the expression of the mutant gene or increase the targeted degradation of the mutant protein would lead to more specific intervention for the disease. Such effort will be extremely significant because aberrant protein aggregation and inadequate PQC implicated in DRM pathogenesis have also been observed in more common forms of life-threatening human disease, such as neurodegenerative disease and congestive heart failure.

References

1. Lazarides E, Hubbard BD (1976) Immunological characterization of the subunit of the 100 A filaments from muscle cells. Proc Natl Acad Sci U S A 73:4344–4348
2. Goldfarb LG, Park KY, Cervenakova L, Gorokhova S, Lee HS, Vasconcelos O, Nagle JW, Semino-Mora C, Sivakumar K, Dalakas MC (1998) Missense mutations in desmin associated with familial cardiac and skeletal myopathy. Nat Genet 19:402–403
3. Munoz-Marmol AM, Strasser G, Isamat M, Coulombe PA, Yang Y, Roca X, Vela E, Mate JL, Coll J, Fernandez-Figueras MT, Navas-Palacios JJ, Ariza A, Fuchs E (1998) A dysfunctional desmin mutation in a patient with severe generalized myopathy. Proc Natl Acad Sci U S A 95:11312–11317
4. Vicart P, Caron A, Guicheney P, Li Z, Prevost MC, Faure A, Chateau D, Chapon F, Tome F, Dupret JM, Paulin D, Fardeau M (1998) A missense mutation in the alphaB-crystallin chaperone gene causes a desmin-related myopathy. Nat Genet 20:92–95
5. Capetanaki YG, Ngai J, Lazarides E (1984) Characterization and regulation in the expression of a gene coding for the intermediate filament protein desmin. Proc Natl Acad Sci U S A 81:6909–6913

6. Li ZL, Lilienbaum A, Butler-Browne G, Paulin D (1989) Human desmin-coding gene: complete nucleotide sequence, characterization and regulation of expression during myogenesis and development. Gene 78:243–254
7. Viegas-Pequignot E, Li ZL, Dutrillaux B, Apiou F, Paulin D (1989) Assignment of human desmin gene to band 2q35 by nonradioactive in situ hybridization. Hum Genet 83:33–36
8. Small JV, Sobieszek A (1977) Studies on the function and composition of the 10-NM(100-A) filaments of vertebrate smooth muscle. J Cell Sci 23:243–268
9. Li ZL, Paulin D (1991) High level desmin expression depends on a muscle-specific enhancer. J Biol Chem 266:6562–6570
10. Li Z, Paulin D (1993) Different factors interact with myoblast-specific and myotube-specific enhancer regions of the human desmin gene. J Biol Chem 268:10403–10415
11. Li Z, Marchand P, Humbert J, Babinet C, Paulin D (1993) Desmin sequence elements regulating skeletal muscle-specific expression in transgenic mice. Development 117:947–959
12. Li H, Capetanaki Y (1994) An E box in the desmin promoter cooperates with the E box and MEF-2 sites of a distal enhancer to direct muscle-specific transcription. EMBO J 13: 3580–3589
13. Kuisk IR, Li H, Tran D, Capetanaki Y (1996) A single MEF2 site governs desmin transcription in both heart and skeletal muscle during mouse embryogenesis. Dev Biol 174:1–13
14. Schaart G, Viebahn C, Langmann W, Ramaekers F (1989) Desmin and titin expression in early postimplantation mouse embryos. Development 107:585–596
15. Babai F, Musevi-Aghdam J, Schurch W, Royal A, Gabbiani G (1990) Coexpression of alpha-sarcomeric actin, alpha-smooth muscle actin and desmin during myogenesis in rat and mouse embryos I. Skeletal muscle. Differentiation 44:132–142
16. Furst DO, Osborn M, Weber K (1989) Myogenesis in the mouse embryo: differential onset of expression of myogenic proteins and the involvement of titin in myofibril assembly. J Cell Biol 109:517–527
17. Hill CS, Duran S, Lin ZX, Weber K, Holtzer H (1986) Titin and myosin, but not desmin, are linked during myofibrillogenesis in postmitotic mononucleated myoblasts. J Mol Biol 103:2185–2196
18. Brown JH, Cohen C, Parry DA (1996) Heptad breaks in alpha-helical coiled coils: stutters and stammers. Proteins 26:134–145
19. Strelkov SV, Burkhard P (2002) Analysis of alpha-helical coiled coils with the program TWISTER reveals a structural mechanism for stutter compensation. J Struct Biol 137:54–64
20. Herrmann H, Strelkov SV, Feja B, Rogers KR, Brettel M, Lustig A, Haner M, Parry DA, Steinert PM, Burkhard P, Aebi U (2000) The intermediate filament protein consensus motif of helix 2B: its atomic structure and contribution to assembly. J Mol Biol 298:817–832
21. Herrmann H, Haner M, Brettel M, Muller SA, Goldie KN, Fedtke B, Lustig A, Franke WW, Aebi U (1996) Structure and assembly properties of the intermediate filament protein vimentin: the role of its head, rod and tail domains. J Mol Biol 264:933–953
22. Heimburg T, Schuenemann J, Weber K, Geisler N (1996) Specific recognition of coiled coils by infrared spectroscopy: analysis of the three structural domains of type III intermediate filament proteins. Biochemistry 35:1375–1382
23. Strelkov SV, Herrmann H, Geisler N, Wedig T, Zimbelmann R, Aebi U, Burkhard P (2002) Conserved segments 1A and 2B of the intermediate filament dimer: their atomic structures and role in filament assembly. EMBO J 21:1255–1266
24. Rogers KR, Eckelt A, Nimmrich V, Janssen KP, Schliwa M, Herrmann H, Franke WW (1995) Truncation mutagenesis of the non-alpha-helical carboxyterminal tail domain of vimentin reveals contributions to cellular localization but not to filament assembly. Eur J Cell Biol 66:136–150
25. Lockard VG, Bloom S (1993) Trans-cellular desmin-lamin B intermediate filament network in cardiac myocytes. J Mol Cell Cardiol 25:303–309
26. Konieczny P, Fuchs P, Reipert S, Kunz WS, Zeold A, Fischer I, Paulin D, Schroder R, Wiche G (2008) Myofiber integrity depends on desmin network targeting to Z-disks and costameres via distinct plectin isoforms. J Cell Biol 181:667–681

27. Poon E, Howman EV, Newey SE, Davies KE (2002) Association of syncoilin and desmin: linking intermediate filament proteins to the dystrophin-associated protein complex. J Biol Chem 277:3433–3439
28. Kaplan SR, Gard JJ, Carvajal-Huerta L, Ruiz-Cabezas JC, Thiene G, Saffitz JE (2004) Structural and molecular pathology of the heart in Carvajal syndrome. Cardiovasc Pathol 13:26–32
29. Bloom S, Lockard VG, Bloom M (1996) Intermediate filament-mediated stretch-induced changes in chromatin: a hypothesis for growth initiation in cardiac myocytes. J Mol Cell Cardiol 28:2123–2127
30. Nikolova V, Leimena C, McMahon AC, Tan JC, Chandar S, Jogia D, Kesteven SH, Michalicek J, Otway R, Verheyen F, Rainer S, Stewart CL, Martin D, Feneley MP, Fatkin D (2004) Defects in nuclear structure and function promote dilated cardiomyopathy in lamin A/C-deficient mice. J Clin Invest 113:357–369
31. McCullagh KJ, Edwards B, Poon E, Lovering RM, Paulin D, Davies KE (2007) Intermediate filament-like protein syncoilin in normal and myopathic striated muscle. Neuromuscul Disord 17:970–979
32. Zhang J, Bang ML, Gokhin DS, Lu Y, Cui L, Li X, Gu Y, Dalton ND, Scimia MC, Peterson KL, Lieber RL, Chen J (2008) Syncoilin is required for generating maximum isometric stress in skeletal muscle but dispensable for muscle cytoarchitecture. Am J Physiol Cell Physiol 294:C1175–C1182
33. Li H, Choudhary SK, Milner DJ, Munir MI, Kuisk IR, Capetanaki Y (1994) Inhibition of desmin expression blocks myoblast fusion and interferes with the myogenic regulators MyoD and myogenin. J Cell Biol 124:827–841
34. Weitzer G, Milner DJ, Kim JU, Bradley A, Capetanaki Y (1995) Cytoskeletal control of myogenesis: a desmin null mutation blocks the myogenic pathway during embryonic stem cell differentiation. Dev Biol 172:422–439
35. Li Z, Colucci-Guyon E, Pincon-Raymond M, Mericskay M, Pournin S, Paulin D, Babinet C (1996) Cardiovascular lesions and skeletal myopathy in mice lacking desmin. Dev Biol 175:362–366
36. Milner DJ, Weitzer G, Tran D, Bradley A, Capetanaki Y (1996) Disruption of muscle architecture and myocardial degeneration in mice lacking desmin. J Cell Biol 134:1255–1270
37. Li Z, Mericskay M, Agbulut O, Butler-Browne G, Carlsson L, Thornell LE, Babinet C, Paulin D (1997) Desmin is essential for the tensile strength and integrity of myofibrils but not for myogenic commitment, differentiation, and fusion of skeletal muscle. J Cell Biol 139:129–144
38. Milner DJ, Taffet GE, Wang X, Pham T, Tamura T, Hartley C, Gerdes AM, Capetanaki Y (1999) The absence of desmin leads to cardiomyocyte hypertrophy and cardiac dilation with compromised systolic function. J Mol Cell Cardiol 31:2063–2076
39. Thornell L, Carlsson L, Li Z, Mericskay M, Paulin D (1997) Null mutation in the desmin gene gives rise to a cardiomyopathy. J Mol Cell Cardiol 29:2107–2124
40. Capetanaki Y, Milner DJ, Weitzer G (1997) Desmin in muscle formation and maintenance: knockouts and consequences. Cell Struct Funct 22:103–116
41. Weisleder N, Soumaka E, Abbasi S, Taegtmeyer H, Capetanaki Y (2004) Cardiomyocyte-specific desmin rescue of desmin null cardiomyopathy excludes vascular involvement. J Mol Cell Cardiol 36:121–128
42. Wede OK, Lofgren M, Li Z, Paulin D, Arner A (2002) Mechanical function of intermediate filaments in arteries of different size examined using desmin deficient mice. J Physiol 540:941–949
43. Milner DJ, Mavroidis M, Weisleder N, Capetanaki Y (2000) Desmin cytoskeleton linked to muscle mitochondrial distribution and respiratory function. J Cell Biol 150:1283–1298
44. Weisleder N, Taffet GE, Capetanaki Y (2004) Bcl-2 overexpression corrects mitochondrial defects and ameliorates inherited desmin null cardiomyopathy. Proc Natl Acad Sci U S A 101:769–774

45. Slats AM, Janssen K, van Schadewijk A, van der Plas DT, Schot R, van den Aardweg JG, de Jongste JC, Hiemstra PS, Mauad T, Rabe KF, Sterk PJ (2008) Expression of smooth muscle and extracellular matrix proteins in relation to airway function in asthma. J Allergy Clin Immunol 121:1196–1202
46. Shardonofsky FR, Capetanaki Y, Boriek AM (2006) Desmin modulates lung elastic recoil and airway responsiveness. Am J Physiol Lung Cell Mol Physiol 290:L890–L896
47. Mohamed JS, Hajira A, Li Z, Paulin D, Boriek AM (2011) Desmin regulates airway smooth muscle hypertrophy through early growth-responsive protein-1 and microRNA-26a. J Biol Chem 286:43394–43404
48. Scott RS, Li Z, Paulin D, Uvelius B, Small JV, Arner A (2008) Role of desmin in active force transmission and maintenance of structure during growth of urinary bladder. Am J Physiol Cell Physiol 295:C324–C331
49. Lazarides E (1980) Intermediate filaments as mechanical integrators of cellular space. Nature 283:249–256
50. Fichna JP, Karolczak J, Potulska-Chromik A, Miszta P, Berdynski M, Sikorska A, Filipek S, Redowicz MJ, Kaminska A, Zekanowski C (2014) Two desmin gene mutations associated with myofibrillar myopathies in polish families. PLoS One 9:e115470
51. Schroder R, Schoser B (2009) Myofibrillar myopathies: a clinical and myopathological guide. Brain Pathol 19:483–492
52. Goldfarb LG, Dalakas MC (2009) Tragedy in a heartbeat: malfunctioning desmin causes skeletal and cardiac muscle disease. J Clin Invest 119:1806–1813
53. Park KY, Dalakas MC, Goebel HH, Ferrans VJ, Semino-Mora C, Litvak S, Takeda K, Goldfarb LG (2000) Desmin splice variants causing cardiac and skeletal myopathy. J Med Genet 37:851–857
54. Iwaki T, Kume-Iwaki A, Liem RK, Goldman JE (1989) Alpha B-crystallin is expressed in non-lenticular tissues and accumulates in Alexander's disease brain. Cell 57:71–78
55. Sacconi S, Feasson L, Antoine JC, Pecheux C, Bernard R, Cobo AM, Casarin A, Salviati L, Desnuelle C, Urtizberea A (2012) A novel CRYAB mutation resulting in multisystemic disease. Neuromuscul Disord 22:66–72
56. Reilich P, Schoser B, Schramm N, Krause S, Schessl J, Kress W, Muller-Hocker J, Walter MC, Lochmuller H (2010) The p.G154S mutation of the alpha-B crystallin gene (CRYAB) causes late-onset distal myopathy. Neuromuscul Disord 20:255–259
57. Inagaki N, Hayashi T, Arimura T, Koga Y, Takahashi M, Shibata H, Teraoka K, Chikamori T, Yamashina A, Kimura A (2006) Alpha B-crystallin mutation in dilated cardiomyopathy. Biochem Biophys Res Commun 342:379–386
58. Selcen D, Engel AG (2003) Myofibrillar myopathy caused by novel dominant negative alpha B-crystallin mutations. Ann Neurol 54:804–810
59. Del Bigio MR, Chudley AE, Sarnat HB, Campbell C, Goobie S, Chodirker BN, Selcen D (2011) Infantile muscular dystrophy in Canadian aboriginals is an alphaB-crystallinopathy. Ann Neurol 69:866–871
60. Forrest KM, Al-Sarraj S, Sewry C, Buk S, Tan SV, Pitt M, Durward A, McDougall M, Irving M, Hanna MG, Matthews E, Sarkozy A, Hudson J, Barresi R, Bushby K, Jungbluth H, Wraige E (2011) Infantile onset myofibrillar myopathy due to recessive CRYAB mutations. Neuromuscul Disord 21:37–40
61. Berry V, Francis P, Reddy MA, Collyer D, Vithana E, MacKay I, Dawson G, Carey AH, Moore A, Bhattacharya SS, Quinlan RA (2001) Alpha-B crystallin gene (CRYAB) mutation causes dominant congenital posterior polar cataract in humans. Am J Hum Genet 69:1141–1145
62. Liu M, Ke T, Wang Z, Yang Q, Chang W, Jiang F, Tang Z, Li H, Ren X, Wang X, Wang T, Li Q, Yang J, Liu J, Wang QK (2006) Identification of a CRYAB mutation associated with autosomal dominant posterior polar cataract in a Chinese family. Invest Ophthalmol Vis Sci 47:3461–3466
63. Xia XY, Wu QY, An LM, Li WW, Li N, Li TF, Zhang C, Cui YX, Li XJ, Xue CY (2014) A novel P20R mutation in the alpha-B crystallin gene causes autosomal dominant congenital posterior polar cataracts in a Chinese family. BMC Ophthalmol 14:108

64. Devi RR, Yao W, Vijayalakshmi P, Sergeev YV, Sundaresan P, Hejtmancik JF (2008) Crystallin gene mutations in Indian families with inherited pediatric cataract. Mol Vis 14:1157–1170
65. Selcen D, Muntoni F, Burton BK, Pegoraro E, Sewry C, Bite AV, Engel AG (2009) Mutation in BAG3 causes severe dominant childhood muscular dystrophy. Ann Neurol 65:83–89
66. Wang X, Osinska H, Dorn GW 2nd, Nieman M, Lorenz JN, Gerdes AM, Witt S, Kimball T, Gulick J, Robbins J (2001) Mouse model of desmin-related cardiomyopathy. Circulation 103:2402–2407
67. Kostareva A, Sjoberg G, Bruton J, Zhang SJ, Balogh J, Gudkova A, Hedberg B, Edstrom L, Westerblad H, Sejersen T (2008) Mice expressing L345P mutant desmin exhibit morphological and functional changes of skeletal and cardiac mitochondria. J Muscle Res Cell Motil 29:25–36
68. Mavroidis M, Panagopoulou P, Kostavasili I, Weisleder N, Capetanaki Y (2008) A missense mutation in desmin tail domain linked to human dilated cardiomyopathy promotes cleavage of the head domain and abolishes its Z-disc localization. FASEB J 22:3318–3327
69. Clemen CS, Stockigt F, Strucksberg KH, Chevessier F, Winter L, Schutz J, Bauer R, Thorweihe JM, Wenzel D, Schlotzer-Schrehardt U, Rasche V, Krsmanovic P, Katus HA, Rottbauer W, Just S, Muller OJ, Friedrich O, Meyer R, Herrmann H, Schrickel JW, Schroder R (2015) The toxic effect of R350P mutant desmin in striated muscle of man and mouse. Acta Neuropathol 129:297–315
70. Wang X, Osinska H, Klevitsky R, Gerdes AM, Nieman M, Lorenz J, Hewett T, Robbins J (2001) Expression of R120G-alphaB-crystallin causes aberrant desmin and alphaB-crystallin aggregation and cardiomyopathy in mice. Circ Res 89:84–91
71. Rajasekaran NS, Connell P, Christians ES, Yan LJ, Taylor RP, Orosz A, Zhang XQ, Stevenson TJ, Peshock RM, Leopold JA, Barry WH, Loscalzo J, Odelberg SJ, Benjamin IJ (2007) Human alpha B-crystallin mutation causes oxido-reductive stress and protein aggregation cardiomyopathy in mice. Cell 130:427–439
72. Andley UP, Hamilton PD, Ravi N, Weihl CC (2011) A knock-in mouse model for the R120G mutation of alphaB-crystallin recapitulates human hereditary myopathy and cataracts. PLoS One 6:e17671
73. Bar H, Mucke N, Kostareva A, Sjoberg G, Aebi U, Herrmann H (2005) Severe muscle disease-causing desmin mutations interfere with in vitro filament assembly at distinct stages. Proc Natl Acad Sci U S A 102:15099–15104
74. Winter DL, Paulin D, Mericskay M, Li Z (2014) Posttranslational modifications of desmin and their implication in biological processes and pathologies. Histochem Cell Biol 141:1–16
75. Cohen S, Zhai B, Gygi SP, Goldberg AL (2012) Ubiquitylation by Trim32 causes coupled loss of desmin, Z-bands, and thin filaments in muscle atrophy. J Cell Biol 198:575–589
76. Caron A, Chapon F (1999) Desmin phosphorylation abnormalities in cytoplasmic body and desmin-related myopathies. Muscle Nerve 22:1122–1125
77. Janue A, Odena MA, Oliveira E, Olive M, Ferrer I (2007) Desmin is oxidized and nitrated in affected muscles in myotilinopathies and desminopathies. J Neuropathol Exp Neurol 66:711–723
78. Li D, Tapscoft T, Gonzalez O, Burch PE, Quinones MA, Zoghbi WA, Hill R, Bachinski LL, Mann DL, Roberts R (1999) Desmin mutation responsible for idiopathic dilated cardiomyopathy. Circulation 100:461–464
79. Maloyan A, Sanbe A, Osinska H, Westfall M, Robinson D, Imahashi K, Murphy E, Robbins J (2005) Mitochondrial dysfunction and apoptosis underlie the pathogenic process in alpha-B-crystallin desmin-related cardiomyopathy. Circulation 112:3451–3461
80. Maloyan A, Osinska H, Lammerding J, Lee RT, Cingolani OH, Kass DA, Lorenz JN, Robbins J (2009) Biochemical and mechanical dysfunction in a mouse model of desmin-related myopathy. Circ Res 104:1021–1028
81. Wang X, Robbins J (2006) Heart failure and protein quality control. Circ Res 99:1315–1328

82. Wang X, Pattison JS, Su H (2013) Posttranslational modification and quality control. Circ Res 112:367–381
83. Wang X, Klevitsky R, Huang W, Glasford J, Li F, Robbins J (2003) AlphaB-crystallin modulates protein aggregation of abnormal desmin. Circ Res 93:998–1005
84. Bova MP, Yaron O, Huang Q, Ding L, Haley DA, Stewart PL, Horwitz J (1999) Mutation R120G in alphaB-crystallin, which is linked to a desmin-related myopathy, results in an irregular structure and defective chaperone-like function. Proc Natl Acad Sci U S A 96:6137–6142
85. Perng MD, Muchowski PJ, van Den IP, Wu GJ, Hutcheson AM, Clark JI, Quinlan RA (1999) The cardiomyopathy and lens cataract mutation in alphaB-crystallin alters its protein structure, chaperone activity, and interaction with intermediate filaments in vitro. J Biol Chem 274:33235–33243
86. Chen Q, Liu JB, Horak KM, Zheng H, Kumarapeli AR, Li J, Li F, Gerdes AM, Wawrousek EF, Wang X (2005) Intrasarcoplasmic amyloidosis impairs proteolytic function of proteasomes in cardiomyocytes by compromising substrate uptake. Circ Res 97:1018–1026
87. Liu J, Chen Q, Huang W, Horak KM, Zheng H, Mestril R, Wang X (2006) Impairment of the ubiquitin-proteasome system in desminopathy mouse hearts. FASEB J 20:362–364
88. Kumarapeli AR, Horak KM, Glasford JW, Li J, Chen Q, Liu J, Zheng H, Wang X (2005) A novel transgenic mouse model reveals deregulation of the ubiquitin-proteasome system in the heart by doxorubicin. FASEB J 19:2051–2053
89. Bence NF, Sampat RM, Kopito RR (2001) Impairment of the ubiquitin-proteasome system by protein aggregation. Science 292:1552–1555
90. Liu J, Tang M, Mestril R, Wang X (2006) Aberrant protein aggregation is essential for a mutant desmin to impair the proteolytic function of the ubiquitin-proteasome system in cardiomyocytes. J Mol Cell Cardiol 40:451–454
91. Li J, Horak KM, Su H, Sanbe A, Robbins J, Wang X (2011) Enhancement of proteasomal function protects against cardiac proteinopathy and ischemia/reperfusion injury in mice. J Clin Invest 121:3689–3700
92. Ranek MJ, Terpstra EJ, Li J, Kass DA, Wang X (2013) Protein kinase g positively regulates proteasome-mediated degradation of misfolded proteins. Circulation 128:365–376
93. Weihl CC, Iyadurai S, Baloh RH, Pittman SK, Schmidt RE, Lopate G, Pestronk A, Harms MB (2015) Autophagic vacuolar pathology in desminopathies. Neuromuscul Disord 25:199–206
94. Tannous P, Zhu H, Johnstone JL, Shelton JM, Rajasekaran NS, Benjamin IJ, Nguyen L, Gerard RD, Levine B, Rothermel BA, Hill JA (2008) Autophagy is an adaptive response in desmin-related cardiomyopathy. Proc Natl Acad Sci U S A 105:9745–9750
95. Zheng Q, Su H, Ranek MJ, Wang X (2011) Autophagy and p62 in cardiac proteinopathy. Circ Res 109:296–308
96. Bhuiyan MS, Pattison JS, Osinska H, James J, Gulick J, McLendon PM, Hill JA, Sadoshima J, Robbins J (2013) Enhanced autophagy ameliorates cardiac proteinopathy. J Clin Invest 123:5284–5297
97. Pattison JS, Osinska H, Robbins J (2011) Atg7 induces basal autophagy and rescues autophagic deficiency in CryABR120G cardiomyocytes. Circ Res 109:151–160
98. Kannan S, Muthusamy VR, Whitehead KJ, Wang L, Gomes AV, Litwin SE, Kensler TW, Abel ED, Hoidal JR, Rajasekaran NS (2013) Nrf2 deficiency prevents reductive stress-induced hypertrophic cardiomyopathy. Cardiovasc Res 100:63–73
99. Rajasekaran NS, Varadharaj S, Khanderao GD, Davidson CJ, Kannan S, Firpo MA, Zweier JL, Benjamin IJ (2011) Sustained activation of nuclear erythroid 2-related factor 2/antioxidant response element signaling promotes reductive stress in the human mutant protein aggregation cardiomyopathy in mice. Antioxid Redox Signal 14:957–971
100. Maloyan A, Gulick J, Glabe CG, Kayed R, Robbins J (2007) Exercise reverses preamyloid oligomer and prolongs survival in alphaB-crystallin-based desmin-related cardiomyopathy. Proc Natl Acad Sci U S A 104:5995–6000

101. Sanbe A, Osinska H, Saffitz JE, Glabe CG, Kayed R, Maloyan A, Robbins J (2004) Desmin-related cardiomyopathy in transgenic mice: a cardiac amyloidosis. Proc Natl Acad Sci U S A 101:10132–10136
102. Zheng H, Tang M, Zheng Q, Kumarapeli AR, Horak KM, Tian Z, Wang X (2010) Doxycycline attenuates protein aggregation in cardiomyocytes and improves survival of a mouse model of cardiac proteinopathy. J Am Coll Cardiol 56:1418–1426
103. Zhang H, Wang X (2015) Priming the proteasome by protein kinase G: a novel cardioprotective mechanism of sildenafil. Future Cardiol 11:177–189
104. McLendon PM, Ferguson BS, Osinska H, Bhuiyan MS, James J, McKinsey TA, Robbins J (2014) Tubulin hyperacetylation is adaptive in cardiac proteotoxicity by promoting autophagy. Proc Natl Acad Sci U S A 111:E5178–E5186

Chapter 13
Possible Functions of Intermediate Filaments in Mammalian Ovarian Follicles and Oocytes

Hiroyuki Suzuki

Introduction

Intermediate filaments (IFs), microfilaments and microtubules comprise the three major cytoskeletal proteins found in most mammalian cells. In contrast to microfilaments and microtubules, various members of the IF protein family are expressed abundantly and differentially in complex patterns during cell growth and differentiation (for reviews, see [1–6]). Depending upon the cell type, IFs are composed of different members of the cytoskeletal IF protein family (Table 13.1).

Type I and type II IFs are the acidic and basic keratins, respectively, which are obligate heteropolymers composed of type I and type II subunits [1, 3, 7–9]. Keratins are the most complex subgroup of the IF family. Vimentin [15, 16], desmin [24], glial fibrillary acidic protein (GFAP) [25, 26] and peripherin [26] form type III IF proteins, that can assemble into filaments on their own, or in combination with type IV and type VI IF proteins [3, 17–19]. For example, vimentin can co-assemble with desmin, GFAP or peripherin (all type III), or with neurofilament light and α-internexin (both type IV) [5, 18]. In addition, vimentin expression precedes the expression of other type III IF proteins during the differentiation and development of neural and muscle cells (later replaced by GFAP and desmin, respectively), suggesting important functions for vimentin as an intracellular scaffold [5].

Neurofilaments, the major IFs found in neurons consist of light (NF-L), medium (NF-M), and heavy (NF-H) subunits, are classified as type IV IF proteins along with α-internexin [27, 28]. Type V proteins are the nuclear lamins, that organize to form the nuclear lamina, a fibrous meshwork of proteins adjacent to the nucleoplasmic face of the inner nuclear membrane [32–36]. The type V nuclear lamins do not

H. Suzuki (✉)
Faculty of Agriculture and Life Science, Hirosaki University,
3 Bunkyo-cho, Hirosaki, Aomori 036-8561, Japan
e-mail: suzuki@hirosaki-u.ac.jp

H. Schatten (ed.), *The Cytoskeleton in Health and Disease*,
DOI 10.1007/978-1-4939-2904-7_13

Table 13.1 Types of intermediate filament (IF) proteins and notable features

Sequence type	IF proteins	Primary tissue distribution	Notable features
I	Acid keratins	Epithelium [7, 8]	Obligatory heteropolymers composed of type I and type II proteins [1, 3, 9] Crosstalk with MT[a] and/or MF[b] [10–14]
II	Neutral-basic keratins		
III	Vimentin	Mesenchymal cells [15, 16]	Homopolymers or in combination with type IV and type VI IF proteins [3, 17–19] Crosstalk with MT and/or MF [20–23]
	Desmin	Muscle fibers [24]	
	GFAP[c]	Glial cells [25]	
	Peripherin	Peripheral neurons and cranial nerves [26]	
IV	NF[d]-L, NF-M and NF-H	Astrocytes and other glial cells [27, 28]	Crosstalk with MT [29] Copolymer with vimentin [30, 31]
V	Lamin A, B, and C	Nuclear lamina [32–36]	MT motors drives interkinetic nuclear migration [37]
VI	Nestin	Neuroepithelial stem cells [38, 39] Pluripotent cells [40, 41] Endocrine cells [42–47] Endothelial cells [48] Metastatic tumors [49]	Crosstalk with MT and MF [18] Copolymer with vimentin [10, 11, 18, 19, 50, 51] Copolymer with vimentin/desmin [52] Transiently expressed during renal development [53]

[a]Microtubules
[b]Microfilaments
[c]Glial fibrillary acidic protein
[d]Neurofilament

co-assemble with members of types I to IV. Nestin is a type VI protein of IFs, as well as tanabin and transitin [54]. Nestin expression occurs in proliferating stem cells of the developing mammalian central nervous system and other pluripotent cells of non-neuronal tissues [3, 18, 38, 40–42, 48, 55]. Nestin is unable to form filaments on its own, but it can readily form copolymer IFs when combined with type III IF proteins such as vimentin both in vitro and in vivo [10, 11, 18, 19, 50–52].

Crosstalk among IFs, microfilaments and/or microtubules via specific linking proteins, such as the plakin family, is also important for stable architecture of the cytoskeletal system [10–14, 18, 20–23, 29–31, 37, 52]. The mechanisms responsible for the bidirectional microtubule-dependent movements of vimentin particles are related to their association with conventional kinesin and cytoplasmic dynein [56, 57].

Moreover, IFs are highly dynamic intracellular structures and new functional and regulatory roles of IFs have been defined, thereby suggesting special physiological capacity besides their mechanical function [1, 3, 18, 58–60]. These include cell growth, organelle distribution, signal transduction, cell polarity, and gene regulation. On the other hand, little information exists concerning the structure of the IF networks performing these functions in the mammalian oocytes. In this chapter, we focus mainly on data obtained from studies in mammalian systems to understand the role of IFs within ovarian follicles and oocytes.

IFs in Ovarian Tissues and Follicles

IF proteins studied in mammalian oocytes and granulosa cells are summarized in Table 13.2. Species differences are noted in the specific IF proteins. In the mammalian ovary, keratin immunoreactivity is consistently demonstrated in the surface epithelium of many species, including cattle [70, 77, 80, 81], pigs ([62]; Suzuki et al. unpublished data), mice [73, 78], rats [62, 67] and humans [62, 71, 75]. Keratin is detected in the mouse [61, 63, 69], hamster [65, 72], sheep [64] and human follicles [68, 75].

Vimentin is the IF protein characteristic of mesenchymal cells, such as fibroblasts and endothelial cells (for reviews, see [5, 15]). Vimentin immunostaining is often observed in follicular epithelial cells maintaining a similar distribution in primary, secondary, and tertiary follicles (Fig. 13.1). Vimentin positivity of follicular cells remains unchanged in the granulosa cell layer and increases in mature follicles during maturation ([70, 80]; Suzuki et al. unpublished data). The theca interna cells and the theca externa cells show a uniformly strong vimentin-positive appearance. The endothelial cells of blood capillaries in stroma, atretic follicles and larger blood vessels were strongly positive for vimentin. Desmin positivity is mainly localized in the wall of blood vessels. Very weak signaling of desmin is noted in the oocytes and granulosa cells of pigs and hamsters (Suzuki et al. unpublished data). Since almost all proteins composed of IFs are able to be located in the mammalian ovarian tissue, the ovary may be suitable for a positive-control tissue in the study of IF proteins.

In the baboon and human ovaries, some dissimilar distribution patterns of IFs are observed, where the surface epithelial cells exhibit keratin staining, whereas vimentin has been primarily localized in the basal regions of these cells [62, 71]. A weak to moderate immunoreactivity for desmin has also been present apically in surface epithelial cells [71].

IFs in Growing Follicles

Vimentin proteins are expressed at all stages of follicular development (Fig. 13.1). In primordial follicles, oocytes are individually surrounded by a single layer of squamous pre-granulosa cells, also referred to as follicular epithelial cells. Cell-to-cell communication between these somatic cells and oocytes is apparent from the formation of primordial follicles onward [82, 83]. A great number of non-growing primordial follicles serve as the source of developing follicles and oocytes until the end of a female's reproductive life. Interestingly, vimentin is detected in the

Table 13.2 Summary of intermediate filament proteins studied in mammalian oocytes and granulosa cells

References	Fixation[a] (methods)[b]	Antibodies used[c]	Response[d]	
			Oocytes	Granulosa cells
Lehtonen et al. [61]	Me-OH/ Acetone (IFT) & (IB)	P, keratin	Mouse, +	Mouse, –
		P/M, vimentin	Mouse, –	ND
		P, GFAP	Mouse, –	ND
		P, neurofilament	Mouse, –	ND
Czernobilsky et al. [62]	Fr (ICT) Acetone (IFT)	P/M, keratin	Human/pig/rat, –	Human, +; pig/rat, –
		P, vimentin	Human/pig/rat, –	Human/pig/rat, +
		P, desmin	Human/pig/rat, –	Human/pig/rat, –
		P, desmoplakin	Human/pig/rat, –	Human, +; pig/rat, ND
Lehtonen [63]	Me-OH (IFT)	M, keratin	Mouse, +	ND
Gall et al. [64]	Fr w/BF (IFT) & (IEM, IB)	P, keratin	Sheep, +	ND
Plancha et al. [65]	Fr or BF (IFT) & (IEM, IB)	M/P, keratin	Hamster, +	ND
van Niekerk et al. [66]	Fr (ICT)	M, keratin	ND	Human, + ⇒ –[e]
Fridmacher et al. [67]	Fr w/ or w/o BF (ICT)	M, keratin	Rat, –	Rat, + ⇒ –[e]
Santini et al. [68]	M (ICT) &(EM)	M, keratin	Human, +	Human, +
		M, vimentin	Human, –	Human, +
		M, actin	Human, –	Human, +
		M, desmin	Human, –	Human, –
Gallicano et al. [69]	G (IFT) & (IEM, IB)	M, keratin	Mouse, +	ND
		P, vimentin	Mouse, –	ND
van den Hurk et al. [70]	B (ICT)	P, keratin	Bovine, –	Bovine, –
		P, vimentin	Bovine, –	Bovine, +
		P, desmin	Bovine, –	Bovine, –
Khan-Dawood et al. [71]	B (ICT)	M, keratin	Baboon/human, –	Baboon/human, –
		M, vimentin	Baboon/human, – (+)[f]	Baboon/human, –
		M, desmin	Baboon/human, –	Baboon/human, –
		M, neurofilament	Baboon/human, +	Baboon/human, –
Plancha [72]	BF (IFT) & (IEM)	M, keratin	Hamster, +	ND
Appert et al. [73]	Fr (IFT)	M, keratin	Mouse, –	Mouse, +
Marettová and Maretta [74]	BF (ICT)	M, vimentin	ND	Sheep, +
		M, desmin	ND	Sheep, –
Bukovsky et al. [75]	Fr w/aceton (ICT)	M, keratin	Human, +	Human, +

(continued)

Table 13.2 (continued)

References	Fixation[a] (methods)[b]	Antibodies used[c]	Response[d]	
			Oocytes	Granulosa cells
Takahashi et al. [48]	Not specified (ICT)	M, nestin	Rat, –	Rat, –
Kabashima et al. [76]	Me-OH (IFT)	M, keratin	Hamster, +	ND
Townson et al. [77]	Fr w/BF (ICT)	M, keratin	Bovine, –	Bovine, – (+)[g]
Mora et al. [78]	BF or Fr (IFT) & (mRNA)	M, keratin	Mouse, –	Mouse, +
		M, vimentin	Mouse, +	Mouse, +
Takahashi and Ishizuka [79]	BF (ICT, IFT)	M, neurofilament	Rat, +	ND
Wendl et al. [80]	B (ICT, IFT)	M/P, keratin	Bovine, –	Bovine, +
		M, vimentin	Bovine, –	Bovine, +
		M, desmin	Bovine, –	Bovine, –
Hummitzsch et al. [81]	Fr (ICT)	M, keratin	Bovine, –	Bovine, + or +/–
Suzuki et al. (unpublished data)	BF (ICT) B (IFT)	M, keratin	Pig, –	Pig, –
		M, vimentin	Pig, –/hamster, –	Pig, +/hamster, –
		M, desmin	Pig, +/hamster, +	Pig, +/hamster, +
		M, GFAP	Pig, +	Pig, +
		M, neurofilament[h]	Pig, +	Pig, –
		P, neurofilaments[i]	Pig, +	Pig, +
		P, nestin	Pig, +/hamster, –	Pig, –/hamster, –

[a] *B* Bouin fixative, *BF* buffered formalin, Me-OH methanol, *Fr* frozen, *G* glutaraldehyde, *M* methacarnoy fixative
[b] *IFT* immunofluorescence technique, *ICT* immunocytochemical technique, *IB* immunoblotting analysis, *IEM* immunoelectron microscopy, *EM* electron microscopy, *mRNA* mRNA analysis. &() shows additional analyses
[c] *M* monoclonal, *P* polyclonal
[d] + positive, – negative, *ND* not determined
[e] Positive signals progressively disappeared in the granulosa cells of growing follicles
[f] Immunoreactivity developed in atretic follicles
[g] Keratin is localized to the cytoplasm of granulosa cells in a growing follicle and to the basal granulosa cells in an antral follicle
[h] Antibody for NF-L
[i] Rabbit anti-pan-neurofilaments polyclonal antibody (Enzo Life Sciences, Inc.), which includes antibodies for NF-L, NF-M and NF, H, was used

follicular epithelial cells even at the early stage of the primordial follicle (Fig. 13.1, Suzuki et al. unpublished data; see also [78, 84]). It is suggested, therefore, that vimentin might have a role in signaling of the cell-to-cell communication, because recent studies have brought light into the role of vimentin that are involved in cell signaling along with adhesion and migration [85–87]. Vimentin is also known to interact with signaling molecules [5].

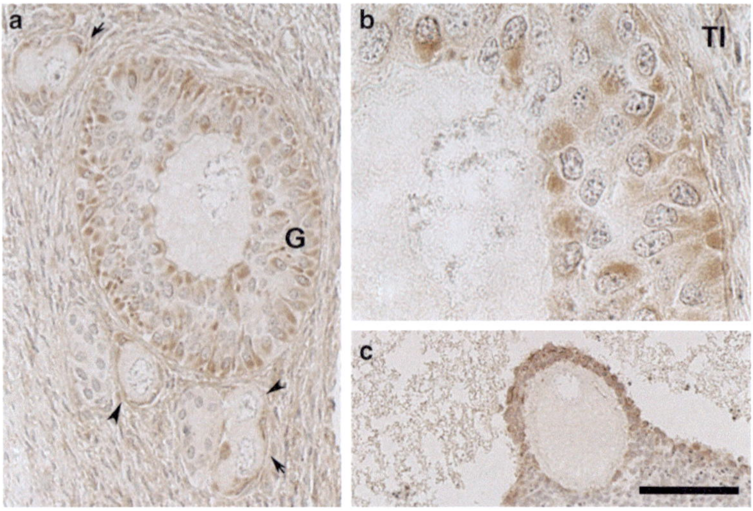

Fig. 13.1 Vimentin immunoreactivity in the pig ovary. Nuclei are stained with hematoxylin. *Bar* represented in (**c**) shows 50 µm in (**a**), 10 µm in (**b**) and 100 µm in (**c**). (**a**) Primordial (*arrowheads*), primary (*arrows*) and secondary follicles. Vimentin reaction is found in follicular epithelial (pre-granulosa) cells of the primordial and primary follicles. The granulosa (G) of a secondary follicle is stained in two-layered fashion. (**b**) Higher magnification of an upper right part of the secondary follicle of (**a**). Vimentin immunoreactivity in the granulosa of the secondary follicles is mainly localized in basal and apical granulosa cells. Theca interna (TI) cells are also weakly positive. (**c**) Cumulus-oocyte complex of a tertiary (antral) follicle. Cumulus oophorus cells are positively stained. Oocytes showed no significant vimentin reaction irrespective of the growing follicular stages

The transition from the primordial to primary follicle is characterized by a morphological change in the surrounding follicular epithelial cells from squamous to cuboidal, where vimentin IF protein was also found ([74, 78, 80]; Suzuki et al. unpublished data).

Secondary follicles contain growing oocytes surrounded by two or more layers of follicular epithelial cells (now called granulosa cells). Preantral follicle development is gonadotropin-independent and is induced by autocrine and paracrine regulatory factors [88, 89]. An additional somatic cell layer, the theca, forms outside the basement membrane of the follicle and differentiates as the theca interna and theca externa [88, 89]. The theca interna cells include numerous mitochondria with tubular cristae, smooth endoplasmic reticulum, and abundant lipid vesicles, corresponding with the endocrine function as a source of androgens for neighboring granulosa cells to convert to estrogens [88, 89]. The theca externa, composed of fibroblasts and smooth muscle-like cells, shows coexistence with actin and myosin and also the desmin antibody occasionally gives positive results (Suzuki et al. unpublished data). During subsequent oocyte-follicular development, surface adhesion molecules are established and maintain contact with appropriate cumulus cells when the zona pellucida is produced [82].

The development of a follicular antrum is clearly dependent on gonadotropins and well-developed antral follicles are called tertiary follicles (often referred to as Graafian follicles). At the transition from preantral to antral follicles, most of the oocytes become meiotically competent and will resume meiosis spontaneously if removed from the follicles and cultured in an appropriate medium [90].

In granulosa cells of the mouse, rat, bovine and human ovary, keratin immunoreactivity is detected [62, 70, 73, 77, 78], whereas in the other species the granulosa cells do not express any keratins (Table 13.2). Furthermore, it has been shown that keratin 8 (K8) and K19 were detected first in primary and secondary follicles in the rat [67] or K8 and K18 in the human [66], but progressively disappear in granulosa cells of growing or mature (Graafian) follicles, respectively. The expression of keratin in the follicular development remains controversial unlike that of vimentin.

IFs in Atretic Follicles

Apoptosis has been implicated in the selective elimination of granulosa cells and oocytes during atresia of ovarian follicles [91–93]. Atretic follicles contain some K18-positive cells with intense cytoplasmic staining [77]. In early antral atretic follicles, keratin-positive cells are present in the most antral layers of the follicle, whereas in advanced atretic follicles, they are distributed throughout the follicle, particularly in the basal atretic follicles, in which the granulosa layer has separated from the basal lamina [77]. Ortega et al. [94] have observed significantly higher intensity of vimentin in the granulosa cell layer of atretic follicles compared to those of healthy antral follicles. Similarly, a greater significant immunostaining for vimentin and keratins is noted in the granulosa cell layer of atretic follicles [70, 71, 95]. The same immunoreactions of these IF proteins are also observed in the granulosa cell layer of cystic follicles in rats [96] and cows [94].

IFs in Mammalian Oocytes

During oocyte maturation, spindle formation, chromosome separation, polar body extrusion and organelle movement occur in the ooplasm for subsequent fertilization and development [97–101]. Cytoskeletons, such as microfilaments and microtubules, are well known to be important for the progression of those events [102–105]. In contrast, research related to IFs remains poorly advanced relative to that of microfilaments and microtubules.

Scanning electron micrographs of the cytoskeleton network just beneath the oolemma are presented in Fig. 13.2, showing the highly ordered filamentous structures of microfilaments, microtubules and IFs as determined by their size. IF proteins studied in mammalian oocytes and granulosa cells are summarized in Table 13.2. Distribution of IFs in the oocytes has been a very controversial issue. The reasons for

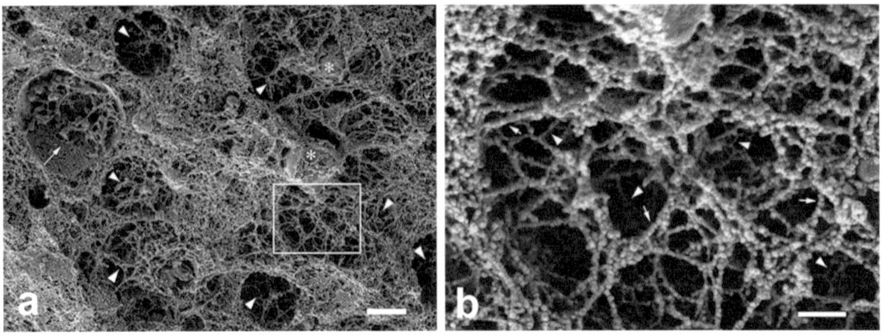

Fig. 13.2 Scanning electron micrograph of the cytoskeleton network of a bovine oocyte. *Bar* represents 1 µm in (**a**) and 250 nm in (**b**). (**a**) The highly ordered filamentous structures just beneath the oolemma. *Arrow* shows a part of membrane debris, which has remained even after proteolytic digestion. *Arrowheads* show the cytoskeleton network in low density. Note the trans-most cisterna of Golgi complex (*asterisks*). (**b**) A high magnification image of a part framed rectangle in (**a**). Note three interconnected filament systems. *Arrows* show the thickest cytoskeleton, microtubules, and *arrowheads* show the thinnest filaments of the cytoskeleton, actin microfilaments. There is very great abundance of intermediate filaments among them

the discrepancies concerning the presence of keratins in oocytes may depend on (1) interspecies differences in the IF protein sequence, expression and organization and (2) the use of different fixatives and/or antibodies. Fridmacher et al. [67] have pointed out that immunoreactivities with keratin monoclonal antibody are affected by fixation, where the results observed from fixed and unfixed tissues has been different. In the course of the study on distribution of IFs in porcine oocytes, we have also noticed that the immunoreactivity of certain secondary antibodies were altered as a result of different fixation methods (buffered formalin vs. methanol).

Keratins in the Oocytes

In mouse and human oocytes, some researchers have reported that the oocytes show keratin positive detection [61, 63, 68, 69, 71]. For example, Balbiani bodies, which contains aggregated mitochondria of the oocytes and persist in resting human primary follicles [106], show immunostaining for K8, K18, and K19 [68, 71, 89]. On the contrary, others have not observed any keratin-positive signals in the oocytes of these species [62, 67, 73, 81]. Furthermore, there are interspecies differences in the literature. Keratin IF protein is observed in sheep [64] and hamsters [65, 72, 76], whereas not in cattle, pigs and rats ([62, 67, 70, 73]; Suzuki et al. unpublished data).

In our previous study [76], non-fibrillar keratin particles have been observed. In germinal vesicle (GV) oocytes, large and oval-shaped aggregates of non-fibrillar keratin have been found in the cortical ooplasm (designated as a 'cortical' pattern). The delicate network of keratin filaments is concentrated in the GV periphery.

The large keratin aggregates begin to divide into small fragments at the pro-MI/MI stage (designated as a 'fragmented' pattern). Some keratin fragments have occasionally been broken down into several granules at the peripheral region. In the MII oocytes, the filament network is extended over the ooplasm and numerous keratin granules are scattered across the oocyte (designated as a 'granular' pattern). It has been suggested, therefore, that non-fibrillar keratin constitute a reservoir of keratin protein that can be recruited into keratin IFs, thereby creating a more effective distribution of IF protein throughout the ooplasm [76].

Vimentin in the Oocytes

Mammalian oocytes show no significant vimentin reaction with any of the antibodies applied (Table 13.2). As mentioned above, vimentin immunoreactivity is found in flattened follicular epithelial cells of primordial follicles and in cuboidal follicular epithelial cells of primary follicles, and numerous granulosa cells of secondary and antral follicles (Fig. 13.1). Furthermore, vimentin-positive protrusions of the corona radiata cells penetrate the zona pellucida and contact the oocyte in cows [70] and pigs (Fig. 13.3, Suzuki et al. unpublished data; see also [84]). In various mammalian species, similar corona cell processes appear to contain IFs at the ultrastructural level [107]. These cytoskeletal components may have a function in various important cellular activities, including aspects of cell–cell adhesions, intercellular transport and mechano-transduction and signaling [3, 5, 108].

Several studies have shown that cell-to-cell communications via gap junctions, as well as other junctional complexes, form the major anchorage between the oocyte and cumulus cells during all stages of follicle development [109–116]. Our previous study has clearly demonstrated by scanning electron microscopy that the cumulus

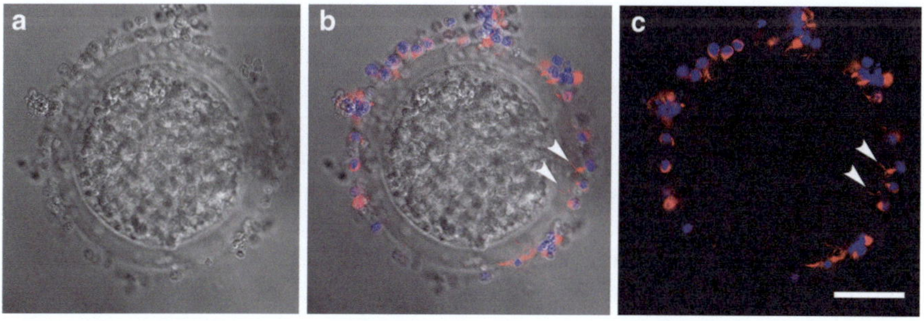

Fig. 13.3 Confocal laser scanning microscopic images of vimentin localization in a porcine oocyte. Sequential differential interference contrast (DIC) and fluorescence imaging. *Bar* represents 50 μm. (**a**) DIC image, (**b**) overlay of DIC and fluorescence images, and (**c**) fluorescence image. Vimentin filaments are *red* and nuclei are *blue*. Note transzonal cumulus cell projections consisting of vimentin (*arrowheads*)

cell projections are directed toward and terminate at the oocyte in the pig [98]. These transzonal projections appear as extremely long and thin extensions at the GV stage and are intermingled with those arising from the adjacent cumulus, which are densely stained for actin but not for tubulin [98].

Desmin in the Oocytes

As shown in Table 13.2, desmin immunoreactivity has not been detected in the granulosa cells and oocytes of the bovine [70, 80], pig [62], sheep [74], rat [62], baboon [71] and human ovary [62, 68, 71]. However, in our immunofluorescent observations on porcine and hamster oocytes, desmin immunoreactivity has been detected in both the oocytes and cumulus cells. Staining with anti-desmin of the oocyte has been very low in intensity, but it has changed during oocyte maturation (Suzuki et al. unpublished data). Figure 13.4 represents confocal laser microscopic images of desmin and actin localization in hamster oocytes. In contrast to the intensive cortical actin staining, the desmin intensity is very weak. At the GV and MI stage of hamster oocytes, the desmin-positive area has been restricted only to the cortical region of the ooplasm, whereas desmin is localized uniformly throughout the ooplasm at the MII stage. The average intensity of desmin is 30 % higher at the MII stage compared to the GV and MI stages ($P<0.05$). These observations suggest that desmin IF protein may play an important role in maintaining the cell architecture during oocyte maturation.

Nestin in the Oocytes

Nestin is widely used as stem cell marker (for review, see [41]). Nestin has been shown to interact with other cytoskeleton proteins, such as vimentin [19, 50, 51] or desmin [52], suggesting a role in regulating cellular cytoskeletal structure. The physiological significance of nestin in the ovary remains unknown. Nestin is expressed during early developmental stages and during regeneration in several tissues such as the brain, pancreas, and testis [41]. This suggests that nestin is necessary in cells with proliferative activity or in cells that are in a dynamic developmental phase, both of which require a high degree of cytoplasmic plasticity [40, 41].

In our unpublished observations, nestin immunoreactivity of GV, MI and MII porcine oocytes has been evaluated by confocal laser scanning microscopy (Fig. 13.5). Fluorescent intensity of nestin is decreased during oocyte maturation. Because the intensity of nestin staining is negatively correlated with the progression of meiosis, it is suggested that nestin may be involved in follicular growth rather than oocyte maturation. Takahashi et al. [48] have reported that nestin is mainly

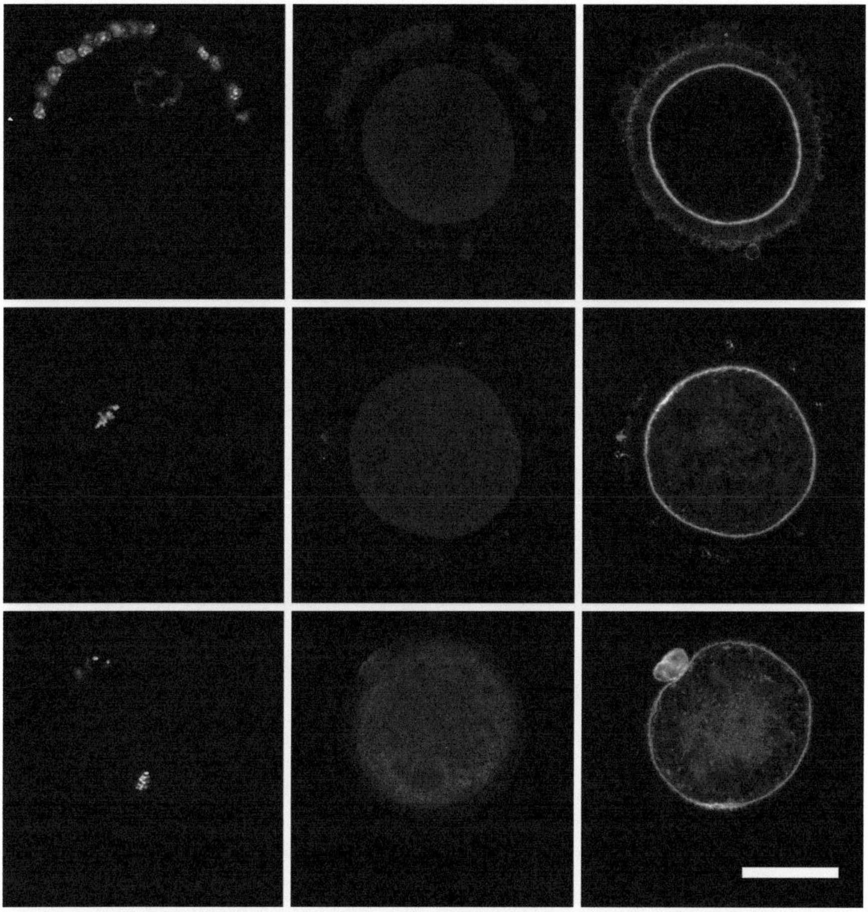

Fig. 13.4 Confocal laser scanning microscopic images of desmin and actin localization in hamster oocytes. *Bar* represents 50 µm. Images of the oocytes at GV (*upper panel*), MI (*middle panel*) and MII (*lower panel*) stages. (A) Chromatin of GV, chromosomes and nuclei of cumulus cells are visualized by DAPI, (B) staining with anti-desmin, and (C) staining with anti-actin. Desmin immunoreactivity (B) is noted in the ooplasm and cumulus cells and staining intensity of the ooplasm is higher at MII stage than at GV and MI stages. Actin microfilaments (C) are strongly stained just beneath the membrane of the oocytes and cumulus cells. Note transzonal cumulus cell projections consisting of actin in the GV oocyte (*upper panel*, C)

expressed in vascular endothelial cells of the theca interna in rat growing follicles and that nestin expression increases with follicular growth or hCG administration, which promote angiogenesis in the ovary [117]. These observations suggest that nestin may be involved in angiogenesis in growing follicles, which is followed by follicle maturation and subsequent ovulation.

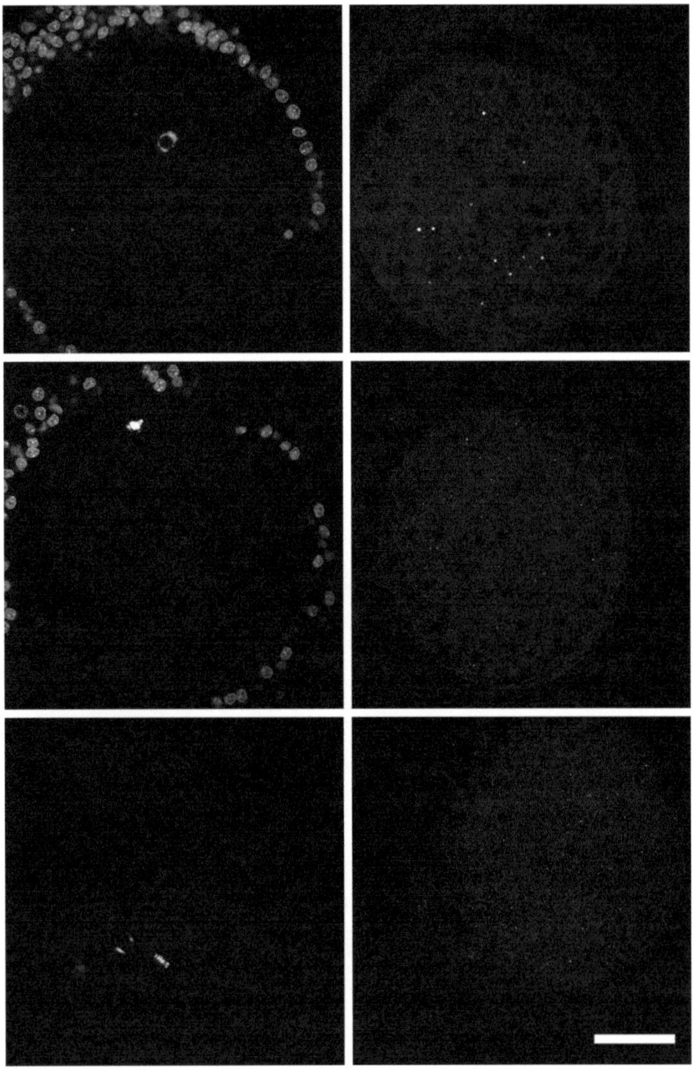

Fig. 13.5 Confocal laser scanning microscopic images of nestin localization in porcine oocytes. *Bar* represents 50 μm. Images of the oocytes at GV (*upper panel*), MI (*middle panel*) and MII (*lower panel*) stages. (A) Chromatin of GV, chromosomes and nuclei of cumulus cells are visualized by DAPI, and (B) staining with anti-nestin. Nestin immunoreactivity (B) is noted in the ooplasm and cumulus cells. Discrete nestin-containing *dots* are strongly stained and the mesh-like structure with weak response is noted at GV stage (*upper panel*). During oocyte maturation, nestin-containing *dots* decrease in size and have become diffuse throughout the ooplasm. Staining intensity of the ooplasm is also decreased until MII stage

Neurofilaments and GFAP in the Oocytes

Neurofilaments (NFs) are the main cytoskeleton elements in neurons. The three types of NFs have different molecular masses and are referred to as NF-L, NF-M, and NF-H. NF proteins synthesized in the neuronal cell body are phosphorylated after transfer to the axon, where they accumulate with other cytoskeletal proteins to help maintain the axonal structure [118]. NF protein is immunohistochemically detected in rat [79] and human oocytes [71], but not in mouse oocytes [61]. Expression of NF-H starts in oocytes at the primary stage of follicles, and continues in fertilized one-cell eggs and vanishes at the two-cell stage [79].

GFAP is the major protein constituent of glial IFs in differentiated fibrous and protoplasmic astrocytes of the central nervous system. Lehtonen et al. [61] have failed to detect GFAP along with NFs in mouse oocytes and early embryos. In our unpublished observations, however, NFs and GFAP have been detected in the GV, MI and MII porcine oocytes. Figure 13.6 shows confocal laser scanning microscopic

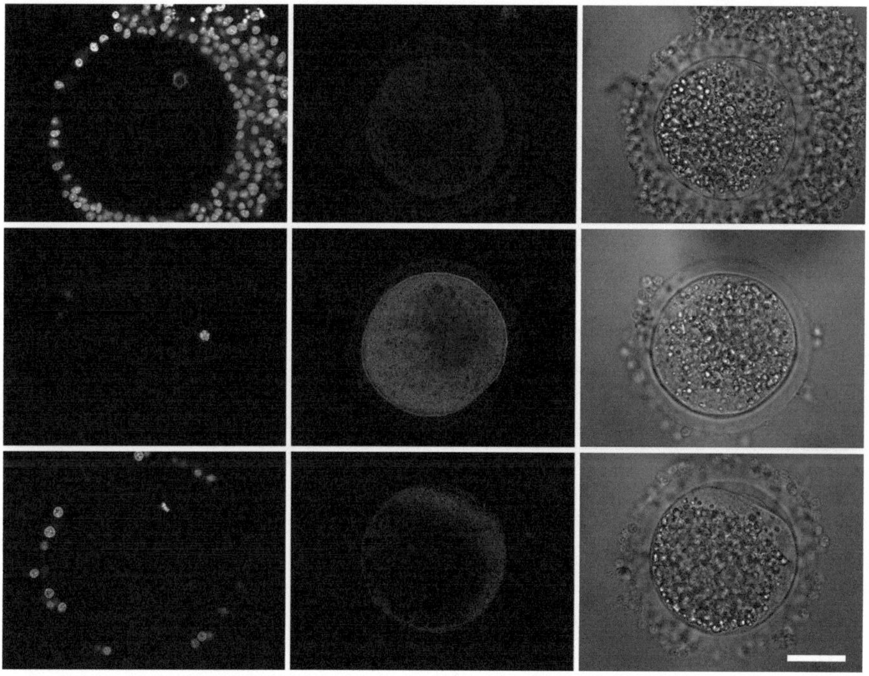

Fig. 13.6 Confocal laser scanning microscopic images of neurofilament localization in porcine oocytes. *Bar* represents 50 μm. Images of the oocytes at GV (*upper panel*), MI (*middle panel*) and MII (*lower panel*) stages. Sequential fluorescence and differential interference contrast (DIC) imaging. (A) A nucleolus of GV, chromosomes and nuclei of cumulus cells are visualized by DAPI, (B) staining with anti-neurofilament, and (C) DIC imaging. Neurofilament immunoreactivity (B) is noted in the ooplasm, but not in the cumulus cells. The lipid droplets appear as small vacuoles under DIC (C). In the area where the lipid droplets are not seen in the oocyte, abundance of neurofilament is detected

images of NF localization in porcine oocytes. NFs are found to be located in inverse proportion to accumulation of lipid droplets in the oocytes.

In the developing nervous system, vimentin is found in both presumptive glial and neural cells, and the tissue-specific NF, GFAP and nestin appear later in development [5, 10, 18]. We have observed similar mesh-like structures stained with antibodies to these neuronal-related IFs. It is suggested, therefore, that they may have been co-assembled and/or replaced often by different IFs even in the mammalian oocytes. In addition, we have observed that transzonal projections of the corona radiata cells are stained with anti-GFAP, anti-desmin and anti-vimentin in porcine oocytes. The physiological significance of NF, GFAP and nestin expressed in the oocytes remains to be determined.

Mechanical and Non-mechanical IF Functions in the Oocytes

IFs support mechanically the structural integrity of tissues and cells and the best example is seen in epithelial cells constituting the epidermis composed of the keratin IFs. Our previous study has shown the increasing complexity of keratin filament network of hamster oocytes during maturation. Keratin IFs which have assembled into extensive cytoskeletal networks in the MII oocytes suggests that keratin may play a specific role in maintaining cell integrity under physical stress during egg transport in the oviduct after ovulation [76]. The extensive distribution of IFs appears to provide the oocytes with important mechanical properties. Keratin filaments (tonofibrils) appear to respond rapidly to shear stresses which are exerted at the surface of epithelial PtK2 cells [14]. IFs contribute to cell adhesion and migration [119, 120].

Recent studies have pointed out the functional significance of cellular mechanotransduction processes in somatic cells [121–123]. Transmission of forces from outside the cell through cell–matrix and cell–cell contacts appears to control the maturation or disassembly of these adhesions and initiates intracellular signaling cascades that ultimately alter many cellular behaviors. In response to externally applied forces, cells actively rearrange the organization and contractile activity of the cytoskeleton and redistribute their intracellular forces. Accumulating evidences suggest that the localized concentration of these cytoskeletal tensions at adhesions is also a major mediator of mechanical signaling [121]. IF networks connect the cell surface with the outer nuclear membrane which connect to components of the nuclear lamina [123, 124], thereby regulating the cellular architecture and also providing an important platform to mediate cellular mechanotransduction processes [121]. Polymerized IF networks also play roles in numerous other signal transduction pathways by providing a scaffold or platform that interacts with signaling molecules including MAP kinases, mTOR, various 14-3-3 protein isoforms, Cdk5, and apoptotic factors [3, 86, 125–130]. Environmental or internal stresses initiate stress signaling cascades, which activate the stress response and transcriptional machineries that induce the expression of the classical stress-induced heat shock protein (HSP) genes [131].

Non-mechanical Physiological Functions of IFs in the Oocytes

IFs provide the cell with a mechanism for resisting mechanical stress and cellular mechanotransduction processes. Furthermore, many studies on different types of somatic cells have revealed that IFs and their precursors are remarkably dynamic and exhibit a complex array of motile activities related to their subcellular assembly and organization. As mentioned above, some IF proteins may locate in the ooplasm (keratins, vimentin, desmin, nestin, GFAP and neurofilaments), but their physiological significances are still to be resolved. We mainly focused on the regulation of organelle positioning and regulation of translation as possible functions of IFs in the oocytes, because the cytoskeleton participates in the spatial organization and regulation of translation [132]. These subcellular events may be crucial for cellular growth, proliferation and function.

Organelle Positioning and IFs

Transport of membranous organelles is mainly mediated by microtubule and microfilament cytoskeletal tracks and their respective molecular motors. In addition, organelle positioning in the cytoplasm seems to involve interactions with IFs [3, 57, 133, 134]. Here a special interest has been paid to the interactions between IFs and membranous organelles, such as mitochondria, the Golgi complex and other membranous components.

Mitochondria

The temporal and spatial dynamic patterns of mitochondrial distribution are important for their biological functions; disruption of their distribution can cause cell death. The morphology and distribution of mitochondria in cells are coordinated by microfilaments and microtubules [135–139]. Evidence that mitochondria associate with IFs has also been obtained [140, 141]. Since then, several IF proteins have been associated with mitochondrial functions in muscle [142–148] and non-muscle cells [149–154]. For example, desmin IFs play a role in mitochondrial positioning and respiratory function in cardiac and skeletal muscle [142, 143] and in smooth muscle [145, 146]. Immunoelectron microscopic studies of chicken skeletal and cardiac muscle have also shown that extensive labeling of desmin is localized to the interfibrillar spaces where mitochondria are located [147, 148]. Furthermore, observations on desmin-deficient mice have revealed the importance of desmin IFs in mitochondrial behavior and function [142, 155–157].

There is also evidence that keratin and vimentin have been implicated in docking mitochondria in muscle cells [144] and hepatocytes [153]. In nerve cells, on the other hand, the subcellular organization and movement of mitochondria are associated with

IFs comprising the NFs [158, 159]. It has also been shown that antibodies against the NF-H subunit disrupt binding between mitochondria and NFs and the interactions between them depend on mitochondrial membrane potential [160].

A linking protein between mitochondria and IFs, termed IEF 24 (MW 56,000), has been extracted from cultured fibroblasts, which is tenaciously associated with a subpopulation of IFs and also correlates closely with mitochondrial distribution [140]. Conserved structures on the mitochondrial surface, such as Mdm10p, Mmm1p and so on, are suggested to be adapted for interaction with different cytoskeletal networks [161]. Mitochondria can also associate with IFs through interactions with the cytolinker protein, plectin [143, 162–164]. Therefore, organization of the cytoskeleton network together with associated protein(s) described above could be essential in regulating mitochondrial function. Furthermore, IFs can directly or indirectly bind the mitochondria, which have been detached from microfilament or microtubule tracks [165].

The Golgi Complex

The Golgi complex plays an important role in the post-translational modifications and sorting of lipids and proteins from the endoplasmic reticulum (ER) and ER-Golgi intermediate compartments [166]. The association of the Golgi complex and microtubules has been demonstrated often in several systems [167–169], whereas in a certain type of cells microfilaments are essential for the Golgi morphology and cytological positioning [170], through various actin-associated proteins [171]. Vimentin IFs are also associated with the Golgi complex, and the Golgi 58K protein (FTCD) is a candidate linker protein connecting the Golgi complex to the vimentin IF cytoskeleton [172, 173]. In addition, a Golgi-associated network surrounding the Golgi complex has been proposed [174], where dense bundles of keratin and actin filaments are observed around the Golgi complex. Interestingly, when NFs are induced to aggregate by microinjection of NF-H into cultured neuronal cells, the Golgi complex is fragmented and dispersed. Such a phenomenon is seen in diseased neuronal IF aggregate containing neurons derived from amyotrophic lateral sclerosis (ALS) patient tissues [159].

Other Membranous Organelles

Autophagosomes are cellular organelles thought to be derived from the membranes of the ER-mitochondria contact sites that engulf organelles targeted for degradation by fusing autophagosomes and lysosomes [175]. The positioning of endosomes and lysosomes and the maturation of autophagosomes have been shown to be tightly associated with the assembly of vimentin and its phosphorylation [176]. It has also been shown that interactions between vimentin IFs and the adaptor complex AP-3 likely control the positioning, content, and subcellular distribution of selected late endosome/lysosome membrane proteins [177].

Lipid droplets, the cellular organelles for the repositories of fatty acids, are thought to arise from the bilayer membrane of endoplasmic reticulum [178]. Vimentin IFs interact with the lipid droplets [179, 180]. Recent studies revealed perilipin as linking protein between lipid droplets and vimentin [181].

Translational Components Associated with IFs

The cytoskeleton acts as a signaling platform that modulates cellular pathways by controlling the activity and/or subcellular localization of signaling proteins and their targets [132]. Polysomes (clustered ribosomes) are observed close to the cytoskeleton in various cell types, such as fibroblasts, epithelial lens cells and sea urchin eggs [132]. Although microfilaments are the main cytoskeletal element that participates in the organization of the translational apparatus [132, 182], there is evidence indicating a physical link between IFs and polysomes [183–185]. Ribonucleoprotein complexes are reported to bind keratin [186] or vimentin IFs, too [187].

Eukaryotic elongation factor-1 (eEF1), composed of 3 subunits (eEF1A, eEF1Bα and eEF1Bγ) is essential for peptide-chain elongation during translation. eEF1A interacts with the actin microfilaments in a wide range of species from yeasts to mammals [188]. It has been shown that eEF1Bγ, a non-catalytic subunit of the eEF1 complex, may be a keratin-binding protein, suggesting an involvement of keratin IF networks in translation [189]. The two other components of the eEF1, eEF1Bα and eEF1A are also associated with keratin IFs in epithelial cells. Thus, there appears to be a remarkable convergence in the reciprocal manner with which two distinct subunits of the eEF1 complex, eEF1A and eEF1Bγ, relate to actin microfilaments and keratin IFs, respectively [132].

Post-translational modifications (PTMs) play important roles in regulating the functional properties of IFs. They include phosphorylation, glycosylation, prenylation, sumoylation, acetylation, and others [130]. Furthermore, PTM studies have revealed important interactions between IFs and other cellular components and structures, such as the interaction of 14-3-3 proteins with multiple IFs [190–192]. PTMs may regulate IF organization and the binding of IFs to IF-associated proteins, thereby regulating numerous cellular processes and cell-specific functions (for review, see [130]).

Concluding Remarks

Unlike microfilaments or microtubules, IFs show a wide range of molecular diversity. Furthermore, IFs have a non-polar structure and therefore have no IF-specific associated motor proteins. IFs show versatile functions and properties, due to an outstanding degree of the molecular diversity. IF proteins are dynamic components of the cytoskeleton characterized by rapid movement and dynamic exchange of the subunits. Non-mechanical IF functions include regulation of the cellular architecture, cell growth, organelle positioning, signaling, and gene expression. IFs are clearly well integrated with the microfilament and microtubule cytoskeletons and their motor proteins: movement of IF proteins likely occur through interactions with the microtubule-based motors kinesin and dynein; IFs are also associated with the microfilament-based motor myosin. Interactions with microfilaments and microtubules are not only restricted to motors, because a family of proteins directly link the microfilaments and microtubules to IFs. Therefore, IFs are dynamically integrated with other cytoskeletons capable of the polarization required for directional movement of organelles and motor cargoes. In addition, certain protein(s) linking between organelle and IFs remains to be clarified. Further studies are clearly needed before the biological significance of IFs and participation in cytoskeletal crosstalk in the oocyte/embryo can be fully assessed.

Acknowledgments This work was supported by JSPS KAKENHI Grant Numbers 17580243, 21580340 and 24580403. The author thanks Mr. Taiki Sato, Mr. Yosuke Shiozaki and Ms. Manami Sumida for technical assistance.

References

1. Albers K, Fuchs E (1992) The molecular biology of intermediate filament proteins. Int Rev Cytol 134:243–279
2. Alberts B, Bray D, Lewis J, Raff M, Roberts K, Watson JD (1994) Intermediate filaments. In: Alberts B et al (eds) Molecular biology of the cell, 3rd edn. Garland Science, New York
3. Chang L, Goldman RD (2004) Intermediate filaments mediate cytoskeletal crosstalk. Nat Rev Mol Cell Biol 5:601–613
4. Lodish H, Berk A, Zipursky SL, Matsudaira P, Baltimore D, Darnell J (2000) Section 19.6, intermediate filaments. In: Lodish H et al (eds) Molecular cell biology, 4th edn. W. H. Freeman, New York
5. Paramio JM, Jorcano JL (2002) Beyond structure: do intermediate filaments modulate cell signalling? Bioessays 24:836–844
6. Steinert PM, Steven AC, Roop DR (1985) The molecular biology of intermediate filaments. Cell 42:411–419
7. Moll R, Divo M, Langbein L (2008) The human keratins: biology and pathology. Histochem Cell Biol 129:705–733. doi:10.1007/s00418-008-0435-6
8. Schweizer J, Bowden PE, Coulombe PA, Langbein L, Lane EB, Magin TM, Maltais L, Omary MB, Parry DA, Rogers MA, Wright MW (2006) New consensus nomenclature for mammalian keratins. J Cell Biol 174:169–174
9. Haines RL, Lane EB (2012) Keratins and disease at a glance. J Cell Sci 125:3923–3928. doi:10.1242/jcs.099655
10. Magin TM, Vijayaraj P, Leube RE (2007) Structural and regulatory functions of keratins. Exp Cell Res 313:2021–2032
11. Windoffer R, Leube RE (2001) De novo formation of cytokeratin filament networks originates from the cell cortex in A-431 cells. Cell Motil Cytoskeleton 50:33–44
12. Green KJ, Talian JC, Goldman RD (1986) Relationship between intermediate filaments and microfilaments in cultured fibroblasts: evidence for common foci during cell spreading. Cell Motil Cytoskeleton 6:408–418

13. Green KJ, Geiger B, Jones JC, Talian JC, Goldman RD (1987) The relationship between intermediate filaments and microfilaments before and during the formation of desmosomes and adherens-type junctions in mouse epidermal keratinocytes. J Cell Biol 104:1389–1402
14. Yoon KH, Yoon M, Moir RD, Khuon S, Flitney FW, Goldman RD (2001) Insights into the dynamic properties of keratin intermediate filaments in living epithelial cells. J Cell Biol 153:503–516
15. Evans RM (1998) Vimentin: the conundrum of the intermediate filament gene family. Bioessays 20:79–86
16. Franke WW, Schmid E, Osborn M, Weber K (1979) Intermediate-sized filaments of human endothelial cells. J Cell Biol 81:570–580
17. Galou M, Colucci-Guyon E, Ensergueix D, Ridet JL, Gimenez y Ribotta M, Privat A, Babinet C, Dupouey P (1996) Disrupted glial fibrillary acidic protein network in astrocytes from vimentin knockout mice. J Cell Biol 133:853–863
18. Herrmann H, Aebi U (2000) Intermediate filaments and their associates: multi-talented structural elements specifying cytoarchitecture and cytodynamics. Curr Opin Cell Biol 12:79–90
19. Steinert PM, Chou YH, Prahlad V, Parry DA, Marekov LN, Wu KC, Jang SI, Goldman RD (1999) A high molecular weight intermediate filament-associated protein in BHK-21 cells is nestin, a type VI intermediate filament protein. J Biol Chem 274:9881–9890
20. Eckes B, Dogic D, Colucci-Guyon E, Wang N, Maniotis A, Ingber D, Merckling A, Langa F, Aumailley M, Delouvée A, Koteliansky V, Babinet C, Krieg T (1998) Impaired mechanical stability, migration and contractile capacity in vimentin-deficient fibroblasts. J Cell Sci 111:1897–1907
21. Esue O, Carson AA, Tseng Y, Wirtz D (2006) A direct interaction between actin and vimentin filaments mediated by the tail domain of vimentin. J Biol Chem 281:30393–30399
22. Goldman RD (1971) The role of three cytoplasmic fibers in BHK-21 cell motility. I. Microtubules and the effects of colchicine. J Cell Biol 51:752–762
23. Helfand BT, Loomis P, Yoon M, Goldman RD (2003) Rapid transport of neural intermediate filament protein. J Cell Sci 116:2345–2359
24. Bär H, Strelkov SV, Sjoberg G, Aebi U, Herrmann H (2004) The biology of desmin filaments: how do mutations affect their structure, assembly, and organisation? J Struct Biol 148: 137–152
25. Kasantikul V, Shuangshoti S (1989) Positivity to glial fibrillary acidic protein in bone, cartilage, and chordoma. J Surg Oncol 41:22–26
26. Eriksson KS, Zhang S, Lin L, Lariviere RC, Julien JP, Mignot E (2008) The type III neurofilaments peripherin is expressed in the tuberomammillary neurons of the mouse. BMC Neurosci 9:26. doi:10.1186/1471-2202-9-26
27. Trojanowski JQ, Lee VM (1985) Expression of neurofilament antigens by normal and neoplastic human adrenal chromaffin cells. N Engl J Med 313:101–104
28. Yuan A, Rao MV, Veeranna, Nixon RA (2012) Neurofilaments at a glance. J Cell Sci 125:3257–3263
29. Shah JV, Flanagan LA, Janmey PA, Leterrier JF (2000) Bidirectional translocation of neurofilaments along microtubules mediated in part by dynein/dynactin. Mol Biol Cell 11:3495–3508
30. Cochard P, Paulin D (1984) Initial expression of neurofilaments and vimentin in the central and peripheral nervous system of the mouse embryo in vivo. J Neurosci 14:2080–2094
31. Shea TB, Beermann ML, Fischer I (1993) Transient requirement for vimentin in neuritogenesis: intracellular delivery of anti-vimentin antibodies and antisense oligonucleotides inhibit neurite initiation but not elongation of existing neurites in neuroblastoma. J Neurosci Res 36:66–76
32. Burke B, Stewart CL (2013) The nuclear lamins: flexibility in function. Nat Rev Mol Cell Biol 14:13–24
33. Gerace L, Burke B (1988) Functional organization of the nuclear envelope. Annu Rev Cell Biol 4:335–374

34. Lin F, Worman HJ (1997) Expression of nuclear lamins in human tissues and cancer cell lines and transcription from the promoters of the lamin A/C and B1 genes. Exp Cell Res 236:378–384
35. Oryschak AF, Ghadially FN, Bhatnagar R (1974) Nuclear fibrous lamina in the chondrocytes of articular cartilage. J Anat 118:511–515
36. Patrizi G, Poger M (1967) The ultrastructure of the nuclear periphery. The zonula nucleum limitans. J Ultrastruct Res 17:127–136
37. Tsai JW, Lian WN, Kemal S, Kriegstein AR, Vallee RB (2010) An unconventional kinesin and cytoplasmic dynein mediate interkinetic nuclear migration in neural stem cells. Nat Neurosci 13:1463–1471. doi:10.1038/nn.2665
38. Lendahl U, Zimmerman LB, McKay RD (1990) CNS stem cells express a new class of intermediate filament protein. Cell 60:585–595
39. Hendrickson ML, Rao AJ, Demerdash ON, Kalil RE (2011) Expression of nestin by neural cells in the adult rat and human brain. PLoS One 6, e18535. doi:10.1371/journal.pone.0018535
40. Michalczyk K, Ziman M (2005) Nestin structure and predicted function in cellular cytoskeletal organisation. Histol Histopathol 20:665–671
41. Wiese C, Rolletschek A, Kania G, Blyszczuk P, Tarasov KV, Tarasova Y, Wersto RP, Boheler KR, Wobus AM (2004) Nestin expression: a property of multi-lineage progenitor cells? Cell Mol Life Sci 61:2510–2522
42. Lobo MV, Arenas MI, Alonso FJ, Gomez G, Bazan E, Paino CL, Fernandez E, Fraile B, Paniagua R, Moyano A, Caso E (2004) Nestin, a neuroectodermal stem cell marker molecule, is expressed in Leydig cells of the human testis and in some specific cell types from human testicular tumours. Cell Tissue Res 316:369–376
43. Abraham EJ, Leech CA, Lin JC, Zulewski H, Habener JF (2002) Insulinotropic hormone glucagon-like peptide-1 differentiation of human pancreatic islet derived progenitor cells into insulin-producing cells. Endocrinology 143:3152–3161
44. Davidoff MS, Middendorff R, Enikolopov G, Riethmacher D, Holstein AF, Muller D (2004) Progenitor cells of the testosterone-producing Leydig cells revealed. J Cell Biol 167:935–944
45. Frojdman K, Pelliniemi LJ, Lendahl U, Virtanen I, Eriksson JE (1997) The intermediate filament protein nestin occurs transiently in differentiating testis of rat and mouse. Differentiation 61:243–249
46. Huang H, Tang X (2003) Phenotypic determination and characterization of nestin-positive precursors derived from human fetal pancreas. Lab Invest 83:539–547
47. Zulewski H, Abraham EJ, Gerlach MJ, Daniel PB, Moritz W, Muller B, Vallejo M, Thomas MK, Habener JF (2001) Multipotential nestin-positive stem cells isolated from adult pancreatic islets differentiate ex vivo into pancreatic endocrine, exocrine, and hepatic phenotypes. Diabetes 50:521–533
48. Takahashi N, Itoh MT, Ishizuka B (2008) Human chorionic gonadotropin induces nestin expression in endothelial cells of the ovary via vascular endothelial growth factor signaling. Endocrinology 149:253–260
49. Dahlstrand J, Collins VP, Lendahl U (1992) Expression of the class VI intermediate filament nestin in human central nervous system tumors. Cancer Res 52:5334–5341
50. Eliasson C, Sahlgren C, Berthold CH, Stakeberg J, Celis JE, Betsholtz C, Eriksson JE, Pekny M (1999) Intermediate filament protein partnership in astrocytes. J Biol Chem 274: 23996–24006
51. Marvin MJ, Dahlstrand J, Lendahl U, McKay RD (1998) A rod end deletion in the intermediate filament protein nestin alters its subcellular localization in neuroepithelial cells of transgenic mice. J Cell Sci 111:1951–1961
52. Sjöberg G, Jiang WQ, Ringertz NR, Lendahl U, Sejersen T (1994) Colocalization of nestin and vimentin/desmin in skeletal muscle cells demonstrated by three-dimensional fluorescence digital imaging microscopy. Exp Cell Res 214:447–458
53. Chen J, Boyle S, Zhao M, Su W, Takahashi K, Davis L, Decaestecker M, Takahashi T, Breyer MD, Hao CM (2006) Differential expression of the intermediate filament protein

nestin during renal development and its localization in adult podocytes. J Am Soc Nephrol 17:1283–1291

54. Guérette D, Khan PA, Savard PE, Vincent M (2007) Molecular evolution of type VI intermediate filament proteins. BMC Evol Biol 7:164. doi:10.1186/1471-2148-7-164
55. Wei R, Yang J, Hou W, Liu G, Gao M, Zhang L, Wang H, Mao G, Gao H, Chen G, Hong T (2013) Insulin-producing cells derived from human embryonic stem cells: comparison of definitive endoderm- and nestin-positive progenitor-based differentiation strategies. PLoS One 8(8), e72513. doi:10.1371/journal.pone.0072513. eCollection 2013
56. Helfand BT, Mikami A, Vallee RB, Goldman RD (2002) A requirement for cytoplasmic dynein and dynactin in intermediate filament network assembly and organization. J Cell Biol 157:795–806
57. Helfand BT, Chang L, Goldman RD (2004) Intermediate filaments are dynamic and motile elements of cellular architecture. J Cell Sci 117:133–141
58. Hyder CL, Pallari HM, Kochin V, Eriksson JE (2008) Providing cellular signposts—post-translational modifications of intermediate filaments. FEBS Lett 582:2140–2148. doi:10.1016/j.febslet.2008.04.064
59. Minin AA, Moldaver MV (2008) Intermediate vimentin filaments and their role in intracellular organelle distribution. Biochemistry (Mosc) 73:1453–1466
60. Oriolo AS, Wald FA, Ramsauer VP, Salas PJ (2007) Intermediate filaments: a role in epithelial polarity. Exp Cell Res 313:2255–2264
61. Lehtonen E, Lehto VP, Vartio T, Badley RA, Virtanen I (1983) Expression of cytokeratin polypeptides in mouse oocytes and preimplantation embryos. Dev Biol 100:158–165
62. Czernobilsky B, Moll R, Levy R, Franke WW (1985) Co-expression of cytokeratin and vimentin filaments in mesothelial, granulosa and rete ovarii cells of the human ovary. Eur J Cell Biol 37:175–190
63. Lehtonen E (1985) A monoclonal antibody against mouse oocyte cytoskeleton recognizing cytokeratin-type filaments. J Embryol Exp Morphol 90:197–209
64. Gall L, Le Guen P, Huneau D (1989) Cytokeratin-like proteins in the sheep oocyte. Cell Differ Dev 28:95–104
65. Plancha CE, Carmo-Fonseca M, David-Ferreira JF (1989) Cytokeratin filaments are present in golden hamster oocytes and early embryos. Differentiation 42:1–9
66. van Niekerk CC, Boerman OC, Ramaekers FC, Poels LG (1991) Marker profile of different phases in the transition of normal human ovarian epithelium to ovarian carcinomas. Am J Pathol 138:455–463
67. Fridmacher V, Locquet O, Magre S (1992) Differential expression of acidic cytokeratins 18 and 19 during sexual differentiation of the rat gonad. Development 115:503–517
68. Santini D, Ceccarelli C, Mazzoleni G, Pasquinelli G, Jasonni VM, Martinelli GN (1993) Demonstration of cytokeratin intermediate filaments in oocytes of the developing and adult human ovary. Histochemistry 99:311–319
69. Gallicano GI, Larabell CA, McGaughey RW, Capco DG (1994) Novel cytoskeletal elements in mammalian eggs are composed of a unique arrangement of intermediate filaments. Mech Dev 45:211–226
70. van den Hurk R, Dijkstra G, van Mil FN, Hulshof SC, van den Ingh TS (1995) Distribution of the intermediate filament proteins vimentin, keratin, and desmin in the bovine ovary. Mol Reprod Dev 41:459–467
71. Khan-Dawood FS, Yusoff Dawood M, Tabibzadeh S (1996) Immunohistochemical analysis of the microanatomy of primate ovary. Biol Reprod 54:734–742
72. Plancha CE (1996) Cytokeratin dynamics during oocyte maturation in the hamster requires reaching of metaphase I. Differentiation 60:87–98
73. Appert A, Fridmacher V, Locquet O, Magre S (1998) Patterns of keratins 8, 18 and 19 during gonadal differentiation in the mouse: sex- and time-dependent expression of keratin 19. Differentiation 63:273–284
74. Marettová E, Maretta M (2002) Demonstration of intermediate filaments in sheep ovary. Acta Histochem 104:431–434

75. Bukovsky A, Caudle MR, Svetlikova M, Upadhyaya NB (2004) Origin of germ cells and formation of new primary follicles in adult. Reprod Biol Endocrinol 2:20. http://www.rbej.com/content/2/1/20
76. Kabashima K, Matsuzaki M, Suzuki H (2010) Intermediate filament keratin dynamics during oocyte maturation requires maturation/M-phase promoting factor and mitogen-activated protein kinase kinase activities in the hamster. Reprod Domest Anim 45:e184–e188
77. Townson DH, Putnam AN, Sullivan BT, Guo L, Irving-Rodgers HF (2010) Expression and distribution of cytokeratin 8/18 intermediate filaments in bovine antral follicles and corpus luteum: an intrinsic mechanism of resistance to apoptosis? Histol Histopathol 25:889–900
78. Mora JM, Fenwick MA, Castle L, Baithun M, Ryder TA, Mobberley M, Carzaniga R, Franks S, Hardy K (2012) Characterization and significance of adhesion and junction-related proteins in mouse ovarian follicles. Biol Reprod 86(153):1–14
79. Takahashi N, Ishizuka B (2012) The involvement of neurofilament heavy chain phosphorylation in the maturation and degeneration of rat oocytes. Endocrinology 153:1990–1998
80. Wendl J, Ebach K, Rodler D, Kenngott RA (2012) Immunocytochemical localization of cytoplasmic and nuclear intermediate filaments in the bovine ovary during folliculogenesis. Anat Histol Embryol 41:190–201
81. Hummitzsch K, Irving-Rodgers HF, Hatzirodos N, Bonner W, Sabatier L, Reinhardt DP, Sado Y, Ninomiya Y, Wilhelm D, Rodgers RJ (2013) A new model of development of the mammalian ovary and follicles. PLoS One 8, e55578. doi:10.1371/journal.pone.0055578
82. Eppig JJ (2001) Oocyte control of ovarian follicular development and function in mammals. Reproduction 122:829–838
83. Matzuk MM, Burns KH, Viveiros MM, Eppig JJ (2002) Intercellular communication in the mammalian ovary: oocytes carry the conversation. Science 296:2178–2180
84. Da Silva-Buttkus P, van den Hurk R, te Velde ER, Taverne MAM (2003) Ovarian development in intrauterine growth-retarded and normally developed piglets originating from the same litter. Reproduction 126:249–258
85. Helfand BT, Chou Y-H, Shumaker DK, Goldman RD (2005) Intermediate filament proteins participate in signal transduction. Trends Cell Biol 15:568–570
86. Ivaska J, Pallari HM, Nevo J, Eriksson JE (2007) Novel functions of vimentin in cell adhesion, migration, and signaling. Exp Cell Res 313:2050–2062
87. Mendez MG, Kojima S, Goldman RD (2010) Vimentin induces changes in cell shape, motility, and adhesion during the epithelial to mesenchymal transition. FASEB J 24:1838–1851
88. Edson MA, Nagaraja AK, Matzuk MM (2009) The mammalian ovary from genesis to revelation. Endocr Rev 30:624–712. doi:10.1210/er.2009-0012
89. Hirshfield AN (1991) Development of follicles in the mammalian ovary. Int Rev Cytol 124:43–101
90. Pincus G, Enzmann EV (1935) The comparative behavior of mammalian eggs in vivo and in vitro I. The activation of ovarian eggs. J Exp Med 62:655–675
91. Bridgham JT, Johnson AL (2001) Expression and regulation of Fas antigen and tumor necrosis factor receptor type I in hen granulosa cells. Biol Reprod 65:733–739
92. Hu CL, Cowan RG, Harman RM, Porter DA, Quirk SM (2001) Apoptosis of bovine granulosa cells after serum withdrawal is mediated by Fas antigen (CD95) and Fas ligand. Biol Reprod 64:518–526
93. Kim JM, Boone DL, Auyeung A, Tsang BK (1998) Granulosa cell apoptosis induced at the penultimate stage of follicular development is associated with increased levels of Fas and Fas ligand in the rat ovary. Biol Reprod 58:1170–1176
94. Ortega HH, Salvetti NR, Müller LA, Amable P, Lorente JA, Barbeito CG, Gimeno EJ (2007) Characterization of cytoskeletal proteins in follicular structures of cows with cystic ovarian disease. J Comp Pathol 136:222–230
95. van Wezel IL, Rodgers RJ, Krupa M (1999) Development of the membrana granulosa of bovine antral follicles: structure, location of mitosis and pyknosis, and immunolocalization of involucrin and vimentin. Reprod Fertil Dev 11:37–48

96. Baravalle C, Salvetti NR, Mira GA, Pezzone N, Ortega HH (2006) Microscopic characterization of follicular structures in letrozole-induced polycystic ovarian syndrome in the rat. Arch Med Res 37:830–839
97. Kim NH, Funahashi H, Prather RS, Schatten G, Day BN (1996) Microtubule and microfilament dynamics during meiotic maturation. Mol Reprod Dev 43:248–255
98. Suzuki H, Jeong B-S, Yang X (2000) Dynamic changes of cumulus-oocyte cell communication during in vitro maturation of porcine oocytes. Biol Reprod 63:723–729
99. Suzuki H, Satoh M, Toyokawa K (2005) Changes in distribution of active mitochondria during oocyte maturation and fertilization in the hamster. J Mamm Ova Res 22:163–169
100. Suzuki H, Satoh M, Kabashima K (2006) Distributions of mitochondria and the cytoskeleton in hamster embryos developed in vivo and in vitro. J Mamm Ova Res 23:128–134
101. Suzuki H, Saito Y (2006) Cumulus cells affect distribution and function of the cytoskeleton and organelles in porcine oocytes. Reprod Med Biol 5:183–194
102. van Blerkom J (1991) Microtubule mediation of cytoplasmic and nuclear maturation during the early stages of resumed meiosis in cultured mouse oocytes. Proc Natl Acad Sci U S A 88:5031–5035
103. Sun QY, Lai L, Park KW, Kuhholzer B, Prather RS, Schatten H (2001) Dynamic events are differently mediated by microfilaments, microtubules, and mitogen-activated protein kinase during porcine oocyte maturation and fertilization in vitro. Biol Reprod 64:879–889
104. Sun QY, Wu GM, Lai L, Park KW, Cabot R, Cheng HT, Day BN, Prather RS, Schatten H (2001) Translocation of active mitochondria during pig oocyte maturation, fertilization and early embryo development in vitro. Reproduction 122:155–163
105. Sun QY, Schatten H (2006) Regulation of dynamic events by microfilaments during oocyte maturation and fertilization. Reproduction 131:193–205
106. Motta PM, Nottola SA, Makabe S, Heyn R (2000) Mitochondrial morphology in human fetal and adult female germ cells. Hum Reprod 15(Suppl 2):129–147
107. Thibault C, Szölosi D, Gérard M (1987) Mammalian oocyte maturation. Reprod Nutr Dev 27:865–896
108. Goldman RD, Grin B, Mendez MG, Kuczmarski ER (2008) Intermediate filaments: versatile building blocks of cell structure. Curr Opin Cell Biol 20:28–34. doi:10.1016/j.ceb.2007.11.003
109. Allworth AE, Albertini DF (1993) Meiotic maturation in cultured bovine oocytes is accompanied by remodeling of the cumulus cell cytoskeleton. Dev Biol 158:101–112
110. Anderson E, Albertini DF (1976) Gap junctions between the oocyte and companion follicle cells in the mammalian ovary. J Cell Biol 71:680–686
111. Gilula NB, Epstein ML, Beers WH (1978) Cell-to-cell communication and ovulation. A study of the cumulus-oocyte complex. J Cell Biol 78:58–75
112. Hyttel P (1987) Bovine cumulus-oocyte disconnection in vitro. Anat Embryol 176:41–44
113. Kruip TAM, Cran DG, van Beneden TH, Dieleman SJ (1983) Structural changes in bovine oocytes during final maturation in vivo. Gamete Res 8:29–47
114. Laurincík J, Kroslak P, Hyttel P, Pivko J, Sirotkin AV (1992) Bovine cumulus expansion and corona-oocyte disconnection during culture in vitro. Reprod Nutr Dev 32:151–161
115. Moor RM, Smith MW, Dawson RM (1980) Measurement of intercellular coupling between oocytes and cumulus cells using intracellular markers. Exp Cell Res 126:15–29
116. Zamboni L (1974) Fine morphology of the follicle wall and follicle cell-oocyte association. Biol Reprod 10:125–149
117. Wulff C, Dickson SE, Duncan WC, Fraser HM (2001) Angiogenesis in the human corpus luteum: simulated early pregnancy by HCG treatment is associated with both angiogenesis and vessel stabilization. Hum Reprod 16:2515–2524
118. Eyer J, Peterson A (1994) Neurofilament-deficient axons and perikaryal aggregates in viable transgenic mice expressing a neurofilament-beta-galactosidase fusion protein. Neuron 12:389–405
119. Chung BM, Rotty JD, Coulombe PA (2013) Networking galore: intermediate filaments and cell migration. Curr Opin Cell Biol 25:600–612. doi:10.1016/j.ceb.2013.06.008

120. Seltmann K, Fritsch AW, Käs JA, Magin TM (2013) Keratins significantly contribute to cell stiffness and impact invasive behavior. Proc Natl Acad Sci U S A 110:18507–18512. doi:10.1073/pnas.1310493110
121. Chen CS, Tan J, Tien J (2004) Mechanotransduction at cell–matrix and cell–cell contacts. Annu Rev Biomed Eng 6:275–302
122. Farge E (2011) Mechanotransduction in development. Curr Top Dev Biol 95:243–265. doi:10.1016/B978-0-12-385065-2.00008-6
123. Herrmann H, Bär H, Kreplak L, Strelkov SV, Aebi U (2007) Intermediate filaments: from cell architecture to nanomechanics. Nat Rev Mol Cell Biol 8:562–573
124. Tzur YB, Wilson KL, Gruenbaum Y (2006) SUN-domain proteins: 'Velcro' that links the nucleoskeleton to the cytoskeleton. Nat Rev Mol Cell Biol 7:782–788
125. Kim S, Coulombe PA (2007) Intermediate filament scaffolds fulfill mechanical, organizational, and signaling functions in the cytoplasm. Genes Dev 21:1581–1597
126. Marceau N, Schutte B, Gilbert S, Loranger A, Henfling ME, Broers JL, Mathew J, Ramaekers FC (2007) Dual roles of intermediate filaments in apoptosis. Exp Cell Res 313:2265–2281
127. Margolis SS, Perry JA, Forester CM, Nutt LK, Guo Y, Jardim MJ, Thomenius MJ, Freel CD, Darbandi R, Ahn JH, Arroyo JD, Wang XF, Shenolikar S, Nairn AC, Dunphy WG, Hahn WC, Virshup DM, Kornbluth S (2006) Role for the PP2A/B56delta phosphatase in regulating 14-3-3 release from Cdc25 to control mitosis. Cell 127:759–773
128. Pallari HM, Lindqvist J, Torvaldson E, Ferraris SE, He T, Sahlgren C, Eriksson JE (2011) Nestin as a regulator of Cdk5 in differentiating myoblasts. Mol Biol Cell 22:1539–1549. doi:10.1091/mbc.E10-07-0568
129. Sahlgren CM, Mikhailov A, Vaittinen S, Pallari HM, Kalimo H, Pant HC, Eriksson JE (2003) Cdk5 regulates the organization of Nestin and its association with p35. Mol Cell Biol 23:5090–5106
130. Snider NT, Omary MB (2014) Post-translational modifications of intermediate filament proteins: mechanisms and functions. Nat Rev Mol Cell Biol 15:163–177
131. Toivola DM, Strnad P, Habtezion A, Omary MB (2010) Intermediate filaments take the heat as stress proteins. Trends Cell Biol 20:79–91. doi:10.1016/j.tcb.2009.11.004
132. Kim S, Coulombe PA (2010) Emerging role for the cytoskeleton as an organizer and regulator of translation. Nat Rev Mol Cell Biol 11:75–81. doi:10.1038/nrm2818
133. Styers ML, Kowalczyk AP, Faundez V (2005) Intermediate filaments and vesicular membrane traffic: the odd couple's first dance? Traffic 6:359–365
134. Toivola DM, Tao GZ, Habtezion A, Liao J, Omary MB (2005) Cellular integrity plus: organelle-related and protein-targeting functions of intermediate filaments. Trends Cell Biol 15:608–617
135. Chan KY, Bunt AH (1978) An association between mitochondria and microtubules in synaptosomes and axon terminals of cerebral cortex. J Neurocytol 7:137–143
136. Couchman JR, Rees DA (1982) Organelle-cytoskeleton relationships in fibroblasts: mitochondria, Golgi apparatus, and endoplasmic reticulum in phases of movement and growth. Eur J Cell Biol 27:47–54
137. Heggeness MH, Simon M, Singer SJ (1978) Association of mitochondria with microtubules in cultured cells. Proc Natl Acad Sci U S A 75:3863–3866
138. Krendel M, Sgourdas G, Bonder EM (1998) Disassembly of actin filaments leads to increased rate and frequency of mitochondrial movement along microtubules. Cell Motil Cytoskeleton 40:368–378
139. Raine CS, Ghetti B, Shelanski ML (1971) On the association between microtubules and mitochondria within axons. Brain Res 34:389–393
140. Mose-Larsen P, Bravo R, Fey SJ, Small JV, Celis JE (1982) Putative association of mitochondria with a subpopulation of intermediate-sized filaments in cultured human skin fibroblasts. Cell 31:681–692
141. Toh BH, Lolait SJ, Mathy JP, Baum R (1980) Association of mitochondria with intermediate filaments and of polyribosomes with cytoplasmic actin. Cell Tissue Res 211:163–169

142. Milner DJ, Mavroidis M, Weisleder N, Capetanaki Y (2000) Desmin cytoskeleton linked to muscle mitochondrial distribution and respiratory function. J Cell Biol 150:1283–1298
143. Reipert S, Steinbock F, Fischer I, Bittner RE, Zeold A, Wiche G (1999) Association of mitochondria with plectin and desmin intermediate filaments in striated muscle. Exp Cell Res 252:479–491
144. Stone MR, O'Neill A, Lovering RM, Strong J, Resneck WG, Reed PW, Toivola DM, Ursitti JA, Omary MB, Bloch RJ (2007) Absence of keratin 19 in mice causes skeletal myopathy with mitochondrial and sarcolemmal reorganization. J Cell Sci 120:3999–4008
145. Stromer MH, Bendayan M (1988) Arrangement of desmin intermediate filaments in smooth muscle cells as shown by high-resolution immunocytochemistry. Cell Motil Cytoskeleton 11:117–125
146. Stromer MH, Bendayan M (1990) Immunocytochemical identification of cytoskeletal linkages to smooth muscle cell nuclei and mitochondria. Cell Motil Cytoskeleton 17:11–18
147. Tokuyasu KT, Dutton AH, Singer SJ (1983) Immunoelectron microscopic studies of desmin (skeletin) localization and intermediate filament organization in chicken skeletal muscle. J Cell Biol 96:1727–1735
148. Tokuyasu KT, Dutton AH, Singer SJ (1983) Immunoelectron microscopic studies of desmin (skeletin) localization and intermediate filament organization in chicken cardiac muscle. J Cell Biol 96:1736–1742
149. Almahbobi G, Williams LJ, Han XG, Hall PF (1993) Binding of lipid droplets and mitochondria to intermediate filaments in rat Leydig cells. J Reprod Fertil 98:209–217
150. Collier NC, Sheetz MP, Schlesinger MJ (1993) Concomitant changes in mitochondria and intermediate filaments during heat shock and recovery of chicken embryo fibroblasts. J Cell Biochem 52:297–307
151. Nekrasova OE, Mendez MG, Chernoivanenko IS, Tyurin-Kuzmin PA, Kuczmarski ER, Gelfand VI, Goldman RD, Minin AA (2011) Vimentin intermediate filaments modulate the motility of mitochondria. Mol Biol Cell 22:2282–2289. doi:10.1091/mbc.E10-09-0766
152. Summerhayes IC, Wong D, Chen LB (1983) Effect of microtubules and intermediate filaments on mitochondrial distribution. J Cell Sci 61:87–105
153. Tao GZ, Looi KS, Toivola DM, Strnad P, Zhou Q, Liao J, Wei Y, Habtezion A, Omary MB (2009) Keratins modulate the shape and function of hepatocyte mitochondria: a mechanism for protection from apoptosis. J Cell Sci 122:3851–3855. doi:10.1242/jcs.051862
154. Tolstonog GV, Belichenko-Weitzmann IV, Lu JP, Hartig R, Shoeman RL, Traub U, Traub P (2005) Spontaneously immortalized mouse embryo fibroblasts: growth behavior of wild-type and vimentin-deficient cells in relation to mitochondrial structure and activity. DNA Cell Biol 24:680–709
155. Capetanaki Y (2002) Desmin cytoskeleton: a potential regulator of muscle mitochondrial behavior and function. Trends Cardiovasc Med 12:339–348
156. Capetanaki Y, Bloch RJ, Kouloumenta A, Mavroidis M, Psarras S (2007) Muscle intermediate filaments and their links to membranes and membranous organelles. Exp Cell Res 313:2063–2076
157. Lindén M, Li Z, Paulin D, Gotow T, Leterrier JF (2001) Effects of desmin gene knockout on mice heart mitochondria. J Bioenerg Biomembr 33:333–341
158. Perez-Olle R, Lopez-Toledano MA, Goryunov D, Cabrera-Poch N, Stefanis L, Brown K, Liem RK (2005) Mutations in the neurofilaments light gene linked to Charcot-Marie-Tooth disease cause defects in transport. J Neurochem 93:861–874
159. Straube-West K, Loomis PA, Opal P, Goldman RD (1996) Alterations in neural intermediate filament organization: functional implications and the induction of pathological changes related to motor neuron disease. J Cell Sci 109:2319–2329
160. Wagner OI, Lifshitz J, Janmey PA, Linden M, McIntosh TK, Leterrier JF (2003) Mechanisms of mitochondria–neurofilament interactions. J Neurosci 23:9046–9058
161. Yaffe MP (1999) The machinery of mitochondrial inheritance and behavior. Science 283:1493–1497

162. Rezniczek GA, Abrahamsberg C, Fuchs P, Spazierer D, Wiche G (2003) Plectin 5′-transcript diversity: short alternative sequences determine stability of gene products, initiation of translation and subcellular localization of isoforms. Hum Mol Genet 12:3181–3194
163. Wiche G (1998) Role of plectin in cytoskeleton organization and dynamics. J Cell Sci 111:2477–2486
164. Winter L, Abrahamsberg C, Wiche G (2008) Plectin isoform 1b mediates mitochondrion-intermediate filament network linkage and controls organelle shape. J Cell Biol 181: 903–911
165. Anesti V, Scorrano L (2006) The relationship between mitochondrial shape and function and the cytoskeleton. Biochim Biophys Acta 1757:692–699
166. Reynders E, Foulquier F, Annaert W, Matthijs G (2011) How Golgi glycosylation meets and needs trafficking: the case of the COG complex. Glycobiology 21:853–863. doi:10.1093/glycob/cwq179
167. Ho WC, Allan VJ, van Meer G, Berger EG, Kreis TE (1989) Reclustering of scattered Golgi elements occurs along microtubules. Eur J Cell Biol 48:250–263
168. Hoshino H, Tamaki A, Yagura T (1997) Process of dispersion and fragmentation of Golgi complex by microtubule bundles formed in taxol treated HeLa cells. Cell Struct Funct 22:325–334
169. Rogalski AA, Singer SJ (1984) Associations of elements of the Golgi apparatus with microtubules. J Cell Biol 99:1092–1100
170. Valderrama F, Babià T, Ayala I, Kok JW, Renau-Piqueras J, Egea G (1998) Actin microfilaments are essential for the cytological positioning and morphology of the Golgi complex. Eur J Cell Biol 76:9–17
171. McCallum SJ, Erickson JW, Cerione RA (1998) Characterization of the association of the actin-binding protein, IQGAP, and activated Cdc42 with Golgi membranes. J Biol Chem 273:22537–22544
172. Gao Y, Sztul E (2001) A novel interaction of the Golgi complex with the vimentin intermediate filament cytoskeleton. J Cell Biol 152:877–894
173. Gao YS, Vrielink A, MacKenzie R, Sztul E (2002) A novel type of regulation of the vimentin intermediate filament cytoskeleton by a Golgi protein. Eur J Cell Biol 81:391–401
174. Iwatsuki H, Suda M (2010) Seven kinds of intermediate filament networks in the cytoplasm of polarized cells: structure and function. Acta Histochem Cytochem 43:19–31. doi:10.1267/ahc.10009
175. Hamasaki M, Furuta N, Matsuda A, Nezu A, Yamamoto A, Fujita N, Oomori H, Noda T, Hara-guchi T, Hiraoka Y, Amano A, Yoshimori T (2013) Autophagosomes form at ER-mitochondria contact sites. Nature 495:389–393. doi:10.1038/nature11910
176. Cogli L, Progida C, Bramato R, Bucci C (2013) Vimentin phosphorylation and assembly are regulated by the small GTPase Rab7a. Biochim Biophys Acta 1833:1283–1293. doi:10.1016/j.bbamcr.2013.02.024
177. Styers ML, Salazar G, Love R, Peden AA, Kowalczyk AP, Faundez V (2004) The endo-lysosomal sorting machinery interacts with the intermediate filament cytoskeleton. Mol Biol Cell 15:5369–5382
178. Martin S, Parton RG (2006) Lipid droplets: a unified view of a dynamic organelle. Nat Rev Mol Cell Biol 7:373–378
179. Brasaemle DL, Dolios G, Shapiro L, Wang R (2004) Proteomic analysis of proteins associated with lipid droplets of basal and lipolytically stimulated 3T3-L1 adipocytes. J Biol Chem 279:46835–46842
180. Franke WW, Hergt M, Grund C (1987) Rearrangement of the vimentin cytoskeleton during adipose conversion: formation of an intermediate filament cage around lipid globules. Cell 49:131–141
181. Heid H, Rickelt S, Zimbelmann R, Winter S, Schumacher H, Dörflinger Y, Kuhn C, Franke WW (2014) On the formation of lipid droplets in human adipocytes: the organization of the perilipin–vimentin cortex. PLoS One 9, e90386. doi:10.1371/journal.pone.0090386

182. St. Johnston D (2005) Moving messages: the intracellular localization of mRNAs. Nat Rev Mol Cell Biol 6:363–375
183. Berciano MT, Andres MA, Calle E, Lafarga M (1995) Age-induced hypertrophy of astrocytes in rat supraoptic nucleus: a cytological, morphometric, and immunocytochemical study. Anat Rec 243:129–144
184. Murti K, Goorha R (1989) Synthesis of frog virus 3 proteins occurs on intermediate filament-bound polyribosomes. Biol Cell 65:205–214
185. Thornell LE, Eriksson A (1981) Filament systems in the Purkinje fibres of the heart. Am J Physiol 241:291–305
186. Grossi de Sa MF, Martins de Sa C, Harper F, Olink-Coux M, Huesca M, Scherrer K (1988) The association of prosomes with some of the intermediate filament networks of the animal cell. J Cell Biol 107:1517–1530
187. Traub P, Bauer C, Hartig R, Grüb S, Stahl J (1998) Colocalization of single ribosomes with intermediate filaments in puromycin-treated and serum-starved mouse embryo fibroblasts. Biol Cell 90:319–337
188. Gross SR, Kinzy TG (2005) Translation elongation factor 1A is essential for regulation of the actin cytoskeleton and cell morphology. Nat Struct Mol Biol 12:772–778
189. Kim S, Kellner J, Lee CH, Coulombe PA (2007) Interaction between the keratin cytoskeleton and eEF1Bγ affects protein synthesis in epithelial cells. Nat Struct Mol Biol 14:982–983
190. Kim S, Wong P, Coulombe PA (2006) A keratin cytoskeletal protein regulates protein synthesis and epithelial cell growth. Nature 41:362–365
191. Ku NO, Michie S, Resurreccion EZ, Broome RL, Omary MB (2002) Keratin binding to 14-3-3 proteins modulates keratin filaments and hepatocyte mitotic progression. Proc Natl Acad Sci U S A 99:4373–4378
192. Wang RC, Wei Y, An Z, Zou Z, Xiao G, Bhagat G, White M, Reichelt J, Levine B (2012) Akt-mediated regulation of autophagy and tumorigenesis through Beclin 1 phosphorylation. Science 338:956–959. doi:10.1126/science.1225967

Part V
Focus on Microfilaments

Chapter 14
Actin Organizing Proteins in Regulation of Osteoclast Function

Brooke K. McMichael and Beth S. Lee

Introduction

Skeletal health requires an ongoing process of bone breakdown and rebuilding, as the cellular components of bone respond to mechanical stresses, hormonal stimuli, and other external signals. The cell type responsible for breakdown, or resorption, of the organic and inorganic components of bone is the osteoclast, a terminally differentiated member of the monocyte–macrophage lineage. Osteoclasts not only resorb bone, but also are intimately involved in regulation of bone formation, hematopoiesis, angiogenesis, and endocrine homeostasis [1, 2]. Further, as members of the monocyte–macrophage lineage, osteoclasts are subject to inflammatory stimuli. Excessive osteoclast function can lead to loss of bone mass and fractures in conditions such as osteoporosis, rheumatoid arthritis, and periodontitis, among others. Conversely, loss of osteoclast function results in a rare condition called osteopetrosis, in which bones are dense, but brittle and prone to fracture. Osteoclasts are formed by a process both of differentiation and precursor fusion, resulting in large polykaryons with the capacity for rapid motility and bone resorption. Resorption is initiated by tight attachment of osteoclasts to the bone surface through a circular, actin-rich gasket-like structure called the sealing zone. Enclosed by this structure is a specialized apical membrane domain termed the "ruffled border", so named for its ruffled appearance in electron micrographs [3]. The ruffled border is responsible for secretion of protons and proteases onto the bone surface, creating a protease-rich, acidic extracellular milieu that degrades the organic component of bone while promoting dissolution of the hydroxyapatite mineral component. Formation of the unique actin-rich sealing zone is essential to the ability of osteoclasts to perform

B.K. McMichael • B.S. Lee (✉)
Department of Physiology and Cell Biology, The Ohio State University College of Medicine, 304 Hamilton Hall, 1645 Neil Avenue, Columbus, OH 43210, USA
e-mail: brooke@brookemcmichael.com; beth.lee@osumc.edu

H. Schatten (ed.), *The Cytoskeleton in Health and Disease*,
DOI 10.1007/978-1-4939-2904-7_14

their primary function of bone resorption. In addition, sealing zones constantly form and dissipate as osteoclasts migrate along bone surfaces, necessitating a uniquely dynamic regulation of the actin cytoskeleton.

The primary structural unit that forms osteoclast sealing zones is the podosome, an adhesion structure expressed in monocytic, endothelial, and smooth muscle cells, among others. Podosomes contain an actin-rich core that undergoes rapid turnover and is surrounded by a plethora of adhesion plaque molecules, including scaffolding proteins, kinases, and small GTPases. Podosomes are abundant in cells of the monocyte/macrophage lineage, including osteoclasts, and perform multiple functions including cell adhesion, sensing of substrate rigidity, and matrix degradation, among others. Recent excellent reviews have detailed the structures, functions, and patterning of podosomes in osteoclasts and other cell types [4–6].

When labeled by fluorescent phalloidin and immunocytochemistry, podosomes present as dot-like structures (about 0.5–1.0 μm in diameter) with a dense F-actin core and a less dense cloud of surrounding F-actin. Under high resolution scanning electron microscopy, the densely packed core is shown to consist of branched filaments that are oriented perpendicular to the plasma membrane, while the cloud consists of less dense linear actin filaments that radiate out from the core and eventually interact with the plasma membrane [7]. This arrangement is shown by schematic in Fig. 14.1, along with many of the actin-binding proteins associated with the podosome. Adhesion to the substrate via the podosome core is mediated by the hyaluronate receptor, CD44 [8]. In contrast, actin filaments within the cloud are linked to the substrate through integrin dimers composed of αv, α2, β1, and β3, and

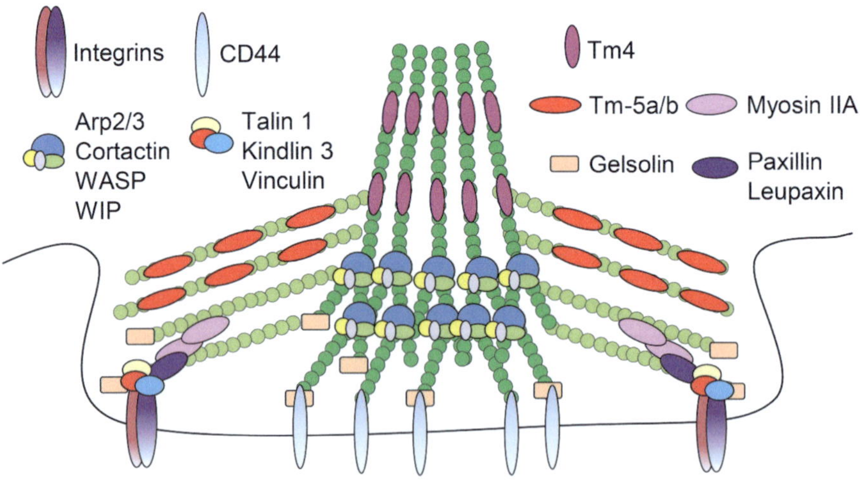

Fig. 14.1 Schematic of a podosome. Actin filaments of the podosome core are indicated in *dark green*, while those of the podosome cloud are indicated in *light green*. Many of the actin-binding proteins discussed in this chapter are also indicated. Those not shown here include proteins only temporarily (e.g. L-plastin, myosin X), or not (e.g. tropomyosins 2/3) associated with podosomes. Others, like the actin bundling protein α-actinin, which are abundant throughout the podosome, are also not shown for the sake of simplicity

associated adaptors such as talin, kindlin, and vinculin. While these proteins traditionally have been viewed as being distributed in a homogeneous ring surrounding the podosome core, super-resolution imaging has instead demonstrated their presence in discrete islets or clusters [9]. Although not discussed here, integrin-mediated signaling and subsequent podosome patterning is highly dependent on tyrosine kinase signaling, with specific kinases (e.g. Src) and phosphatases (e.g. PTP-PEST) playing key roles. Further, the small GTPases Rho and Rac are critical in patterning of podosomes. Recent reviews have provided excellent discussions of these proteins in osteoclast differentiation and function [10, 11].

While there are many similarities between the podosomes of osteoclasts and other cells, some differences exist [4], most notably the ability of osteoclast podosomes to adopt circular structures such as the sealing zone (when on a mineralized bone-like substrate) or a peripheral belt of podosomes (when on a smooth surface like glass or plastic). Recent interesting studies have begun to define the chemical and physical properties that induce cell "choice" in generating podosome belts or sealing zones [12–15]. The ability of osteoclast podosomes to generate these circular patterns is a function of cellular maturity, as illustrated in Fig. 14.2. As osteoclasts

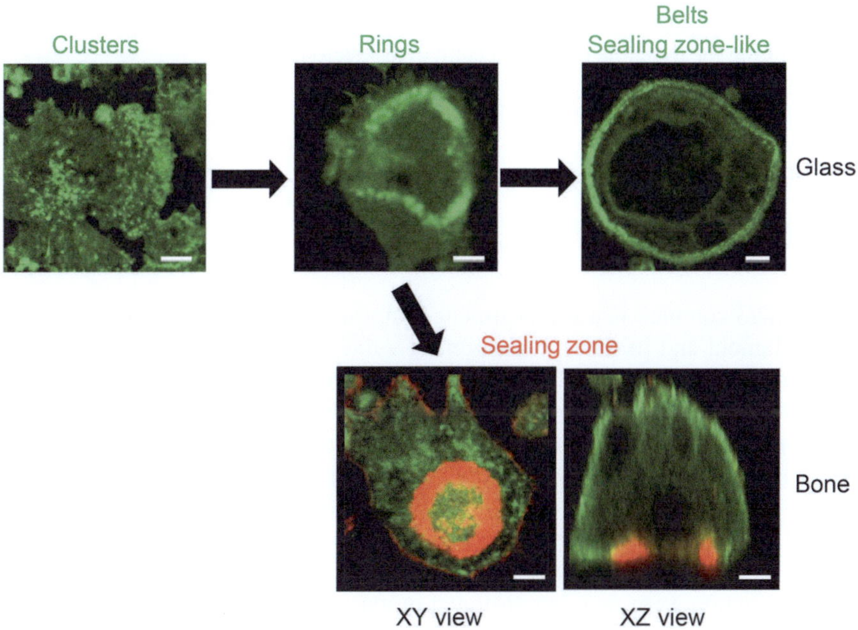

Fig. 14.2 Maturation of osteoclast podosome superstructures. This figure demonstrates the stages of podosome distribution, from clusters to internal rings, and then to peripheral belts when osteoclasts are on a smooth, non-mineralized surface, or to dense sealing zones when osteoclasts are on mineralized substrates like bone. In the upper row, cells are labeled with *fluorescent green phalloidin* to indicate cellular F-actin. In the lower row, cells are labeled with a *green membrane stain* and with *fluorescent red phalloidin* to demonstrate the sealing zone. The XZ view of an osteoclast on bone illustrates its dome-like shape; in contrast, osteoclasts on glass or plastic are much more flattened (not shown). Scale bars = 10 μm

differentiate from monocytic precursors to the large polykaryons capable of resorbing bone, they first develop the ability to pattern podosomes into clusters. As the cell matures, podosomes then adopt a distribution of short-lived internal rings. If the osteoclast is on a mineralized substrate such as bone, the ring matures into a sealing zone in which the podosome cores become more densely packed and tightly interconnected to form a thickened, stable annular structure [7]. If the osteoclast is on a smooth, non-mineralized substrate, the rings continue to expand to the cell periphery where the podosomes also become denser and more interconnected, though not to the same degree as those found in sealing zones. These peripheral structures in cells not on bone are referred to as podosome belts or sealing zone-like structures. The expansion of these annular structures and their stabilization is a microtubule-dependent process, and indeed, loss of microtubule integrity results in a collapse of podosome patterning [16].

In this chapter, we will focus not on the enzyme regulators of actin organization in osteoclasts (e.g. small GTPases, tyrosine kinases), but rather on discussion of regulatory actin binding proteins as previously summarized [17, 18]. These classes include actin nucleators, F-actin binding proteins, F-actin depolymerizing/severing proteins, actin bundling/cross-linking proteins, linkers that promote actin association with the plasma membrane and with integrins, and actin motor proteins and their regulators. We will discuss their distribution in osteoclasts and current knowledge of their roles in generating the dynamic osteoclast actin cytoskeleton.

Actin Nucleators

Arp2/3 Complex and Regulators

The Arp2/3 complex is a seven-protein composite that binds an existing mother actin filament and initiates nucleation of a daughter filament at a ~70° angle. The complex consists of Arp2, Arp3, and five additional subunits termed ARPC1-5. On its own, the Arp2/3 complex is poor at nucleating formation of new filaments, so this process requires regulators known as nucleation promoting factors (NPFs). NPFs include WASP (Wiskott-Aldrich syndrome protein), N- (neural-)WASP, WAVE (WASP family verprolin-homologous protein), and cortactin, among others [19, 20]. This machinery plays a critical role in creating the branched, high F-actin density of podosome cores. The essential roles of these proteins in podosome formation were first demonstrated in macrophages and dendritic cells [21, 22]. In osteoclasts, the Arp2 and 3 proteins were demonstrated to increase in abundance during differentiation and to be very closely associated both with podosome cores and with sealing zones [23]. Notably, this complex is enriched at the membrane surfaces of these structures, where actin monomers enter the podosomes [16]. SiRNA-mediated knockdown of Arp2 strongly diminishes the appearance of podosome-like structures and almost completely abolishes sealing zone formation, demonstrating the essential role of the Arp2/3 complex in podosome core formation [23].

The NFP cortactin binds both to actin filaments and the Arp2/3 complex and can consequently enhance nucleation of daughter filaments and prevent debranching [24]. Although cortactin is not expressed in hematopoietic cells, including the monocytic precursors of osteoclasts, its expression is rapidly induced upon the onset of osteoclast differentiation [25, 26]. Additionally, cortactin expression is further induced when osteoclasts are exposed to bone [27]. As expected, cortactin strongly co-localizes with Arp2/3 in the core of podosomes. Time-lapse microscopy demonstrates that cortactin is notably more enriched in nascent, rather than established, podosomes, suggesting its role in formation of these structures [26]. Indeed, cortactin is one of the first actin-binding proteins detected in nascent podosomes [28]. Consistent with this hypothesis, loss of cortactin induced by lentivirally-expressed shRNA results in an apparent complete loss of podosomes [26].

Another NFP, Wiskott-Aldrich syndrome protein, similarly is required for functional osteoclast actin organization. WASP is expressed primarily in hematopoietic cells and is composed of functional domains that interact with F-actin, Arp2/3, and the signaling molecules Cdc42 and phosphatidylinositol 4,5-bisphosphate (PIP_2). Indeed, both podosome and sealing zone formation require interactions of PIP_2 and Cdc42 with WASP [29]. Osteoclasts from WASP-null mice demonstrate a greater degree of multinucleation than wild-type osteoclasts, but they also show a distinct lack of podosomes and sealing zones, and have a modest deficiency in bone resorption. WASP-null mice do not demonstrate an abnormal bone volume relative to wild-type mice; however, when mice are ovariectomized to stimulate osteoclast-mediated bone loss, the bone resorption in WASP-null mice is less pronounced, indicating functional osteoclast deficiency [30]. WASP function also is dependent on its interaction with the WASP interacting protein, WIP [31]. While WIP is essential for WASP-mediated actin polymerization, WIP also can bind to F- and G-actin on its own to promote the formation and stabilization of actin filaments. WIP is a critical component of the podosome F-actin core; indeed, WIP-null osteoclasts lack the core structure while still maintaining the podosome cloud [8]. As might be expected, podosome cores in these cells also lack Arp2/3, cortactin, and WASP. Further, the cellular expression of WASP is markedly reduced in $WIP^{-/-}$ cells, consistent with previous studies suggesting that WIP protects WASP from degradation in podosomes [32]. Outside-in activation of CD44, through attachment of $WIP^{-/-}$ cells to substrates such as collagen I or osteopontin, or reaction with an anti-CD44 antibody, is able to restore normal podosome core formation, suggesting that outside-in signaling through CD44 may have WASP stabilizing functions on its own [8]. However, CD44 activation cannot fully compensate for a lack of WIP in the bone resorptive process. While $WIP^{-/-}$ cells are capable of forming sealing zones, resorption is diminished by ~30 %.

Formins

Formins are a family of conserved proteins that both nucleate actin and act as elongation factors, but unlike the Arp2/3 complex, formins elongate filaments in a linear direction. In general, formins nucleate actin monomer assembly, then processively move in the direction of the growing filament barbed end while preventing binding of actin capping proteins. Many formins have additional activities that can include actin filament severing, depolymerization, and bundling, as well as regulation of microtubule dynamics [33]. Fifteen formins are expressed in mammals, including humans, and can be subdivided into eight families [34]. The best studied of these, the Dia formins (named because of similarity to the Drosophila gene *diaphanous*) act as Rho effector proteins and have been implicated in polarity and migration of both normal and cancer cells [35]. They also have been shown to bind microtubules and regulate their organization and dynamics. Formins have not been well-studied in osteoclasts; however, the mouse Dia protein mDia2 was demonstrated to alter osteoclast podosome patterning through post-translational modification of microtubules [36]. Briefly, it was demonstrated that Rho activity, which is disruptive to the organization of mature podosome belts, activates mDia2. mDia2, in turn, binds and activates the microtubule deacetylase activity of histone deacetylase 6 (HDAC6), resulting in loss of microtubule stability and podosome organization. Therefore, mDia2's role in podosome dynamics appears to depend more on its activity as a regulator of microtubules than its function as a regulator of actin nucleation and elongation. Another formin, FRL1, has been shown to reside atop the actin core of macrophage podosomes and to be necessary for their stability [37]. However, it has not yet been demonstrated whether the elongation function of FRL1 contributes to this activity, and any role for FRL1 in osteoclast podosomes has not yet been identified.

F-Actin Binding Proteins

VASP

VASP, or vasodilator-stimulated phosphoprotein, is a member of the Ena/VASP family of actin polymerases that also includes Mena and EVL. Like formins, Ena/VASP proteins promote linear actin filament elongation at barbed ends; however, unlike formins, they do not nucleate formation of new filaments. Ena/VASP proteins stimulate elongation of existing filaments by processively binding barbed ends, thereby protecting these ends from capping protein, and by promoting actin monomer addition. It is also suggested that Ena/VASP may suppress filament branching triggered by Arp2/3 [38]. Although formins and Ena/VASP proteins have similar functions at barbed ends, their roles are not interchangeable [39, 40]. In osteoclasts, VASP is demonstrated to be a downstream mediator of nitric oxide

(NO)-stimulated motility. NO is a potent downregulator of osteoclastic resorptive activity, due to its stimulation of cytoskeletal rearrangements that promote motility (and, at high concentrations, detachment), but not bone resorption [41]. In non-motile cells, VASP associates with the integrin $\alpha_v\beta_3$, but upon NO stimulation of motility, cGMP-dependent protein kinase I causes phosphorylation of VASP, resulting in its separation from the integrin [42]. Further, VASP also is necessary for activation of μ-calpain, a Ca^{2+}-activated protease that is essential for osteoclast motility [43, 44]. Therefore, VASP's involvement in NO-stimulated cell motility is clear, but its precise associations with podosomal components in resting and motile osteoclasts remain to be determined.

Actin Depolymerizing/Severing Proteins

Cofilin

The ADF (actin depolymerizing factor)/cofilin family of proteins consists of three members: ADF, cofilin-1, and cofilin-2. ADF and cofilins exhibit qualitative similarities, but differ quantitatively in their abilities to bind actin monomers and nucleate assembly. While cofilin has numerous functions in cell biology, it is best understood as a regulator of actin filament non-equilibrium assembly and disassembly [45]. At low cofilin/actin ratios, filament severing occurs, while at high ratios cofilin can stabilize the severed filaments and even nucleate new assembly.

Cofilin is encoded by one of the most abundantly expressed genes in osteoclasts [46]. Cofilin does not appear to be necessary for podosome formation, but rather, for organization of mature podosome belts [47]. Although cofilin in present in both podosome cores and clouds, the fraction present in clouds is inactive due to phosphorylation at serine-3, a modification that prevents cofilin from binding to actin and initiating its severing activity. Indeed, experiments with phospho-mimetic or phospho-deficient cofilin mutants suggest that the phosphorylation state of cofilin directs its distribution in podosomes. Cofilin activation (dephosphorylation) was found to be triggered by the osteoclast differentiation factor RANKL, and the phosphatase SSH1 was implicated in this process [47]. It is of interest that cofilin was found to be a component of mature, and not nascent, podosomes, since mature podosomes found in belts are more dynamically active (i.e. less stable) than the individual structures [48]. Cofilin's function as a modulator of actin dynamics may be key to the lifespan of mature podosomes.

Two more modulators of cofilin phosphorylation recently have been demonstrated to play roles in osteoclast activity. LIM kinase 1 (LIMK1) phosphorylates and inactivates cofilin, and mice that lack LIMK1 have reduced bone mass, due in part to the increased resorptive activity of *Limk1*$^{-/-}$ osteoclasts [49]. In addition, the small GTPase RhoE is necessary for osteoclast motility and resorptive capacity due to its inhibition of ROCK1, a serine-threonine kinase downstream of RhoE that phosphorylates and inactivates cofilin. RhoE-deficient osteoclasts have impaired migration velocity and directionality, and far less resorptive capacity. Further, in keeping with previous studies showing the role of cofilin in podosome core dynamics, RhoE-depleted osteoclasts demonstrate fewer but thicker podosome cores, with slower actin turnover, and smaller sealing zones [50]. These results further underscore cofilin's role as a key regulator of actin dynamics in osteoclasts.

Gelsolin Superfamily

Gelsolin is the founding member of a superfamily of proteins that bind to and sever microfilaments [51, 52]. Following severing, gelsolin remains as a cap on the new barbed end, preventing elongation or reannealing of the shortened filaments. However, severing by gelsolin amplifies the number of filaments, so regulated gelsolin activity followed by uncapping can promote actin polymerization and rebuilding of the cytoskeleton as needed. Gelsolin may also nucleate new actin filament formation through its binding of actin monomers. In osteoclasts, gelsolin is present at high levels in both podosome cores and clouds [53]. Binding of osteopontin to its integrin receptor $\alpha v\beta 3$, which causes formation of podosomal structures in osteoclasts, results in gelsolin's association with phosphoinositides [54]. This association, which is regulated by proline-rich tyrosine kinase 2 (PYK2) [55], causes uncapping of actin filament barbed ends and subsequent filament formation [54, 56]. At the same time, gelsolin also associates with numerous podosome structural proteins and kinases in the podosome signaling complex, including vinculin, talin, focal adhesion kinase, and Src kinase [57]. Gelsolin's presence in these complexes is necessary, since gelsolin-deficient osteoclasts fail to make podosomes, have defective $\alpha v\beta 3$-stimulated signaling, are hypomotile, and resorb bone poorly. Further, mice lacking gelsolin expression demonstrate an age-dependent thickening of both cortical and trabecular bone, resulting in increased mechanical strength [58]. This finding is consistent with gelsolin's important role in osteoclast function, although the phenotype is subtle relative to the loss of other podosomal proteins that result in osteopetroses and accompanying poor bone quality. Interestingly, one human study demonstrated seemingly contradictory results by showing that low expression of gelsolin in peripheral blood monocytes (which contain osteoclast precursors) was associated with low bone mineral density in Caucasian women [59]. The discrepancy between the murine and human studies may be due to cell type or species specificity; nonetheless, these results further suggest the importance of gelsolin in osteoclast function. Indeed, gelsolin may be more critical to the function of osteoclasts than to other podosome-containing cells, since a lack of gelsolin has no effect on podosomes of dendritic cells [60].

Recently, another member of the gelsolin superfamily, adseverin (previously called scinderin, *SCIN*), was examined for its role in osteoclast function [61]. Adseverin is the superfamily member most closely related to gelsolin, and like gelsolin, it binds, severs, and caps actin filaments. However, it differs from gelsolin in its regulation by calcium and phosphoinositides [62]. In osteoclasts, adseverin was found to be strongly upregulated during the early phases of osteoclast differentiation, while gelsolin levels remained constant, indicating a specific role for adseverin in the differentiation process. Indeed, shRNA-mediated knockdown in osteoclast precursors showed that diminished levels of adseverin resulted in failure to express tartrate-resistant acid phosphatase, a marker of osteoclastogenesis, and failure to fuse into polykaryons. Knockdown precursors also exhibited altered morphology and F-actin distribution, and increased migration [61]. Thus, adseverin plays an important role in osteoclast activity, but one that is distinct from gelsolin.

Actin Bundling/Crosslinking Proteins

Alpha-Actinin

Alpha-actinin is a ubiquitously expressed actin binding protein that crosslinks actin filaments and serves as a platform for initiating protein–protein interactions with numerous cytoskeletal and regulatory proteins [63]. It was one of the first actin binding proteins demonstrated to be associated with the adhesion structures of osteoclasts [64]. In osteoclasts, α-actinin is most strongly present in podosome cores, but also is present in the cloud domain [48]. Live-cell microscopy has indicated that during de novo assembly of podosomes, α-actinin becomes present in podosomes simultaneously with the appearance of the actin core. Further, the abundance and distribution of α-actinin in the podosome core over time parallels that of F-actin [28]. These results suggest that α-actinin's filament bundling properties are critical to consolidation of the core during podosome formation.

Plastins/Fimbrins

Plastins (also known as fimbrins) similarly were identified as components of podosomes in the earliest studies of the osteoclast actin cytoskeleton [64]. Plastins are generally distributed in regions of high actin filament turnover such as podosomes, focal adhesions, ruffled membranes, and filopodia, where they crosslink filaments into tight bundles [65]. Three plastin isoforms are encoded by distinct genes that appear to have evolved from a common ancestor [66], and are called I-plastin (*PLS1*), L-plastin (*LCP1*), and T-plastin (*PLS3*). These are expressed in a cell-restricted manner, with osteoclasts expressing both L- and T-plastin [67]. A role for L-plastin was later determined when expression of this protein was demonstrated to be markedly decreased upon osteoclast attachment to bone, suggesting a role in sealing zone formation [27]. L-plastin was found to be a component of actin aggregates that form at membrane extensions at the earliest stages of osteoclast attachment to bone. Its complete loss resulted in reductions of these extensions and of sealing zones. These results suggest that L-plastin may be involved in an initial actin bundling step that is required for actin aggregates that act as precursors to sealing zone formation. Interestingly, mutations in the gene that encodes T-plastin, or plastin 3 (*PLS3*), have been associated with X-linked childhood osteoporosis and increased bone fracture risk [68, 69]. Skeletal examination of four boys from two families with nonsense mutations in *PLS3* revealed low lumbar spine bone mineral density, low trabecular bone volume and low osteoid maturation time but a normal bone formation rate [69]. Any potential contribution of osteoclasts to the bone phenotype is yet to be determined. These patients presented with normal numbers of osteoclasts; however, bone mineral density was improved in the two boys treated with oral alendronate (bisphosphonate inhibitors of osteoclast activity), suggesting that the osteoclasts may be overactive. Therefore, additional study of the cellular basis of bone fragility in patients with *PLS3* mutations is required.

Filamins

Filamins are hinged, flexible homodimers that crosslink actin into orthogonal networks. Filamin acts as a bracket to arrange actin filaments perpendicular to each other in T-, X-, and L-shaped junctions that adapt to morphological changes such as cell migration [70]. Three filamin isoforms, A, B, and C, are expressed in humans, with filamin A being the most broadly expressed. Filamin A deficiency in osteoclast precursors results in defective osteoclastogenesis. This is due to impairment of the cellular migration that allows precursors to come in contact and undergo fusion into polykaryons, since the defect can be rescued in vitro by plating precursors at higher densities [71]. This impaired migration is due to attenuated signaling by the Rho family GTPases RhoA, Rac, and Cdc42, which bind directly to filamin homodimers. The loss of signaling by these GTPases ultimately results in diminished actin polymerization at the leading edge of cells, a process required for cellular migration. Mice that lack filamin A are osteoporotic, a somewhat surprising finding given that these mice have diminished numbers of osteoclasts. However, the mice also show decreased bone formation, demonstrating filamin A's importance in cells of the osteogenic lineage in addition to osteoclasts [71].

Plasma Membrane Association

Annexins

Annexins are a family of proteins that bind acidic phospholipids in a calcium-dependent manner, and thereby control numerous aspects of cellular membrane dynamics, including membrane organization, vesicle trafficking, and intracellular signaling [72]. In addition, a number of annexins have been demonstrated to bind F-actin in a calcium-dependent manner, suggesting their potential roles in mediating cytoskeletal-membrane contacts. One of the actin binding members of this family is annexin A2, which can bind and contribute to bundling of F-actin in the presence of calcium [73]. Annexin A2 can function intracellularly to organize membrane domains, but it also can work extracellularly to effect intracellular signaling. Annexin A2 (AnxA2) has been well studied in osteoclasts, with its major role described as a secreted autocrine/paracrine factor that promotes osteoclastogenesis through stimulating expression of RANKL, the cytokine most critical for osteoclast differentiation and function [74]. Indeed, elevated levels of AnxA2 in human peripheral blood monocytes (which contain osteoclast precursors) are associated with low bone mineral density, suggesting increased osteoclast activity [75]. In contrast, another F-actin-binding annexin family member, annexin A8 (AnxA8) [76], is suggested to work intracellularly to regulate sealing zone formation. AnxA8 was identified in a profile of genes that are upregulated when osteoclasts are cultured on bone versus plastic. SiRNA-mediated knockdown of AnxA8 results in osteoclasts that have a markedly reduced ability to spread on plastic and are unable to form sealing zones on bone [77]. The mechanism by which this deficiency occurs remains to be discovered.

MARCKS

MARCKS (myristoylated alanine-rich C kinase substrate) is a broadly expressed mediator of membrane-actin interactions [78]. MARCKS binds acidic phospholipids through a basic effector domain, and binds the sides of actin filaments to crosslink them into a rigid meshwork at the membrane. The interaction of MARCKS with membranes is disrupted by protein kinase C (PKC)-mediated phosphorylation, while MARCKS interactions with actin are disrupted both by PKC-mediated phosphorylation and by binding calcium/calmodulin [79]. Because of MARCKS' ability to organize actin at membranes, it plays key roles in cell motility and membrane trafficking. In bone resorbing osteoclasts, MARCKS is present not in the sealing zone, but at the membrane of the ruffled border. Osteoclasts deficient in MARCKS demonstrate excessive exocytosis of cathepsin K, a proteolytic enzyme secreted from the ruffled border that is particularly important to osteoclastic bone resorption [80]. Not unexpectedly, this excessive release of cathepsin K results in enhanced resorptive activity. This finding suggests a mechanism by which MARCKS crosslinks actin in the ruffled border to act as a barrier to exocytosis of cathepsin K-containing vesicles. Further, MARCKS' presence at the membrane is regulated by PKCδ. Mouse osteoclasts lacking PKCδ have defective cathepsin K exocytosis due to their limited ability to phosphorylate MARCKS and remove it from the membrane. This defect can be corrected by silencing MARCKS expression, demonstrating that the PKCδ-MARCKS pathway is a major regulator of cathepsin K secretion [80].

Cell-Extracellular Matrix Junction

Talin

Talin is a cytosolic adaptor protein that links β integrins to the actin cytoskeleton and thereby acts as a critical mediator of inside-out integrin activation. This activation is a process by which intracellular signal transduction pathways result in conformational alterations of integrins, thus increasing the binding affinity of integrins toward their substrates. The resulting increased binding subsequently promotes downstream cellular processes that include reorganization of the cytoskeleton. Humans express two talins, talin1 (TLN1) and talin2 (TLN2), but only the former protein is present in osteoclasts. Conditional knockout of mouse talin1 in either osteoclast precursors or mature osteoclasts (using the LysM-Cre and Ctsk-Cre systems, respectively) results in increased bone density and osteopetrosis, an indicator of osteoclast dysfunction [81]. Ex vivo, talin1-deficient precursors produce a dearth of mature osteoclasts when exposed to osteoclastogenic cytokines, and these precursors demonstrate poor adhesion to substrate and impaired directional migration. Osteoclast-specific deletion of Rap1, a small GTPase that promotes the association of talin with β integrins, similarly promotes dysfunctional osteoclasts and osteopetrosis. Notably, loss of talin, which would affect all osteoclast integrins, produces a more severe phenotype in mice than loss of $\alpha_v\beta_3$ only. This is in line with other findings demonstrating that all integrin classes, and not just a single class, play important roles in osteoclast function [82].

Kindlin

Kindlins are a family of proteins that, like talin, bind to and activate β integrins via inside-out signaling. In spite of the similar overall functions of talin and kindlins, they are not functionally redundant, since kindlin mutations are associated with human disease, and knockouts in mice elicit severe phenotypes [83]. The three family members are designated kindlin-1, -2, and -3. Mutations of kindlin-3, which is expressed specifically in hematopoietic cells [84], are associated with the rare inherited disease leukocyte adhesion deficiency type III (LAD-III). This disease is characterized by severe bleeding and recurrent bacterial and fungal infections [85–87]. Mice deficient for kindlin-3 similarly have severe bleeding, and their platelets and leukocytes demonstrate a failure of integrin activation, leading to defective platelet aggregation and leukocyte adhesion [88, 89]. A relationship between kindlin-3 and bone health became evident when it was observed that some LAD-III patients have increased bone density [86, 90, 91]. Further, kindlin-3-deficient mice develop a severe osteopetrosis [82], illustrative of osteoclast defects. Osteoclasts from these mice undergo normal, though slightly delayed, differentiation, but have an attenuated ability to adhere to the bone matrix protein osteopontin. Integrin-mediated signaling is severely compromised in these cells. Further, although kindlin-3-deficient osteoclasts produce podosome cores, they are unable to arrange them into podosome belts or sealing zones, and cannot resorb bone. Loss of kindlin-3 produces greater osteoclast dysfunction and osteopetrosis than loss of a single integrin, demonstrating this protein's role in activation of the entire inventory of osteoclast integrins [82].

Vinculin

Vinculin is a ubiquitously expressed protein that localizes to integrin-based adhesion sites and contains binding motifs for numerous proteins, including F-actin and talin [92]. Loss of vinculin from focal adhesions reduces cell adhesion and alters motility. In osteoclasts, vinculin is associated with the podosome cloud [64]. Although vinculin levels increase during osteoclastogenesis, its absence in osteoclast precursors does not alter the differentiation process ex vivo [93]. However, mature osteoclasts lacking vinculin are smaller than wild-type osteoclasts in size, and they correspondingly possess smaller sealing zones and resorb bone poorly. Reflective of the cell culture findings, mice lacking vinculin in the osteoclast lineage have increased bone mass relative to wild-type animals, with unchanged osteoclast numbers. Vinculin's activity in promoting osteoclast cytoskeletal rearrangements appears to be distinct from integrin activation, since signaling through $\alpha_v\beta_3$ is unimpaired in vinculin-deficient osteoclasts. Rather, vinculin's association with talin was shown to be a key feature of its function, since vinculin molecules with mutated talin binding sites were unable to rescue the defects in vinculin-deficient cells [93]. Thus, vinculin's primary role in osteoclast function is as an adaptor to link integrin–talin complexes to the actin cytoskeleton.

Paxillin Superfamily

Paxillin (PAX) is a widely expressed adaptor protein that binds actin and integrins, thereby distributing into the podosome cloud. Paxillin is also one the first actin regulatory proteins to associate with nascent podosomes [28]. Although deletion of many cytoskeletal regulators in osteoclasts causes poor cellular spreading, a complete lack of paxillin results in a "superspread" phenotype that is not caused by accelerated differentiation [94]. Nonetheless, PAX-null osteoclasts demonstrate poor bone resorption that is likely to be caused by dysfunction in sealing zone formation. Unlike wild-type osteoclasts, which tend to form small sealing zones that are smaller than the cell circumference, PAX-null osteoclasts form actin structures that are reminiscent of podosome belts in that they encompass the cell periphery. In spite of these large actin structures, resorption pits are shallow and lack well-defined margins. The "superspread" phenotype of PAX-null osteoclasts is likely to be related to paxillin's physical association with myosin IIA [94]. As described below, loss of myosin IIA in osteoclasts also results in increased cell spreading and expansion of the sealing zone [95], suggesting that this myosin–paxillin interaction is involved in constricting the actin structures of osteoclasts into a bone-resorptive morphology.

Leupaxin (LPXN) is a member of the paxillin superfamily whose distribution is preferentially expressed in hematopoietic cells, including cells of the monocyte–macrophage lineage [96]. Like paxillin, leupaxin is strongly associated with podosomal molecules, including the signaling kinase Pyk2. Overexpression of leupaxin via TAT-mediated transduction resulted in osteoclasts adopting lamellipodia-like projections, although it is unclear whether this is due to cellular retraction or expansion, since cell size was not measured [97]. Further, knockdown of leupaxin with the use of antisense oligonucleotides resulted in increased spreading and decreased bone resorptive capacity, similar to the results obtained from paxillin deficiency. Therefore, leupaxin is likely to work in conjunction with paxillin to mediate organization of actin structures in podosomes and sealing zones.

Motor Proteins and Regulators

Myosins

The myosin superfamily can be divided into at least 35 classes based on their heavy chain structure, resulting in expression of about 40 different myosins expressed in human tissues [98]. Class II nonmuscle myosins, considered to be "conventional" because of their structural homology to muscle myosins, were the first to be identified in osteoclasts. Their importance to osteoclast function was first revealed in a study showing that injection of anti-myosin II antibodies into these cells inhibits bone resorptive activity [99]. In addition, an F-actin/myosin II complex binds and aids in trafficking of V-ATPases, proton-translocating complexes that are required for acid secretion and resorption of bone by osteoclasts [100]. Later immunocytochemical labeling of osteoclasts for the specific myosin isoforms IIA and IIB demonstrated that myosin IIA is enriched in osteoclast podosome belts and sealing zones, while myosin IIB is absent from these regions and instead is distributed in a fibrillar pattern throughout the cell. Rearrangement of podosomes induced by addition of soluble osteopontin results in redistribution of MyoIIA and heightened MyoIIA presence in lateral F-actin fibers that connect individual podosomes [101]. A similar distribution of MyoIIA was demonstrated in A7r5 smooth muscle cells induced to form podosomes by stimulation with phorbol esters [102]. These and other studies in macrophages suggest that myosin IIA-mediated contractility may not be required for the integrity of individual podosomes, but rather, for condensation of podosomes by regulating the tension of radial actin filaments in the podosome cloud [28, 103]. This is consistent with a study in osteoclasts in which it was demonstrated that siRNA-mediated loss of MyoIIA results both in increased cell spreading and expansion of the sealing zone [95]. Unexpectedly, this study also demonstrated a role for MyoIIA in regulation of osteoclast precursor fusion. MyoIIA heavy chain (MYH9) expression temporarily decreases by about 50 % during the osteoclast differentiation process, with its nadir at days 3–4, at the onset of osteoclast precursor fusion. This loss of expression is due to proteolytic degradation of MYH9 and is both necessary and sufficient for the onset of fusion, suggesting that MyoIIA-mediated contractility of podosomes can serve as a barrier to cell fusion [95]. As described above, the association of MyoIIA with paxillin is likely to be key to its contractile function in osteoclasts [94].

The functions of two unconventional nonmuscle myosins similarly have been studied in osteoclasts. The first of these, myosin X (Myo10) is a double-headed myosin that contains in its tail three pleckstrin homology (PH) domains, a myosin tail homology 4 (MyTH4) domain, and a FERM domain, and is the sole member of its class [104]. While PH domains are involved in protein–protein and protein–lipid interactions, MyTH4 domains bind microtubules, and FERM domains link cell membrane proteins to the cytoskeleton. MyTH4 domains in this myosin are particularly of interest since it is well established that maturation of podosome belts and sealing zones from internal podosome rings is a microtubule-dependent process [16, 105–108]. Immunocytochemical studies showed that Myo10 is absent from the podosome belts and sealing zones of mature osteoclasts, but in cells containing internal podosome rings, Myo10 is positioned between, and binds to, both F-actin and microtubules [109]. This finding suggests that Myo10 acts as a linker between the two cytoskeletal systems during maturation of rings into podosome belts and sealing zones (as shown in Fig. 14.2). Indeed, osteoclasts in which Myo10 is knocked down do not make podosome belts, but only podosome clusters and rings; sealing zones also do not expand properly. Further, overexpression of the MyTH4 or the MyTH4/FERM tail domains specifically inhibits both podosome belt and sealing zone formation in a dominant negative manner [109]. These findings demonstrate that Myo10 is critical for expansion of podosome rings into functional peripheral belts and sealing zones. Interestingly, studies in endothelial cells demonstrated that Myo10 is stimulated by bone morphogenetic protein (BMP) in a positive feedback loop [110]. Because BMPs are well established regulators of bone morphogenesis, this further suggests a role for Myo10 in skeletal development and health. Recently, genetic screens showed a coding region variant of Myo10 to be associated with hind limb conformation in Swiss Large White boars [111]. Therefore, although Myo10 has not yet been associated with variation in humans, it seems likely that this motor protein may play an as-yet undiscovered role in skeletal health.

Most recently, the unconventional myosin IXB (Myo9B) was demonstrated to be critical to osteoclast function [112]. Class IX myosins (Myo9A and Myo9B) are unusual in containing RhoGAP domains in their tails [113]. RhoGAPs, numbering near 70 in mammalian genomes, accelerate hydrolysis of Rho-bound GTP, thereby promoting Rho toward an inactive, GDP-bound state [114]. These myosins appear to be able to regulate the local Rho activity of distinct F-actin pools in cells [115]. While Myo9A is restricted primarily to brain, testis, and spleen, Myo9B is most highly expressed in leukocytes, including osteoclast precursors and differentiated osteoclasts [116]. Notably, Myo9B is localized to F-actin structures in osteoclasts that require low Rho activity (podosome belts), while being absent from structures that require high Rho activity (sealing zones) [112]. Further, siRNA-mediated knockdown of Myo9B alters podosome patterning and diminishes bone resorptive capacity in a Rho-dependent manner. Inhibition of excessive Rho activity caused by loss of Myo9B is able to restore normal podosome formation and bone resorption. Interestingly, the *MYO9B* gene, which is also expressed in the growth plates of long bones, has been associated with human height [117, 118], so it will be of interest to determine how alteration of Myo9B expression or function in bone and cartilage cells may directly affect skeletal health.

Tropomyosins

Tropomyosins (Tms) are coiled-coil homo- or heterodimers that bind along the α-helical groove of actin polymers and regulate access of other regulatory proteins (e.g. myosins, actin nucleators, actin depolymerizing/severing proteins) to the actin filament [119]. Approximately 40 tropomyosins have been identified in mammals, with the bulk of these expressed in nonmuscle cells. They are encoded by four genes, *TPM1-4*, and are generated by multiple promoters and alternate splicing to create a large molecular and functional diversity. Tropomyosins may be structurally divided into high molecular weight and low molecular weight forms of about 284 and 248 amino acids respectively, based on the presence or absence of sequences coded by N-terminal exons. Initial screening of osteoclast tropomyosins by Western analysis and RT-PCR revealed the presence of at least seven isoforms from genes *TPM1*, *TPM3*, and *TPM4* [120]. Because of the similarity in sequence of many tropomyosin isoforms, it is often not possible to distinguish between related isoforms using antibody reagents; nonetheless, available antibodies demonstrated markedly different distributions of the isoforms by immunocytochemistry.

Tm-2 and 3 are high molecular weight tropomyosins encoded by *TPM1* that differ only by the presence of exons 2a or 2b, and are distributed rather diffusely throughout the osteoclast, with only loose association with podosome belts and sealing zones [120, 121]. Notably, these tropomyosins are gradually upregulated from undetectable levels in osteoclast precursors to high levels late in osteoclastogenesis, suggesting osteoclast-specific functions for these proteins. SiRNA-mediated knockdown of Tm-2/3 causes marked flattening and spreading of osteoclasts, along with diminished motility and altered bone resorption. This was determined to be caused by gelsolin redistributing from podosomes to the cell interior, where its actin severing activity increases to cause collapse of the osteoclast support scaffolding. Conversely, overexpression of Tm-2 induces cells to adopt a highly rounded morphology, caused by the failure of gelsolin to bind and sever Tm-2-associated filaments. Overexpression of Tm-2 also causes aberrant effects on osteoclast motility and bone resorption. These studies demonstrate that high molecular weight tropomyosins stabilize the internal scaffolding of osteoclasts that is required for these very large cells to maintain a functional morphology [121].

In contrast to the high molecular weight Tms -2 and -3 that show little presence in osteoclast adhesion structures, the low molecular weight tropomyosin-4 (Tm-4), encoded by *TPM4*, is highly enriched in podosome cores and sealing zones. Specifically, Tm-4 is present on the tops of these structures, away from the membrane faces where new actin filament formation occurs [120]. This finding suggests that Tm-4 stabilizes the "older" end of podosome cores. Indeed, a ~50 % reduction in Tm-4 expression results in thinning of the sealing zone actin ring to half its normal height. This suppression also results in dramatic losses of osteoclast motility and bone resorptive capacity. Conversely, only a modest overexpression of Tm-4 causes increased podosome height, but thickened, aberrant sealing zone formation and an even greater loss of bone resorption [122]. These findings demonstrate that minimal changes in Tm-4 expression can severely alter osteoclast adhesion structures and provide evidence for Tm-4's role in regulating stability of podosome cores.

Osteoclasts also express the low molecular weight tropomyosins Tm-5a and Tm-5b, which, like Tm-2 and Tm-3, are expressed from the *TPM1* gene and differ only by inclusion of either exon 2a or 2b. However, where Tm-2 and -3 are 284 amino acid residues in length, Tm-5a and -5b contain only 248 residues and differ from their longer counterparts in the choice of N-terminal exon. These few differences between Tm-2/3 and Tm-5a/5b are sufficient to result in completely different subcellular distributions for these tropomyosins. Whereas Tm-2/3 are distributed throughout the osteoclast, Tm-5a/5b are strongly present in podosome clouds and sealing zones. Like Tm-4, these tropomyosins were present on the basal sides of the attachment structures, suggesting that they stabilize the more mature filaments, leaving the newly forming filaments at the apical membrane free to undergo formation and turnover [120]. SiRNA-mediated knockdown of Tm-5a/5b to about 20 % of normal levels resulted in more densely packed podosomes, suggesting that loss of actin stability in the podosome cloud caused aggregation of the core structures. Further, the resulting sealing zones were flattened relative to normal controls. On a functional level, loss of Tm-5a/5b resulted in osteoclasts with increased motility that generated more, but shallower, resorption pits. Preliminary studies of podosome dynamics suggested that loss of Tm-5a/5b resulted in podosome instability, consistent with the accompanying increase in cell motility (P. Kotadiya and B.S. Lee, unpublished results).

Conclusions

While the basic nature of the osteoclast actin cytoskeleton has been appreciated for several decades, understanding of its intricacies has grown rapidly in this century. Advanced imaging techniques, including high resolution scanning electron microscopy and live-cell fluorescence microscopy, have contributed greatly to elucidating the structure and dynamics of the unique podosome-based adhesion structures of osteoclasts. Advances in the study of podosomes in other cell types such as macrophages also have contributed to this understanding. However, osteoclasts, due to their status as syncytial cells and their ability to generate sealing zones and other distinct podosomal patterns, present special challenges in cytoskeletal organization. The effects of deletions and mutations of actin-regulatory proteins in murine and human osteoclasts underscore the importance of the actin cytoskeleton in bone health. Although great progress has been made in identifying the roles for numerous regulators of actin organization, there still exist a variety of these proteins whose functions have not been discerned. As we continue to expand our appreciation of the role of individual proteins in regulating osteoclast adhesion, motility, and bone resorption, future challenges will include trying to understand how podosomal complexes perform on a three dimensional level to interact with the extracellular environment, both chemically and mechanically. In addition, there is ample room for exploration of actin-based functions in osteoclasts beyond the podosome. The mechanics of cell–cell fusion in osteoclast differentiation are poorly understood, and understanding of intracellular trafficking of secretory vesicles and other internal components is still limited. Nonetheless, the unique nature of the osteoclast has made its study of great interest, both for understanding its role in skeletal and organismal health, and for the lessons it provides about regulation of the actin cytoskeleton.

References

1. Cappariello A, Maurizi A, Veeriah V, Teti A (2014) The Great Beauty of the osteoclast. Arch Biochem Biophys 558:70–78. doi:10.1016/j.abb.2014.06.017
2. Charles JF, Aliprantis AO (2014) Osteoclasts: more than 'bone eaters'. Trends Mol Med 20(8):449–459. doi:10.1016/j.molmed.2014.06.001
3. Gonzales F, Karnovsky MJ (1961) Electron microscopy of osteoclasts in healing fractures of rat bone. J Biophys Biochem Cytol 9:299–316
4. Georgess D, Machuca-Gayet I, Blangy A, Jurdic P (2014) Podosome organization drives osteoclast-mediated bone resorption. Cell Adh Migr 8(3):191–204
5. Linder S, Wiesner C (2014) Tools of the trade: podosomes as multipurpose organelles of monocytic cells. Cell Mol Life Sci. doi:10.1007/s00018-014-1731-z
6. Schachtner H, Calaminus SD, Thomas SG, Machesky LM (2013) Podosomes in adhesion, migration, mechanosensing and matrix remodeling. Cytoskeleton (Hoboken) 70(10):572–589. doi:10.1002/cm.21119
7. Luxenburg C, Geblinger D, Klein E, Anderson K, Hanein D, Geiger B, Addadi L (2007) The architecture of the adhesive apparatus of cultured osteoclasts: from podosome formation to sealing zone assembly. PLoS One 2(1), e179
8. Chabadel A, Banon-Rodriguez I, Cluet D, Rudkin BB, Wehrle-Haller B, Genot E, Jurdic P, Anton IM, Saltel F (2007) CD44 and beta3 integrin organize two functionally distinct actin-based domains in osteoclasts. Mol Biol Cell 18(12):4899–4910. doi:10.1091/mbc.E07-04-0378
9. Meddens MB, van den Dries K, Cambi A (2014) Podosomes revealed by advanced bioimaging: what did we learn? Eur J Cell Biol 93(10–12):380–387. doi:10.1016/j.ejcb.2014.09.002
10. Touaitahuata H, Blangy A, Vives V (2014) Modulation of osteoclast differentiation and bone resorption by Rho GTPases. Small GTPases 5, e28119. doi:10.4161/sgtp.28119
11. Granot-Attas S, Elson A (2008) Protein tyrosine phosphatases in osteoclast differentiation, adhesion, and bone resorption. Eur J Cell Biol 87(8–9):479–490. doi:10.1016/j.ejcb.2008.01.006

12. Anderegg F, Geblinger D, Horvath P, Charnley M, Textor M, Addadi L, Geiger B (2011) Substrate adhesion regulates sealing zone architecture and dynamics in cultured osteoclasts. PLoS One 6(12), e28583
13. Geblinger D, Addadi L, Geiger B (2010) Nano-topography sensing by osteoclasts. J Cell Sci 123(Pt 9):1503–1510. doi:10.1242/jcs.060954
14. Geblinger D, Zink C, Spencer ND, Addadi L, Geiger B (2010) Effects of surface microtopography on the assembly of the osteoclast resorption apparatus. J R Soc Interface 9(72): 1599–1608
15. Saltel F, Destaing O, Bard F, Eichert D, Jurdic P (2004) Apatite-mediated actin dynamics in resorbing osteoclasts. Mol Biol Cell 15(12):5231–5241
16. Destaing O, Saltel F, Geminard JC, Jurdic P, Bard F (2003) Podosomes display actin turnover and dynamic self-organization in osteoclasts expressing actin-green fluorescent protein. Mol Biol Cell 14(2):407–416
17. Siripala AD, Welch MD (2007) SnapShot: actin regulators I. Cell 128(3):626. doi:10.1016/j.cell.2007.02.001
18. Siripala AD, Welch MD (2007) SnapShot: actin regulators II. Cell 128(5):1014. doi:10.1016/j.cell.2007.02.021
19. Rotty JD, Wu C, Bear JE (2013) New insights into the regulation and cellular functions of the ARP2/3 complex. Nat Rev Mol Cell Biol 14(1):7–12. doi:10.1038/nrm3492
20. Pollard TD (2007) Regulation of actin filament assembly by Arp2/3 complex and formins. Annu Rev Biophys Biomol Struct 36:451–477. doi:10.1146/annurev.biophys.35.040405.101936
21. Burns S, Thrasher AJ, Blundell MP, Machesky L, Jones GE (2001) Configuration of human dendritic cell cytoskeleton by Rho GTPases, the WAS protein, and differentiation. Blood 98(4):1142–1149
22. Linder S, Nelson D, Weiss M, Aepfelbacher M (1999) Wiskott-Aldrich syndrome protein regulates podosomes in primary human macrophages. Proc Natl Acad Sci U S A 96(17):9648–9653
23. Hurst IR, Zuo J, Jiang J, Holliday LS (2004) Actin-related protein 2/3 complex is required for actin ring formation. J Bone Miner Res 19(3):499–506
24. Weaver AM, Karginov AV, Kinley AW, Weed SA, Li Y, Parsons JT, Cooper JA (2001) Cortactin promotes and stabilizes Arp2/3-induced actin filament network formation. Curr Biol 11(5):370–374
25. Hiura K, Lim SS, Little SP, Lin S, Sato M (1995) Differentiation dependent expression of tensin and cortactin in chicken osteoclasts. Cell Motil Cytoskeleton 30(4):272–284. doi:10.1002/cm.970300405
26. Tehrani S, Faccio R, Chandrasekar I, Ross FP, Cooper JA (2006) Cortactin has an essential and specific role in osteoclast actin assembly. Mol Biol Cell 17(7):2882–2895. doi:10.1091/mbc.E06-03-0187
27. Ma T, Sadashivaiah K, Madayiputhiya N, Chellaiah MA (2010) Regulation of sealing ring formation by L-plastin and cortactin in osteoclasts. J Biol Chem 285(39):29911–29924. doi:10.1074/jbc.M109.099697
28. Luxenburg C, Winograd-Katz S, Addadi L, Geiger B (2012) Involvement of actin polymerization in podosome dynamics. J Cell Sci 125(Pt 7):1666–1672. doi:10.1242/jcs.075903
29. Chellaiah MA (2005) Regulation of actin ring formation by rho GTPases in osteoclasts. J Biol Chem 280(38):32930–32943
30. Calle Y, Jones GE, Jagger C, Fuller K, Blundell MP, Chow J, Chambers T, Thrasher AJ (2004) WASp deficiency in mice results in failure to form osteoclast sealing zones and defects in bone resorption. Blood 103(9):3552–3561. doi:10.1182/blood-2003-04-1259
31. Fried S, Matalon O, Noy E, Barda-Saad M (2014) WIP: more than a WASp-interacting protein. J Leukoc Biol. doi:10.1189/jlb.2RU0314-162R
32. Tsuboi S (2007) Requirement for a complex of Wiskott-Aldrich syndrome protein (WASP) with WASP interacting protein in podosome formation in macrophages. J Immunol 178(5):2987–2995

33. Breitsprecher D, Goode BL (2013) Formins at a glance. J Cell Sci 126(Pt 1):1–7. doi:10.1242/jcs.107250
34. Schonichen A, Geyer M (2010) Fifteen formins for an actin filament: a molecular view on the regulation of human formins. Biochim Biophys Acta 1803(2):152–163. doi:10.1016/j.bbamcr.2010.01.014
35. DeWard AD, Eisenmann KM, Matheson SF, Alberts AS (2010) The role of formins in human disease. Biochim Biophys Acta 1803(2):226–233. doi:10.1016/j.bbamcr.2009.11.006
36. Destaing O, Saltel F, Gilquin B, Chabadel A, Khochbin S, Ory S, Jurdic P (2005) A novel Rho-mDia2-HDAC6 pathway controls podosome patterning through microtubule acetylation in osteoclasts. J Cell Sci 118(Pt 13):2901–2911
37. Mersich AT, Miller MR, Chkourko H, Blystone SD (2010) The formin FRL1 (FMNL1) is an essential component of macrophage podosomes. Cytoskeleton (Hoboken) 67(9):573–585. doi:10.1002/cm.20468
38. Skoble J, Auerbuch V, Goley ED, Welch MD, Portnoy DA (2001) Pivotal role of VASP in Arp2/3 complex-mediated actin nucleation, actin branch-formation, and Listeria monocytogenes motility. J Cell Biol 155(1):89–100. doi:10.1083/jcb.200106061
39. Barzik M, McClain LM, Gupton SL, Gertler FB (2014) Ena/VASP regulates mDia2-initiated filopodial length, dynamics, and function. Mol Biol Cell. doi:10.1091/mbc.E14-02-0712
40. Bilancia CG, Winkelman JD, Tsygankov D, Nowotarski SH, Sees JA, Comber K, Evans I, Lakhani V, Wood W, Elston TC, Kovar DR, Peifer M (2014) Enabled negatively regulates diaphanous-driven actin dynamics in vitro and in vivo. Dev Cell 28(4):394–408. doi:10.1016/j.devcel.2014.01.015
41. Yaroslavskiy BB, Li Y, Ferguson DJ, Kalla SE, Oakley JI, Blair HC (2004) Autocrine and paracrine nitric oxide regulate attachment of human osteoclasts. J Cell Biochem 91(5):962–972. doi:10.1002/jcb.20009
42. Yaroslavskiy BB, Zhang Y, Kalla SE, Garcia Palacios V, Sharrow AC, Li Y, Zaidi M, Wu C, Blair HC (2005) NO-dependent osteoclast motility: reliance on cGMP-dependent protein kinase I and VASP. J Cell Sci 118(Pt 23):5479–5487. doi:10.1242/jcs.02655
43. Marzia M, Chiusaroli R, Neff L, Kim NY, Chishti AH, Baron R, Horne WC (2006) Calpain is required for normal osteoclast function and is down-regulated by calcitonin. J Biol Chem 281(14):9745–9754
44. Yaroslavskiy BB, Sharrow AC, Wells A, Robinson LJ, Blair HC (2007) Necessity of inositol (1,4,5)-trisphosphate receptor 1 and mu-calpain in NO-induced osteoclast motility. J Cell Sci 120(Pt 16):2884–2894. doi:10.1242/jcs.004184
45. Bernstein BW, Bamburg JR (2010) ADF/cofilin: a functional node in cell biology. Trends Cell Biol 20(4):187–195. doi:10.1016/j.tcb.2010.01.001
46. Sakai D, Tong HS, Minkin C (1995) Osteoclast molecular phenotyping by random cDNA sequencing. Bone 17(2):111–119
47. Blangy A, Touaitahuata H, Cres G, Pawlak G (2012) Cofilin activation during podosome belt formation in osteoclasts. PLoS One 7(9), e45909. doi:10.1371/journal.pone.0045909
48. Luxenburg C, Addadi L, Geiger B (2006) The molecular dynamics of osteoclast adhesions. Eur J Cell Biol 85(3–4):203–211. doi:10.1016/j.ejcb.2005.11.002
49. Kawano T, Zhu M, Troiano N, Horowitz M, Bian J, Gundberg C, Kolodziejczak K, Insogna K (2013) LIM kinase 1 deficient mice have reduced bone mass. Bone 52(1):70–82. doi:10.1016/j.bone.2012.09.024
50. Georgess D, Mazzorana M, Terrado J, Delprat C, Chamot C, Guasch RM, Perez-Roger I, Jurdic P, Machuca-Gayet I (2014) Comparative transcriptomics reveals RhoE as a novel regulator of actin dynamics in bone-resorbing osteoclasts. Mol Biol Cell 25(3):380–396. doi:10.1091/mbc.E13-07-0363
51. Silacci P, Mazzolai L, Gauci C, Stergiopulos N, Yin HL, Hayoz D (2004) Gelsolin superfamily proteins: key regulators of cellular functions. Cell Mol Life Sci 61(19–20):2614–2623. doi:10.1007/s00018-004-4225-6
52. Sun HQ, Yamamoto M, Mejillano M, Yin HL (1999) Gelsolin, a multifunctional actin regulatory protein. J Biol Chem 274(47):33179–33182

53. Hu S, Biben T, Wang X, Jurdic P, Geminard JC (2011) Internal dynamics of actin structures involved in the cell motility and adhesion: modeling of the podosomes at the molecular level. J Theor Biol 270(1):25–30. doi:10.1016/j.jtbi.2010.11.014
54. Chellaiah M, Hruska K (1996) Osteopontin stimulates gelsolin-associated phosphoinositide levels and phosphatidylinositol triphosphate-hydroxyl kinase. Mol Biol Cell 7(5):743–753
55. Wang Q, Xie Y, Du QS, Wu XJ, Feng X, Mei L, McDonald JM, Xiong WC (2003) Regulation of the formation of osteoclastic actin rings by proline-rich tyrosine kinase 2 interacting with gelsolin. J Cell Biol 160(4):565–575
56. Janmey PA, Stossel TP (1987) Modulation of gelsolin function by phosphatidylinositol 4,5-bisphosphate. Nature 325(6102):362–364. doi:10.1038/325362a0
57. Chellaiah MA, Biswas RS, Yuen D, Alvarez UM, Hruska KA (2001) Phosphatidylinositol 3,4,5-trisphosphate directs association of Src homology 2-containing signaling proteins with gelsolin. J Biol Chem 276(50):47434–47444
58. Chellaiah M, Kizer N, Silva M, Alvarez U, Kwiatkowski D, Hruska KA (2000) Gelsolin deficiency blocks podosome assembly and produces increased bone mass and strength. J Cell Biol 148(4):665–678
59. Deng FY, Zhu W, Zeng Y, Zhang JG, Yu N, Liu YZ, Liu YJ, Tian Q, Deng HW (2014) Is GSN significant for hip BMD in female Caucasians? Bone 63:69–75. doi:10.1016/j.bone.2014.02.015
60. Hammarfjord O, Falet H, Gurniak C, Hartwig JH, Wallin RP (2011) Gelsolin-independent podosome formation in dendritic cells. PLoS One 6(7), e21615. doi:10.1371/journal.pone.0021615
61. Hassanpour S, Jiang H, Wang Y, Kuiper JW, Glogauer M (2014) The actin binding protein adseverin regulates osteoclastogenesis. PLoS One 9(9), e109078. doi:10.1371/journal.pone.0109078
62. Nag S, Larsson M, Robinson RC, Burtnick LD (2013) Gelsolin: the tail of a molecular gymnast. Cytoskeleton (Hoboken) 70(7):360–384. doi:10.1002/cm.21117
63. Sjoblom B, Salmazo A, Djinovic-Carugo K (2008) Alpha-actinin structure and regulation. Cell Mol Life Sci 65(17):2688–2701. doi:10.1007/s00018-008-8080-8
64. Marchisio PC, Cirillo D, Naldini L, Primavera MV, Teti A, Zambonin-Zallone A (1984) Cell-substratum interaction of cultured avian osteoclasts is mediated by specific adhesion structures. J Cell Biol 99(5):1696–1705
65. Delanote V, Vandekerckhove J, Gettemans J (2005) Plastins: versatile modulators of actin organization in (patho)physiological cellular processes. Acta Pharmacol Sin 26(7):769–779. doi:10.1111/j.1745-7254.2005.00145.x
66. Lin CS, Park T, Chen ZP, Leavitt J (1993) Human plastin genes. Comparative gene structure, chromosome location, and differential expression in normal and neoplastic cells. J Biol Chem 268(4):2781–2792
67. Babb SG, Matsudaira P, Sato M, Correia I, Lim SS (1997) Fimbrin in podosomes of monocyte-derived osteoclasts. Cell Motil Cytoskeleton 37(4):308–325. doi:10.1002/(SICI)1097-0169(1997)37:4<308::AID-CM3>3.0.CO;2-0
68. Fahiminiya S, Majewski J, Al-Jallad H, Moffatt P, Mort J, Glorieux FH, Roschger P, Klaushofer K, Rauch F (2014) Osteoporosis caused by mutations in PLS3 – Clinical and bone tissue characteristics. J Bone Miner Res. doi:10.1002/jbmr.2208
69. van Dijk FS, Zillikens MC, Micha D, Riessland M, Marcelis CL, de Die-Smulders CE, Milbradt J, Franken AA, Harsevoort AJ, Lichtenbelt KD, Pruijs HE, Rubio-Gozalbo ME, Zwertbroek R, Moutaouakil Y, Egthuijsen J, Hammerschmidt M, Bijman R, Semeins CM, Bakker AD, Everts V, Klein-Nulend J, Campos-Obando N, Hofman A, te Meerman GJ, Verkerk AJ, Uitterlinden AG, Maugeri A, Sistermans EA, Waisfisz Q, Meijers-Heijboer H, Wirth B, Simon ME, Pals G (2013) PLS3 mutations in X-linked osteoporosis with fractures. N Engl J Med 369(16):1529–1536. doi:10.1056/NEJMoa1308223
70. Popowicz GM, Schleicher M, Noegel AA, Holak TA (2006) Filamins: promiscuous organizers of the cytoskeleton. Trends Biochem Sci 31(7):411–419. doi:10.1016/j.tibs.2006.05.006

71. Leung R, Wang Y, Cuddy K, Sun C, Magalhaes J, Grynpas M, Glogauer M (2010) Filamin A regulates monocyte migration through Rho small GTPases during osteoclastogenesis. J Bone Miner Res 25(5):1077–1091. doi:10.1359/jbmr.091114
72. Gerke V, Creutz CE, Moss SE (2005) Annexins: linking Ca2+ signalling to membrane dynamics. Nat Rev Mol Cell Biol 6(6):449–461. doi:10.1038/nrm1661
73. Filipenko NR, Waisman DM (2001) The C terminus of annexin II mediates binding to F-actin. J Biol Chem 276(7):5310–5315. doi:10.1074/jbc.M009710200
74. Li F, Chung H, Reddy SV, Lu G, Kurihara N, Zhao AZ, Roodman GD (2005) Annexin II stimulates RANKL expression through MAPK. J Bone Miner Res 20(7):1161–1167. doi:10.1359/JBMR.050207
75. Deng FY, Lei SF, Zhang Y, Zhang YL, Zheng YP, Zhang LS, Pan R, Wang L, Tian Q, Shen H, Zhao M, Lundberg YW, Liu YZ, Papasian CJ, Deng HW (2011) Peripheral blood monocyte-expressed ANXA2 gene is involved in pathogenesis of osteoporosis in humans. Mol Cell Proteomics 10(11). doi:10.1074/mcp.M111.011700
76. Goebeler V, Ruhe D, Gerke V, Rescher U (2006) Annexin A8 displays unique phospholipid and F-actin binding properties. FEBS Lett 580(10):2430–2434. doi:10.1016/j.febslet.2006.03.076
77. Crotti TN, O'Sullivan RP, Shen Z, Flannery MR, Fajardo RJ, Ross FP, Goldring SR, McHugh KP (2011) Bone matrix regulates osteoclast differentiation and annexin A8 gene expression. J Cell Physiol 226(12):3413–3421. doi:10.1002/jcp.22699
78. Aderem A (1995) The MARCKS family of protein kinase-C substrates. Biochem Soc Trans 23(3):587–591
79. Hartwig JH, Thelen M, Rosen A, Janmey PA, Nairn AC, Aderem A (1992) MARCKS is an actin filament crosslinking protein regulated by protein kinase C and calcium-calmodulin. Nature 356(6370):618–622. doi:10.1038/356618a0
80. Cremasco V, Decker CE, Stumpo D, Blackshear PJ, Nakayama KI, Nakayama K, Lupu TS, Graham DB, Novack DV, Faccio R (2012) Protein kinase C-delta deficiency perturbs bone homeostasis by selective uncoupling of cathepsin K secretion and ruffled border formation in osteoclasts. J Bone Miner Res 27(12):2452–2463. doi:10.1002/jbmr.1701
81. Zou W, Izawa T, Zhu T, Chappel J, Otero K, Monkley SJ, Critchley DR, Petrich BG, Morozov A, Ginsberg MH, Teitelbaum SL (2013) Talin1 and Rap1 are critical for osteoclast function. Mol Cell Biol 33(4):830–844. doi:10.1128/MCB.00790-12
82. Schmidt S, Nakchbandi I, Ruppert R, Kawelke N, Hess MW, Pfaller K, Jurdic P, Fassler R, Moser M (2011) Kindlin-3-mediated signaling from multiple integrin classes is required for osteoclast-mediated bone resorption. J Cell Biol 192(5):883–897. doi:10.1083/jcb.201007141
83. Meves A, Stremmel C, Gottschalk K, Fassler R (2009) The Kindlin protein family: new members to the club of focal adhesion proteins. Trends Cell Biol 19(10):504–513. doi:10.1016/j.tcb.2009.07.006
84. Ussar S, Wang HV, Linder S, Fassler R, Moser M (2006) The Kindlins: subcellular localization and expression during murine development. Exp Cell Res 312(16):3142–3151. doi:10.1016/j.yexcr.2006.06.030
85. Kuijpers TW, van de Vijver E, Weterman MA, de Boer M, Tool AT, van den Berg TK, Moser M, Jakobs ME, Seeger K, Sanal O, Unal S, Cetin M, Roos D, Verhoeven AJ, Baas F (2009) LAD-1/variant syndrome is caused by mutations in FERMT3. Blood 113(19):4740–4746. doi:10.1182/blood-2008-10-182154
86. Malinin NL, Zhang L, Choi J, Ciocea A, Razorenova O, Ma YQ, Podrez EA, Tosi M, Lennon DP, Caplan AI, Shurin SB, Plow EF, Byzova TV (2009) A point mutation in KINDLIN3 ablates activation of three integrin subfamilies in humans. Nat Med 15(3):313–318. doi:10.1038/nm.1917
87. Svensson L, Howarth K, McDowall A, Patzak I, Evans R, Ussar S, Moser M, Metin A, Fried M, Tomlinson I, Hogg N (2009) Leukocyte adhesion deficiency-III is caused by mutations in KINDLIN3 affecting integrin activation. Nat Med 15(3):306–312. doi:10.1038/nm.1931
88. Moser M, Bauer M, Schmid S, Ruppert R, Schmidt S, Sixt M, Wang HV, Sperandio M, Fassler R (2009) Kindlin-3 is required for beta2 integrin-mediated leukocyte adhesion to endothelial cells. Nat Med 15(3):300–305. doi:10.1038/nm.1921

89. Moser M, Nieswandt B, Ussar S, Pozgajova M, Fassler R (2008) Kindlin-3 is essential for integrin activation and platelet aggregation. Nat Med 14(3):325–330. doi:10.1038/nm1722
90. Kilic SS, Etzioni A (2009) The clinical spectrum of leukocyte adhesion deficiency (LAD) III due to defective CalDAG-GEF1. J Clin Immunol 29(1):117–122. doi:10.1007/s10875-008-9226-z
91. Sabnis H, Kirpalani A, Horan J, McDowall A, Svensson L, Cooley A, Merck T, Jobe S, Hogg N, Briones M (2010) Leukocyte adhesion deficiency-III in an African-American patient. Pediatr Blood Cancer 55(1):180–182. doi:10.1002/pbc.22386
92. Carisey A, Ballestrem C (2011) Vinculin, an adapter protein in control of cell adhesion signalling. Eur J Cell Biol 90(2–3):157–163. doi:10.1016/j.ejcb.2010.06.007
93. Fukunaga T, Zou W, Warren JT, Teitelbaum SL (2014) Vinculin regulates osteoclast function. J Biol Chem 289(19):13554–13564. doi:10.1074/jbc.M114.550731
94. Zou W, Deselm CJ, Broekelmann TJ, Mecham RP, Vande Pol S, Choi K, Teitelbaum SL (2012) Paxillin contracts the osteoclast cytoskeleton. J Bone Miner Res 27(12):2490–2500. doi:10.1002/jbmr.1706
95. McMichael BK, Wysolmerski RB, Lee BS (2009) Regulated proteolysis of nonmuscle myosin IIA stimulates osteoclast fusion. J Biol Chem 284(18):12266–12275
96. Lipsky BP, Beals CR, Staunton DE (1998) Leupaxin is a novel LIM domain protein that forms a complex with PYK2. J Biol Chem 273(19):11709–11713
97. Gupta A, Lee BS, Khadeer MA, Tang Z, Chellaiah M, Abu-Amer Y, Goldknopf J, Hruska KA (2003) Leupaxin is a critical adaptor protein in the adhesion zone of the osteoclast. J Bone Miner Res 18(4):669–685
98. Foth BJ, Goedecke MC, Soldati D (2006) New insights into myosin evolution and classification. Proc Natl Acad Sci U S A 103(10):3681–3686
99. Sato M, Grasser W (1990) Myosin II antibodies inhibit the resorption activity of isolated rat osteoclasts. Cell Motil Cytoskeleton 17(3):250–263
100. Lee BS, Gluck SL, Holliday LS (1999) Interaction between vacuolar H(+)-ATPase and microfilaments during osteoclast activation. J Biol Chem 274(41):29164–29171
101. Krits I, Wysolmerski RB, Holliday LS, Lee BS (2002) Differential localization of myosin II isoforms in resting and activated osteoclasts. Calcif Tissue Int 71(6):530–538
102. Burgstaller G, Gimona M (2004) Actin cytoskeleton remodelling via local inhibition of contractility at discrete microdomains. J Cell Sci 117(Pt 2):223–231
103. Labernadie A, Thibault C, Vieu C, Maridonneau-Parini I, Charriere GM (2010) Dynamics of podosome stiffness revealed by atomic force microscopy. Proc Natl Acad Sci U S A 107(49):21016–21021. doi:10.1073/pnas.1007835107
104. Berg JS, Derfler BH, Pennisi CM, Corey DP, Cheney RE (2000) Myosin-X, a novel myosin with pleckstrin homology domains, associates with regions of dynamic actin. J Cell Sci 113(19):3439–3451
105. Evans JG, Correia I, Krasavina O, Watson N, Matsudaira P (2003) Macrophage podosomes assemble at the leading lamella by growth and fragmentation. J Cell Biol 161(4):697–705
106. Jurdic P, Saltel F, Chabadel A, Destaing O (2006) Podosome and sealing zone: specificity of the osteoclast model. Eur J Cell Biol 85(3–4):195–202
107. Linder S, Hufner K, Wintergerst U, Aepfelbacher M (2000) Microtubule-dependent formation of podosomal adhesion structures in primary human macrophages. J Cell Sci 113(Pt 23):4165–4176
108. Okumura S, Mizoguchi T, Sato N, Yamaki M, Kobayashi Y, Yamauchi H, Ozawa H, Udagawa N, Takahashi N (2006) Coordination of microtubules and the actin cytoskeleton is important in osteoclast function, but calcitonin disrupts sealing zones without affecting microtubule networks. Bone 39(4):684–693
109. McMichael BK, Cheney RE, Lee BS (2010) Myosin X regulates sealing zone patterning in osteoclasts through linkage of podosomes and microtubules. J Biol Chem 285(13):9506–9515
110. Pi X, Ren R, Kelley R, Zhang C, Moser M, Bohil AB, Divito M, Cheney RE, Patterson C (2007) Sequential roles for myosin-X in BMP6-dependent filopodial extension, migration, and activation of BMP receptors. J Cell Biol 179(7):1569–1582

111. Becker D, Luther H, Hofer A, Leeb T (2014) A variant in MYO10 is associated with hind limb conformation in Swiss Large White boars. Anim Genet 45(2):308. doi:10.1111/age.12111
112. McMichael BK, Scherer KF, Franklin NC, Lee BS (2014) The RhoGAP activity of myosin IXB is critical for osteoclast podosome patterning, motility, and resorptive capacity. PLoS One 9(1), e87402
113. Bahler M (2000) Are class III and class IX myosins motorized signalling molecules? Biochim Biophys Acta 1496(1):52–59
114. Tcherkezian J, Lamarche-Vane N (2007) Current knowledge of the large RhoGAP family of proteins. Biol Cell 99(2):67–86
115. Omelchenko T, Hall A (2012) Myosin-IXA regulates collective epithelial cell migration by targeting RhoGAP activity to cell–cell junctions. Curr Biol 22(4):278–288. doi:10.1016/j.cub.2012.01.014
116. Wirth JA, Jensen KA, Post PL, Bement WM, Mooseker MS (1996) Human myosin-IXb, an unconventional myosin with a chimerin-like rho/rac GTPase-activating protein domain in its tail. J Cell Sci 109(Pt 3):653–661
117. Lango Allen H, Estrada K, Lettre G, Berndt SI, Weedon MN, Rivadeneira F, Willer CJ, Jackson AU, Vedantam S, Raychaudhuri S, Ferreira T, Wood AR, Weyant RJ, Segre AV, Speliotes EK, Wheeler E, Soranzo N, Park JH, Yang J, Gudbjartsson D, Heard-Costa NL, Randall JC, Qi L, Vernon Smith A, Magi R, Pastinen T, Liang L, Heid IM, Luan J, Thorleifsson G, Winkler TW, Goddard ME, Sin Lo K, Palmer C, Workalemahu T, Aulchenko YS, Johansson A, Zillikens MC, Feitosa MF, Esko T, Johnson T, Ketkar S, Kraft P, Mangino M, Prokopenko I, Absher D, Albrecht E, Ernst F, Glazer NL, Hayward C, Hottenga JJ, Jacobs KB, Knowles JW, Kutalik Z, Monda KL, Polasek O, Preuss M, Rayner NW, Robertson NR, Steinthorsdottir V, Tyrer JP, Voight BF, Wiklund F, Xu J, Zhao JH, Nyholt DR, Pellikka N, Perola M, Perry JR, Surakka I, Tammesoo ML, Altmaier EL, Amin N, Aspelund T, Bhangale T, Boucher G, Chasman DI, Chen C, Coin L, Cooper MN, Dixon AL, Gibson Q, Grundberg E, Hao K, Juhani Junttila M, Kaplan LM, Kettunen J, Konig IR, Kwan T, Lawrence RW, Levinson DF, Lorentzon M, McKnight B, Morris AP, Muller M, Suh Ngwa J, Purcell S, Rafelt S, Salem RM, Salvi E, Sanna S, Shi J, Sovio U, Thompson JR, Turchin MC, Vandenput L, Verlaan DJ, Vitart V, White CC, Ziegler A, Almgren P, Balmforth AJ, Campbell H, Citterio L, De Grandi A, Dominiczak A, Duan J, Elliott P, Elosua R, Eriksson JG, Freimer NB, Geus EJ, Glorioso N, Haiqing S, Hartikainen AL, Havulinna AS, Hicks AA, Hui J, Igl W, Illig T, Jula A, Kajantie E, Kilpelainen TO, Koiranen M, Kolcic I, Koskinen S, Kovacs P, Laitinen J, Liu J, Lokki ML, Marusic A, Maschio A, Meitinger T, Mulas A, Pare G, Parker AN, Peden JF, Petersmann A, Pichler I, Pietilainen KH, Pouta A, Ridderstrale M, Rotter JI, Sambrook JG, Sanders AR, Schmidt CO, Sinisalo J, Smit JH, Stringham HM, Bragi Walters G, Widen E, Wild SH, Willemsen G, Zagato L, Zgaga L, Zitting P, Alavere H, Farrall M, McArdle WL, Nelis M, Peters MJ, Ripatti S, van Meurs JB, Aben KK, Ardlie KG, Beckmann JS, Beilby JP, Bergman RN, Bergmann S, Collins FS, Cusi D, den Heijer M, Eiriksdottir G, Gejman PV, Hall AS, Hamsten A, Huikuri HV, Iribarren C, Kahonen M, Kaprio J, Kathiresan S, Kiemeney L, Kocher T, Launer LJ, Lehtimaki T, Melander O, Mosley TH Jr, Musk AW, Nieminen MS, O'Donnell CJ, Ohlsson C, Oostra B, Palmer LJ, Raitakari O, Ridker PM, Rioux JD, Rissanen A, Rivolta C, Schunkert H, Shuldiner AR, Siscovick DS, Stumvoll M, Tonjes A, Tuomilehto J, van Ommen GJ, Viikari J, Heath AC, Martin NG, Montgomery GW, Province MA, Kayser M, Arnold AM, Atwood LD, Boerwinkle E, Chanock SJ, Deloukas P, Gieger C, Gronberg H, Hall P, Hattersley AT, Hengstenberg C, Hoffman W, Lathrop GM, Salomaa V, Schreiber S, Uda M, Waterworth D, Wright AF, Assimes TL, Barroso I, Hofman A, Mohlke KL, Boomsma DI, Caulfield MJ, Cupples LA, Erdmann J, Fox CS, Gudnason V, Gyllensten U, Harris TB, Hayes RB, Jarvelin MR, Mooser V, Munroe PB, Ouwehand WH, Penninx BW, Pramstaller PP, Quertermous T, Rudan I, Samani NJ, Spector TD, Volzke H, Watkins H, Wilson JF, Groop LC, Haritunians T, Hu FB, Kaplan RC, Metspalu A, North KE, Schlessinger D, Wareham NJ, Hunter DJ, O'Connell JR, Strachan DP, Wichmann HE, Borecki IB, van Duijn CM, Schadt EE, Thorsteinsdottir U, Peltonen L, Uitterlinden AG, Visscher PM, Chatterjee N,

Loos RJ, Boehnke M, McCarthy MI, Ingelsson E, Lindgren CM, Abecasis GR, Stefansson K, Frayling TM, Hirschhorn JN (2010) Hundreds of variants clustered in genomic loci and biological pathways affect human height. Nature 467(7317):832–838
118. Lui JC, Nilsson O, Chan Y, Palmer CD, Andrade AC, Hirschhorn JN, Baron J (2012) Synthesizing genome-wide association studies and expression microarray reveals novel genes that act in the human growth plate to modulate height. Hum Mol Genet 21(23):5193–5201
119. Gunning P, O'Neill G, Hardeman E (2008) Tropomyosin-based regulation of the actin cytoskeleton in time and space. Physiol Rev 88(1):1–35
120. McMichael BK, Kotadiya P, Singh T, Holliday LS, Lee BS (2006) Tropomyosin isoforms localize to distinct microfilament populations in osteoclasts. Bone 39(4):694–705
121. Kotadiya P, McMichael BK, Lee BS (2008) High molecular weight tropomyosins regulate osteoclast cytoskeletal morphology. Bone 43(5):951–960
122. McMichael BK, Lee BS (2008) Tropomyosin 4 regulates adhesion structures and resorptive capacity in osteoclasts. Exp Cell Res 314(3):564–573

Chapter 15
The Role of Drebrin-Binding Stable Actin Filaments in Dendritic Spine Morphogenesis

Tomoaki Shirao and Noriko Koganezawa

Actin Cytoskeletons in Dendritic Spines

Actin Governs Dynamic Spine Motility

Dendritic spines are the postsynaptic receptive regions of most excitatory synapses in the brain [1]. They are small protrusions from the parent dendritic shaft and typically consist of a head (about 5.1×10^{-20} m^3) and a neck (about 1.2×10^{-20} m^3) [2]. They form various shapes and marked abnormalities in spine morphology in human children with mental retardation [3], suggesting that the differences between shapes reflect functional differences. Furthermore, high-frequency synaptic activity induces changes in the population of spine shapes [4] and the balance among various shapes is changed in a close correlation with learning and memory. Time lapse imaging analysis of hippocampal slices has shown that dendritic spines rapidly changed their shapes [5]. However, neither blockade nor induction of neuronal activity affect spine motility [6]. This suggests that the basal motility of dendritic spines is intrinsic to the neuron, but it does not directly link to the molecular mechanism of learning and memory.

Dendritic spines contain high concentrations of actin [7]. In general, high concentrations of actin plays a central role in supporting cell motility, suggesting that the morphological change of dendritic spines is governed by actin. In 1998, Fischer et al. showed using hippocampal neurons expressing actin labeled with green fluorescent protein that the motility of dendritic spines was completely inhibited when the neurons were treated by latrunculin A, a G-actin-sequestering agent [8]. This clearly indicates that actin plays a pivotal role in spine motility.

T. Shirao (✉) • N. Koganezawa
Department of Neurobiology and Behavior, Gunma University Graduate School of Medicine, 3-39-22 Showa-machi, Maebashi, Gunma 371-8511, Japan
e-mail: tshirao@gunma-u.ac.jp

H. Schatten (ed.), *The Cytoskeleton in Health and Disease*,
DOI 10.1007/978-1-4939-2904-7_15

Polymerization of F-actin (filamentous actin) against cellular membranes is thought to provide the force for cell membrane dynamics, such as the formation of plasma membrane protrusions of migrating cells, suggesting that the polymerization of dynamic F-actin in the spine head regulates spine motility. However, the molecular mechanism on how F-actin polymerization is regulated in the spine head has still not yet been clarified.

Studies of actin filaments in spines suggest the presence of two kinds of actin filaments. On the one hand, actin filaments appear to be stable over many hours and exhibit marked resistance to actin depolymerizing drugs such as cytochalasin D [9], which implies that actin filaments in spines are extremely stable and fulfill a purely structural role. On the other hand, individual spines undergo shape changes within a timespan of seconds or minutes as was described earlier. Actually the spines rapidly and continuously change shapes, but this motility seldom involves changes in spine size under basal conditions [8]. Halpain [10] addressed that dynamic and stable configurations of actin filaments fulfil their needs for motility versus structural integrity of dendritic spines. Two pools of actin filaments compose the spine head under normal circumstances. 'Core' of stable actin filaments form the structural foundation of the spine, while the peripheral population of actin filaments in the spine is dynamic.

Two Kinds of Actin Filaments in Mature Dendritic Spines

Dynamic and Stable Actins

In accordance with Halpain's prediction, Kasai and co-workers have reported the presence of dynamic and stable actins in dendritic spines using PAGFP-actin [11]. They photoactivated PAGFP-actin protomers in F-actin of dendritic spines. As the filaments treadmilled, activated PAGFP-actin protomers reached the end of the filament, depolymerized, and diffused away. The fluorescence from activated molecules decayed in two phases with time constants of 40 s and 17 min, indicating that there are two pools of F-actin in the spine: a dynamic one with a fast treadmilling rate and a stable one with a much slower treadmilling rate. Larger spines have a greater proportion of stable F-actin, although the proportion of stable F-actin is usually less than the dynamic one. While the dynamic F-actin is observed in the spine tip, the stable one is largely restricted to the base core of spine heads. The treadmilling of dynamic F-actin is observed from the apex to the base, but the actin protomer in the dynamic F-actin pool did not flow into the stable F-actin pool. This suggests that the two F-actin pools are differentially regulated.

It is known that polymerization of dynamic F-actin in spine heads regulates the spine motility; however, the spine size under basal conditions does not seem to be regulated by F-actin polymerization because, as stated previously, the spine size remains rather constant in spite of rapid spontaneous motility.

Mathematical Model for Lamellipodium Protrusion

It is suggested that protrusion formation is a consequence of the dynamics downstream from nucleation promoting factors (NPFs), with signaling setting the dynamic regime but not initiating the formation of individual protrusions [12]. Zimmermann and Falcke developed a mathematical model for lamellipodium protrusion. The model lamellipodium consists of an actin gel in the bulk and a highly dynamic range at the leading edge, called semiflexible region (SR). Signaling cascades that lead to the activation of NPFs, activate the actin related protein complex Arp2/3. Arp2/3 initiates the growth of a new filament branch from an existing filament in SR, pushing the lamellipodial membrane forward. Individual lamellipodia form due to random supercritical filament nucleation events amplified by autocatalytic branching. This model can be applied to the incessant lamellipodia formation in many cells with a constant state of the signaling pathways upstream from NPFs.

Model of Spontaneous Spine Motility

Applying Zimmermann and Falcke's model to the actin cytoskeleton in dendritic spine, spontaneous rapid spine motility (periodic lamellipodium formation) may be determined by the autocatalytic nature of branching of actin filaments of the dynamic F-actin pool, and by the length dependence of bundling, capping and severing of them. The stable F-actin core provides a stiff substrate for actin filaments in the lamellipodium to push back against to extend the postsynaptic membrane [13]. Although the nature of the stable F-actin core has not been well elucidated, a spine-resident side-binding protein of F-actin named drebrin, which is localized at the core region of dendritic spines [14], is thought to be an important element of the stable F-actin pool [15].

Although spine is too small to directly observe the two kinds of F-actins, we can observe fast and slow treadmilling of actin filaments in axonal growth cones. In the lamellipodia at the periphery of the growth cone, actin filaments are similar to the dynamic F-actin in dendritic spines in terms of the absence of drebrin. They flow retrogradely at rates of approximately 4 μm/min [16]. F-actin in the actin arc at the base of the lamellipodia contains drebrin, which flows transversely rather than longitudinally. This retrograde flow of drebrin-binding F-actin occurs more slowly (approximately 1 μm/min) [17]. This indicates that peripheral F-actin that does not bind to drebrin shows more rapid treadmilling than drebrin-binding actin filaments. In addition, the drebrin-binding F-actin is resistant to the actin depolymerizing agent cytochalasin D, suggesting that they are stable actin filaments.

Together it is suggested that dendritic spines consist of dynamic and stable F-actin pools. Basal motility of dendritic spines occurs due to random supercritical filament nucleation events amplified by autocatalytic branching in the dynamic F-actin pool. Drebrin-binding actin filaments in the stable F-actin pool form a cross-linked gel, serving as the structural element for treadmilling of dynamic F-actin to push back against the spine membrane.

Drebrin Governs the Formation of Stable F-Actin

Modulation of Helical Structure of F-Actin by Drebrin

F-actin consists of a double helix of actin protomers decorated with its binding proteins. The helical structure plays an important role in modifying the relationship (binding activity) between F-actin and actin-regulating proteins [18]. Variations in the helical structure of F-actin are modulated by several side-binding proteins of F-actin (the double helix of actin protomers).

Tropomyosin is a typical side-binding protein of F-actin found in virtually all eukaryotic cells. Brain tropomyosin binds to F-actin with a stoichiometry of 1:7 (tropomyosin: actin protomer) with a dissociation constant (Kd) of 2.2×10^{-7} M [19]. Similarly drebrin binds to F-actin with a stoichiometry of 1:5 (drebrin : actin protomer) with a dissociation constant (Kd) of 1.2×10^{-7} M [20]. In spite of their similarity in the biochemical actin-binding property, atomic force microscopy analysis shows the significant differences in the helical structure. Tropomyosin forms a helix pitch of 36.5 nm, which is similar to the pitch of bared double helix of actin protomers. In contrast, drebrin forms the 40.0 nm pitch of actin filaments [21]. This difference makes it possible that the drebrin-binding F-actin and the other F-actins respond differently to the same signal within small dendritic spines.

Inhibition of F-Actin Depolymerization by Drebrin

Mikati et al. [22] reported that drebrin significantly decreased the depolymerization rates of uncapped filaments, reaching 88 % inhibition at full saturation, and 50 % inhibition is achieved at a low binding density of drebrin (~18 %). Drebrin causes stronger inhibition of barbed-end depolymerization compared to pointed-end depolymerization at the same binding density. Even in the presence of latrunculin A, drebrin inhibits the full depolymerization of actin filaments. Furthermore, differential scanning calorimetry (DSC) study shows that the T_m of F-actin was increased by 0.5 °C in the presence of saturating amounts of drebrin. Taken together, it is indicated that drebrin forms stable actin filaments.

Spine Morphogenesis

Drebrin Clustering in Dendritic Filopodia Mediates Spine Morphogenesis

Dendritic spines have two major structural elements, the postsynaptic density (PSD) and the actin cytoskeleton. Although PSD scaffold proteins such as PSD-95, Shank, and Homer are known to play pivotal roles in spine morphogenesis [23–25], the initiation of spine morphogenesis precedes synaptic assembly of PSD-95 [26]. Moreover mutant mice which lack PSD-95 expression exhibit standard spine morphology [27], suggesting that molecules other than PSD scaffold proteins govern spine morphogenesis.

There are two models for the formation of dendritic spines. One model is that dendritic filopodia serve as the precursor of dendritic spines, and the other is that dendritic spines emerge from shaft synapses. The former model predominates during neuronal development. Developmental changes of the actin cytoskeleton within filopodia during spine morphogenesis have been intensely studied because the actin cytoskeleton regulates the morphology of both filopodia and spines.

In vitro study shows that filopodia from the dendrites are classified into two types in terms of the presence of drebrin clusters: diffuse-type filopodia and cluster-type filopodia [28]. Most cluster-type filopodia appose presynaptic terminals, but diffuse-type filopodia do not. This indicates that cluster-type filopodia are more matured than diffuse-type. On the other hand, the half of cluster-type filopodia do not contain PSD-95, while most mature spines contain PSD-95 [28], indicating that cluster-type filopodia are not mature spines but their precursors. Similarly, drebrin has been already observed at the nascent contact site of the dendrite by the axon in vivo [29]. Thus it is indicated that dendritic spines develop via cluster-type filopodia that have been transformed from diffuse-type filopodia.

Drebrin-binding stable F-actin seems to play a pivotal role for the establishment of postsynaptic structures. Drebrin content in dendritic spines correlates with spine head size, suggesting that the proportion of stable F-actin in the spine head seems to regulate the spine size [14]. During development, clustering of drebrin with F-actin occurs at postsynaptic sites in dendritic filopodia. In parallel with this change, drebrin changes its isoform from embryonic-type (drebrin E) to adult-type (drebrin A) [29, 30]. Interestingly, synaptic clustering of PSD-95 and NMDARs partially depend on drebrin [28, 31]. Additionally, drebrin is involved in the regulation of AMPAR trafficking to the postsynaptic site [32].

AMPA Receptor Facilitates the Drebrin Clustering in Dendritic Spines

How is drebrin clustered at postsynaptic sites? Although the aforementioned studies suggest that the conversion of drebrin isoform expression from drebrin E to drebrin A is involved in the drebrin clustering, the premature expression of drebrin A induces abnormally large headless protrusions with the unrestricted accumulation of F-actin, PSD-95 and drebrin [33], indicating that the conversion of drebrin isoform plays a role for the targeting mechanism of postsynaptic molecules, but is not a sufficient condition for postsynaptic formation.

The synchronous development of drebrin clustering and functional turnover of synaptic vesicles indicates that synaptic activity is involved in drebrin clustering at postsynaptic sites. Inhibition of action potentials with TTX decreases drebrin cluster density, while inhibition of GABAA-receptor with picrotoxin, which enhances the excitatory component of synaptic transmission, increases drebrin clustering [34]. Thus spontaneous synaptic activity is involved in the drebrin clustering. Moreover, the study using subtype-specific blockers of glutamate receptors has shown that AMPA receptor, but neither NMDA receptor nor metabotropic glutamate receptor, regulates the clustering of drebrin at the postsynaptic site [34].

Then how does AMPAR regulate drebrin clustering? Using the fluorescence recovery after photobleaching (FRAP) analysis we have explored a cellular basis for activity-dependent drebrin clustering and have demonstrated that AMPARs specifically regulates drebrin dynamics within dendritic spines. Neurons were transfected with vectors that encoded drebrin A fused to enhanced green fluorescent protein (eGFP). Individual eGFP molecules can be rendered nonfluorescent, or 'bleached', with high-intensity laser pulses. Such pulses darken the target area until new, unbleached eGFP-drebrin replaces the bleached molecules during normal protein turnover. Under normal physiological conditions that allow spontaneous neuronal activity, about a quarter of total drebrin within a single spine is stabilized. Applications of CNQX or AP5 show that the activity of AMPARs, but not that of NMDARs, significantly decreases the level of stable drebrin in spines. Together it is indicated that activated AMPAR accumulates the stable F-actin bound by drebrin at the postsynaptic site, facilitating the recruitment of PSD-95, NMDAR and other postsynaptic proteins, including AMPARs themselves as suggested in the above section, into dendritic spines during development.

Spikar Is Involved in the Drebrin-Mediated Spine Formation

Drebrin initiates spine formation and the decrease of drebrin results in the decrease of spine density [31, 35]. However, an increased amount of drebrin does not raise the number of normal spines, but forms the large number of small protrusions from the dendritic shaft [36, 37]. These facts suggest that there is an unidentified protein which mediates the drebrin-dependent spine formation.

To explore a drebrin-binding molecule mediating spine formation, we performed a yeast two-hybrid screen using drebrin as bait and found a novel drebrin binding protein [38]. This protein localizes in neuronal nuclei as well as in dendritic spines, and this is why we named it spikar (localizes in *spi*ne and *kar*yoplasm). Unlike drebrin, the up-regulation and down-regulation of spikar expression results in the increase and decrease of the spine density, respectively. Interestingly spikar does not affect the spine morphology different from drebrin [38]. The localization of spikar depends on drebrin whereas that of drebrin does not depend on spikar. In addition, spine formation activity of spikar depends on drebrin. Together it is suggested that drebrin might function to include spikar to the stable F-actin complex at postsynaptic sites, resulting in the spine formation.

Conclusion and Perspective

Dendritic spines are formed from dendritic filopodia in parallel with the appearance of stable F-actin instead of dynamic F-actin at postsynaptic sites. Stable F-actin consists of drebrin-binding actin double helix polymers, which shows the slow treadmilling and the increase of heat stability as well as the elongation of the helix pitch. Although the developmental conversion of drebrin isoforms, drebrin E to drebrin A, is involved in the accumulation of stable F-actin, which facilitates further accumulation of postsynaptic scaffold proteins and neurotransmitter receptors, AMPA receptor activation seem to be needed for more precise accumulation of stable F-actin at postsynaptic sites.

It is believed that the motility of actin filaments is of importance for synaptic plasticity but further investigation, particularly focusing on the stable and dynamic F-actin, is needed to reveal the actual role of the actin filaments. Furthermore, since the appearance of the stable actin pool is a good marker of synaptic maturation, we suggest drebrin as an appropriate surrogate marker of synaptic function. Recently, it has been recognized that mislocalization and dysregulation of postsynaptic cytoskeletons are crucial events regarding pathophysiology of so-called "synaptopathies" such as Alzheimer disease. In addition, human induced pluripotent stem cells (hiPSCs) provide new possibilities for drug discoveries because human specific side effects could be tested easily using those cells. Thus, drebrin can be used as the surrogate marker in hiPSCs-derived neurons as well. For this reason we expect drebrin be widely used in drug discovery and developmental fields.

Referencces

1. Harris KM, Kater SB (1994) Dendritic spines: cellular specializations imparting both stability and flexibility to synaptic function. Annu Rev Neurosci 17:341–371
2. Harris KM, Stevens JK (1989) Dendritic spines of CA 1 pyramidal cells in the rat hippocampus: serial electron microscopy with reference to their biophysical characteristics. J Neurosci 9:2982–2997
3. Purpura DP (1974) Dendritic spine "dysgenesis" and mental retardation. Science 186: 1126–1128
4. Desmond NL, Levy WB (1983) Synaptic correlates of associative potentiation/depression: an ultrastructural study in the hippocampus. Brain Res 265:21–30
5. Dailey ME, Smith SJ (1996) The dynamics of dendritic structure in developing hippocampal slices. J Neurosci 16:2983–2994
6. Dunaevsky A, Tashiro A, Majewska A, Mason C, Yuste R (1999) Developmental regulation of spine motility in the mammalian central nervous system. Proc Natl Acad Sci U S A 96:13438–13443
7. Matus A, Ackermann M, Pehling G, Byers HR, Fujiwara K (1982) High actin concentrations in brain dendritic spines and postsynaptic densities. Proc Natl Acad Sci U S A 79:7590–7594
8. Fischer M, Kaech S, Knutti D, Matus A (1998) Rapid actin-based plasticity in dendritic spines. Neuron 20:847–854

9. Allison DW, Gelfand VI, Spector I, Craig AM (1998) Role of actin in anchoring postsynaptic receptors in cultured hippocampal neurons: differential attachment of NMDA versus AMPA receptors. J Neurosci 18:2423–2436
10. Halpain S (2000) Actin and the agile spine: how and why do dendritic spines dance? Trends Neurosci 23:141–146
11. Honkura N, Matsuzaki M, Noguchi J, Ellis-Davies GC, Kasai H (2008) The subspine organization of actin fibers regulates the structure and plasticity of dendritic spines. Neuron 57:719–729
12. Zimmermann J, Falcke M (2014) Formation of transient lamellipodia. PLoS One 9:e87638
13. Burnette DT, Manley S, Sengupta P, Sougrat R, Davidson MW, Kachar B, Lippincott-Schwartz J (2011) A role for actin arcs in the leading-edge advance of migrating cells. Nat Cell Biol 13:371–381
14. Kobayashi C, Aoki C, Kojima N, Yamazaki H, Shirao T (2007) Drebrin a content correlates with spine head size in the adult mouse cerebral cortex. J Comp Neurol 503:618–626
15. Asada H, Uyemura K, Shirao T (1994) Actin-binding protein, drebrin, accumulates in submembranous regions in parallel with neuronal differentiation. J Neurosci Res 38:149–159
16. Medeiros NA, Burnette DT, Forscher P (2006) Myosin II functions in actin-bundle turnover in neuronal growth cones. Nat Cell Biol 8:215–226
17. Mizui T, Kojima N, Yamazaki H, Katayama M, Hanamura K, Shirao T (2009) Drebrin E is involved in the regulation of axonal growth through actin-myosin interactions. J Neurochem 109:611–622
18. Sharma S, Grintsevich EE, Hsueh C, Reisler E, Gimzewski JK (2012) Molecular cooperativity of drebrin1-300 binding and structural remodeling of F-actin. Biophys J 103:275–283
19. Yamashiro-Matsumura S, Matsumura F (1988) Characterization of 83-kilodalton nonmuscle caldesmon from cultured rat cells: stimulation of actin binding of nonmuscle tropomyosin and periodic localization along microfilaments like tropomyosin. J Cell Biol 106:1973–1983
20. Ishikawa R, Hayashi K, Shirao T, Xue Y, Takagi T, Sasaki Y, Kohama K (1994) Drebrin, a development-associated brain protein from rat embryo, causes the dissociation of tropomyosin from actin filaments. J Biol Chem 269:29928–29933
21. Sharma S, Grintsevich EE, Phillips ML, Reisler E, Gimzewski JK (2011) Atomic force microscopy reveals drebrin induced remodeling of f-actin with subnanometer resolution. Nano Lett 11:825–827
22. Mikati MA, Grintsevich EE, Reisler E (2013) Drebrin-induced stabilization of actin filaments. J Biol Chem 288:19926–19938
23. El-Hussein AE, Schnell E, Chetkovich DM, Nicoll RA, Bredt DS (2000) PSD-95 involvement in maturation of excitatory synapses. Science 290:1364–1368
24. Marrs GS, Green SH, Dailey ME (2001) Rapid formation and remodeling of postsynaptic densities in developing dendrites. Nat Neurosci 4:1006–1013
25. Sala C, Piech V, Wilson NR, Passafaro M, Liu GS, Sheng M (2001) Regulation of dendritic spine morphology and synaptic function by Shank and Homer. Neuron 31:115–130
26. Okabe S, Miwa A, Okado H (2001) Spine formation and correlated assembly of presynaptic and postsynaptic molecules. J Neurosci 21:6105–6114
27. Migaud M, Charlesworth P, Dempster M, Webster LC, Watabe AM, Makhinson M, He Y, Ramsay MF, Morris RGM, Morrison JH et al (1998) Enhanced long-term potentiation and impaired learning in mice with mutant postsynaptic density-95 protein. Nature 396:433–439
28. Takahashi H, Sekino Y, Tanaka S, Mizui T, Kishi S, Shirao T (2003) Drebrin-dependent actin clustering in dendritic filopodia governs synaptic targeting of postsynaptic density-95 and dendritic spine morphogenesis. J Neurosci 23:6586–6595
29. Aoki C, Sekino Y, Hanamura K, Fujisawa S, Mahadomrongkul V, Ren Y, Shirao T (2005) Drebrin A is a postsynaptic protein that localizes in vivo to the submembranous surface of dendritic sites forming excitatory synapses. J Comp Neurol 483:383–402
30. Hayashi K, Suzuki K, Shirao T (1998) Rapid conversion of drebrin isoforms during synapse formation in primary culture of cortical neurons. Brain Res Dev Brain Res 111:137–141

31. Takahashi H, Mizui T, Shirao T (2006) Down-regulation of drebrin A expression suppresses synaptic targeting of NMDA receptors in developing hippocampal neurones. J Neurochem 97(Suppl 1):110–115
32. Kato K, Shirao T, Yamazaki H, Imamura K, Sekino Y (2012) Regulation of AMPA receptor recruitment by the actin binding protein drebrin in cultured hippocampal neurons. J Neurosci Neuroeng 1:153–160
33. Mizui T, Sekino Y, Takahashi H, Yamazaki H, Shirao T (2005) Overexpression of drebrin A in immature neurons induces the accumulation of F-actin and PSD-95 into dendritic filopodia, and the formation of large abnormal protrusions. Mol Cell Neurosci 30:149–157
34. Takahashi H, Yamazaki H, Hanamura K, Sekino Y, Shirao T (2009) Activity of the AMPA receptor regulates drebrin stabilization in dendritic spine morphogenesis. J Cell Sci 122: 1211–1219
35. Ivanov A, Esclapez M, Pellegrino C, Shirao T, Ferhat L (2009) Drebrin A regulates dendritic spine plasticity and synaptic function in mature cultured hippocampal neurons. J Cell Sci 122:524–534
36. Hayashi K, Shirao T (1999) Change in the shape of dendritic spines caused by overexpression of drebrin in cultured cortical neurons. J Neurosci 19:3918–3925
37. Biou V, Brinkhaus H, Malenka RC, Matus A (2008) Interactions between drebrin and Ras regulate dendritic spine plasticity. Eur J Neurosci 27:2847–2859
38. Yamazaki H, Kojima N, Kato K, Hirose E, Iwasaki T, Mizui T, Takahashi H, Hanamura K, Roppongi RT, Koibuchi N et al (2014) Spikar, a novel drebrin-binding protein, regulates the formation and stabilization of dendritic spines. J Neurochem 128:507–522

Chapter 16
The Role of the Actin Cytoskeleton in Cancer and Its Potential Use as a Therapeutic Target

Simon Brayford, Galina Schevzov, Julien Vos, and Peter Gunning

Actin; The Building Block of a Functionally Diverse Cytoskeleton

Actin is the most abundant protein found in eukaryotic cells and represents the basic subunit of the microfilament system. As well as the role of actin filaments in muscle cells as part of the contractile apparatus, non-muscle actin plays a role in almost all cellular functions including cell division, polarity, maintenance of cell shape and cell motility [1]. This 42 kDa globular protein undergoes cycles of polymerisation and disassembly between its globular and filamentous forms. The filaments formed are helical polymers with an overall molecular polarity and exist in a constant state of flux with new monomers being added to the barbed end, and removed at the pointed end to produce free monomers in an ATP dependent process [2]. This cyclical process, known as tread-milling, gives rise to the dynamic nature of actin filaments, allowing precise cytoskeletal structures to be quickly and efficiently assembled as needed. Actin filaments are found as organised assemblies localised in discrete areas of the cytoplasm where they perform their functions in response to different stimuli. As well as individual filaments, actin is often found bundled into thick cables which traverse the cell, known as stress fibers. The biochemistry of actin alone cannot explain the complexity of function observed under physiological conditions. Actin achieves its vast array of functions via its interaction with a rich variety of actin binding proteins, of which more than 60 classes have been described [3]. A handful of these are relevant to the role of the actin cytoskeleton in disease, and their potential as therapeutic targets will be discussed here.

S. Brayford • G. Schevzov • J. Vos • P. Gunning (✉)
Oncology Research Unit, School of Medical Sciences, UNSW Australia,
Room 229, Wallace Wurth East, Sydney 2052, NSW, Australia
e-mail: p.gunning@unsw.edu.au

H. Schatten (ed.), *The Cytoskeleton in Health and Disease*,
DOI 10.1007/978-1-4939-2904-7_16

Actin Nucleating Factors

Spontaneous assembly of pure actin monomers is energetically unfavourable due to the instability of actin dimers and trimers, but once polymerisation has initiated, filaments grow rapidly. Subunit addition at the barbed end is diffusion limited, meaning that the rate of growth is determined by the probability of subunits colliding with the end [4]. Although ADP-actin subunits dissociate faster from the barbed end than ATP-actin subunits, the resulting tread-milling would still be very slow [5]. Under physiological conditions, actin assembly proteins or nucleating factors mediate filament assembly by bringing monomers within close proximity to one another, promoting rapid polymerisation. The actin-related protein 2 and 3 complex (Arp2/3) was the first actin nucleating factor to be described [6]. Upon activation, Arp2/3 binds to the side of a pre-existing actin filament, generating a stable trimer for the growth of a daughter filament, forming a branch at a 70° angle from the mother filament [7]. Subsequent branching creates a dendritic actin network crucial for the production of rapid membrane protrusions associated with cell motility such as lamellipodia. Unsurprisingly, the Arp2/3 complex is abundant at the leading edge of migrating cells where it functions downstream of WAVE and N-WASp activators, respectively, in response to rac signalling [8]. Furthermore, it has been demonstrated that cells lacking Arp2/3 expression are unable to produce lamellipodia. Interestingly however, Arp2/3-depleted cells respond normally to shallow gradients of PDGF, indicating that lamellipodia are not required for chemotaxis [9]. Cortactin acts synergistically with the Arp2/3 complex to stabilise branch junctions [10]. Another family of actin filament nucleating factors are the formins which can mediate both actin assembly and disassembly to produce a vast array of cytoskeletal structures. Formins have been demonstrated to be capable of nucleating, polymerising, bundling and severing actin filaments in vitro [11]. To date, there are 15 known mammalian formins, grouped into 7 families [12]. Unlike the Arp2/3 complex, formins nucleate actin filaments and remain bound to the barbed end, generating unbranched actin filaments through a processive capping mechanism [13]. Generally, formins exist in an auto-inhibited state between their N-terminal Diaphanous Inhibitory Domain (DID) and C-terminal Diaphanous Activating Domain (DAD). Activation occurs when an active Rho-GTPase disrupts the interaction between the DID and DAD domains [14]. It is important to note that, while plants express many isoforms of actin [15] suggesting an evolutionary need to diversify function, only two cytoskeletal isoforms of actin exist in mammalian cells. Given the functional diversity of actin observed in higher organisms, it begs the question of how a putatively more complex system can achieve this with a seemingly simpler array of building blocks at its disposal. The answer lies in the fact that, unlike plants, actin filaments in animal and yeast cells are not a homogeneous system but rather consist of compositionally distinct filaments arising from the inclusion of various isoforms of tropomyosin [16].

Tropomyosins as Master Regulators of Actin Filament Form and Function

Tropomyosin (Tm) is an alpha-helical, coiled-coil protein, dimers of which associate head-to-tail to form a continuous polymer which lies along the major groove of actin filaments [17]. The role of tropomyosin in muscle is very well understood where it is responsible for regulating the interaction between actin and myosin [18]. Non-muscle cells also contain tropomyosin where its proposed role is to stabilise actin filaments by modulating their interactions with actin binding proteins [19]. Recent studies in a variety of systems have shown that the diversity of actin cytoskeletal function is paralleled by a diversity of tropomyosin isoforms. Isoforms of tropomyosin are generated by the alternative splicing of four mammalian genes resulting in over 40 isoforms, the majority of which are cytoskeletal. The use of alternative promoters at the amino terminus gives rise to either high molecular weight (HMW) isoforms or low molecular weight (LMW) isoforms [20]. Different tropomyosin isoforms can differentially regulate actin filament function. Actin filaments decorated by different isoforms of tropomyosin have been shown to recruit different actin binding proteins and myosin motors, leading to a difference in filament stability [21, 22]. Functionally distinct sub-populations of actin filaments can therefore be defined on the basis of their tropomyosin isoform composition. Tropomyosin isoforms also display extensive intracellular sorting, resulting in spatially distinct actin filament populations. Sorting of Tm isoforms has been observed in a number of cell types, including fibroblasts, epithelial cells, osteoclasts, neurons and muscle cells [23]. The exact mechanism underlying the way in which different isoforms of tropomyosin are targeted to specific actin structures within the cell remains unknown but a few independent experiments have provided some insights. On one hand, a molecular sink model has been proposed whereby isoforms accumulate in actin-based structures where they have the highest affinity, rather than the presence of an intrinsic sorting signal that directs particular isoforms to a single geographical location [24]. Other groups' work in yeast has revealed the importance of the N-terminal acetylation of tropomyosin in effecting its cellular localisation and function. Acetylated tropomyosin (Cdc8) is found predominantly in the contractile cytokinetic actomyosin ring (CAR) whereas the un-acetylated form is seen exclusively on interphase actin filaments [25, 26]. In addition, the acetylated state of yeast tropomyosin can regulate the motility of myosin, with the motility of class II myosin being affected but not that of class I and V myosin [27, 28]. Johnson et al. [29] have recently demonstrated that the formin isoform used to build an actin filament in yeast determines which tropomyosin is incorporated into the filament and that in turn dictates the functional characteristics of the filament including binding of specific myosin motors.

Data so far provides us with a working model for tropomyosin-directed regulation of actin filament function (Fig. 16.1). Cell culture studies have shown that actin filaments decorated with different Tm isoforms recruit different actin binding proteins

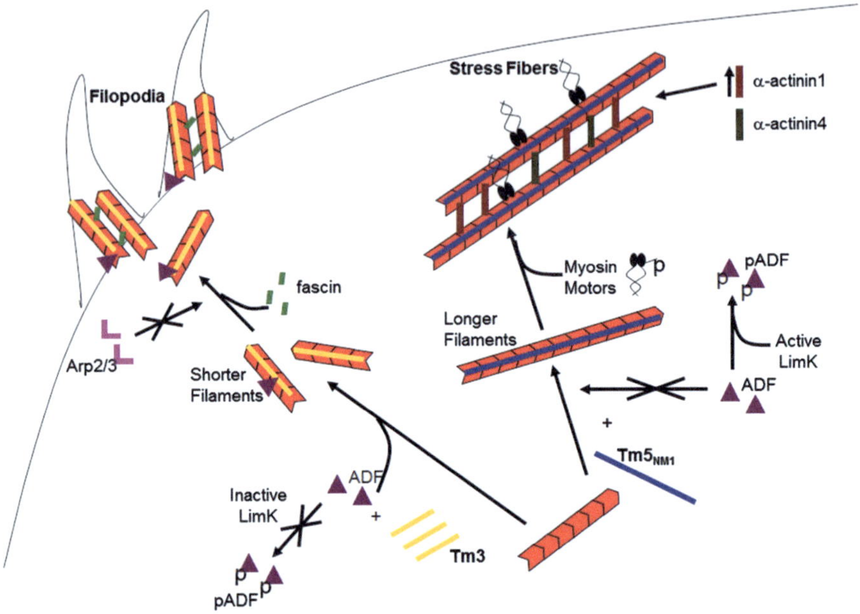

Fig. 16.1 Working model for tropomyosin-directed regulation of actin filament function. When Tm5NM1 binds to actin filaments, they become permissive for myosin II interaction, and this can lead to myosin II recruitment to these filaments. Simultaneously, Tm5NM1 eliminates ADF binding to the filaments, which results in more-stable filaments engaged in contractile activity [22]. By contrast, binding of Tm3 is permissive for ADF binding, which promotes severing of the filaments and greater turnover. In parallel, fascin promotes filament bundling and the result is non-contractile, stable filopodia forming filaments [21]. In both cases, multiple mechanisms reinforce the final outcome

ultimately regulating the organisational and functional properties of the filaments [21, 22]. Finally, there is the issue of whether a single actin filament is restricted to being decorated by only one isoform of tropomyosin or if hetero-polymers can lie along the same filament. With the advancement of super-resolution microscopy techniques, answers to these questions are likely on the horizon.

Cellular Motility Driven by the Assembly and Disassembly of Actin Filaments

Directional motility is a fundamental cellular process essential for embryonic morphogenesis, wound healing, immune surveillance and tissue repair. Dysregulation of this process resulting in aberrant cell movement is a hallmark feature of metastatic cancer cells. The development of metastases accounts for more than 90 % of cancer related mortality [30], highlighting the need for an increased understanding of the regulatory mechanisms underlying cell motility. Almost universally, eukaryotic cell migration involves a series of four highly orchestrated steps including protrusion of the leading plasma membrane, formation of cell-substrate adhesions, generation of acto-myosin contractile force and release of substrate adhesions at the trailing cell rear [31]. Dynamic remodelling of the actin cytoskeleton underlies cell migration, with many essential cytoskeletal proteins conserved across eukaryotes. This may explain why similar motility phenotypes are observed across a broad range of cells such as fibroblasts, neuronal and epithelial cells. As mentioned earlier, rapid assembly of actin filaments by addition of monomers at the barbed end and disassembly at the pointed end results in a tread-milling effect. This retrograde flow of the actin filament in a particular direction underlies the principal step in cell migration, being protrusion of the leading edge in the direction of movement. Although this process has been widely studied, a detailed mechanism underlying how this process is regulated and coupled to the rest of the cell migration cycle to translate a persistent protrusive force into whole cell translocation remains to be fully understood. Cells extend four different plasma membrane protrusions at the leading edge; lamellipodia which can extend long distances through the extracellular matrix to pull cells through tissues [32], filopodia which explore the cell's surroundings, blebs which have been described to drive directional migration during development [33] and invadopodia are protrusions which allow invasion through tissues via metalloprotease mediated degradation of the extracellular matrix [34]. Each of these structures uniquely contributes to migration and, depending on the specific circumstances can also co-exist at the leading edge as has been previously observed in migrating zebrafish cells during gastrulation [35]. Perhaps the most well-studied of these structures is the lamellipodium. These thin, sheet-like projections were first described and named by Abercrombie in 1970 who observed them at the leading edge of fibroblasts in culture [36]. Through electron microscopy studies, he and colleagues identified the dominant structural component of lamellipodia to be branched actin filaments [37]. Later, it was demonstrated that the Arp2/3 complex was localised to the branch junctions in these networks and that the branches adhered very closely to the 70° angle observed in vitro [38].

Early Views on the Composition of the Leading Edge

As the lamellipodium continues to advance, it leaves behind a region known as the lamella, where actin filaments are unbranched, and the Arp2/3 complex is absent. It has previously been demonstrated in vitro that tropomyosin competes with Arp2/3 for binding sites on actin filaments [39]. Due to the enrichment of Arp2/3 and the branched organization of actin filaments observed in lamellipodia it was proposed that tropomyosin must be absent from this region. This model was supported by DesMarais et al. [40] who examined cells stained with anti-tropomyosin antibodies and found that none stained the lamellipodium. However, more recent work in neuroblastoma cells has shown that, in contrast to Tm5NM1, TmBr3 (previously shown to induce lamellipodia formation in non-neuronal cells [22]) actually supported the activity of the Arp2/3 complex rather than inhibiting it, indicating that TmBr3 may work simultaneously with the Arp2/3 complex in neuronal cells [41]. Finally, through the use of antibodies and fluorescently-tagged isoforms it has been demonstrated that tropomyosins are indeed present in significant amounts in the lamellipodia and filopodia of spreading normal and transformed cells [42]. These observations indicate that the view of the role of tropomyosin in regulating actin filaments within the lamellipodium needs to be significantly revised.

Two Distinct Actin Filament Networks to Drive the Protrusion of Migrating Cells

For many years, the Arp2/3 complex was thought to be the sole mediator of actin filament assembly in lamellipodia. More recently however, other actin nucleators have been found to contribute to lamellipodium extension, including several members of the formin family. As mentioned earlier, formins promote filament elongation without branching through a processive capping mechanism and the formin mDia1, a RhoA target, has been shown to localise to the leading edge [43]. It has been proposed that two structurally distinct populations of actin filaments may actually overlap at the leading edge to occupy the same space [44]. In other words, the actin cytoskeleton is organised into two molecularly distinct yet collaborating filament networks wherein the narrow lamellipodium undergoing fast, Arp2/3-mediated treadmilling is superimposed by a more stable, linear array of actin filaments that reach all the way to the leading edge [45]. A critique of this model has raised the question of how these two structures, both composed of actin and occupying the same cellular region, are differentially regulated and coordinated in time and space to maintain a state of persistent protrusion coupled to cell translocation [46]. With an ever-increasing understanding of the biology of formins and tropomyosins, the answer to this key regulatory mechanism will likely be revealed in the near future.

Stress Fibers Interact with Non-muscle Myosins to Generate Contractile Force

For coordinated cell migration to occur, persistent membrane protrusions must be coupled to contraction of the cell body to generate force. Much like the contractile apparatus of the sarcomere in muscle, this is achieved in non-muscle cells via the interaction between actin and myosin motor proteins. The non-muscle myosin II is a ubiquitous molecular motor which, upon phosphorylation by Rho Kinase, binds and contracts actin bundles in an ATP dependent manner [47]. The tension induced by myosin II depends upon the tethering of actin stress fibers to the substratum via focal adhesions. Focal adhesions are integrin-based, large multimolecular assemblies that form a structural bridge between the substrate and the actin cytoskeleton [48]. Through fluorescence microscopy studies it has been established that the HMW Tropomyosins; Tms 1, 2, 3 & 6 as well as the LMW Tms 5NM1 & 2 clearly localise to actin stress fibers in a variety of cell types. Studies show that the Tm5NM1 isoform is able to promote isoform-specific recruitment of a myosin II motor to stress fibres containing this Tm. By contrast, a Tm that induces lamellipodia, TmBr3, leads to a reduction in active myosin II levels [22]. This is compatible with the observation that Tm isoforms can differentially regulate myosin mechanochemistry in a cell-free system and suggests a possible mechanism to explain the effects of Tm5NM1 and TmBr3 on myosin location and activity [49].

Aberrant Signalling Leads to Alterations to the Actin Cytoskeleton in Cancer

Most of the alterations that lead to tumour formation and metastasis can be described as being associated with several hallmarks of cancer cells, which represent the properties that are necessary for cancer cell survival and tumour spreading. These genomic or epigenetic changes target pathways that lead to uncontrolled proliferation, disruption of apoptotic mechanisms, initiation of angiogenesis, evasion of immune surveillance and the ability to invade into surrounding tissues to form metastases [50]. Owing to the critical role of invasion and migration in metastasis, there has been considerable interest in targeting cancer cells' migration machinery as a novel therapeutic approach. However, it has become clear that cancer cells use a range of motility phenotypes to migrate and invade [51]. Transitions between epithelial and mesenchymal phenotypes of cells are required for normal morphogenetic processes and tissue remodelling during embryogenesis. However, sustained signalling by oncogenic Ras may result in morphological transformation to a mesenchymal phenotype, which is associated with changes in gene expression, loss of cell–cell adhesions and increased invasiveness of tumour cells [52]. Alterations to the actin based cytoskeleton are also an established characteristic of transformed cells. Oncogenic signalling pathways directly target the actin cytoskeleton leading to disruption of stress fibers (Fig. 16.2) [53], and associated adhesive structures which in turn leads to enhanced motility and invasiveness of tumour cells along with anchorage-independent growth and cellular tumourigenicity [54]. These findings are supported by studies in which oncogenic-mediated changes to the actin cytoskeleton were able to be reversed by ectopic expression of specific actin filament stabilising proteins such as tropomyosin [55–60]. The precise mechanisms by which these changes to the actin cytoskeleton contribute to signalling events that provide a tumour cell with a selective growth advantage remain to be fully understood. Consequently, the observed aberrant organisation of the actin cytoskeleton in transformed cells has made it an attractive target for early chemotherapeutic strategies [61].

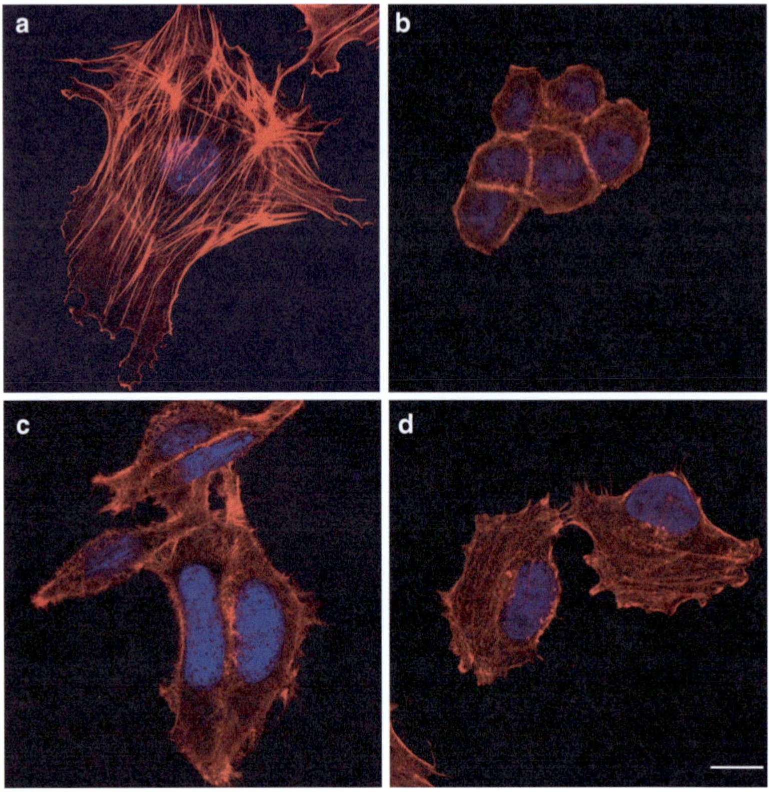

Fig. 16.2 The unique organisation of the actin cytoskeleton of cancer cells. Fluorescent staining of the actin cytoskeleton with phalloidin (*red*) and DAPI, a nuclear stain (*blue*) of (**a**) primary fibroblasts, (**b**) HT29 colon cancer, (**c**) SW480 colon cancer and (**d**) 131W1 astrocytoma cancer cells. Scale bar: 10 μm (Schevzov, 2014, unpublished)

The Problem with Targeting the Actin Cytoskeleton for Chemotherapy

To date, little progress has been made with compounds that disrupt the organisation of actin filaments, mainly due to the essential role of actin in the composition of the functional unit of muscle contraction, the sarcomere, universal disruption of which results in intolerable toxicity to cardiac and respiratory muscle. Furthermore, selective drug targeting has been hampered by the plasticity of the actin cytoskeleton [61]. As mentioned earlier, the actin cytoskeleton is not a single, uniform system but rather a series of unique filament populations with distinct functions arising from their inclusion of various isoforms of tropomyosin. This opens up the exciting possibility that certain filament populations may be indispensable for tumour cell function, yet molecularly distinct from those found in the contractile apparatus of heart and skeletal muscle.

Tropomyosin as a Regulator of Cancer Cell Transformation

The changes in rearrangement of microfilament bundles seen in transformed cells appear to correlate with alterations in tropomyosin expression. Decreased expression of non-muscle tropomyosins is commonly associated with the transformed phenotype. In addition, these changes in Tm expression occur in cells of all species examined including chicken, rodents and human, indicating that alterations of Tm expression is a common feature of the transformed phenotype [62]. In particular, the expression of HMW Tm isoforms (Tm1, 2 and 3) is decreased during oncogenic transformation [63–65]. The drive behind this shift in tropomyosin expression in transformed cells is not well understood but may reflect a requirement for the cell to eliminate certain functions associated with HMW Tm containing actin filaments [66].

Down-Regulation of HMW Tropomyosins; A Crucial Step in Oncogenic Transformation

It has been demonstrated that Tm2 expression is reduced in malignant ovarian carcinomas compared to benign ovarian tumours [67] and that Tm1 and Tm2 are reduced in prostate carcinoma compared to prostate hyperplasia [68]. Although it remains unclear whether suppression of HMW Tm contributes directly to the disruption in cyto-architecture, or the loss of these Tms is simply associated with the formation of aberrant filaments, decreased expression of HMW Tms correlates with the disruption of stress fibers seen in transformed cells and is supported by studies which also showed that forced expression of HMW Tm1 reverses transformation-associated changes by restoring the structural components of the cell and abolishing anchorage-independent growth [58, 59]. Comparison between low- and highly-metastatic lung carcinoma cells showed that a decrease in Tm2 is associated with a higher level of metastasis [69, 70]. A comprehensive summary of the current literature on Tm expression in cancer cells can be found in Table (Appendix). Figure 16.3 shows representative western blot analysis of the expression of Tm isoforms in a range of cancer cell lines. All cancer cell lines analysed exhibited expression of Tm5NM1 whereas varied expression of HMW isoforms Tm1,2,3,6 and the LMW isoform Tm4 was detected.

While further studies will be required to determine how changes in Tm expression contribute to tumour growth and if Tm expression can be utilised as a diagnostic tool [71], these observations have sparked interest in tropomyosins as a potential new target for chemotherapy. While traditionally, the approach has been to try and restore the expression of HMW Tms in transformed cells in the hope of reversing tumourigenicity, another more exciting possibility is that an increased reliance on LMW Tms may in fact make tumour cells more vulnerable and therefore a better target for cancer chemotherapy.

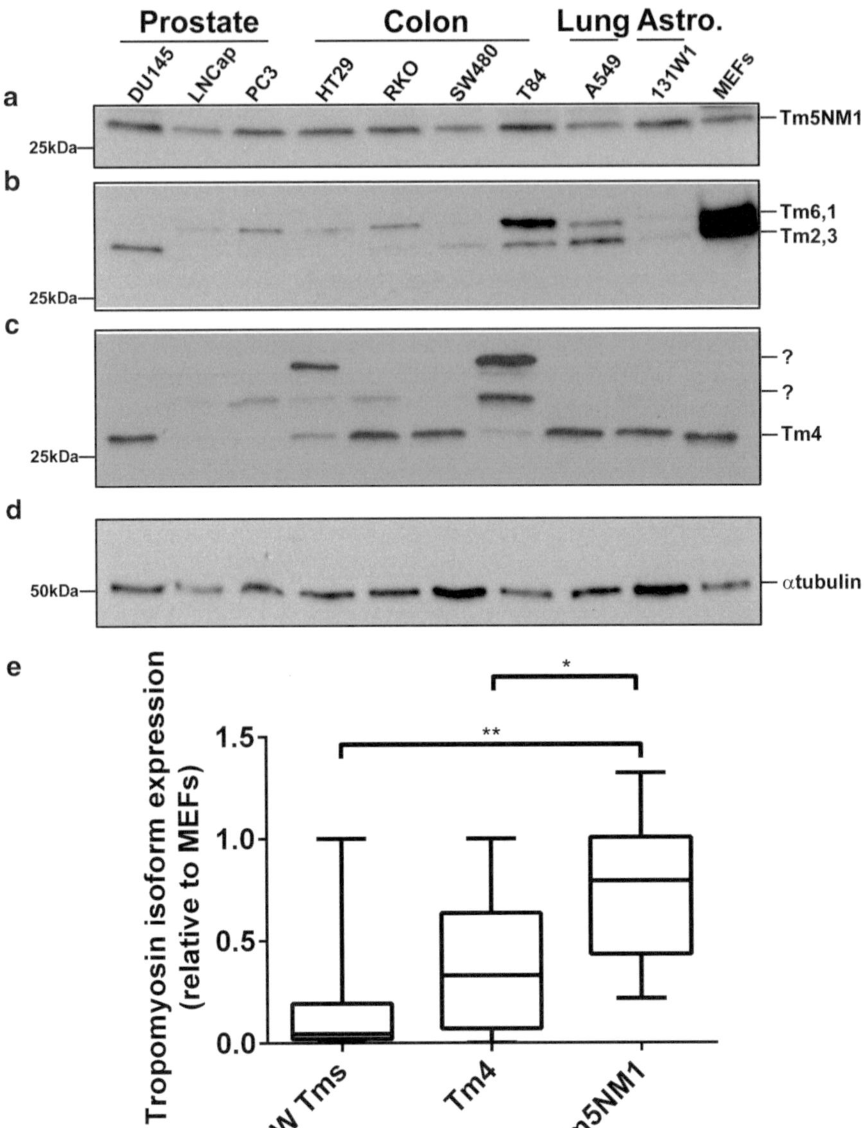

Fig. 16.3 Consistent expression of Tm5NM1 in a range of cancer cell lines. 10 μg of total protein was extracted from various cancer cells lines including DU145, LNCaP, PC3 (prostate), HT29, RKO, SW480, T84 (colon), A549 (lung), 131W1 (astrocytoma) and primary fibroblasts (MEFs) and SDS PAGE electrophoresis followed by western blotting was conducted. Blots were probed with tropomyosin antibodies that detect the high molecular weight tropomyosin isoforms (Tm1,2,3,6) and the low molecular weight isoforms Tm4 and Tm5NM1 as previously described [19]. All cancer cell lines expressed high levels of the LMW isoform Tm5NM1 (**a**). Apart from two prostate cell lines, all cancer cell lines expressed visible amounts of LMW Tm4 (**c**). However, a significant down-regulation of HMW Tm isoforms, specifically Tm6/1 and Tm 2/3 (**b**), was seen in all cancer cells compared to the MEF control cell line. Densitometry was performed for quantitation with bands normalised to α-tubulin (**d**) and the data were analysed by One-way ANOVA nonparametric test and Tukey's multiple comparison test was used to compare medians. (**e**) The data are presented as a box plot whose endpoints are the first and third quartile, with the center line corresponding to the median. $^{**}P<0.01$ and $^{*}P<0.05$ (Vos, 2014, unpublished)

Novel Strategies for Anticancer Compounds to Target the Actin Cytoskeleton

Recently, Stehn et al. (2013) described a novel class of anti-tropomyosin compounds which preferentially target Tm5NM1-containing actin filaments in cancer cells. The lead compound, TR100, has been shown to be effective against a panel of neural crest-derived tumour cell lines in both 2D and 3D cultures, with minimal impact on the contractile properties of isolated rat adult cardiomyocytes. Furthermore, using xenograft models, Stehn and colleagues showed that TR100 is effective in reducing tumour cell growth in vivo (Fig. 16.4) without compromising cardiac function [72], showing that it is indeed possible to selectively target actin filaments fundamental to tumour cell viability based on their tropomyosin isoform composition. This improvement in specificity provides a pathway for the development of a novel class of anti-actin compounds for the potential treatment of a wide variety of cancers.

Concluding Remarks

Actin microfilaments are core constituents of the cytoskeletal network fundamental to all eukaryotic cells. The actin cytoskeleton is essential for many biological processes including cell motility, intracellular organisation, cytokinesis and endocytosis. Structural alterations to the actin cytoskeleton are an established characteristic of transformed cancer cells. Despite the disruption of their internal architecture, transformed cells retain, or even increase many actin-based functional properties such as enhanced motility, invasiveness and metastasis. The actin cytoskeleton therefore represents a point for chemotherapeutic intervention. To date, little progress has been made with compounds that universally disrupt actin filaments due to their essential role in the function of cardiac and skeletal muscle. Tropomyosins are actin-associated polymers which form an integral component of the actin filament. Mammals have over 40 isoforms of tropomyosin which sort to spatially distinct actin filament populations and differentially regulate the interaction of various actin binding proteins. Changes in the expression of tropomyosin isoforms play a crucial role in the onset of oncogenic properties of the cell. While most transformed cells display a decrease in expression of high molecular weight tropomyosins, one isoform, the low molecular weight Tm5NM1 is consistently expressed. This tropomyosin is sufficiently different from those found in muscle cells and therefore represents a novel way to target the actin cytoskeleton of cancer cells without damaging the contractile apparatus of the heart or diaphragm. Part of a novel class of anti-tropomyosin compounds, TR100 selectively targets Tm5NM1 containing actin filaments and has been shown to be effective in vitro and in vivo in reducing tumour cell growth in neuroblastoma and melanoma models. Importantly, TR100 shows no adverse impact on cardiac structure and function. This shows that it is possible to target specific actin filament populations fundamental to tumour cell viability based on their tropomyosin isoform composition and represents a novel approach to potentially treat a wide variety of cancers.

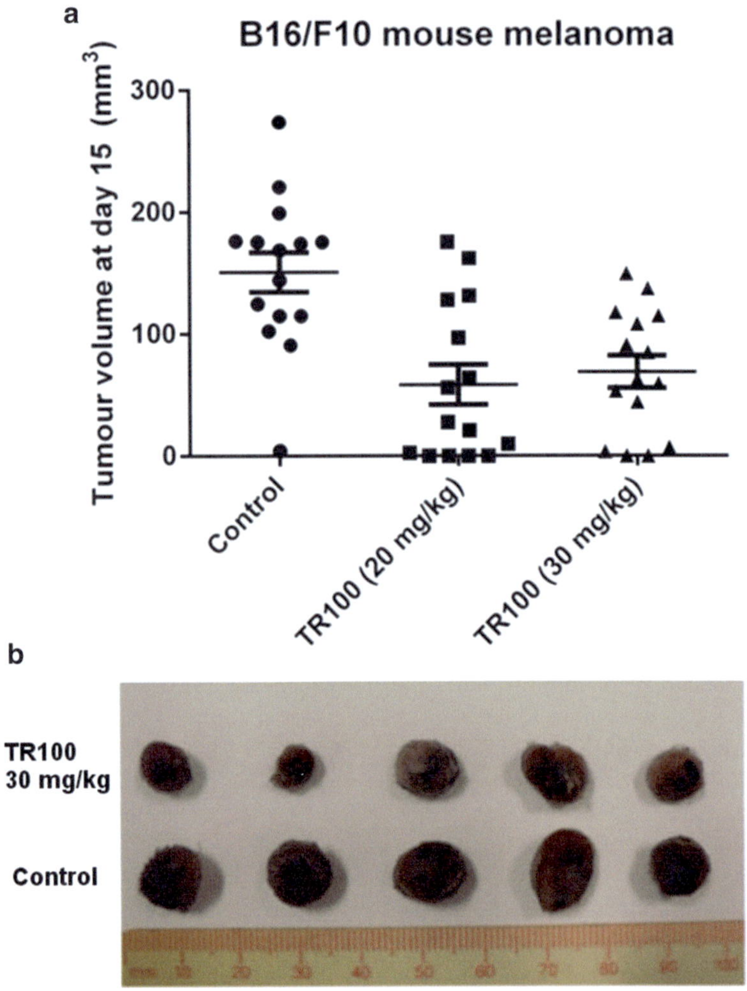

Fig. 16.4 TR100 inhibits tumour growth in a melanoma mouse model. (**a**) C57/Bl6 animals were injected with B16/10F cells and n = 15 per group were treated with DMSO, 20 or 30 mg/kg of TR100 for 5 days/week for 15 days. *Graph points* represent tumour volumes at day 15 of treatment ± SEM ($P = 0.0005$ for both treatment groups compared to control group). (**b**) Photo shows five representative tumours from control (*bottom*) and 30 mg/kg TR100 treatment group (*top*) after 15 days (figure generated using raw data from [72])

Appendix: Table

Cell type	Tm isoform expression[a]			References
	Decreased	Increased	Unaltered	
Experimentally transformed cells:				
Jun-transformed rat fibroblasts	Tm2			[73]
Ras-transformed NIH3T3	Tm1–3		Tm4,5	[59, 74]
Ras-transformed rat intestinal epithelial cell	α-Tm			[75]
REF-52 transformed with DNA or RNA virus	Tm1	Tm3,5		[76]
RSV-transformed NRK	Tm1,2	Tm4,5		[76]
RSV-transformed chick embryo fibroblasts	(α and β) Tm1			[77, 78]
Src-transformed NIH3T3	Tm1			[76]
Transformed/tumorigenic 267B1 prostate cell	Tm1,3			[79]
Transformed HUT-12 fibroblasts	Tm1,2,6		Tm4,5	[80]
Transformed HUT-14 fibroblasts	Tm1,2,3,4,6		Tm5	[80]
Tumorigenic HUT-14T fibroblasts	Tm1,2,3,4,6		Tm5	[80]
Tumor derived HOS	Tm1,2,6	Tm5	Tm4	[80]
Cultured cancer cell lines:				
Breast carcinoma cell lines:				
BT-20	Tm1	Tm5, Tm32[b]		[81]
BT-474	Tm1, Tm38[b]			[81]
MCF7	Tm1, Tm38[b]	Tm3,4,32[b]		[81]
MDA-MB-231	Tm1	Tm4,5,32[b]		[81]
Novel MCF7 cisplatin resistant	Tm1[c]			[82]
T47D	Tm1	Tm3,36[b]		[81]
ZR-75.1	Tm1,38[b]	Tm5,32[b]		[81]
Cholangiocarcinoma cell lines:				
HuCCT1	Tm1			[83]
QBC939		Tm1		[83]
Esophageal carcinoma cell lines:				
Novel esophagus squamous cancer cell line		Tm3		[84]
TE15	Tm1–3			[85]
Gastric carcinoma cell lines:				
OCUM-1		Tm4		[86]
OCUM-2D		Tm4		[86]
OCUM-2M		Tm4		[86]
OCUM-2MLN		Tm4		[86]
OCUM-D3		Tm4		[86]
OCUM-9		Tm4		[86]
OCUM-12		Tm4		[86]
Neuroblastoma cell lines:				
IMR32	Tm1–3, 5a,5b			[58]
BE(2)-C	Tm1–3, 5a,5b			[58]

(continued)

Cell type	Tm isoform expression[a]			References
	Decreased	Increased	Unaltered	
Prostate cell lines:				
DU-145	Tm1	Novel β-Tm[f]		[64, 87]
LNCaP	Tm1	Novel β-Tm[f]		[64, 87]
PC3	Tm1	Novel β-Tm[f]		[64, 87]
DLD-1 human colon cancer cell line	α-Tm			[75]
Tumor derived HT1080 fibrosarcoma	Tm2,6	Tm5	Tm4	[75, 80]
Lewis lung carcinoma cell line	Tm2			[69]
PLA801D non-small cell lung carcinoma cell line		Tm3[d]		[88]
QRsP-11 fibrosarcoma cell line		Tm1[e]		[89]
Patient tumor material:				
Astrocytoma (high grade)	HMW Tm			[90]
Breast carcinoma:	Tm1–3			[63, 91]
Infiltrating ductal breast carcinoma		Tm4		[92]
Cervical carcinoma	Tm1,2,4	Tm3		[93–95]
Colon cancer	β-Tm[g]	TC22[f], Tm2[h]		[95–97]
Esophageal cancer	β-Tm, Tm1	α-Tm, Tm4		[98–101]
Fibrous histiocytoma		Tm3,4		[65, 102]
Gastric carcinoma	α-Tm	LMW Tm		[65, 103]
Hepatocellular carcinoma		Tm5		[104, 105]
Leiomyosarcoma:				
Pleomorphic leiomyosarcoma	Tm1,2	Tm3,4		[102, 106]
Conventional leiomyosarcoma	Tm3,4	Tm1,2		[102, 106]
Lung carcinoma (high grade)	Tm3[i]			[107]
Oral squamous cell carcinoma		Tm2		[108]
Oral tongue squamous cell carcinoma	Tm1	LMW Tm		[103]
Ovarian carcinoma	Tm2,4			[67, 109]
Prostate cancer	Tm1			[64, 110][j]
Renal cell carcinoma	Tm4			[111]
Transitional bladder cell carcinoma	Tm1–3	Tm5		[112]
Vaginal carcinoma	Tm1			[113]
Patient plasma material:				
Ovarian carcinoma		Tm4[k]		[114]

[a]Decreased expression refers to down-regulation or loss of isoform specific protein or mRNA and increased expression correspondingly refers to the gain of protein or mRNA

[b]None of the isoforms corresponded to previously identified isoforms expressed in fibroblasts

[c]Down-regulation of Tm1 compared to the MCF7 breast cancer cell line

[d]Up-regulation of Tm3 in the highly metastatic PLA801D subline compared to the poorly metastatic PLA801C subline

[e]Up-regulation of Tm1 in the progressive cancer cell line QRsP-11 compared to the regressive cell line QR-32

[f]A novel Tm isoform

[g]Down-regulation of β-Tm was also found in colorectal adenoma tissue

[h]Tm2 was found to be increased in patients with poor outcome compared to patients with good outcome

[i]Tm3 showed a steady decline with the malignant progression of squamous cell lung carcinoma from stage I to stage IV

[j]Down-regulation of β-Tm was observed in only one of the tumors (high grade) assayed

[k]Expression of Tm4 showed an early increase in the first stages of ovarian cancer, followed by a steady decrease in later stages

References

1. Dominguez R, Holmes KC (2011) Actin structure and function. Annu Rev Biophys 40:169–186
2. Pollard TD (1986) Rate constants for the reactions of ATP- and ADP-actin with the ends of actin filaments. J Cell Biol 103(6 Pt 2):2747–2754
3. Pollard TD, Blanchoin L, Mullins RD (2000) Molecular mechanisms controlling actin filament dynamics in nonmuscle cells. Annu Rev Biophys Biomol Struct 29:545–576
4. Drenckhahn D, Pollard TD (1986) Elongation of actin filaments is a diffusion-limited reaction at the barbed end and is accelerated by inert macromolecules. J Biol Chem 261(27): 12754–12758
5. Carlier MF, Pantaloni D (1986) Direct evidence for ADP-Pi-F-actin as the major intermediate in ATP-actin polymerization. Rate of dissociation of Pi from actin filaments. Biochemistry 25(24):7789–7792
6. Goley ED, Welch MD (2006) The ARP2/3 complex: an actin nucleator comes of age. Nat Rev Mol Cell Biol 7(10):713–726
7. Mullins RD, Heuser JA, Pollard TD (1998) The interaction of Arp2/3 complex with actin: nucleation, high affinity pointed end capping, and formation of branching networks of filaments. Proc Natl Acad Sci U S A 95(11):6181–6186
8. Pollitt AY, Insall RH (2009) WASP and SCAR/WAVE proteins: the drivers of actin assembly. J Cell Sci 122(Pt 15):2575–2578
9. Wu C et al (2012) Arp2/3 is critical for lamellipodia and response to extracellular matrix cues but is dispensable for chemotaxis. Cell 148(5):973–987
10. Weaver AM et al (2002) Interaction of cortactin and N-WASp with Arp2/3 complex. Curr Biol 12(15):1270–1278
11. Pruyne D et al (2002) Role of formins in actin assembly: nucleation and barbed-end association. Science 297(5581):612–615
12. Higgs HN, Peterson KJ (2005) Phylogenetic analysis of the formin homology 2 domain. Mol Biol Cell 16(1):1–13
13. Otomo T et al (2005) Structural basis of actin filament nucleation and processive capping by a formin homology 2 domain. Nature 433(7025):488–494
14. Seth A, Otomo C, Rosen MK (2006) Autoinhibition regulates cellular localization and actin assembly activity of the diaphanous-related formins FRLalpha and mDia1. J Cell Biol 174(5):701–713
15. McDowell JM et al (1996) Structure and evolution of the actin gene family in *Arabidopsis thaliana*. Genetics 142(2):587–602
16. Gunning P, O'Neill G, Hardeman E (2008) Tropomyosin-based regulation of the actin cytoskeleton in time and space. Physiol Rev 88(1):1–35
17. Li XE et al (2011) Tropomyosin position on F-actin revealed by EM reconstruction and computational chemistry. Biophys J 100(4):1005–1013
18. Holmes KC, Lehman W (2008) Gestalt-binding of tropomyosin to actin filaments. J Muscle Res Cell Motil 29(6-8):213–219
19. Schevzov G et al (2005) Tissue-specific tropomyosin isoform composition. J Histochem Cytochem 53(5):557–570
20. Gunning PW et al (2005) Tropomyosin isoforms: divining rods for actin cytoskeleton function. Trends Cell Biol 15(6):333–341
21. Creed SJ et al (2011) Tropomyosin isoform 3 promotes the formation of filopodia by regulating the recruitment of actin-binding proteins to actin filaments. Exp Cell Res 317(3):249–261
22. Bryce NS et al (2003) Specification of actin filament function and molecular composition by tropomyosin isoforms. Mol Biol Cell 14(3):1002–1016
23. Martin C, Gunning P (2008) Isoform sorting of tropomyosins. Adv Exp Med Biol 644:187–200
24. Martin C, Schevzov G, Gunning P (2010) Alternatively spliced N-terminal exons in tropomyosin isoforms do not act as autonomous targeting signals. J Struct Biol 170(2):286–293

25. Skoumpla K et al (2007) Acetylation regulates tropomyosin function in the fission yeast *Schizosaccharomyces pombe*. J Cell Sci 120(Pt 9):1635–1645
26. Coulton AT et al (2010) The recruitment of acetylated and unacetylated tropomyosin to distinct actin polymers permits the discrete regulation of specific myosins in fission yeast. J Cell Sci 123(Pt 19):3235–3243
27. Clayton JE et al (2010) Differential regulation of unconventional fission yeast myosins via the actin track. Curr Biol 20(16):1423–1431
28. Skau CT, Kovar DR (2010) Fimbrin and tropomyosin competition regulates endocytosis and cytokinesis kinetics in fission yeast. Curr Biol 20(16):1415–1422
29. Johnson M, East DA, Mulvihill DP (2014) Formins determine the functional properties of actin filaments in yeast. Curr Biol 24(13):1525–1530
30. Sporn MB (1996) The war on cancer. Lancet 347(9012):1377–1381
31. Lauffenburger DA, Horwitz AF (1996) Cell migration: a physically integrated molecular process. Cell 84(3):359–369
32. Friedl P, Gilmour D (2009) Collective cell migration in morphogenesis, regeneration and cancer. Nat Rev Mol Cell Biol 10(7):445–457
33. Charras G, Paluch E (2008) Blebs lead the way: how to migrate without lamellipodia. Nat Rev Mol Cell Biol 9(9):730–736
34. Buccione R, Caldieri G, Ayala I (2009) Invadopodia: specialized tumor cell structures for the focal degradation of the extracellular matrix. Cancer Metastasis Rev 28(1-2):137–149
35. Diz-Munoz A et al (2010) Control of directed cell migration in vivo by membrane-to-cortex attachment. PLoS Biol 8(11), e1000544
36. Abercrombie M, Heaysman JE, Pegrum SM (1970) The locomotion of fibroblasts in culture. II. "Ruffling". Exp Cell Res 60(3):437–444
37. Abercrombie M, Heaysman JE, Pegrum SM (1971) The locomotion of fibroblasts in culture. IV. Electron microscopy of the leading lamella. Exp Cell Res 67(2):359–367
38. Svitkina TM, Borisy GG (1999) Arp2/3 complex and actin depolymerizing factor/cofilin in dendritic organization and treadmilling of actin filament array in lamellipodia. J Cell Biol 145(5):1009–1026
39. Blanchoin L, Pollard TD, Hitchcock-DeGregori SE (2001) Inhibition of the Arp2/3 complex-nucleated actin polymerization and branch formation by tropomyosin. Curr Biol 11(16):1300–1304
40. DesMarais V et al (2002) Spatial regulation of actin dynamics: a tropomyosin-free, actin-rich compartment at the leading edge. J Cell Sci 115(Pt 23):4649–4660
41. Kis-Bicskei N et al (2013) Purification of tropomyosin Br-3 and 5NM1 and characterization of their interactions with actin. Cytoskeleton (Hoboken) 70(11):755–765
42. Hillberg L et al (2006) Tropomyosins are present in lamellipodia of motile cells. Eur J Cell Biol 85(5):399–409
43. Chesarone MA, DuPage AG, Goode BL (2010) Unleashing formins to remodel the actin and microtubule cytoskeletons. Nat Rev Mol Cell Biol 11(1):62–74
44. Ponti A et al (2004) Two distinct actin networks drive the protrusion of migrating cells. Science 305(5691):1782–1786
45. Danuser G (2005) Coupling the dynamics of two actin networks—new views on the mechanics of cell protrusion. Biochem Soc Trans 33(Pt 6):1250–1253
46. Vallotton P, Small JV (2009) Shifting views on the leading role of the lamellipodium in cell migration: speckle tracking revisited. J Cell Sci 122(Pt 12):1955–1958
47. Bresnick AR (1999) Molecular mechanisms of nonmuscle myosin-II regulation. Curr Opin Cell Biol 11(1):26–33
48. Burridge K, Chrzanowska-Wodnicka M (1996) Focal adhesions, contractility, and signaling. Annu Rev Cell Dev Biol 12:463–518
49. Fanning AS et al (1994) Differential regulation of skeletal muscle myosin-II and brush border myosin-I enzymology and mechanochemistry by bacterially produced tropomyosin isoforms. Cell Motil Cytoskeleton 29(1):29–45
50. Hanahan D, Weinberg RA (2011) Hallmarks of cancer: the next generation. Cell 144(5):646–674

51. Friedl P (2004) Prespecification and plasticity: shifting mechanisms of cell migration. Curr Opin Cell Biol 16(1):14–23
52. Behrens J et al (1989) Dissecting tumor cell invasion: epithelial cells acquire invasive properties after the loss of uvomorulin-mediated cell–cell adhesion. J Cell Biol 108(6):2435–2447
53. Pollack R, Osborn M, Weber K (1975) Patterns of organization of actin and myosin in normal and transformed cultured cells. Proc Natl Acad Sci U S A 72(3):994–998
54. Shin SI et al (1975) Tumorigenicity of virus-transformed cells in nude mice is correlated specifically with anchorage independent growth in vitro. Proc Natl Acad Sci U S A 72(11):4435–4439
55. Braverman RH et al (1996) Anti-oncogenic effects of tropomyosin: isoform specificity and importance of protein coding sequences. Oncogene 13(3):537–545
56. Gimona M, Kazzaz JA, Helfman DM (1996) Forced expression of tropomyosin 2 or 3 in v-Ki-ras-transformed fibroblasts results in distinct phenotypic effects. Proc Natl Acad Sci U S A 93(18):9618–9623
57. Janssen RA, Mier JW (1997) Tropomyosin-2 cDNA lacking the 3′ untranslated region riboregulator induces growth inhibition of v-Ki-ras-transformed fibroblasts. Mol Biol Cell 8(5):897–908
58. Yager ML et al (2003) Functional analysis of the actin-binding protein, tropomyosin 1, in neuroblastoma. Br J Cancer 89(5):860–863
59. Prasad GL, Fuldner RA, Cooper HL (1993) Expression of transduced tropomyosin 1 cDNA suppresses neoplastic growth of cells transformed by the ras oncogene. Proc Natl Acad Sci U S A 90(15):7039–7043
60. Takenaga K, Masuda A (1994) Restoration of microfilament bundle organization in v-raf-transformed NRK cells after transduction with tropomyosin 2 cDNA. Cancer Lett 87(1):47–53
61. Bonello TT, Stehn JR, Gunning PW (2009) New approaches to targeting the actin cytoskeleton for chemotherapy. Future Med Chem 1(7):1311–1331
62. Helfman DM et al (2008) Tropomyosin as a regulator of cancer cell transformation. Adv Exp Med Biol 644:124–131
63. Franzen B et al (1996) Expression of tropomyosin isoforms in benign and malignant human breast lesions. Br J Cancer 73(7):909–913
64. Wang FL et al (1996) Two differentially expressed genes in normal human prostate tissue and in carcinoma. Cancer Res 56(16):3634–3637
65. Jung MH et al (2000) Identification of differentially expressed genes in normal and tumor human gastric tissue. Genomics 69(3):281–286
66. O'Neill GM, Stehn J, Gunning PW (2008) Tropomyosins as interpreters of the signalling environment to regulate the local cytoskeleton. Semin Cancer Biol 18(1):35–44
67. Alaiya AA et al (1997) Phenotypic analysis of ovarian carcinoma: polypeptide expression in benign, borderline and malignant tumors. Int J Cancer 73(5):678–683
68. Alaiya AA et al (2001) Identification of proteins in human prostate tumor material by two-dimensional gel electrophoresis and mass spectrometry. Cell Mol Life Sci 58(2):307–311
69. Takenaga K, Nakamura Y, Sakiyama S (1988) Differential expression of a tropomyosin isoform in low- and high-metastatic Lewis lung carcinoma cells. Mol Cell Biol 8(9):3934–3937
70. Takenaga K et al (1988) Isolation and characterization of a cDNA that encodes mouse fibroblast tropomyosin isoform 2. Mol Cell Biol 8(12):5561–5565
71. Stehn JR et al (2006) Specialisation of the tropomyosin composition of actin filaments provides new potential targets for chemotherapy. Curr Cancer Drug Targets 6(3):245–256
72. Stehn JR et al (2013) A novel class of anticancer compounds targets the actin cytoskeleton in tumor cells. Cancer Res 73(16):5169–5182
73. Ljungdahl S et al (1998) Down-regulation of tropomyosin-2 expression in c-Jun-transformed rat fibroblasts involves induction of a MEK1-dependent autocrine loop. Cell Growth Differ 9(7):565–573
74. Cooper HL et al (1985) Suppression of tropomyosin synthesis, a common biochemical feature of oncogenesis by structurally diverse retroviral oncogenes. Mol Cell Biol 5(5):972–983

75. Shields JM et al (2002) Opposing roles of the extracellular signal-regulated kinase and p38 mitogen-activated protein kinase cascades in Ras-mediated downregulation of tropomyosin. Mol Cell Biol 22(7):2304–2317
76. Matsumura F et al (1983) Differential expression of tropomyosin forms in the microfilaments isolated from normal and transformed rat cultured cells. J Biol Chem 258(22):13954–13964
77. Hendricks M, Weintraub H (1981) Tropomyosin is decreased in transformed cells. Proc Natl Acad Sci U S A 78(9):5633–5637
78. Lin JJ et al (1985) Tropomyosin isoforms in chicken embryo fibroblasts: purification, characterization, and changes in Rous sarcoma virus-transformed cells. J Cell Biol 100(3):692–703
79. Prasad SC et al (1997) Protein expression changes associated with radiation-induced neoplastic progression of human prostate epithelial cells. Electrophoresis 18(3-4):629–637
80. Leavitt J et al (1986) Tropomyosin isoform switching in tumorigenic human fibroblasts. Mol Cell Biol 6(7):2721–2726
81. Bhattacharya B et al (1990) Tropomyosins of human mammary epithelial cells: consistent defects of expression in mammary carcinoma cell lines. Cancer Res 50(7):2105–2112
82. Smith L et al (2007) The proteomic analysis of cisplatin resistance in breast cancer cells. Oncol Res 16(11):497–506
83. Yang W et al (2013) Genetic and epigenetic alterations are involved in the regulation of TPM1 in cholangiocarcinoma. Int J Oncol 42(2):690–698
84. Moghanibashi M et al (2012) Proteomics of a new esophageal cancer cell line established from Persian patient. Gene 500(1):124–133
85. Zare M et al (2012) Downregulation of tropomyosin-1 in squamous cell carcinoma of esophagus, the role of Ras signaling and methylation. Mol Carcinog 51(10):796–806
86. Kuramitsu Y et al (2010) Proteomic differential display analysis shows up-regulation of 14-3-3 sigma protein in human scirrhous-type gastric carcinoma cells. Anticancer Res 30(11):4459–4465
87. Assinder SJ et al (2010) A novel splice variant of the beta-tropomyosin (TPM2) gene in prostate cancer. Mol Carcinog 49(6):525–531
88. Jiang D et al (2003) Identification of metastasis-associated proteins by proteomic analysis and functional exploration of interleukin-18 in metastasis. Proteomics 3(5):724–737
89. Hayashi E et al (2005) Proteomic profiling for cancer progression: differential display analysis for the expression of intracellular proteins between regressive and progressive cancer cell lines. Proteomics 5(4):1024–1032
90. Hughes JA et al (2003) High-molecular-weight tropomyosins localize to the contractile rings of dividing CNS cells but are absent from malignant pediatric and adult CNS tumors. Glia 42(1):25–35
91. Raval GN et al (2003) Loss of expression of tropomyosin-1, a novel class II tumor suppressor that induces anoikis, in primary breast tumors. Oncogene 22(40):6194–6203
92. Kabbage M et al (2013) Tropomyosin-4 correlates with higher SBR grades and tubular differentiation in infiltrating ductal breast carcinomas: an immunohistochemical and proteomics-based study. Tumour Biol 34(6):3593–3602
93. Lomnytska MI et al (2011) Differential expression of ANXA6, HSP27, PRDX2, NCF2, and TPM4 during uterine cervix carcinogenesis: diagnostic and prognostic value. Br J Cancer 104(1):110–119
94. Bae SM et al (2005) Two-dimensional gel analysis of protein expression profile in squamous cervical cancer patients. Gynecol Oncol 99(1):26–35
95. Luo YX et al (2009) Identification of cancer-associated proteins by proteomics and downregulation of beta-tropomyosin expression in colorectal adenoma and cancer. Proteomics Clin Appl 3(12):1397–1406
96. Lin JL et al (2002) Isolation and sequencing of a novel tropomyosin isoform preferentially associated with colon cancer. Gastroenterology 123(1):152–162
97. Kim HJ et al (2012) Profiling of differentially expressed proteins in stage IV colorectal cancers with good and poor outcomes. J Proteomics 75(10):2983–2997

98. Adami R et al (2003) On the stiffness of the natural actin filament decorated with alexa fluor tropomyosin. Biophys Chem 104(2):469–476
99. Harada T et al (2007) Expression of tropomyosin alpha 4 chain is increased in esophageal squamous cell carcinoma as evidenced by proteomic profiling by two-dimensional electrophoresis and liquid chromatography–mass spectrometry/mass spectrometry. Proteomics Clin Appl 1(2):215–223
100. Zhang J et al (2011) Using proteomic approach to identify tumor-associated proteins as biomarkers in human esophageal squamous cell carcinoma. J Proteome Res 10(6):2863–2872
101. Qi Y et al (2005) Comparative proteomic analysis of esophageal squamous cell carcinoma. Proteomics 5(11):2960–2971
102. Kawai A et al (2008) Global protein-expression analysis of bone and soft tissue sarcomas. Clin Orthop Relat Res 466(9):2099–2106
103. He QY et al (2004) Identification of tumor-associated proteins in oral tongue squamous cell carcinoma by proteomics. Proteomics 4(1):271–278
104. Kim TM et al (2008) Clinical implication of recurrent copy number alterations in hepatocellular carcinoma and putative oncogenes in recurrent gains on 1q. Int J Cancer 123(12):2808–2815
105. Choi HS et al (2010) Tropomyosin3 overexpression and a potential link to epithelial–mesenchymal transition in human hepatocellular carcinoma. BMC Cancer 10:122
106. Suehara Y et al (2006) Proteomic signatures corresponding to histological classification and grading of soft-tissue sarcomas. Proteomics 6(15):4402–4409
107. Deng B et al (2005) Proteomics analysis of stage-specific proteins expressed in human squamous cell lung carcinoma tissues. Cancer Biomark 1(6):279–286
108. Lo WY et al (2007) Identification of over-expressed proteins in oral squamous cell carcinoma (OSCC) patients by clinical proteomic analysis. Clin Chim Acta 376(1-2):101–107
109. Chow SN et al (2010) Analysis of protein profiles in human epithelial ovarian cancer tissues by proteomic technology. Eur J Gynaecol Oncol 31(1):55–62
110. Ahram M et al (2002) Proteomic analysis of human prostate cancer. Mol Carcinog 33(1):9–15
111. Hwa JS et al (2005) Identification of proteins differentially expressed in the conventional renal cell carcinoma by proteomic analysis. J Korean Med Sci 20(3):450–455
112. Pawlak G et al (2004) Alterations in tropomyosin isoform expression in human transitional cell carcinoma of the urinary bladder. Int J Cancer 110(3):368–373
113. Hellman K et al (2004) Protein expression patterns in primary carcinoma of the vagina. Br J Cancer 91(2):319–326
114. Bailey MJ et al (2013) Stage-specific analysis of plasma protein profiles in ovarian cancer: difference in-gel electrophoresis analysis of pooled clinical samples. J Carcinog 12:10

Index

H. Schatten (ed.), *The Cytoskeleton in Health and Disease*,
DOI 10.1007/978-1-4939-2904-7